AGRICULTURAL PRAIRIES

Natural Resources and Crop Productivity

AGRICULTURAL PRAIRIES
Natural Resources and Crop Productivity

K. R. Krishna, PhD

Apple Academic Press Inc.	Apple Academic Press Inc.
3333 Mistwell Crescent	9 Spinnaker Way
Oakville, ON L6L 0A2	Waretown, NJ 08758
Canada	USA

©2015 by Apple Academic Press, Inc.

First issued in paperback 2021

Exclusive worldwide distribution by CRC Press, a member of Taylor & Francis Group
No claim to original U.S. Government works

ISBN 13: 978-1-77463-363-2 (pbk)
ISBN 13: 978-1-77188-050-3 (hbk)

Library of Congress Control Number: 2014955225

Library and Archives Canada Cataloguing in Publication

Krishna, K. R. (Kowligi R.), author
Agricultural prairies: natural resources and crop productivity / K.R. Krishna, PhD.

(Advances in hospitality and tourism book series)
Includes bibliographical references and index.
ISBN 978-1-77188-050-3 (bound)
1. Prairies. 2. Agriculture. 3. Natural resources. 4. Crop yields. 5. Agricultural productivity.
I. Title.

| QK938.P7K75 2014 | 630.915'3 | C2014-907284-8 |

Apple Academic Press also publishes its books in a variety of electronic formats. Some content that appears in print may not be available in electronic format. For information about Apple Academic Press products, visit our website at **www.appleacademicpress.com** and the CRC Press website at **www.crcpress.com**

ABOUT THE AUTHOR

K. R. Krishna, PhD

K. R. Krishna received his PhD in Agriculture from the University of Agricultural Sciences, Bangalore, India. Retired from the International Crops Research Institute for the Semi-Arid Tropics (ICRISAT) in India, he has been a cereals scientist in India and a visiting professor and research scholar at the Soil and Water Science Department at the University of Florida, Gainesville, USA. Dr. Krishna is a member of several professional organizations, including the International Society for Precision Agriculture, the American Society of Agronomy, the Soil Science Society of America, the Ecological Society of America, and the Indian Society of Agronomy.

CONTENTS

LIST OF ABBREVIATIONS

ANPP	above-ground net productivity
ADB	Asian Development Bank
BNPP	below ground net productivity
BMP	best management practices
BNF	biological nitrogen fixation
CEC	cation exchange capacity
CGMS	crop growth monitoring systems
ENSO	El Nino southern oscillation
Ec	electrical conductivity
FYM	farm yard manure
GEMS	general ensemble biogeochemical modeling system
GE	genetically engineered
GM	genetically modified
GHG	greenhouse gas
GDP	gross domestic product
GW	ground water
ICAR	Indian Council of Agricultural Research
ICLS	integrated crop-livestock systems
INM	integrated nutrient management
ISFM	integrated soil fertility management
ICRISAT	International Crops Research Institute for the Semi-arid Tropics
IFDC	International Fertilizer Development Centre
IITA	International Institute for Tropical Agriculture
LGP	lower gangetic plains
MGP	middle gangetic plains
MWR	Ministry of Water Resources
MP	moldboard plow
NASA	National Aeronautic and Space Agency
NPP	net primary product
NT	no-tillage
OC	organic carbon
RT	ridge tillage
SSNM	site specific nutrient management
SAR	state agency recommendations
SW	surface water

TAGB	total above-ground biomass
TBGB	total below-ground biomass
TGP	trans gangetic plains
UGP	upper gangetic plains
WPE	water production efficiency
WUE	water use efficiency

PREFACE

Natural prairies have evolved and thrived on the earth's surface since several million years ago. They have adapted to diverse physiographic conditions that occurred at various times during history. Several plant species have evolved, peaked at certain periods, and perished. Natural species that we witness today are the resultant situation derived through the biological evolutionary process that has occurred in response to environmental conditions.

Agricultural cropping is said to have begun sometime during the 12th millennium B.C. Human ingenuity has consistently played its role in domesticating crops, mending soils, and augmenting water to crop species. In due course, human intellect was responsible for developing exclusive crop lands—that is, "agricultural prairies." Through the ages, agricultural prairies and human species have coevolved into interdependence. However, it may not be so. Human dependence on food grain/forage generating agricultural prairies seems to be relatively more obligated than we usually acknowledge. During the past few centuries, we have excessively focused on very few cereals, legumes, and oilseeds, and preferentially accentuated them in the croplands. Agricultural prairies occupy more than 80 percent of crop land on earth. It is a lopsided preference of prairie vegetation. Further, we may realize that together, three major cereals—maize, wheat, and rice—dominate the cropland and master the human dietary preference. Such excessive preference and dependence, after all, exposes the crops and humans alike to detrimental factors with potentially large-scale disasters. Perhaps, there is a strong need to diversify the food generating vegetation pattern.

This book, titled *Agricultural Prairies: Natural Resources and Crop Productivity*, deals with several aspects of agricultural prairies such as geology, geographical settings, natural resources such as soil types, water and irrigation, and crop genetic stocks. Human resource, migration, and settlement patterns within each prairie zone have been described lucidly. Land use change and environmental concerns such as climate change have also been discussed in detail for each prairie zone. Fluctuation in expanse and productivity of grain generating systems provides insight into performance of prairies. There are eight chapters in this book. The first three deal with agricultural vegetation of the Americas. They deal with large agricultural prairies of North America, in particular the Great Plains of North America, the Cerrados of Brazil, and the Pampas of South America. Chapter 4 deals with the agricultural prairies of Europe, especially those traced in the French and German Plains, Central Europe, and Russian Steppes. Detailed and lucid discussions on cereal-dominated prairies of Sahelian, Sudanian, and Guinea-Savannah zones of West Africa are presented in Chapter 5. The South Asian agricultural prairies are important food grain suppliers to the over 1.4 billion human population. Several aspects of the croplands in India, Pakistan, and Bangladesh are found in Chapter 6. Chapter 7 deals exclusively with relevant discussions on North China Steppes. These are among the most intensely cultivated zones with high food grain output and biomass turnover. Apart from the above-mentioned prairie regions, there are a few other agricultural prairies that flourish in Southern Africa, West Asia, and Southern Australia. They serve a large population of humans in situ with food grains and are important in terms of financial resources. However, they have not been dealt in this book.

Agricultural prairies are almost entirely a human effort to answer his requirements of food grains and few other items. Human involvement in the prairies is varied. In nature, there are many facets to prairie versus human coexistence and interaction. Agricultural crops have induced large-scale migrations, altered human dwelling and settlement patterns, and influenced his food habits, affected monetary funds, and several other aspects of daily life. In a way, agricultural crops, especially a few grain cereals, have driven humans to subordination and enslavement. There are no tangible alternatives to agricultural prairies as yet. However, admixtures of prairies and plantations could be envisioned to overcome lopsided dependence on prairies for food grains and other goods. Evidences and arguments for many of the above aspects have been listed, and lucid discussions are presented in Chapter 8, titled "Agricultural Prairies and Man."

This book on agricultural prairies is a scholarly edition useful to those interested in studying earth systems, biomes, vegetation, ecology, environment, and food crops. It is apt as a textbook for students in geography, natural sciences, ecology, environment, and agriculture.

ACKNOWLEDGMENT

During preparation of this book, several of my colleagues, friends, researchers, and professors from different institutions have helped me directly or indirectly. They have been a source of inspiration. I wish to thank the following scientists/administrators and their institution for allowing use of photographs and other research publications: Lindsay Kennedy, Director, External Affairs, Sorghum Checkoff, Lubbock, Texas, USA, for research reports and pictures on sorghum prairies in Southern Great Plains; United States Department of Agriculture (USDA), at Beltsville in Maryland, USA, for maps and pictures; EMBRAPA, Sao Paola, Brazil, for research reports on cereals and picture of natural vegetation of Cerrado and its clearing; International Crops Research Institute for the Semi-Arid Tropics (ICRISAT) at Hyderabad in India and ICRISAT Sahelian Centre at Niamey in Niger for pictures depicting Niger River and Savannah crop in West Africa; International Rice Research Institute (IRRI), Manila in Philippines for maps showing distribution of rice crop; and University of Agricultural Sciences (UAS), GKVK campus, Bangalore, South India.

I wish to thank Dr. Uma Devi Krishna, Sharath Kowligi, Roopashree Kowligi and offer my best wishes to Tara Kowligi.

CHAPTER 1

THE GREAT PLAINS OF NORTH AMERICA: NATURAL RESOURCES, PRAIRIE CROPS, AND AGRICULTURAL PRODUCTIVITY

CONTENTS

1.1   GREAT PLAINS: AN INTRODUCTION

The Great Plains of North America is geographically a vast expanse of land with fertile soils, natural prairies, agricultural cropping regions, forests, mountains, un-dulating and flat regions, rivers, and lakes. It is located in the central part of North American continent and extends for about 1.4 million mi². It stretches for 3,870 km in length from Alberta, Saskatchewan, and Manitoba in the North through to Texas coast line in the South and on to small region in Mexico. The width of this ecoregion is estimated at 1,612 km starting from foothills of Rocky Mountains to Indiana in the East (Figure 1.1). Precipitation depreciates from wet regions in east (1,050 mm year⁻¹) to dry arid belts of West (720 mm year⁻¹). Rocky Mountains create a rain shadow region. The Great Plains encompass wide variety of vegetation. It supports predominantly tall grass prairies in the Eastern zone, short grass prairies in the West, and mixed pastures and cropland all around. Great Plains is among the top most food grain producing regions of the world. It has large wheat, corn, and soybean belts that contribute grains and forage. Great Plains produce 60 percent wheat, 87 percent sorghum and 36 percent of cotton harvested in the entire USA. Annual grain production in the Great Plains is greater than 334 million tons. It amounts to 25 percent of global harvest of major dry land cereals such as wheat, barley, maize, sor-ghum, legume soybean, and so on. However, Great Plains is one of the least densely populated agricultural belts. It has a population of 10 million human beings. It also supports a large population of cattle and other farm animals that has to be supplied with food (IISD, 2012a).

1.1.1   GEOLOGICAL ASPECTS

Trimble (1980) believes that for most part of half a billion years, approximately 570 million, the un-glaciated parts of North American continent was least complicated geologically. Shallow seas covered the zone until about 70 million years ago. Later, a sequence of layered sediments of 5,000–10,000 ft thick was deposited on to sub-siding ocean floor. These layers consolidated into rocks comparable to ancient rocks of the superior upland. During the Cretaceous period, Great Plains was engulfed by a shallow sea called the Western Interior Seaway. During late Cretaceous to Paleo-cene, some 55–65 million years ago, the seaway began to recede, thus creating a flat plain with dried marine deposits (Table 1.1).

During Eocene that is 45 million years ago, it seems Great Plains experienced a long period of stability lasting over 10 million years. Later, there was uplift of mountains. Soil formation occurred during this stable period. Large quantities of sediment were carried through streams and spread into Northern Great Plains. Some of these sediments gave rise to White river group and South Dakota Bad lands. Lush vegetation developed in the semiarid grass lands, where herds of large and small animals thrived in the Great Plains.

FIGURE 1.1   The Great Plains of North America.
Source: Courtesy University of Nebraska, Extension Services, Lincoln, Nebraska, USA
Note: Great Plains and Canadian Prairies, as they are commonly known lies within the marked area. It extends into three Prairie Provinces of Canada namely Alberta, Saskatchewan, and Manitoba. In the United States of America, the plains region extends into 10 different states namely Montana, North Dakota, Wyoming, South Dakota, Nebraska, Colorado, Kansas, New Mexico, Oklahoma, and Texas.

About 20–30 million years ago, streams transported large quantities of gravel, sand, and silt. They deposited these eroded material into Arikaree and Ogallala formations. This process of deposition, it seems continued for another 10 million years in the entire Great Plains from Texas to Canada except in mountainous regions. During past 5–10 million years, Great Plains became lush green with vegetation, sloping Westward due to depositional plain. Further, during past 5–10 million years, streams played prominent role in down cutting and excavating the plains region, removing sediments, and depositing elsewhere. These processes created, features currently known as Missouri Plateau, Colorado Piedmont, Pecos Valley, Edwards Plateau, and Plains Border section.

Geologic history of Great Plains region in Oklahoma and adjoining states is varied. It has rock features pertaining to every geologic period from Precambrian to Quaternary. During past 4 billion years, this region has undergone gradual as well as drastic changes in spurts. It has supported shallow oceans, mountains, and basin formations (Table 1.1). The geologic region of Central Plains is dominated with sandstones, limestones, and shales. There are three major basins and drainage systems namely the Canadian, Arkansas, and Red river. There are several man-made lakes with long shore lines.

Geologically, Great Plains region within United States has been classified into 10 physiographic subdivisions. They are (1) Coteau du Missouri or Missouri Plateau, Glaciated Eastern region of South Dakota, Eastern regions of North Dakota, and Northeastern Montana; (2) Cotueau du Missouri, unglaciated western region of South Dakota, Northeastern Wyoming, Southwestern North Dakota, and Southeastern Montana; (3) Black Hills—western South Dakota; (4) High plains—Eastern New Mexico, Northwestern Texas (including Llano Estacado and Texas panhandle), Kentucky, eastern Colorado, Western Kansas, Nebraska (including sandhills) and Southeastern Wyoming; (5) Plains border—central Kansas and Northern Oklahoma; (6) Colorado Piedmont—eastern Colorado; (7) Raton section—Northeastern New Mexico; (8) Pecos valley—eastern New Mexico; (9) Edwards Plateau—Southcentral Texas; and Central Texas section-central Texas; and (10) Plains Border section (Fenneman, 1917; Trimble, 1980; Figure 1.2).

Land forms and geologic features that are apparent currently are mostly those created through various geologic processes during past 2 million years—that is, since beginning of Pleistocene. Let us discuss these features of Great Plains in slightly greater detail.

Black Hills: It is situated in Northwest Dakota and adjoining regions in Wyoming and covers an area 200 km in length and 105 km in width. The dome in the Black hills region is caused by granite and metamorphic rock in the core and marine sedimentary deposits. The central area of Black hills is 3,000 ft above sea level. The Black Hills is therefore an uplifted zone and that has been carved out differentially by the streams (Trimble, 1980).

Central Texas Uplift: A close examination of topography shows that incessant erosion of uplifted areas has exposed granites, gneiss, and schists. Further, erosion has caused a basin. It is a low plateau that gently slopes eastward. Balcone fault zone determines eastern edge of Great Plains in this part. The Colorado River flows in a low land zone about 160 km. The Northern edge of Texas uplift is a sand stone and limestone zone that divides Brazos and Colorado rivers.

Raton Section: This geologic zone is characterized by volcanic rocks that have induced formation of mesas and cones protected by sedimentary rocks that avoid erosion. It is located close to Colorado Piedmont and Pecos valley. In the South, it is marked with Dakota sandstones. In the North, it is a flat plateau on Cretaceous rocks showing dispersed volcanic cones vents and lava flow. It seems huge lava flow that occurred sometime 8–2 million years ago rests above Ogallala formation. The Ogallala in this region actually rests on Cretaceous rock formations. High mesas called Raton Mesa and Mesa de Maya dissects the Arkansas and Canadian rivers.

TABLE 1.1   A generalized chart describing geologic evolution and rocks of Great Plains of North American continent

Geologic Era and Age (Million Years Ago)	Missouri Plateau— Black Hills	High Plains— Colorado Plateau	Pecos Valley, Edwards Plateau— Central Texas
Quaternary Era (<2)			
Pleistocene (<2)	Glacial deposits, alluvium and terrace deposits	Alluvium, sand dunes, and loess	Piedmont formation, terraces, and sand deposits
Tertiary Era (65–2)			
Pliocene (5–2)	Gravel, Ogallala formation	Ogallala formation	Erosional surface, no depositions
Miocene (23–5)	Arikare formation	-	-
Oligocene (37–24)	White river group	White river group	-
Eocene (54–38)	Waswatch, golden valley	Dawson Arkose	-
Paleocene (65–55)	Fort union formation	Denver, poison canyon, and Raton formations	-
Cretaceous Era (136–66)			
	Hell creek, lance formation fox hills sandstones	Vermejo and Laramie formations Trinidad and Fox hills sandstones	
-	Shales, sandstones, limestone deposits in late cretaceous sea		-
-		Dakota sandstone and Lakota formation	Glen rose and Edwards's limestone
Jurassic Era (195–137)			
	Sundance formation	Morrison formation Ellis, Unkpapa sandstone	Jurassic rocks absent

TABLE 1.1 *(Continued)*

Geologic Era and Age (Million Years Ago)	Missouri Plateau— Black Hills	High Plains— Colorado Plateau	Pecos Valley, Edwards Plateau— Central Texas
Triassic era (225–296)	-	Red rocks	-
Paleozoic era (570–226)	-	Paleozoic rocks	-
Precambrian Era (>570)	-	Precambrian rocks	-

Source: Trimble (1980), Pierson (2012), Geological Survey Bulletin No 1493

High Plains: Great Plains is a gentle sloppy terrain from mountains to Missouri river and beyond. Rivers such as Missouri, Platte, Arkansas, and Pecos strip, the uplifted zones through constant erosion. The High Plains extends from Pin Ridge escarpment near South Dakota-Nebraska border to Edwards Plateau in Texas. A large central plain is still preserved unaffected by streams. This Ogallala preserved zone is known as "High Plains." High Plains region South of Canadian River is known as "Southern High Plains" or "Llano Estacado." Llano Estacada is flanked by Mescalero escarpment in the west and Caprock escarpment in the east. Llano Estacado is a flat surface with dispersed pitted patches and sinks in the Ogallala formation. Dissolution of limestone and wind erosion causes such sinks. The Northern High Plains is a vast stretch of sand dunes and wind-blown silt deposits that rest on the Ogallala formation. Nebraska sandhills, for example, is a large well-stabilized sandy high plains area. It extends from White River in South Dakota to Platte River in Western Nebraska. Loess region extends from western high plains to southward up to Arkansas River and South fork. This loess region is currently a rich wheat production belt. Paleozoic rocks and deposits make this zone rich in petroleum oil.

Missouri Plateau: Rivers that had been depositing sediments into Great Plains for 30 million years changed course of their flow due to regional uplift of terrain. The region dissected much by the Missouri river is known as the "Missouri Plateau." During past 2 million years, continental ice sheets moved from Canada to United States several times as temperatures fluctuated (Trimble, 1980). Regions covered by glaciers are known as "Glaciated Missouri Plateau" and that presently not covered by ice is known as "Unglaciated Missouri Plateau."

Glaciated Missouri Plateau: Movement of continental ice sheets to South into Montana and Dakotas causes the "Glaciated Missouri Plateau." The rolling upland in North Dakota covered by dead-ice moraine and ridges of terminal moraines form last glacial advance known as "Coteau du Missouri." This Missouri escarpment is continuous with Dakotas. Little Missouri flows eastward into narrow valleys. This stream carved terrain is also deposited with thick layer of rock debris. Currently, it supports large expanses of cereals such as wheat, barley, and canola.

Unglaciated Missouri Plateau: This region occurs beyond the margin of glaciation and displays large diversification in terms of landforms, vegetation, and agricultural expanses. Missouri river and its tributaries, such as Sun, Judith, Smith, Musseleh, and Yellow stone have dissected this zone into broad upland surfaces and flood plains. Most of these rivers currently flow into valleys established some 2 million years ago at the end of last glaciation. Bad Lands of South Dakota are created by streams and rivulets flowing through soft and steep regions.

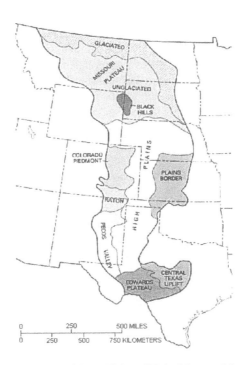

FIGURE 1.2   Geologic regions of Great Plains of United States of America.
Source: University of Nebraska, Lincoln, Nebraska

Colorado Piedmont: This geologic region occupies area between foot of Rocky Mountains, South Platt, and Arkansas River. These two rivers have excavated deeply into Tertiary sedimentary rock layers. They have removed large quantities of sediments and moved them across into plains region. In the western Colorado Piedmont, hard sand stone and limestone deposits make prominent ridges. There are well-formed terraces and flood plains in this zone. Arkansas River has excavated much of Tertiary deposits and cut deeply into Cretaceous rocks near Kansas border (Trimble, 1980). The terrain in Colorado Piedmont is made of wind-blown sand and silt. The northwesterly winds bring large deposits of sand and dust and it may blow up into hurricanes. Gentle ridges of silt, sand, and fertile belts are prominent.

Pecos Valley: It occurs in South of Raton section. Pecos River has cut broad valley in New Mexico. It has removed the piedmont cover of Ogallala formation. Rocks found underlying in the Pecos valley are limestone that belongs to Paleozoic era. The limestone is prone to dissolve. It is responsible for some of the spectacular geologic features of this zone. Large expanses of silt and sand deposits are common and they give rise semiarid plains, useful for crop production.

Edwards Plateau: This region occurs in South of Pecos Valley and is characterized by Cretaceous limestone. Rivers have eroded most the of rock layers except for thin tertiary cover of Ogallala formation. Rivers such as West Neuces and Neuces have sharply eroded the Eastern part of Edwards Plateau. The Southern part of plateau has been dissected intricately by Frio, Sabinal, Guadalupe, and Pedernales rivers. Sink holes and limestone plateau are frequently encountered in this area. Ancient oceans deposited large amounts of limestone that currently cap the plateau. This is also known for oil and gas fields. This region is semiarid with low fertility soils, supports scattered natural vegetation and useful to cultivate drought-tolerant cereals and pasture species.

Plains Border Section: This region consists of several East flowing rivers and valleys created by them. Rivers such as Republican, Solomon, Saline, Smoky Hill, Arkansas, Cimmaron, and North Canadian cut the terrain intricately. North of the Arkansas, rivers have incised into Tertiary Plains and developed closely placed tributaries. Sand dunes are common in many of river valleys. Oil and gas fields are found in this region. Soil formation is accentuated and it allows good vegetation and crop growth.

The Mexican region of Great Plains possesses flat areas and small undulating hills. It rests on Cenozoic sedimentary rocks with recent continental deposits (McGinley, 2008). In addition to crop production, mining for oil and gas is a major occupation in this region. Rio Grande valley currently supports scrub lands and natural vegetation. It irrigates major crops such as corn, sorghum, canola, and beans.

1.1.2 PHYSIOGRAPHY AND AGROCLIMATE OF NORTH AMERICAN PRAIRIES

Great Plains agricultural zones and natural prairies have been classified using several criteria and purposes. For example, Hudson (2012) demarcated the cropping zone of Great Plains in more convenient nine zones. They are Parkland belt, Canadian Prairies, Northern Spring wheat region, Un-glaciated Missouri Plateau, Sand hills; Eastern Feed Grain and Livestock area, Winter Wheat region, Irrigated High Plains, Upland Cotton region, Irrigated Valley, and Range lands. This system allows us to concentrate crop production pattern with priority. Physiographic demarcation of each of the major regions, namely Northern Plains, Western Plains, Central, and

Southern Plains has been done by using several different criteria. As an example, let us consider Northern Great Plains exclusively in greater detail in the following paragraphs.

Padbury et al. (2002) describe and classify Northern Plains based on land resources and climatic conditions. They have utilized parameters such as annual precipitation, frost-free days, diurnal pattern, thermal units, and daily temperatures to demarcate the agricultural prairies into 14 different agroecoregions. The soil types, fertility status, and crop species that dominate each agroecoregion varies. The vastness of each agroecoregion varies from 3.2 to 22.4 m ha. Crops such as wheat, oat, barley, and canola dominate the landscape in most agroecoregions. Following is a summary of characteristics of all 14 agroecoregions of Northern Great Plains:

Agroecoregion 1: It encompasses regions in Southwest of North Dakota state and northeast region of South Dakota. This region occurs between 45°4–46°9 latitude and 100°9–110°0 longitude. Annual precipitation ranges from 300 to 475 mm. Soils are classified as Orthents, Argids, and Psamments that are frigid, deep, and well drained due to sand fraction. The region experiences 112–156 frost-free days. The growing season is short and often confines farmers to practice wheat-fallow systems.

Agroecoregion 2: It covers areas in east of Red river valley. It extends from South Dakota through North Dakota, Southwest Manitoba, and Southeast Saskatchewan. It is a hammocky glacial till region. Mean annual temperature ranges from >5°C in South to 1.5°C in North. Frost-free days are restricted to 120–140 days annually. Yearly precipitation ranges from 375 to 500 mm. Major soil types are grouped as Black Chernozems. Spring wheat dominates the prairie land scape. Other crops are oat, barley, brassicas, and peas.

Agroecoregion 3: It includes areas in a narrow band along Red river covering portion of North Dakota-Minnesota Border to Winnipeg and Manitoba in Canada. It is a region with glacial till deposits. Soil types encountered are known as Rego Black Chernozems, Cyrels (Vertisols) with imperfect clays, and Aquerts. Annual precipitation is about 500 mm. Mean annual temperature is 5°C and frost-free days are 125. Prairies support both tall grass and short statured species. Major crop species are spring wheat, oat, barley, canola, flax, potato, and sugar beet.

Agroecoregion 4: This ecoregion covers areas in South Dakota and small sections in North Dakota along the border and Nebraska. Soils are classified as frigid Ustolls. Annual precipitation ranges from 350 to 425 mm. Precipitation decreases as we move from East to West of this zone. Subzero temperatures and short growing period are common. Major crops are dryland cereals such as wheat, oat, and barley.

Agroecoregion 5: It covers region east of river Missouri within the state of South Dakota. It comprises of short prairies, but several tall grasses such as *Andropogan* do occur frequently in patches. Soils are Ustolls. They are deep and loamy. Annual precipitation is about 375–430 mm. The topography is undulated with gentle slopes. About 50–60percent of this agroecoregion is filled with crop species such as sorghum, corn, wheat, oats, and lentils. Productivity is moderate, but crop growth is luxuriant in patches that receive irrigation from Missouri river.

Agroecoregion 6: This region occurs west of Pierre in North Dakota. Soils encountered are Orthids, Orthents, Ustolls, and Fluvents. These are actually plains formed on shales and bedrocks. Most portion of this agroecoregion is not utilized for cropping. They are grouped as Bad Lands. They are mostly non arable rangelands. Cereals such as sorghum and forage are grown in 10 percent of the land of this region. Temperature reach subzero and frost-free days are restricted to 140. Mean annual precipitation is 475 mm.

Agroecoregion 7: This agroecoregion extends from Wyoming in the East to Eastern Montana. Vegetation is predominantly grassland. Soils are shallow and loamy. They are classified as Orthents, Orthids, Argids, and fluvents. Topography ranges from plains to undulated terrain to steep hills in some locations. Grain production is localized to areas near river and plains. Dry land wheat is predominant. Irrigated alfalfa and legumes are common near riverine regions.

Agroecoregion 8: This region occurs in South Dakota. Soils are classified as Ustolls and Torriorthents. Soils in the western part of this agroecoregion are shallow, excessively drained, and sandy. Soils are basically derived from soft bedrock, alluvial, and eroded material. They are often termed as loess in regions where crop production is predominant. Nearly 75 percent of this agroecoregion supports native range land and pastures. Cropland is confined to riverine zones. Alfalfa and grain crops occur near river Platte. Annual precipitation ranges from 440 to 620 mm depending on location. Frost-free period is about 140–170 days and it influences cropping systems.

Agroecoregion 9: This area occurs as a narrow stretch along the foothills of Rocky Mountains to Southern Alberta in Canada. There are regions with elevation above 900–1,200 m above sea level and undulated topography. Soils in the rolling hills are derived from shale and sand stones. Soils are mostly classified as Ustolls, Orthents, Fluvents, and frigid sandy soils. In Alberta, soils are glacial. Native vegetation is largely grass land and range. Average annual temperature is 6–7°C. Precipitation ranges from 300 to 600 mm, but it can be higher in hilly terrain. Temperature ranges from -5°C to 12°C. Cold tolerance is a required trait for crops. Frost-free period is 120–130 days. Wheat and canola are chief crops in the plains and undulating hills. About 20 percent of the land produces wheat. Wheat is predominant south of 49th parallel.

Agroecoregion 10: This agroecoregion extends from Southcentral Saskatchewan to Central Alberta. Soils are relatively loamy. They are mostly black to brown Chernozems (Udic Ustolls). Coarse textured sandy soils are also common in this region. About 80 percent of the areas are filled with arable cereals such as Red Wheat, Oats, and Barley. Oilseeds and forage are other species that occur frequently in the prairies. Frost-free period ranges from 95 to 115 days.

Agroecoregion 11: It encompasses prairies that occur within parts of North Dakota, Saskatchewan, Alberta, and Montana. It supports large expanses of moist mixed-grassland. Natural prairies are conspicuous. Landscape is basically glacial characterized by undulated terrain, steep hills, and plains. Soils are classified as Ustolls (Dark Brown Chernozems) and Cryerts (Vertisols). These soils are endowed with good water holding capacity due to clay and organic matter. About 90percent of the plains region is cropped with cereals, such as spring Red Wheat, oats, barley, canola, and vegetables. Frost-free period ranges from 110 to 120 days. Precipitation ranges from 380 to 420 mm.

Agroecoregion 12: It is located in the southwest of Saskatchewan, eastern Alberta, and Northern Montana. This ecoregion is characterized by glacial undulated terrain, hilly regions, and rolling plains that support crops and native prairie grasses. About 85 percent of the land is cropped. Cereals include Red wheat and Durum wheat. Oilseeds such as mustard are common. Chickpeas and vegetables are other crops grown in this region. Soils are classified as Ustolls and frigid sandy loams. Frost—free period ranges from 120 to 140 days and restricts crop rotations. Crop fallow is most common. Precipitation ranges from 380 to 475 mm annually.

Agroecoregion 13: This region comprises of Northern most parts of Great Plains in Alberta. Forests are dominant. Soils are classified as Gray Luvisols (HaplUdalfs) and Eutrochrepts. Histosols that are peaty and rich in organic matter are also common. Native prairie with perennial grass is frequently encountered. Frosty conditions restrict crop sequences. Crop season is shortened. Wheat is not a dominant cereal. Barley and canola is preferred. Precipitation ranges from 320 to 380 mm.

Agroecosystem 14: Area comprising Northwestern Alberta and adjacent British Columbia along the Peace River marks this agroecoregion. Soils are clayey or fine textured silts. Grey Luvisol is most frequent. Mollic Hapludals, Natrudalfs, and Hapludalfs are other soil types encountered here. Native vegetation is mostly forest. Spring red wheat, barley, and canola are important crops. Forage and flax are other species found in the plains. Mean annual temperature is low for crop production. It is 0.5°C. Summer temperature reaches 13°C. Cold tolerance is mandatory for crop genotypes. Mean annual precipitation is 350–600 mm.

1.1.3 AGROCLIMATE OF GREAT PLAINS OF NORTH AMERICA

General climatic conditions and local variations have played a key role in molding the crop production trends in the entire Great Plains. Climate has largely decided the cropping systems, sequences, planting dates, and periodicity of various agronomic procedures. The Canadian Prairies experiences dry semiarid climate in the Brown soil region and slightly humid climate in the dark Brown and Black soil areas. The Canadian Prairies receive 300–375 mm precipitation annually in the drier tracts and 400–500 mm in wetter continental regions. Crops and natural vegetation in the Canadian Prairies are also served by water bodies, such as Lake Winnipeg and rivers that flow into the Plains. The Great Plains provinces such as Alberta, Saskatchewan, and Manitoba receive precipitation mainly during April to June. The region is affected by frost during October to April. Frost restricts cropping systems to cereal-fallow that is, to just one crop season. Prairie Provinces are relatively humid regions. Canadian Prairies experiences relatively cooler climate throughout the year. During summer (July), temperature ranges from 15–18°C and in winter (January) it dips to −12°C to −17°C. Regarding diurnal pattern, it is said that excessively long day length during summer is useful for crop production.

The Great Plains in United States of America receives moisture from several sources that emanate from atmosphere and ground. Moist tropical winds that move into plains from Central America are the major source of moisture for vegetation that thrives in the plains. The moist winds actually sweep through Southern plains covering areas in Texas and Oklahoma. Moisture laden wind later takes a curve in the Mississippi region and heads on to Northern regions. This causes a dry warm zone in the Western Great Plains, especially in New Mexico and Arizona. The effect of movement of moist wind is easily discernible. For example, in Kansas, regions that experience moist winds receive over 115 cm rain fall and in western part, which the wind misses, receives only 40 cm rainfall annually. This phenomenon results in westward decline of precipitation and actually sets the rainfall distribution pattern in the Great Plains. Western areas of Great Plains receive progressively lower quantities of precipitation. There are periods of high rainfall in the Great Plains caused by movement of tropical air mass. However, farmers usually depend on precipitations that occur during seeding time from April to August, to initiate crops.

Precipitation that occurs during spring and summer takes the form of strong thunderstorms. Hail storms too are common in this period. They are frequent in Western and Southern Plains. Further, Nebraska and Wyoming are known to receive high amounts of hail storms in a year (United States Department, 2013). "Tornadoes," identified easily by the funnel shaped wind systems are frequent in Central Great Plains. Tornadoes with wind swirling at 350 km h^{-1} speed occur in the Oklahoma, Northern Texas, Kansas, and Arkansas. Such tornadoes may bring in moisture, but they do create havoc to standing crops if any.

Winds that bring moisture into Central Plains move with high velocities. Windmills work efficiently, but it may also induce rapid movement of moisture away from crops and induce rapid evaporation. "Chinook" is a winter wind system that emanates from Western Rocky Mountains. Chinook descends into Great Plains and induces warmer climate. It helps in overcoming severe winter cold. Such winds also keep Eastern regions warmer. As a consequence of warmer climate, frost-free period may stretch slightly more than normally feasible. Clearly, snow, blizzards, wind, cold breeze, warm air, and moisture all add to natural elements that either induce or restrict crop season and productivity in the Great Plains (United States Department, 2013).

Prairie based crop production is a major preoccupation of a large fraction of farmers in the Southern Plains of North America. Major cereals that dominate the prairies are wheat, maize, and sorghum. Legumes and pastures are also conspicuous in this semiarid region. Regarding physiography, Southern Plains encompasses a region situated South of Arkansas River (Southern Kansas and Colorado), and east of Rocky Mountains foothills (eastern New Mexico). It extends from Central Oklahoma to Coastal prairies of Texas. The altitude of landscape ranges from 200 to 300 m above sea level in Texas and Oklahoma to 2,000 m above sea level along the hills of New Mexico (Soil Survey Staff, 1981). Growing season in Southern Plains is affected by frost and cool temperatures. Frost-free period ranges from 150 to 300 days depending on location. Typically, cotton thrives well in regions with 180–200 frost free days that occur in areas up to Southern Kansas. Soybean and Sorghum genotypes with short duration to maturity are preferred in entire Southern plains (Baumhardt, 2003). The average annual wind speed is 5.5 ms^{-1}. The mean precipitation does not vary much as we travel from North to South within the Southern Great Plains. However, annual precipitation is 900 mm in Central Oklahoma and decreases to 400 mm as we move toward New Mexico. The evaporation ranges from 1,800 mm in Oklahoma to 2,600 mm in dry belts of Texas. This results in a water deficit (precipitation–evaporation) of 900–1,500 mm. Overall, precipitation increases from West to East, varies little North to South. Precipitation occurs primarily during summer months and is bimodal (Baumhardt and Salinas—Garcia, 2005).

Southern region of plains is relatively warm and experiences temperature of slightly over 90°F for 70–100 days in a year. Northern prairies face bitter cold and temperatures reach above 90°F for only 10–20 days in a year. Average temperature of Northern Great Plains during past 30 years has been 70°F. Reports by USEPA (2012) suggest that mean annual temperature in the entire Great Plains regions has increased steadily. Summers are clearly getting drier and hotter. It seems average temperature in the Great Plains has increased by 6°F during past 30 years (USGCRP, 2009a). Clearly, cold days are proportionately fewer, but hot days are encountered frequently. Temperatures are forecasted to increase further during this century. Some of the reasons attributed are enhanced emissions that trap heat. Northern plains may get wetter and Southern plains drier with time. Heat wave conditions and high-intensity rainfall days may increase in the South. Forecasts by USGCRP (2009a,b) suggest that by the fourth quarter of this century (2080–2099) temperature might increase by 5–7°F over the present level depending on emissions. In the Northern and Central Great Plains, ambient temperatures have increased by 1°C during the past century. In some parts of Montana and Dakota increase in temperature has been 3°C (USGCRP, 2012).

Precipitation is actually projected to change in the Great Plains due to climate change phenomenon. Once again, depending on trends in emissions, precipitation may fluctuate to ± 22 percent over the present levels. These climatic changes may affect water resources and crop productivity. According to USGCRP (2012) reports, annual precipitation has decreased by 10 percent during past 100 years in the major cereal farming zones of Montana, Dakotas, Wyoming, and Colorado.

Precipitation pattern during past has varied drastically in some locations and marginally in others. Average precipitation in the plains is generally calculated by using a vast network of weather stations located in the entire plains. Ross and Anderson (2008) has reported that over past 50 years, total precipitation has not fluctuated much, but precipitation pattern in the Northern and Southern plains have been consistently different. Precipitation pattern in the Northern Great Plains is well spread out and rainfall events are small each time. However, in the Southern Plains, rain fall events are intense or it is a long stretch of no rain.

Precipitation pattern in the Great Plains prairies varies between 34 cm in regions with short prairies and up to 57 cm in regions with tall grass prairies. Precipitation is feeble or nil during summer months. It induces dryland conditions. Natural prairies and cropland experience long-term cyclical patterns of rainfall. Drought spells get accentuated once in 30 years. Dust bowl of 1930 is said to be the most prominent drought spell during recent period. It is said that climate in the Great Plains is also influenced by sheltering effects of Rocky Mountains.

Natural Prairies and crops are exposed to sprees of fire. They occur once in 5 years and may destroy above ground biomass of large patches of short or tall prairies, depending on location and flora. Natural prairie in the Great Plains is a mixture of great number of plant species. The below-ground portion of Great Plains is actually tangled mass of roots, rhizomes, stolon, bulbs, and root stocks. The natural vegetation may die back each year as summer intensifies and water is scarce. However, below-ground rhizomes, bulbs, and dry seeds may allow regeneration of grass lands. In case of croplands, we normally regenerate vegetation by sowing at the onset of rains.

There are computer-based simulations about influence of climate, weather pattern, soils, and land use on biomass/grain productivity in the Great Plains (Bradford et al., 2006). Firstly, temperature fluctuations affect net primary production of prairies. In water limiting areas, higher temperature range restricts its availability to crops and natural vegetation of the prairies. It is believed that seasonal variations in temperature, precipitation, and other weather parameters have major say in net biomass generation in the Great Plains. Soil properties and inherent fertility status may affect natural vegetation marginally. Cropping intensity and its interactive effects with climate affects biomass generation perceptibly. Monocrops may result in abrupt end to crop season. It affects the influence of weather on crop growth. Cropping systems and weather affect several ecosystematic functions, both in the soil profile and above-ground canopy (Bradford et al., 2005, 2006).

1.2   VEGETATION AND CROPPING HISTORY OF GREAT PLAINS OF NORTH AMERICA

North American Prairies is a geographic region with relatively huge expanse of short and tall grasses, legumes, and other species of plants and crops. It also supports a variety of other flora and fauna that has acclimatized to the prairies (see Weaver, 1927). Trees are generally sparse and grow scattered.

Canadian Prairies support diverse vegetation constituted by tall grass prairie, mixed-grass prairie, and fescue prairie. The grass lands are composed of several species of plants. Tall grass prairie regions have been mostly converted into crop-lands that support wheat and canola. Currently, it is said only 6,000 km^2 of Red river valley supports scattered tall grass prairie. Mixed grass prairie is common to the dry regions of Canadian Great Plains. Native grass in all three Great Plains provinces is mixed and good for grazing. About 24 percent of original native grass lands is unaffected by cropping patterns. Moist regions of Saskatchewan and Manitoba supports fescue prairie (Pieper, 2002; Wikipedia Foundation, 2013). Natural vegetation in the Canadian Provinces is constituted by grass species such as spear grass, wheat grass, and blue grama grass. Saline zones are occupied by alkali grass, wild barley, greasewood, and red samphire (McGinley, 2008). In Canada, wetland prairies are integral to cropping zones. They are not vast stretches of natural wet grass or crops. Wetland inventory suggests that in Canada about 80 percent of wetland ecosystem is lost due to drainage of water that is required to initiate an arable crop such as wheat or barley. Wetland fields are usually drained using surface drainage (Haak, 2008).

Some of the earliest efforts to cultivate the Canadian regions of Great Plains occurred during 1800s. The British Government and Canadian Government tested vast expanses of Rupert's Land (i.e., Manitoba, Saskatchewan, Alberta, and Northern Territories) for agricultural crop production and permanent settlement of farming families (Hubner, 1998). Hudson Bay Company was one among the pioneers to develop agricultural farms. Crops such as wheat, barley, hemp, and pastures were grown during early part of 1800s. Productivity of wheat was low. Earliest record-ed wheat harvest in the Canadian Great Plains dates to 1814 and it amounted to 2,310-gallon kegs. During 1820s, wheat germplasm was brought from Wisconsin and Minnesota to replenish lost stocks. During mid-1800s, Red wheat with superior seed traits was grown in Manitoba. Grain elevators were used to store excess har-vests. During the same period in mid-1800s, ware houses to store larger stocks of wheat were constructed and rail heads were in operation to transport grains across to different regions (Hubner, 1998). Each flat warehouse could store 10–15,000 bush-els. Development of pastures and ranches soon became popular with Canadians. Generally, 1881–1897 is termed as "Golden age of Ranching in the Canadian Prai-ries." It covered regions in Alberta, Saskatchewan, Manitoba, and Southern Quebec.

Western Great Plains occurs in the Western half of Great Plains covering ar-eas East of Rocky Mountains and constitutes parts of Nebraska, Texas, and New Mexico. Western Great Plains supports short grass prairies to a relatively higher ex-tent. Some areas in Northwest have mixed grass prairies (CNHP, 2013). *Boutelouea gracilis* and its variants form the dominant short grass vegetation. *Boutelouea eri-opoda* and *Aristida purpurea* are other herbaceous vegetation traced. *Buchloe dac-tyloides, Hespirostipa comate, Agropyron smithi*, and *Sporobolus cryptandrus* are prominent graminoids found in the Western plains. In addition, agricultural crops and pastures are common. Western Great Plains is made up of stretches of rolling terrain with loamy and rustic soils. Clayey and sandy fields are also frequent. It seems crested wheat grass that was introduced into Great Plains in 1900s established rapidly and helped in reclaiming abandoned wheat fields. However, a species such as leafy spurge (*Euphorbia esula*) became prolific and spread like a weed. Several other grass species such as Japanese Brome (*Bromus japanicus*) also had its impact on perennial flora in the Great Plains (Pieper, 2002).

According to Patton and Marston (1996), Central Plains were dominated by short and tall grass prairies interspersed with slush and wetlands, forests, and grassy savannahs. The vegetation and fauna were used since the Paleo-Indian occupation in 25000 B.C. Central Great Plains has been colonized and utilized by a series of migrants that includes early hunters, foragers, and farmers/settlers. Indian tribes such as Osage, Wichita, Kiowa, Comanche, and Apaches had a major share in nurturing natural vegetation that had different grasses, legumes, and other species. Further, they utilized the Central Great Plains for crop production; mainly they grew crops such as maize, barley, beans, and vegetable. During Medieval and recent history, Spanish and French explorers and settlers did change the vegetation pattern, especially the crop production region to a great extent. It seems relocation of certain major Indian tribes from Oklahoma and other vicinities in Central plains did affect vegetation and crop species/genotypes that dominate the region. The Anglo-American settlements grew wheat and barley in great proportion. Mixed farming that supported cereals, cotton, and vegetables spread rapidly in regions with European settlers. The Osage and other Indian tribes had ranches and grew crops such as maize, legumes, and vegetables. During the last century, dominant crops were winter wheat, rye, maize, pecans, grain sorghum, cotton, soybean, and peanuts (Patton and Marston, 1996).

In Mexico, natural vegetation in the Plains consists of scrub species such as Acacia, Paloverde, Silver leaf, Hackberry, Texas olive, Barreta, Corbagallina, and Ocotillo (McGinley, 2008). According to Hudson (2012), domestication of crops and their cultivation began much later compared to southwestern region of North American continent. Maize, amaranth, chenopods, sunflower, and tobacco were among the earliest of crops grown by Native Americans situated in the plains. Archaeological excavations suggest that at least 13–18 different types of maize cultivars were in use in the Northern Great Plains as early as fifth century A.D. Sedentary village and settlements that produced corn and other crops occurred in North Dakota by 850 AD. Prairie crops such as wheat, sorghum, and cotton were introduced into plains around sixteenth century by the European settlers. Soybean and lentil production became wide spread much later during eighteenth century. Regarding wheat, it seems settlers from Wisconsin, Ontario, and Minnesota transshipped spring wheat into North Great Plains and Canadian Prairies. Marquis wheat, a hard spring wheat, was most preferred in the Canadian plains. Wheat monoculture became prominent in the Dakotas and Red river valley. Montana supported wheat and barley production in larger scale. Minnesota and Dakota farmers also produced sugar beet and potato in large expanses. Corn production became prominent in Kansas, Nebraska,

South Dakota, and Texas. In fact crops such as corn, sorghum, and soybean became a mainstay for farmers in Central Plains region. Large-scale wheat production in Kansas and other locations of Central Great Plains was initiated by migrants from Pennsylvania and Ohio. They grew winter wheat. Turkey red wheat was introduced into Kansas Plains by Germans who migrated in 1870s. This hard winter wheat produced better bread. Winter wheat production in large farms is common in Kansas, Oklahoma, Texas, and Colorado. The Upland cotton region in Southern Plains occurs in Texas and Oklahoma. The Texas high plain is an important cotton growing region in the plains (see Hudson, 2012).

1.2.1 DEVELOPMENT OF FARMS AND FARMING IN THE GREAT PLAINS: RECENT HISTORY

Initiation of large-scale farming was an event of immense influence to natural vegetation of the Great Plains. It converted natural prairies to agricultural prairies with vast expanses of cereal, legumes, cotton, and soybean as we perceive it today. Several factors related to human preferences, economics, and natural resources affected the development of agricultural prairies in the Plains area. During past two centuries, human migration and farming are two crucial factors that drastically changed the diversity of Great Plains in terms of plant species and related aspects.

Farming was encouraged by the US government agencies as a matter of policy. Farmer was provided with 160 acres of dryland per family under Homestead Act of 1862. Farmers had to stay on the farm for 5 years and develop land to acquire a farm. Farmers could also claim land immediately after 6 months of stay, if they paid 1.25 US$ ac^{-1}. Generally, small farms were not efficient and sustainable, given the semiarid and drought prone conditions that prevailed in the Great Plains. Farmers were provided a further 160 acres of land under Timber act of 1873, if they opted to raise tree and developed forest plantations. The Desert Land act of 1877 allowed settlers to obtain a further large patch of land of 640 acres, if they opted to develop irrigation facilities within 3 years after acquiring the land. Farmers and settlers also bought land from other agencies and initiated crop production in the Great Plains (Cliff-notes, 2012; Library of Congress, 2012). There were also vast stretches of land in Great Plains that were converted to agricultural crop production under system known as "Bonanza farms." It allowed over 10,000 acres of native prairie land to be converted agricultural prairies. "Bonanza farms" were developed to produce wheat. The "bonanza farms," it seems disappeared by 1890s. A few that were symbolic of the beginning of large-scale farming enterprise in US Great Plains (Rodrigues, 2004) are still operational.

It seems European settlers and farming community soon realized that crop productivity in the Great Plains is comparatively lower than what they achieved in the Northeastern zone. Great Plains experienced semiarid climate and was drought prone. Further, soil fertility too was moderate. Hence, crop production, at times was a costly enterprise. The dry farming depended more on farm machinery, deep tillage of soil, irrigation to protect crop from wilting, sod busters, seed drills, wind mills, pumps, and irrigation channels. By 1874, barbed wires and fences were used by farmers to protect the crop zone. Wheat was the staple crop produced in the Great Plains and exported. Wheat production increased from 170 million bushels in 1860s to 700 million bushels by 1900. This was attributed to large-scale use of farm machinery, irrigation, and organic manures. Overproduction of wheat occurred in several other agricultural belts of the world, such as in Europe, Argentina, Australia, and Canada. It led to depreciation of price for wheat grain. To offset depreciated profits, farmers in Great Plains expanded farming zones. They increased farm size to increase total grain harvest. This further led to decrease in wheat grain price, because supply far exceeded demand (Cliff-Notes, 2012).

Following is a chronological list of Farming practices that affected crop production in the Great Plains:

1833—Initiation of wheat seeding, use of reapers to collect wheat grains;

1846—Use of steel plow to break tough sod and loosen surface soil;

1860—sod grass cultivation, beginning of cattle farming and grazing stretches;

1866—Development of railroads and rail heads to transport surplus wheat grains;

1874—Use of barbed wire to protect wheat farming stretches and cattle ranches;

1880—Development of standard procedures for dryland faming;

1880—Use of wind mills to provide irrigation;

1890—Harvesting wheat using combined harvesters. Large-scale introduction of tractors fitted with internal combustion engine and that could plow using fuel oil. Tractors were introduced in 1914, first at Omaha in the plains of Nebraska (See Krishna, 2002);

1946—Great Plains Agricultural Council was founded in 1946. It merged the previous organization such as Northern Great Plains Regional Agricultural Council of 1938 and Southern Regional Agricultural Council of 1939. The main aim was to coordinate agricultural cropping in the Great Plains region. Further, it aimed to build cooperation among producers and marketing agencies (Lange, 2008).

North American Agriculture experienced a boom in food crop production during 1880s till mid-1920s. This period is sometimes referred as "Golden Age of American Agriculture." The Great Plains had its share of success to contribute. The spurt in farming actually resulted from rapid expansion wheat production zones, enhanced productivity, and mechanization. For example, in North Dakota alone, crop production zone doubled from 4.3 to 9.4 m ac. Wheat grain yield in North Dakota increased from 69 to 159 m bu (IRSNDSU, 2013). It seems gross income of family farms and large companies doubled or even tripled. Large-scale migration into Northern Great Plains, especially Dakotas, and Midwest further added to agricultural boom. Reports suggest that over 250,000 farm families migrated into North Dakota to initiate wheat production. Development of infrastructure such as roads and railways helped farm produce to be transported rapidly to sea ports. Homestead Act made it easy for farmers to acquire 320 acres of land to develop the wheat belt. During the same period farm mechanization and advent of tractors helped in plowing large expanses relatively easily and rapidly.

1.2.2 POPULATION, SETTLEMENTS, AND MIGRATIONS AS RELATED TO AGRICULTURAL CROPPING

The human inhabitation of Great Plains, it seems began several thousand years ago with movement of Paleo-Indians through Beringia or Bering Strait. Great Plains was inhabited by native Indians, whose population was scattered into all regions on North American plains and mountainous terrain. Their dwellings and citadels, if any, were found all through the plains. Some of the Indian tribes who evolved in the Great Plains are Blackfoot, Camanche, Arapaho, Sioux, Cheyenne, Crow, Wichita, Pawnee, and Mandan. They farmed maize, vegetables, and beans for their sustenance. According to estimates by USGCRP (2009a), there are still about 65 Native American-Indian tribes who inhabit the Great Plains at different population intensities. These Native tribes may not have resource to negotiate changes in climate and crop productivity. Many reservations, it seems often try to migrate as they experience water scarcity and reduction in soil fertility conditions.

Native Indians first encountered European settlers in 1540s in the plains of Nebraska, Kansas, and Texas. The Great Plains region was then colonized by people of European origin during eighteenth to nineteenth century with a rapid and perceptible spree (Powell, 1878; Webb, 1931). It is generally opined that political, sociological, and economical reasons proved to be good for people to migrate to semiarid plains. In due course, agriculture and cropping systems practiced in the plains had a major impact on the extent of migration from Northeastern region of the continent (Gutmann et al., 2005). In the Eastern Plains, possibility of consistent cropping on arable and fertile soils was a major factor for European settlers. This region also allowed better economic advantages through farming. The intensity of European settlements depended on precipitation pattern, soil fertility, and aridity. The arid Western regions of Great Plains allowed only, proportionately feeble settlements on farm land. Yet, population pressure, political reasons, and hope for better farming enterprises induced steady migration of people from North and Northeast to Great Plains. No doubt, years with good rainfall and better crop harvests must have induced proportionately greater people to migrate to plains. Agricultural enterprises were initiated in the Great Plains in large number during 1910–1930 (Gutmann and Sample, 1995). Initially, fertile prairie locations attracted farmers from Northeast to migrate and initiate new farms. Migrated population must have in turn aided spread of cropping zones into vast stretches of Great Plains. A phenomenon interrupted only by drought, loss of soil fertility, natural calamity, and economic pit falls. Plowing large stretches of hitherto natural grasslands to cultivate wheat and other field crops induced loss of topsoil and its structure. This combined with frequent droughts and dust storms induced "dust bowls" of 1930s. The dust bowl condition plus repeated crop loss caused migration of farmers from Great Plains. During 1910–1930, population growth in Great Plains far exceeded national average. Those farmers who remained *in situ* had actually protected their farms using improved varieties, erosion control, irrigation, and organic manure supply. Gutmann et al. (2005) suggest that agricultural cropping trends, migration, and population changes had impacted environment. Conversely, we should note that droughts and dust bowl condition indeed affected demography and agricultural cropping practices. For example, during 1917–1920, migration of farm population from states like North Dakota and Nebraska occurred due to drought. According to Popper and Popper (2004), dust bowls had ecological consequences. They affected crop productivity. Great Plains actually went through depressed precipitation pattern and low grain yield for dozens of years prior to facing severe dust bowl. The short grass prairies and crop fields in many places were severely deteriorated due to rampant cattle grazing and removal of top grass. It is said that due to dust bowls, farmers in Great Plains have more frequently operated in the fields and farms with constraints not common to those in other regions of the North America.

In the Southern plains, migrations related to farmers and farm workers are attributable to variety of reasons. Factors that induced migrations in different regions/directions have occurred at different intensities and for lengths of periods. During mid-1800 (1865–70), agricultural workers from South migrated to North and Midwestern States to pursue different professions. It was caused by crop failures and massive loss of exchequer to farm owners and share croppers. It is said that 6 million farmer workers who otherwise depended on farm activity left their farms in search of farm work elsewhere (Lewis, 2012).

Influx of European settlers into Great Plains did alter the landscape, agricultural area, demography, and distribution of human population. It seems Great Plains extended into 390 million hectare, but a mere 19 million hectare were cultivated with cereals and other crops. Migration into plains was rapid and it induced expansion of farming enterprises. Agricultural cropping covered 190 million hectare by 1910 and 288 million hectare by 1930. No doubt, conversion of natural grass land and vegetation into crop fields was marked and rapid. During early 1960s, 90 percent of the cultivable land in Great Plains had been occupied by crops. Gutmann et al. (2005) and Cunfer (2005) state that rampant plowing, soil disturbance, and much of conversion of natural grass stands into crop fields occurred during first three decades of twentieth century. Conversion of natural prairies to cropping expanses was linked to demography.

During twentieth century, it said that agricultural cropping and manufacturing sectors experienced rapid fluctuations. We can classify the period into premechanization and postmechanization era. Population increase and agricultural expansion in Great Plains was marked during premechanization. White (2008) points out that farm mechanization had occurred well-before mid-twentieth century. For example, John Deere Company had been producing steam engine tractors, grain threshing devices, and combines. These improvements had perceptibly enhanced agricultural productivity and economic status of farmers. It induced migrations into Great Plains. Farmers produced more bushels of cereals per acre. Farm labor requirement decreased perceptibly, almost by 80 percent of original levels in 45 years, due to mechanization. Further, Cochrane (1993) points out that from 1830 to 1930, mechanization reduced requirement of farm laborers by 90percent to produce 100 bushels. There was a further spurt in crop production efficiency during 1900–1945. Farmers in Great Plains reduced labor needs by another 45 percent over initial levels. Albrecht (1993) has reported that proportion of population depending on farms decreased, if county population increased rapidly. In fact, population readjustments were rampant during twentieth century. It was caused due to migrations to urban zones, farm mechanization, and farm economics.

According to White (2008), researchers do believe that farming enterprises, their productivity, population change, and migrations were also affected by global agricultural production trends and marketing. For example, wheat market in Great Britain had perceptible influence on farming enterprises, their expansion, mechanization levels, and farm labor usage. Also, small family-based farms soon changed to larger farm production/business companies. This aspect again affected farm settlements, their viability, and it actually induced migrations in different directions within Great Plains (Friedmann, 1978; White, 2008). Grazing and development of forage farms were important aspects of Great Plains farming enterprises (Cunfer, 2005). In areas afflicted with poor soil fertility, soil maladies, and lack of proper irrigation, higher profitability of cattle-induced establishment of farms and settlements. Livestock production added richness to farms in terms of economics. We should note that Great Plains experienced rapid improvement in population during 1950s and 1960s. It amounted to an increase by 12 percent over initial levels in 1900. White (2008) stated that between 1990 and 2000, population in the Great Plains increased by 16 percent, while that of USA improved by 13 percent. Migrations were largely to counties with urban proliferation.

1.3 NATURAL RESOURCES WITH PARTICULAR REFERENCE TO AGRICULTURAL CROP PRODUCTION IN GREAT PLAINS

1.3.1 SOIL RESOURCES OF GREAT PLAINS

Canadian Provinces:—Soils in the Canadian Prairies range from rich fertile Brown Chernozems in the dry mixed grass lands and wheat belts of Alberta to Black Chernozems in the Aspen Park land and Gray Luvisols in the forest zones. Soils are continuously exposed to aridity and cold temperatures. Canadian Prairies possess vast stretches of fertile Chernozems that support agricultural crop production. Gleysolic patches are traced throughout the agricultural zones of Canadian prairie province. They are found associated with Chernozems and Luvisols (Bedard-Haughn, 2010).

Anderson and Cerkowniak (2010) opine that Canadian Prairies are a special region with regard to forces that aid soil formation. It is clear that gradual and consistent changes in climate and vegetation have resulted in distinct soil types, peculiar to cold regions. Major processes that aid soil formation are additions, removals, transfers, and translocation of soil material and physicochemical transformation. Additions refer to organic matter inputs, N input through atmospheric N fixation, and supply of soil amendments. Removal refers to leaching of soil material through erosion, snow melts, wind, and physical removal of surface soil. Soil components such as silt, clay, nutrients, and organic matter are translocated from one location to another. This process is called translocation. Clay translocation is a dominant process in the Canadian Prairies. Soil components such as organic matter and nutrients are transformed through variety of soil chemical reactions and microbial activity. This refers to transformations. These soil formation processes also affect availability of nutrients to crop roots. Several factors such as climate, parent material, vegetation and other biota, topography, time, groundwater, and human activity affects soil formation trends.

Soils in Dakotas, Montana, Wyoming, and glaciated regions of Missouri are classified as deep black Chernozems rich in organic matter and nutrient supply capacity. Northern Spring wheat belt, in fact, flourishes on such soils known as Borolls, which is a type of Mollisol. Such deep black soils also support vast stretches of pastures and small millets in the Dakotas (Hudson, 2012).

Great Plains in the Kansas region harbors a large variety of soils that serve natural prairies and vast cropland with anchorage, nutrients, and water. Cropland that extends into 29 million acres flourishes on soil types such as Mollisols, Alfisols, and Inceptisols. They are loamy and support major cereals such as wheat, corn, and sorghum, in addition to vast stretches of cotton and soybean. Pastures thrive luxuriously on brown Chernozems of Kansas. The fertile soils are, however, prone to erosion due to wind and water. They are deep, well drained, and occur on slopes and flat ridges found in the Central plains. Silt loams are common in the wheat belt. They are 25–30 cm deep with topsoil that is grayish brown and rich in organic matter (USDA, 2012a).

There are about seven soil orders that are easily traced in the Southern Plains. They are Alfisols, Aridisols, Entisols, Inceptisols, Mollisols, Ultisols, and Vertisols. Soil orders such as Histosols and Spodosols do occur but are sparsely distributed. Andisols, Gelisols, and Oxisols are not traceable in Southern Great Plains (USDA, 2012b). Alfisols of the Duval series with subsurface accumulation of clay occur in the Rio Grande Plains. Aridisols are predominant in desert and dry belts of western Texas and New Mexico. They do possess horizons that accumulate salts such as carbonates and chlorides. Gypsum may be conspicuous. The Upton series Aridisols occur in Trans-Pecos zone. Entisols are prominent in the Texas Coastal Plains. Histosols with relatively higher amounts of organic matter are traced near the marshes in the Southern Coastal and Delta zones. Mollisols of the Pullman series that have dark colored horizon are enriched with organic matter. They are common in natural prairies of North Texas, Oklahoma, and New Mexico. High Plains and gentle sloppy terrain in Southern Plains is rich in Mollisols. Inceptisols with weakly developed horizons and shallow depth occur in the rolling slopes of Southern United States of America. Weswood Inceptisols are traced in the flood plains of rivers Brazo and Colorado in Central Texas. Ultisols of Bowie series are highly weathered and possess subsurface horizon enriched with clay. They occur in the East Texas timber land area. Vertisols of black series are predominant in the prairies of Bad Lands and San Antonio area. These soils are known for swell-shrink traits and good water holding capacity. Spodosols of Babco series are acidic and are marked by subsurface accumulation of iron and organic matter. Spodosols are extensively traced in the woods of Texas.

1.3.2 SOIL TILLAGE SYSTEMS IN THE PRAIRIES—IMPACT ON SOILS AND CROP PRODUCTIVITY

Plow is an important implement that has influenced soils and cropping expanses of the Great Plains. It has imparted wide range of advantages in terms use of soil profile. Its influence has been marked if considered on a long term. Plowing actually disturbs and turns up the field soil before it is sown with seeds. The aeration that results in the upper soil induces organic matter decomposition. The soil nutrients become easily available to crop roots. Plowing, especially, deep plowing on a long run induces rapid loss of SOM, and diminishes several aspects related to SOM such as soil structure, moisture holding capacity, tilth, and biological activity.

Historically, Great Plains was subjected to rapid and intense plowing during eighteenth and nineteenth century. Sod breaking and converting natural prairies to cropped land was rampant as settlers converged into more areas in the Great Plains. It is said that intense plowing led to depreciation of soil organic carbon from 2–3 to 1 percent in soil. In some locations, rich prairie loams that had 8–10 percent SOC lost C rapidly and it decreased to 1 percent SOC by weight. This loss in SOM and related effects reduced the crop productivity of Great Plains to 71 percent of original levels. Replenishment of SOC through FYM became a necessity. All these effects were originally attributable to intense plowing. During recent years, concerted efforts have been made to reduce soil disturbance and shift to agronomic procedures such as conservation or reduced tillage and no-tillage (Cole, 1990; Flach, 1997; and Allamaras, 2000). One of the earliest comparisons of different tillage practices adopted in the Great Plains suggests that, soil disturbance was least if no-till or stubble mulch tillage is practiced (Woodruff et al., 2011). In several locations within Central Great Plains, Smika (1976) found that soil water retention ranged from 42 to 203 mm in fields under stubble-mulch tillage system compared to 28–115 mm in case of conventional tillage systems. It was a clear demonstration of advantages of adopting conservation tillage or no-tillage systems over conventional deep tillage. Soon, no-tillage systems became popular in the Great Plains. Farmers gained in terms of reduction in loss of surface soil and moisture, whenever they practiced no-tillage systems.

Soil tillage systems adopted through the ages had its own impact on soil structure, soil fertility, moisture holding capacity, crop growth, and productivity. Tillage, especially, conservation tillage or no-tillage systems has been effectively used during past two decades to thwart loss of soil fertility and quality. Land surface management techniques, such as flat-bed formation, ridges-furrows, and broad bed-furrow, too have been utilized to manage soil fertility. They are useful in regulating soil moisture. They help in improving irrigation efficiency and water use by crops. We should take note that in the vast plains, even marginal advantages in conserving moisture has its valuable effect, upon extrapolation.

Flat land surface and leveled fields are most common in the plains. Ridge tillage, which has ridges and furrows, is used to guide irrigation water and conserve it just around the crop roots. According to Unger and Musick (1990), cereal farmers in Southern Great Plains have often adopted ridge tillage to manage irrigation water and improve crop yield. Ridge tillage is useful in soils with appropriately low infiltration and allow free surge of water in furrows.

Ridge-furrow tillage systems with small earthen dams and basins that collect water are important procedures in the plains. Small basins formed behind the dikes help to reduce rainstorm effects. It also reduces irrigation runoff and allows better water infiltration. Such small dikes allow us to retain water for a certain period of time (Jones and Stewart, 1990). Tillage equipment used has its impact on crop residue management. For example, in the Great Plains only 5–10 percent of maize residue may remain after use of moldboard plow. About 30–50 percent crop residue may still remain after use of disk, about 70–85 percent after chisel plow, and 80–90 percent after para plow. We should note that mulches have an impact on soil erosion.

1.3.3   ADOPTION OF NO-TILLAGE AND CONSERVATION OR RESTRICTED TILLAGE SYSTEMS IN THE GREAT PLAINS

According to Triplett and Dick (2008), since several centuries, agricultural cropping in any area of the world was synonymous with tillage. Crops were never grown without tillage. No-tillage became possible only with advent of herbicides that effectively eradicated weeds. No-tillage, which is defined as sowing seeds without preparing soil but applying about 30 percent residues, was suggested in 1950s, but it was considered seriously by the Great Plains farmers only by 1980s. Historically, several variations of no-tillage were adopted in Great Plains. For example, Native Indians planted maize without tillage. They dibbled seeds in a trough or hole along with dead fish that added organic matter.

Long-term studies indicate that no-tillage system adopted in the Canadian Prairies has performed advantageously to farmers. For example, Smith et al.

(2012) reported results from a study initiated in 1968, when no-tillage concept was still in its rudiments. It suggests that management of crop residue, weeds, and use of proper seeding equipment is required. They compared three different tillage system, namely wide-blade deep tillage, heavy-duty cultivator that incorporates large amounts of crop residues, and no-tillage system with chemical weed control. Soil compaction could be a problem, if soil is not loosened once periodically in 3–5 years. During the 40 years, grain yield under wheat-fallow systems were slightly lower in no-tillage plots. It was compensated by decreased costs on tillage. However, during past 10 years, wheat grain productivity has not been affected due to adoption of no-tillage system. Further, a 3-year evaluation of fallow–wheat–wheat–flax rotation has shown that wheat grain productivity, if sown as no-till after fallow was equal to grain yield under conventional tillage and manure schedules. Utilization of a hoe opener was efficient compared to disks and heavy tillage. Indeed, there are now several reports that suggest no-tillage with chemical weeding as cost effective and profitable alternative in the Canadian Prairies. In the Brown soil region around Alberta, McConkey et al. (2012) found that minimum or no-tillage did not reduce grain yield and residue. Aspects such as energy efficiency, soil compaction, quality, net economic returns were not adversely affected by adoption of no-tillage systems. In fact, during the past 30 years, since the first trials in 1981, no-tillage system has consistently improved C sequestration in fields, compared to conventional deep tillage with disks. Selection of appropriate crop rotations, fallows, and no-tillage systems has often helped farmers in the Great Plains to conserve soil fertility, sequester C better, and still obtain optimum crop yield. No-tillage reduces costs on tillage equipment and labor.

There are several evaluations about the influence of no-tillage summer fallow systems on water storage efficiency in the soil profile. The Mollisols on which major cereals, soybean, and cotton thrive has deep profile with good water holding capacity generated by organic matter and clay fractions. For example, Peterson et al. (1996) have shown that water storage capacity actually varies depending on geographic location, soil type, and cropping patterns. In the Northern Prairie region, storage capacity of fallows is 18 percent, in North Dakota 26–36 percent, Montana 32–42 percent, Colorado 18–22 percent, Kansas 20–37 percent and in Texas it is 10–12 percent.

During past 2 years, farmers in the Midwest and North Great Plains have shown a tendency to replace no-tillage systems with one-pass shallow vertical tillage and seed drilling systems. It is said that use of shallow vertical tillage helps in deriving better impact on soil conservation, moisture retention, better seed germination, and soil nutrient management (Klingberg and Weisenbeck, 2011). The shallow vertical tillage machines and implements cause least disturbance to soil and therefore, as such minimize SOC loss. Following is an example:

Tillage Equipment	No of Fields Tilled	Soil Depth Tilled (Inches)	Crop Residue Retained on Surface (%)
Great plains turbo till equipment (one pass, shallow till)	8	0–1.5	90–94
Summers super coulter (one-pass, No till)	6	2.5–3.0	76–88

Source: Klingberg and Weisenbeck (2011)
Note: Figures are average of 6–8 trials in the plains region and Midwest.

Firstly, there is net gain in crop residue retention in soil. It means SOC sequestration is enhanced. Soil moisture too is retained better for use by the crops. Soil disturbance is least; hence, oxidative loss of CO_2 and microbial respiration will be at threshold. Now, if we extrapolate large expanses of the Great Plains, such a modification to one-pass shallow tillage system will have enormous impact on soil quality and crop productivity. We may also reduce environmental problems that may emanate due to tillage.

Baumhardt (2003) points out that; no-tillage is a good alternative to series of soil and conservation methods including tillage, ridging, formation of dikes, and so on. Historically, there has been a progressive decrease in deep tillage and dust mulching practiced in 1920 to almost no-tillage or restricted tillage by 1970s. As a consequence, precipitation use efficiency and fallow efficiency has increased. Water stored in soil has increased from 20 to 50 percent due to no-tillage systems.

According to Padbury et al. (2002), farmers in the Northern Great Plains have consistently practiced agronomic procedures that suit the dryland conditions. They have practiced conventional tillage and grown wheat, mostly water efficient and drought-tolerant varieties. However, during past two decades advent of moisture conserving no-tillage systems has allowed farmers to cultivate a series of other crop species, instead of fallows. Crops such as oilseeds, pulses, forage, and vegetables have been grown in sequence with wheat. Clearly, modification of tillage system has influenced crop pattern in the North plains region.

1.4 SOIL FERTILITY ASPECTS OF GREAT PLAINS

1.4.1 SOIL NITROGEN DYNAMICS IN GREAT PLAINS AND ITS IMPACT ON CROP PRODUCTIVITY

Natural prairies suffered due to depreciated soil fertility and water scarcity as and when they got manifested at higher intensity. Further, we should note that, establishment of agricultural crops in place of natural vegetation or sod has been typically constrained by the same two factors, namely water and soil nitrogen availability. Peterson et al. (1996) believes that plowing and supply of extraneous organic matter altered water and nutrient (N) dynamics of Great Plains. Water storage and mineral-ization of N were accentuated. Farmers of Great Plains continued with crop (wheat or maize)-fallow system for several years, perhaps more than 30–50 years without addition of fertilizer-N or using legume derived-N after conversion of natural prai-rie to cropping expanse. Since, 1970s, soils in the Great Plains have been replen-ished with N through chemical fertilizer, because it resulted in perceptible increase in productivity. Clearly, soils were exhausted of essential nutrient-N and hence, crops responded to N supply. Further, Peterson et al. (1996) observes that cropping intensification necessitated greater use of water, fertilizer-N, and other nutrients. Fertilizer-N inputs further helped farmers with better options such as wheat-maize-fallow; wheat-oilseed fallow, maize-maize fallow, and so on.

Great Plains area has been incessantly cultivated for past two–three centuries, may not be intensely all through the area, but at different levels of planting den-sity and productivity. Replenishments of nutrients, especially N and P, scavenged by crops have occurred. Yet, it seems insufficient. In case of K and certain mi-cronutrients, farmers try to neglect its supply thinking that soils are endowed with sumptuous levels and that routine recycling of crop residues is enough to answer their depletion. We should realize that since the advent of chemical fertilizer tech-nology and its adoption in the Great Plains, grain crops, oilseeds, and forages have all responded with enhanced productivity. There are too many reports about influ-ence of fertilizer supply trends, especially quantity, formulations, timing, method of application, and so on. We have to take note that grain crops respond to nutrients inherently held in soil and available to roots plus that supplied through chemicals and organic manures. Stewart (2002) has summarized innumerable research reports and concluded that about 16–41 percent of grain formation in cereals grown in Great Plains could be attributed to extraneous N supply. He has examined the crop yield with and without N supply to fields. Following are examples:

Crop Species	Wheat	Maize	Sorghum	Barley	Soybean	Peanut	Cotton
Reduction in Grain Yield (%)	16	41	19	19	0	0	37

Due to absence of fertilizer-N
Source: Stewart (2002)
Note: Lack of fertilizer-N affects maize productivity with greater impact, compared to other cereals. Legumes such as soybean or peanut do not suffer much because of their ability to derive benefits from *Bradyrhizobium* that fixes atmospheric-N (see Herridges et al., 2008).

Schlegal et al. (2005) have classified crops grown in the Great Plains into those demanding high fertilizer-N supply such as corn, grain sorghum, canola, spring, and winter wheat; those needing medium levels of N supply such as flax, oat, sunflower, and buck wheat; those requiring only low amounts of fertilizer-N inputs such as dry beans. There are several legumes that need no fertilizer-N supply, perhaps except-ing a starter-N in some locations to induce rapid rooting. They are pea, faba bean, soybean, and forage legumes. Such information supplied to farmers and extensions agents can be helpful in distributing and channeling fertilizer to specific areas within Great Plains more accurately.

It is interesting to note that among various agronomic procedures and soil amendments practiced in Great Plains, staple crop such as wheat depends percep-tibly on fertilizer-N supply. Data accrued since 1930s till past decade suggests that about 27–50 percent of wheat growth rate and grain formation is attributable to fertilizer-N and P supply (OSU, 2000). As stated above, maize too is relatively more dependent on fertilizer-N supply and it responds to N better. For example, a simi-lar analysis of data since 1950 indicates that 57–68 percent of biomass and grain formation by maize grown in the prairies is attributable to fertilizer, N, P, and lime inputs (Stewart, 2002). No doubt, fertilizer technology, especially, N supply has enhanced biomass and grain productivity of Great Plains. It has actually primed the Great Plains ecosystem with nutrients and led to better turnover and recycling of nutrient elements and organic fraction of ecosystem. Grain harvests and export to far off consumers have of course affected *in situ* recycling. It therefore, necessities repeated replenishment of chemical and organic manures.

About 40 percent of forage production zones in Canada are located in the Prairie Provinces. Forages are often localized to fields with low fertility soils. However, they do respond to extraneous supply of fertilizers, especially, N. Soil type may affect the extent of biomass increase due to fertilizer-N. For example, in Saskatch-ewan, forage crop thriving on Black Chernozems responded to fertilizer-N better with 2.25 t biomass increase for 50 kg N input. Forage grass on Gray Luvisol of-fered 1.35 t ha^{-1} and Brown Chernozems 0.96 t ha^{-1} for a supply of 50 kg N ha^{-1} (Malhi et al., 2004). The extent of biomass response depended on forage species. For example, response of wheat grass, smooth brome grass, crested wheat grass, and Russian wild rye responded with marginal increase in biomass. Generally, supply of fertilizer-N has enhanced protein content of forage grass, and has also allowed accumulation NO$_3$-N in soil. Excessive supply of fertilizer-N needs to be curtailed, since it is prone to loss via seepage and percolation.

Schlegel et al. (2005) opine that cropping systems practiced in the Great Plains has gradually changed from cereal-fallow to intensification and diversified crop se-quences. Monocropping of cereals and conventional tillage systems have given way to no—tillage with multiple cropping systems. Crops stay in the field throughout the year. They argue that many of the fertilizer-N recommendations are based on models that consider crop-fallow systems. There is actually need to change over to fertilizer-N schedules that are tailored to suit higher cropping intensity, better precipitation-use efficiency, and diversified cereal dominated crop rotations that include legumes, oilseeds, and forages. There is also a need to achieve better syn-chrony between fertilizer-N supply, availability of N to roots, and absorption pattern of crops. It is believed that use of handheld leaf-N measuring devises that use optical sensors and measure chlorophyll contents will enable farmers to distribute fertilizer-N more accurately. Satellite mediated measurements of crop reflectance, crop growth rate, and yield maps will be highly relevant in multiple cropping systems. Supplying fertilize-N using spectral data/sensors is effective for several crops grown in Great Plains (Barker and Sawyer, 2013). Actually, Great Plains farmers are now able to schedule fertilizer inputs based on yield goals and economic returns. Currently, maize crop is supplied with a preplanting rate of 180 kg N ha^{-1} and in-season N needs are answered using spectral data, rainfall data, and crop growth stages. It has allowed farmers to channel nutrients exactly when response to fertilizer-N is greatest. Usually for crops such as wheat, maize, sorghum, or cotton, the window of opportunity for best response occurs during rapid growth phase that begins after seedling stage and ends after seed-fill. Decisions on quantity of fertilizer-N are arrived at using computer models and simulations. For example, "Maize-N" is a decision support system for fertilizer-N supply to corn (Setiyono et al., 2011). They consider large set of factors related to soil, environment, and crop growth rates in addition to economic advantages. These agronomic techniques also avoid undue fertilizer-N supply to crops during late grain maturity or senescence period, when crops do not absorb nutrients in larger quantities. It avoids undue accumulation of N in subsoil, its erosion or emission. Great Plains farmers are therefore able to enhance fertilizer-N use efficiency.

Currently, it is common to calculate fertilizer-N requirements for the entire sequence of crops grown, instead of wheat alone or canola alone. Crop rotations too are tailored after carefully considering the N demand by crops at different stages of growth, residual-N available in soil, N derived from mineralization process, and N lost to environment. Generally, fertilizer schedules adopted on the immediate previous crop has a marked effect on wheat grown in the Great Plains. For example, Schlegal et al. (2005) have shown that spring wheat yield after a legume such as fababean, lentil, or pea averaged 2.23–2.5 t grain per hectare, but if the previous crop is yet another cereal or wheat itself, grain yield is low at 1.82–1.86 t ha^{-1}. Residual-N that is carried forward should be utilized effectively. It can then reduce problems related to soil-N accumulation and groundwater contamination in the Great Plains.

Currently, precision techniques that utilize within field soil-N fertility maps, crop reflectance patterns, and accurately supply fertilizer-N using variable-rate suppliers are in vogue in many locations within Great Plains. The variable-rate supply involves use of several crop growth models, previous data on crop's response to fertilizer-N, soil-N maps, and yield goals. Clearly, intensification of cropping and productivity using precision techniques, matching fertilizer-N supply accurately, rapid removal of N from soil will be a better proposition. It avoids excessive accumulation of N in soil profile, avoids undue loss via erosion, and emission. It also minimizes movement of soil-N into groundwater resource such as Ogallala.

Nitrogen management techniques in North America, including Great Plains region revolves around matching N inputs with crop's need, minimizing N loss via erosion, percolation, seepage, and more importantly as N_2, NO_2, and N_2O emissions. Agronomic procedures adopted by Great Plains farmers often aim at reducing NO_3 accumulation in soil profile that otherwise is vulnerable to loss through natural processes. Application of Nitrification inhibitor is an important procedure adopted by Great Plains Farmers. It supposedly improves soil-N dynamics. There are indeed too many reports from different states within Great Plains that suggest that application of nitrification inhibitors such as "Nitrapyrin" improves fertilizer-N recovery and consequently grain yield of cereals. Let us consider a few examples. Randall and Vetsch (2005) have found that despite erratic rainfall pattern that may affect availability of fertilizer-N in the vicinity of crop roots, application of Nitrapyrin improved fertilizer-N recovery to 47 percent. Supply of fertilizer-N and Nitrapyrin during spring increases N recovery to 61 percent. Maize productivity was 10.8 t grains per hectare, if fertilizer-N was supplied during spring season, along with Nitrapyrin. Application of fertilizer-N during fall along with Nitrapyrin yielded only 10.01 t grains per hectare. Clearly, supply of fertilizer-N at appropriate quantities, addition of nitrification inhibitors, and timing them accurately is important in the semiarid cropping zones.

Tremblay et al. (2012) have made detailed analysis of influence of soil properties, especially texture, water holding capacity, soil-N content, and their interactions with yearly weather pattern on crop yield in the Central Great Plains. They have pointed that most recent computer-based crop growth models and recommendations do not consider effects of weather and soil properties, while deciding on fertilizer-N supply. Weather pattern during crop season is indeed a major factor that affects various soil biological activities, mineralization, nutrient release rates, several chemical reactions, crop growth rate, and differentiation, grain-fill, and final yield. According to Tremblay et al. (2012) accumulated heat units, fine textured soil, and well-distributed rainfall is essential in Great Plains, to obtain high crop yield. Crop's response to fertilizer-N increased by 4.5-folds due to in-season N inputs. Fine texture of soil enhanced grain productivity by 2.7-folds and medium texture by 1.6-folds. Analysis of soil N distribution maps suggest that interactions of three factors, namely soil texture, precipitation pattern, and fertilizer-N supply schedules affects soil-N availability to crop roots. It has direct consequences on grain/forage yield maps. In due course, as precision techniques become common across all of Great Plains regions, farmers should be able to tackle soil fertility variation periodically even within a crop season more effectively. There are innumerable studies that depict the influence of several factors related to soil; crops, agronomic procedures adopted, and fertilizer-N supply levels on it use efficiency. Simplest of goals is to enhance N-use efficiency and reduce its application to soil. During past 6–7 decades, the N-use efficiency of crops such as maize, wheat, or soybean grown in Great Plains has increased steadily. It is revision of yield goals upwards each year that makes farmers add more of fertilizer-N and organic manures.

1.4.2 SOIL PHOSPHORUS MANAGEMENT IN THE GREAT PLAINS

Soil phase of the Great Plains of North America is the main repository of phosphorus that is utilized by crops and natural prairie. Soil P occurs in different forms and at different concentrations based on soil type, organic fraction, cropping pattern, precipitation pattern, and its influence on soil erosion, percolation, and seepage into irrigation channels, and so on. Phosphorus levels reported for soils (top 15 cm) increase from South to North in the Great Plains region (Haas et al., 1960). Crops absorb soil-P that occurs in available form that is, soluble form. The available—P is usually replenished through diffusion and conversion of labile forms. Soil P bound to matrix is not freely available. Soil-P levels depleted due to crop production is usually replenished through residue recycling, farm yard manure supply, or inorganic fertilizers such as single or triple super phosphate. Mollisols found in Great Plains do accumulate sufficient levels of inorganic and organic—P. Yet, available-P levels need to be above critical concentrations stipulated for each crop species/genotype. Generally, soil with available-P (Bray-1, Mehlich, or Troug-P) above 15–20 ppm is taken as optimum. Fertilizer-P sources are often replenished to reach a mark above critical level. Fertilizer or manure supply to wheat/maize or soybean is also decided based on soil type, its P buffering capacity, residual-P level, and yield goals fixed. Nutrient ratios between N:P, S:P, and C:P is also monitored in some farms. According to Grant et al. (2002), cropping intensity and diversity has increased in the Great Plains. Plus, no-tillage systems have become popular in Great Plains. These two phenomena have a major impact on soil-P dynamics. Intense cropping increases residue recycling. It enhances SOM and organic-P levels in soil. Soil P depletion rates could be high but replenishments often keep pace with removal rates. Soil test crop response studies are conducted periodically in the Great Plains at many locations. They aim at prescribing fertilizer-P supply as accurately as possible (Johnston, 2000; Grant, et al., 2002; Mohr et al., 2004). Wortmann et al. (2009) have reviewed the soil P needed and response of major crops grown in the Central Plains. They suggest that critical Bray-1 P for maize is 20 ppm, if maize is repeatedly grown in the field and 10 ppm if maize is followed by soybean.

In the Canadian Prairies, crop rotations have major say in the amount of P needed to be replenished. Rotations may include crops that deplete soil-P in higher quantities or those that allow sizeable residual-P to be utilized by the crop species that succeeds in the sequence. Long-term trials were conducted from 1967 to 2005 at Swift Current in Saskatchewan using 12 different crop rotations that included continuous wheat, wheat-rye-fallow, wheat-fallow, and so on. Soil fertility treatments included supply of nitrogen and phosphorus. Lemke et al. (2012a) reported that grain yield was better in fields with rotations involving fallows and provided with both N and P. Phosphorus recovery was 72 percent of that applied to soil during the crop period, in case of continuous wheat. Following is an example that depicts the effect of crop rotation and P inputs on grain yield:

Crop Rotation	Grain Yield	Phosphorus Input	P Recovered into Crop	P-Recovery Efficiency
-	t ha^{-1}	kg ha^{-1}	kg ha^{-1}	%
Continuous Wheat (N and P)	67.1	375	269.3	72
Fallow-Wheat-Wheat (N and P)	64.3	259	253.1	98

Source: Lemke et al. (2012a)

1.4.3 POTASSIUM DYNAMICS IN THE GREAT PLAINS

Potassium occurs in soil in different forms such as nonexchangeable or chemically bound form, labile form, and exchangeable/available form. Nonexchangeable form is the portion bound to soil matrix chemically. Crops and natural grasses are incapable of extracting this fraction. Exchangeable-K is utilized by crop roots/natural vegetation. Charts that explain the minimum exchangeable-K level required for different crops and fertilizer-K that needs to be replenished are available in each of the Great Plains States. Soil K is derived from weathering of feldspars, micas, and other ores. Soils in the Great Plains that are utilized for crop production possess abundant amounts of K, both in nonexchangeable and exchangeable form. Deficiency of K, if it occurs, restricts growth and grain fill of crops. Soils containing 250–300 lb K ac^{-1} are considered sufficient in K for crop production. According to Johnston (2004), periodic soil testing for K has suggested that plant-available-K and ratios of K versus other nutrients are crucial aspects that need attention by farmers. A general survey of soil K and trends in fertilizer-K supply in the Northern Plains has shown a deficit. Soil K removed into crops and transported out of the fields is greater than that applied seasonally under wheat-fallow system (Table 1.2). Crops may draw K inherent in soil as exchangeable or that supplied through complex fertilizers. Basically, K$^+$ is required at a level above critical concentration and in exchangeable form to support crop growth. State agencies in the Great Plains have general recommendations regarding critical values for exchangeable-K. Fertilizer-K that needs to be applied based on soil-K test value and yield goals is calculated using computer models. Johnston (2004) highlights the following facts about soil K in the Great Plains, they are: (a) K nutrition is critical for crop productivity. Most crop species respond with grain/forage yield increase in soils deficient for exchangeable-K. However, most soil types found in Great Plains are abundantly endowed with K and may not need repeated application of fertilizer-K each season; (b) Crop species grown in Great Plains remove relatively large quantities of soil-K, just as much as N; (c) Surveys have shown that crops such as barley, canola, and wheat respond appreciably to fertilizer-K supply.

TABLE 1.2 Average Annual Budget for Potassium during 2000–2010 in the Northern Great Plains region

Region/State	K Removal by Crop	Fertilizer-K Supply to Soil K$_2$O Million lb year^{-1}	Residual-K in Soil	Soil-K Balance
Alberta	607	128	136	−343
Saskatchewan	640	59	43	−538
Manitoba	332	92	45	−195
Montana	352	42	9	−301
North Dakota	609	52	13	−544

Source: Johnston (2004)

1.4.4 SOIL ORGANIC CARBON AND ORGANIC MANURE SUPPLY TRENDS IN THE GREAT PLAINS

Soil organic matter is an important fraction that has a series of advantages to Prairie soil ecosystem. Briefly, it helps in improving soil structure, improves water holding capacity and infiltration, reduces soil compaction, and provides essential nutrients upon mineralization. It is said that large-scale conversion of grass/legume filled prairies of the Great Plains to the famously known wheat-fallow system did reduce soil organic carbon levels. Several factors related to wheat-fallow systems induced loss of SOC. They are; conversion of sod, repeated plowing that enhances mineralization of SOM, highly oxidative conditions in soil that induce greater microbial activity, and loss of C as CO_2, soil erosion, and other natural process.

Reports by IISD (2012b) suggest that practice of summer fallow has enhanced loss of SOM. Tillage systems other than no-tillage/restricted tillage have further accentuated loss of SOM as CO_2. The magnitude of SOM loss as CO_2 or through soil erosions is no doubt dependent on soil type and crop species that flourishes in the fields. Firstly, conversion of natural prairies/grass lands to crop fields induces changes in SOM component. Next, crop species such as wheat or barley may increase loss of SOC as CO_2. In the Canadian Prairies, adoption of cereal-based cropping sequences has caused a loss of 20–50 percent of SOC during the past five decades. We should note that, in addition to loss of SOC, organic mineralization induces loss of N also, which is the other important constituent of SOM. In the Brown Chernozem region of Canada, SOM loss from topmost layer of soil has been estimated at 41 percent of original levels recorded during early 1900s. The decrease in SOM is dependent on soil type. Dark Brown Chernozems showed 41 percent loss of SOM, Black soil showed 49 percent loss of SOM, and Grey soil zone showed 36 percent loss of SOM in five decades. It seems careful selection of cropping systems and suitable agronomic measures, such as mulching; recycling of residues and cover crops may thwart SOC loss in the Canadian prairies. During past century, since 1900s, production of wheat and other crops in the Canadian prairies has shown marginal decrease during 1915–1930 and has steadied later. This trend coincides with similar pattern for SOC content (IISD, 2012b).

Residue generated by a single crop or crop sequence has its impact on potential recyclable organic matter. As a thumb rule, ratio of residue (stover:grain) generated by different crops that dominate the Great Plains are fallows: Corn—1:1; Sorghum—1:1; Soybeans—1.5:1; Spring Wheat—1.3:1; Winter Wheat—1.7:1; Durum Wheat—1.0:1; Oats—2:1 and Barley—1.5:1. In addition, crop residue left on the ground after harvest is of concern, since it has direct impact on organic carbon recycled into soil and erosion control. The ratio of crop residue still remaining in the field for crops is follows: Continuous wheat—0.75; Fallow-Wheat system—0.50; Barley—0.60; Corn—0.80; and Sorghum—0.80 (Lindstorm et al., 1979). On a wider scale, aspects such as crop residue generated and left on the field has marked effect on C sequestration and emissions in the Great Plains.

A few recent studies in the Central Great Plains conducted by USDA, at Akron Research Station in Colorado indicate that carbon accumulation in soil per unit time is better, in case of continuous cropping systems compared to crop-fallow systems. Continuous cropping with wheat-cotton-maize-fallow and variations that have sorghum or a legume interjection has showed a 20 percent increase in SOC in the upper layer of soil (0–5 cm depth). At depths greater than 5–15 cm, soil organic matter was not markedly different between continuous rotations and wheat-fallow (Bowman et al., 2005). Long-term evaluations suggest that SOM accumulation is similar in fields under no-tillage or restricted tillage but higher than in fields conventional tillage.

1.4.5 CARBON SEQUESTRATION IN NORTH AMERICAN GREAT PLAINS

Carbon sequestration into soils of Great Plains that have consistently supported production of cereals and other crops is an important environmental aspect. There are several instances where conversion of natural prairies and repeated fallows have decreased SOC contents and induced degradation of topsoil. Carbon sequestration is mathematically a difference between effect of process that aids carbon storage and those which trigger loss of C from soil profile or crop biomass (Table 1.3). Campbell et al. (2005) have reviewed several factors such as cropping systems, cropping frequency, tillage systems, soil type, and manure supply that affect both C sequestration and loss in the Canadian Prairies. They have argued that inability of farmers in the Great Plains to continue a second crop affects net biomass production. Summer fallows are actually practiced to improve soil moisture and refurbish its quality. It improves soil moisture status in the profile, rather it conserves. However, there are several short comings that are attributable to fallows. It lessens C fixation through photosynthesis, since there is absence of a second crop. Soil C losses occur mainly due to decomposition, restricted recycling of crop residues, lack of weeds that could be incorporated, soil erosion, and loss of topsoils, loss through emission as CO_2 from soil, microbial respiration, and so on.

Further, Campbell et al. (2005) examined the influence of fallow on C storage in Canadian Prairies. In most of the 17 locations, soils accrued SOC if farmers adopted continuous cropping. The gain in SOC due to continuous cropping in fields kept under no-tillage amounted to 250 kg ha^{-1} year^{-1} greater than, if fields were plowed. Crops had specific influence on C sequestration in the Great Plains. For example, replacing wheat with low yielding flax resulted in low levels of C sequestration in soil. Replacing wheat with Rye (*Secale cereale*) that prevented soil erosion enhanced C sequestration. Replacing wheat with lentil did not affect C sequestration trends in soil. In many locations, if crop-fallow gave 50 kg C ha^{-1} year^{-1} advantage, continuous cropping gave over 260 kg C ha^{-1} year^{-1}. Hence, famers in the prairies are currently advised about cropping sequence and possible C gains to soil using computer-based simulations. Reports suggest that net primary productivity due to conversion to agricultural cropping depends on the intensity of cropping systems. The change in net primary product (NPP) ranges from nil to 410 g C m^{-2}. The net primary productivity of agricultural prairies has direct impact on C fixed in ecosystem. It includes both above-ground net productivity (ANPP) and below ground net productivity (BNPP). Bradford et al. (2005) have reported that net ANPP for cropping zones in Great Plains increased by 0.066 Pg C year^{-1}, but BNPP decreased by 0.020 Pg C year^{-1}. It is equivalent to 26 percent increase in ANPP and 10 percent decrease in BNPP due to conversion to croplands. They have also pointed out that NPP is higher in the Great Plains counties that are endowed with better irrigation facilities and optimum precipitation pattern for crops to flourish. In other words, irrigation accounts for marked change in C fixation in the Great Plains. Generally, about 50 percent of biomass generated is harvested and removed out of fields. They have further argued that 17 percent of NPP, that is, C fixed into crop portion is removed and transported out of the Great Plains. This amounts to substantial amount of loss of C and it reduces C storage potential of Great Plains. No doubt *in situ* recycling is essential.

TABLE 1.3   Relative soil organic carbon change due to cropping frequency. A comparison of fallow-wheat and continuous cropping system in several locations of Great Plains of North America

Location	Cropping System	Relative SOC Change kg ha^{-1} year^{-1}
Soil Type: Grey Luvisol		
Edmonton, Alberta	Fallow-Wheat (none)	−261
	Fallow-Wheat (manure)	−43

	Continuous Wheat-Oat-Barley	0
Soil Type: Brown Chernozems		
Sterling, Colorado	Fallow-Wheat	−100
	Fallow-Wheat (+N)	−133
	Continuous Cropping	0
Mandon, North Dakota	Fallow-Wheat	−500
	Fallow-Wheat (N+P)	−383
	Continuous Wheat	0
Soil Type: Dark Brown Chernozems		
Bushland, Texas	Fallow-Wheat (+N)	−175
	Fallow-Wheat-Sorghum (+N)	−338
	Continuous—Wheat	0
	Continuous—Sorghum	−100

Source: Campbell et al. (2005)
Note: There are several other locations in the Canadian Prairies, mainly in Saskatchewan, that were evaluated for C sequestration trends. In most instances, fallow-wheat has resulted in C loss ranging from 125 to 291 kg SOC ha^{-1} year^{-1} compared to continuous cropping.

Restriction of soil loss via erosion and improvement in C sequestration within soil are major concerns to farmers in many locations within the Great Plains. Soil loss is almost nil, if farmers adopt 100 percent mulch and keep entire field mulched. Crop residue cover available after harvest of previous crop has its effect on soil erosion and C sequestration. Factors such as crops species grown, quantity, and type of residue generated need due consideration. According to Fore (2008), residue cover available after harvest of maize crop (100 bu ac^{-1}) is 70–95 percent, for small grain cereals (40 bu ac^{-1}), it is 70–80 percent and for legumes like soybean, it is 40–65 percent. Residue left on the field has direct effect on SOC content, loss as CO_2, and soil nutrients. In the Northern Great Plains, there are crops such as potato and sugar beets that generate very low quantities of crop residues for recycling. Managing soil erosion and improving C sequestration is therefore difficult. Crop species that fills the landscape and its ability to generate organic residue is a factor that controls C sequestration trend.

Review of Greenhouse Gas (GHG) production and loss of soil C in different locations within Great Plains suggests that implementation of ecological intensification of crop production is an appropriate suggestion to enhance C sequestration. Snyder et al. (2009) have suggested few methods that enhance C sequestration in the Great Plains. They are: (a) Fertilizer-N inputs help in generating higher biomass that can be recycled into soil. It also increases SOC content and improves quality; (b) Adoption of "Best Management Practices" for fertilizer-N reduces residual-N that would otherwise become vulnerable for loss via emissions; (c) Tillage practices that reduce soil disturbance and maintain SOC content should be adopted; and (d) Intensive crop production could be adopted with due care to avoid GHG emissions. Techniques that reduce loss of soil-C as CO_2 and CH_4 should be sought. In fact, intensification of crop production need not always induce proportionately higher quantities of greenhouse gas and result in SOC loss. It has been argued that high productivity may offset production loss in other regions affected by soil degradation. Further, a sizeable area of cropland and conversion of natural prairies could be spared, if intensification is adopted.

1.5   WATER RESOURCES, IRRIGATION, AND CROP PRODUCTIVITY IN THE GREAT PLAINS

The Great Plains agricultural belt is served by variety of water sources. Glacial melt during spring and summer results in regular flow of water via rivers, rivulets, and canals. It also leaves sizeable amounts of stored moisture in the soil profile. Major rivers that feed these agricultural prairies, either directly or through dams and irrigation channels are Missouri, Mississippi, Red, Platt, Arkansas, and Rio Grande. About 17 percent of cropping zones in the Great Plains is irrigated, although irrigation potential is much higher at >40 percent. Irrigation is predominantly utilized on staple cereals and cash crops such as groundnut and cotton. Much of the Great Plains crop production occurs under dryland condition and is dependent on annual rainfall pattern. Rainfall is higher in the Eastern flanks (500–550 mm year^{-1}) and decreases as we move to west and reaches a low at 300 mm year^{-1} in Western Colorado and parts of New Mexico. Farmers match their crop species with rainfall pattern. Lakes scattered all around the plains zone feed the crops in surrounding regions. Lift irrigation from groundwater (aquifer) is the major provider of water to crops in most parts of plains.

The high plains Aquifer also known as "Ogallala aquifer" is the major water source for people and farming enterprises. It supposedly serves 80 percent of Great Plains human population and irrigates 13 million acres. The Ogallala aquifer is replenished with water through precipitation and long standing stored water in the underground, perhaps since last Ice age. Reports by USEPA (2012) caution that population increase, agricultural needs, and general economic development of the plains has enhanced demand for water supply from the Ogallala aquifer. The average level of the aquifer has depreciated by 13 ft. This amounts to a decrease of 9 percent decrease in groundwater storage (USGCRP, 2009a). In some large cropping zones within Great Plains, the water level of this aquifer has dropped by more than 100–250 ft. Therefore, it involves greater investment on lift irrigation. In the absence of a good water source, such as Ogallala aquifers, impending droughts and high ambient temperatures have resulted in depreciated crop production (USCCSP, 2008).

Moore and Rojstaczer (2001) point out that irrigation projects and their effect on cropping expanse are some of the most conspicuous effects created in the North American continent. It was engineered by Native American tribes and European settlers. The high plains aquifer irrigated an estimated $2 \times 1,010$ m^3 in 1990 annually. The area irrigated covered about 60,000 km^2 of farm land. Further, they suggest that irrigation and conversion of natural prairies to intensified agricultural farms might have affected certain climatic parameters like precipitation patterns, although feebly. It is possible that effect of irrigation on precipitation pattern is related to soil moisture status and convection currents induced (Harding and Snyder, 2012).

Ranchers and early European settlers developed irrigation and diverted water from streams into farm land. Water was free of charge to farmers in the Great Plains. Irrigation was not levied and it proved profitable for any farmer who could develop irrigation facilities. It generally involved construction of crude channels, flumes, and storage facilities. Private irrigation constructors were enthusiastic and often estimates of profits were exaggerated. Failures of irrigation enterprises, it seems did not dampen Great Plain farmers from expanding cropland. Federal government aid helped them to sustain their enterprises (Johnson, 1936).

Irrigation is an important procedure in the Southern Plains, especially in locations prone to droughts and low water table conditions. For example, Unger and Musick (1990) opine that in states such as Texas, 66 percent of cropland is irrigated. Crop fields are regularly provided with irrigation supplements or in many cases protective irrigation is adopted. Farmers in Texas also employ most appropriate tillage and land management techniques to improve irrigation efficiency and crop yield. Irrigation using major source, that is, "Ogallala Aquifer" reached a peak during 1970s. Over 2.43 m ha were irrigated in just Texas plains. It decreased marginally initially to 1.82 m ha during 1980s. It has further decreased since then. Groundwater depletion has been the major cause in reduction of use of irrigation in the prairies of Southern Plains. During past two decades, share of irrigated area has depreciated to 40 percent of total cropped in the Texas alone. Major crops that are irrigated are cotton, sorghum, winter wheat, and corn. Interestingly, most farms with irrigation utilized ridge-furrow system to spread water in the fields (Unger and Musick, 1990).

Reports by USGCRP (2009a, b) suggest that irrigation in the Great Plains became common and it induced enhanced water removal from groundwater resources. Currently, on an average, 19 billion gallons of water is pumped out of Ogallala aquifer to feed crops, serve residing humans and cattle. Ogallala aquifer actually feeds 13 million acres of cropland and provides water to 80 percent of human population in Great Plains. Since, 1950s, water level in the aquifer has depreciated by 13 ft each year (USGCRP, 2009a,b). As stated earlier, in some of the highly irrigated regions of Texas, Oklahoma, South Dakota, and Nebraska water levels have, at times gone below 250 ft. However, we should note that as rainy season sets in, Ogallala aquifer is periodically recharged with large quantities of water.

Great Plains receives only 375–500 mm precipitation annually. Therefore, devising techniques and storing 3–5 cm more moisture in soil could make a significant impact on water resources (Mogel, 2013) and crop productivity. Soil moisture holding capacity is influenced by SOM content. Reports by USGCRP (2012) scientists suggest that a soil block of 15 cm depth can hold 2.5 cm of precipitation, if the SOM content is 1 percent. However, at 4–5 percent SOM, same block of soil can hold 3–5 cm precipitation. Firstly, increasing SOM content could reduce risks of flooding and runoff loss.

Precipitation pattern in the Great Plains is classified as erratic, especially, in the drier zones. Fluctuations are discernible in terms of intensity, quantity, and length of precipitation events. Yet, precipitation is sufficient to support a dryland wheat and forage. The wheat—fallow system that has 14 months crop free period allows storage of water received during rainfall events. This system preserves water enough to support a cropping sequence that offers two crops in 3 years to farmers in the Great Plains. It also supports sequences such as wheat-sorghum-fallow, wheat-cotton-fallow, or sorghum-cotton-fallow even in the drier Southern Great Plains. Length of fallow and fallow efficiency is crucial. According to Baumhardt (2003), measures that help farmers in the Great Plains to improve water storage and its efficiency during crop production are as follows:

(i)   Crop sequence with fallows that match rainfall pattern,
(ii) Crops that produce residues that can serve as mulch,
(iii) Tillage systems such as conservation or no-tillage that helps in reducing soil disturbance and loss of moisture,
(iv) Cover crops that reduce loss of water during fallow,
(v)   Irrigation, using furrows and ridges or flat systems may create runoffs and soil loss through channels. Such water loss is prevented effectively by using furrow dikes at small interval within furrow-ridge systems. This helps in retaining water at various spots in the field, avoids runoff and improves water storage. Conservation or no-tillage with mulches, reduce loss of water received through precipitation events. It reduces evaporation of soil water directly from soil surface. Clearly, integration of several of these agronomic measures improves water storage and fallow efficiency. Whatever be the crop and its productivity, since Great Plains is mostly a drier agricultural tract, aspects such as precipitation pattern, its storage, and use-efficiency need attention. Intensification, no doubt, requires larger quantities of water that has to be augmented through irrigation and matching nutrient supply to support biomass production.

Peterson et al. (1996) have studied the effects of tillage and rainfall pattern on precipitation use efficiency and cereal production in the Great Plains during twentieth century. Their observations suggest that since 1916, tillage methods have been modified by farmers mainly to restrict excessive soil disturbance. During 1940–1960s, in most areas within Great Plains, hitherto maximum tillage practice got replaced with conservation tillage. In the next phase, beginning in 1980s, No-tillage systems have been practiced widely in the Plains. Tillage intensity, annual precipitation, fallow efficiency, and wheat production during 1900s are as follows:

Period during 1900s	1916–1945	1946–1960	1961- mid 1980s	1985- till date
Tillage systems/ intensity	Maxi-mum/7–10	Convention-al/5–7	Minimum/Re-stricted/5–7	Zero/No-tillage/2
Annual precipitation (inches)	17.3	15.8	16.4	16.2
Fallow water storage (inches)	4.0	4.4	6.2	7.2
Fallow efficiency (%)	19.0	24.0	33.0	40.0
Wheat grain yield (bu ac^{-1})	15.9	25.7	32.2	42.1
Precipitation-use Efficiency (bu ac-in^{-1})	0.46	0.54	1.05	1.23

Source: Excerpted from Peterson et al. (1996); Nielsen et al. 2005
Note: Intensity refers to number of passes the tillage equipment makes in a field.

Generally, no-tillage has resulted in better water storage during fallows. However, advantage due to no-tillage has varied based on geographic location within the Great Plains and cropping pattern adopted. A compilation of facts about fallow storage systems by Peterson et al. (1996) suggests that mean water storage efficiency is about 18 percent in Canadian Prairies, 32–42 percent in Montana and North Dakota, 17–23 percent in Colorado, and 25–35 percent in Kansas.

According to Elgaai and Garcia (2005), aspects of climate change, especially, higher ambient temperature created due to CO_2 emissions may further stress water supply and affect crop productivity in the Great Plains detrimentally. Agroclimate, in general, may be severely affected by global warming process. It may actually alter precipitation pattern, irrigation resources, evapotranspiration, and water use efficiency of crops in the vast expanses of Great Plains. Forecasts derived out of weather models and computer-based simulations suggest that water supplies in the Great Plains are decreasing since past two decades. It has caused about 12 percent of Great Plains farmers to migrate to better pastures.

Farmers in the Great Plains have often ensured that irrigation systems and precipitation together answer crop's peak demand during its growth cycle. During recent years, irrigation schedules have been tailored considering both rainfall pattern and inherent soil moisture content in the surface and subsurface horizons. Maize is among crops that demand high amounts of water for definite period. Maize has a critical period extending from 45th to 85th day after emergence. The evapotranspiration during this period is 0.3 inches (75 mm) day^{-1}. Short-term demand for water in certain locations within Great Plains reaches 0.4–0.5 inches day^{-1} (Lamm, 2005). Evapotranspiration rates for maize fluctuate between 0.33 and 0.26 inches day during the rapid growth phase from early July to late August. Therefore, farmers schedule irrigation event in such a way that water supply sufficiently meets with crop's demand. Now, if we extrapolate consumptive demand for water within high-intensity maize farms in Northern Great Plains and Corn belt, it is a large increase in water requirement, which the agencies has to meet at least during critical growth phases. Farmers in the Great Plains do adopt different methods of water discharge into fields. For example, Lamm (2005) reports that "In-canopy Sprinkler system" provides 200–230 bu ac^{-1} during wet years, but during dry years same amount of water discharge gives only 150–185 bu ac^{-1}. Subsurface irrigation techniques are supposedly efficient. In the Great Plains, subsurface drip applied during wet years gives 200–260 bu ac^{-1}. However, in dry years grain yield fluctuates from 80 to 240 bu ac^{-1}.

Several oilseed crops such as sunflower, canola, soybeans, and groundnut are grown with the irrigation support. Reports suggest that about 12–25 percent of oilseeds grown in the prairie zones are irrigated (Aiken and Lamm, 2006). They are actually important cash crops to farmers in the Great Plains. Standardized irrigation schedules to match canopy development and grain formation are in place in almost all states of Great Plains. Such a prescription often considers entire crop sequence that includes oilseed crops. It helps in gaining accuracy and efficiency of irrigation.

There are several aspects of irrigation, such as methods, equipment, timing, flow quantity, and frequency, etc. that affect water use efficiency in the Great Plains. There are several reports that emphasize need for enhanced water-use efficiency and crop productivity. Maintaining optimum environmental parameters, reducing water loss via runoff, seepage, or percolation is of utmost importance in many parts of semiarid Great Plains. Periodic evaluation of crops, soil types, and irrigation frequency are made at different locations within the Plains. For example, Jabro et al. (2013) stated that for major crops, such as wheat, maize, sugar beet, malt barley, potato, and other vegetables, supplying water through conventional low frequencies served better than high frequency systems. Currently, most farm companies and farmers adopt irrigation systems such as furrow/ridge, flat field, sprinkler, and drip systems. Each has its advantage for a given crop and in a given location. State agency recommendations (SAR) based on several location trials and long-term data are also in vogue in Great Plains. Clearly, these irrigation measures do influence vegetative growth, biomass, and grain formation in the Plains. Excessive moisture is held in soil horizon but it is not uncommon to lose precipitation to groundwater sources.

There are other concepts such as deficit irrigation, which again minimize water use and yet try to maintain crop productivity. Crops are irrigated at optimum levels during critical sages such as rapid vegetative, flowering, and grain fill stages. During other periods, irrigation is either curtailed or done at reduced levels. It still does not affect final grain/biomass productivity. Such techniques are most useful to farmers in the Great Plains who are exposed to water shortage or improper precipitation pattern (Nielsen, 2005).

1.5.1 DROUGHTS AND DUST BOWL SITUATIONS IN GREAT PLAINS

Reports by Government of Canada indicate that during past 5 years from 2005 to 2010, CN$3 billion have been spent to study impacts of drought and poor precipitation patterns on agricultural crop production in the Northern Great Plains. Computer simulations and forecasts based on previous crop production trends suggest that agricultural prairies in Canada may shift gradually, but slightly to Northern regions from present locations. This may happen due to uncongenial precipitation patterns. Periodic droughts may not allow for higher crop production that has to satisfy the extra 5 percent needed to support biofuel production (NRC, 2009).

In the Central and Southern Great Plains, desertification is severe in areas that were originally affected by dust bowl conditions during 1930s. About 2.8 m ha are affected by erratic rainfall, soil erosion, and severe desertification. Generalized environmental stress, blizzards, tornadoes, droughts, and neglect or isolation of large areas of landscape can result in dust bowl conditions. Sand drifts, sand blasting, disruption of infrastructure relevant to agricultural operations, and impediments to human movement in the field are also detrimental to agricultural cropping systems. This happens despite active intervention and measures that mitigate soil degradation. Heathcote (1980) states that desertification is impoverishment of soil fertility and water resource. It is attributable to natural factors and interactive activities of humans. Droughts and desertification process in the Great Plains is accentuated by human activity and natural deprivation of precipitation. Soil erosion is an important index of dust bowl or desertification trends in a given cropping zone. During 1930s, dust bowl was construed as a situation that results through soil erosion, natural degradation, and lapses in social and economic activities of humans dwelling in the Great Plains (Heathcote, 1980). In fact, during past several years, the official governmental response in USA has been to first subsidize the farmers for natural loss of crops and then induce them to adopt soil erosion control measures effectively (see Helms, 1990; Loring, 2004). However, soil conservation programs do face limitations. In the Great Plains, it is said that only 40–56 percent of cereal cropping zones produce enough crop residues to serve erosion control measures (see Onstad and Otterby, 1979). Deep plowing that creates large clods and avoids dust formation is an important agronomic procedure advised to farmers. At the bottom line, dust bowl conditions are disastrous. They severely alter affect soil nutrient dynamics, moisture availability, crop growth pattern, and yield formation. They enhance loss of soil C and nutrients.

1.5.2 CROP PRODUCTION IN GREAT PLAINS

1.5.3 AGRONOMIC PRACTICES IN VOGUE IN GREAT PLAINS

According to Cash and Johnson (2003), there are at least five different production systems that are in vogue in the Great Plains region. Over 90 percent of the production systems are related to crop production and ranching. Crop production zones are predominant and occupy 75 percent of Great Plains. The five major production systems identified are Crop-fallows, Groundwater (aquifer) irrigation supported cropping, River valley irrigation supported crop production, that is dependent on snow-melt phenomenon, Forage production and confined livestock feeding, and Range and large pasture dependent livestock farming. Within the context of this book, we are concerned more about agronomic procedures and techniques adopted during crop production in large expanses supported by natural precipitation, various irrigation sources, and fertilizer-based nutrient supply. Agronomic procedures adopted by Great Plains farmers have actually evolved through the ages. Each set of procedures has imparted its effect on soils, crop fields, and land scape in general. For example, conventional tillage induced more of oxidative reactions, decomposition of SOC, and evolution of CO_2. No-tillage systems that are recently popular conserve SOC, reduce soil erosion, and loss of nutrients.

1.5.4 TRADITIONAL/CONVENTIONAL SYSTEMS

Native Indians practiced subsistence farming in the Great Plains for a long period. Land and soil management practices were feeble. Tillage, if any, was negligible or nil. Seeds were hand dibbled. Inter-culturing to remove weeds was done using handheld implements. The advent of European settlers induced large-scale farming, tractors, and other mechanical implements. Conventional tillage using disks or chisel plow that disturbed deeper layers of soil became popular. Land surface is often mended into furrows/ridges or flat beds or broad beds and seeds were mechanically sown. The intensity of conventional plowing depends on the crop, soil type, and purpose. Conventional seeding devices, intercropping, weeding twice, or thrice during early stages of crop are practiced under conventional farming systems. Fertilizer supply is based on soil tests and crop's demand. Yet crops may not reach potential yield possible in the region or sometimes it may lead to nutrient accumulation. Factors such as yield goals, previous data on productivity of individual fields, and general information decides the extent of chemical fertilizers and organic manures added to soil. Nutrient supply may or not be balanced. Nutrient ratios are not considered stringently. Regular supply of FYM at the beginning of the crop season adds micronutrients to soil. Conventional systems disturb soil to a great extent, causing loss of SOC, and erosion. The impact of conventional farming on the ecosystem varies. It may result in soil nutrient dearth and deterioration in the long run. Soil nutrient accumulations, if any, may induce emissions and/or contamination of channels and underground aquifers. Fallows are required periodically to refurbish soil fertility.

1.5.5   STATE RECOMMENDATIONS

Great Plains region supports cultivation of several different cereals, legumes, oilseeds, and fiber crops. They thrive on different soil types, in different soil moisture, and fertility regimes. The crop genotypes preferred in a zone also varies. The state agricultural programs recommend the most suitable crop, its genotype, soil fertility, irrigation, and pest control measures based on counties, soil type, precipitation pattern, or economic goals for the entire region. For example, in North Dakota, farmers are advised to supply exact quantities of N, P, and K to different crop species. In case of corn, state fertilizer recommendations are prepared using soil test values and yield goals. Corn, basically needs 1.25 lb N, 0.6 lb P_2O_5 and 1.4 lb K to achieve 1 bu grain. Hence, governmental agencies consider this information along with soil fertility status and develop blanket recommendations to the crop grown in different zones (NDSU, 2012a)

Following is an example about recommendation of N under a program called "PROCROP" to be applied to winter wheat crop based on yield goals set by the state (NDSU, 2012b):

Yield Goal	Total N Required to Achieve the Targeted Yield
Bu ac^{-1}	lb ac^{-1}
30	75
40	100
50	125
60	150
70	175
80	200

Source: NDSU (2012a, b)

In case of legume crops, such as soybean, peanuts, or lentils, state recommendation considers inherent soil fertility, extent of nodulation, and nodulation pattern.

It then suggests inoculation with *Bradyrhizobium* and fertilizer-N supply appropriately. Now, let us consider state recommendations made in a state situated in the Southern Great Plains. In Texas, state agencies consider nutrient and water needs of major cereals such as wheat, sorghum, and maize at different growth stages. For example, nutrient absorption pattern of sorghum at different growth stages and yield goal of 7,500 lb grains per ac^{-1} is taken into consideration, while deciding on fertilizer supply (Stichler and McFarland, 2012). Similar recommendations are available for crops such as cotton with a yield goal of 3,000 lb seeds, and 1,000 lb lint. Blanket recommendations for major and micronutrients required by a wheat crop that yields 60 bu grain and 5,000 lb forages per ac^{-1} is available. In general, blanket recommendation of fertilizer-N in Texas plains is 1.1 lb bu^{-1} grain for corn, 2 lb N for bu^{-1}sorghum grain, 10 lb N 100 lb^{-1} lint for cotton, 1.5 lb N for un-grazed wheat, 5.0 lb N t^{-1} for forage crops (Stichler and McFarland, 2012).

We should note that State Agency Recommendations (SAR) does not consider within field variations in soil fertility and water needs. Yield goals are stipulated for a region and not based on individual farm. Fertilizer supply may not meet exact needs, since it is a blanket recommendation for entire region. Accumulation of soil nutrients is a clear possibility. Micronutrients are supplied in quantities based on soil test values and critical threshold values and not based on depletion by previous crops or the needs of present crop. Crops are decided based on state agency decisions and not based on soil type or agroclimatic variations.

1.5.6 BEST MANAGEMENT PRACTICES

Farmers in the Great Plains have periodically modified and upgraded soil and crop management procedures to match yield goals and answer environmental concerns. Traditional systems that involve conventional tillage, cereal-fallow rotations, and application of fertilizers have been in vogue for several decades. Chemical and organic fertilizer supplies were revised based on soil fertility tests and state recommendations. During recent years, agronomic procedure known as "Best Management Practices" (BMP) have been devised for each region, rotation, and yield goal. According to Johnston (2006), BMP involves application of chemical and organic manure, using integrated approaches and site-specific techniques. In fact, he suggests that fertilizers applied under BMP allow farmers to harvest best possible grain/forage yield. We should realize that reaping profits is the main stay for farming enterprises in Great Plains. Further, BMPs involve three fertilizer and crop management principles. They are: (a) matching nutrient supply with crop's demand in space and time. It requires use of proven soil testing procedures, consulting soil-test crop response studies, knowledge about growth, and yield history of the fields, a thorough understanding of nutrient requirements at different stages of growth, plant analysis to ascertain nutrient needs periodically, establishing realistic yield goals, and most importantly obtaining estimates of nutrient budgets in fields and adopting procedures that do not deplete nutrients nor restrict crop growth and yield; (b) Adoption of efficient fertilizer application system based on soil analysis. This aspect of BMP is easily explained by "four rights" such as right amounts and ratios of fertilizer-based on nutrients, right fertilizer form, right placement, and right timing of fertilizer inputs; and (c) minimize nutrient loss through erosion, percolation, or emissions, so that fertilizer-use efficiency is not affected adversely. This aspect involves an understanding about soil erosion and emission pattern during crop season. Reduction of nutrient leaching and seepage. Adoption of conservation tillage or no-tillage systems and C sequestration methods, establishing buffer strips.

1.5.7 INTEGRATED NUTRIENT MANAGEMENT

Farmers in the Great Plains are currently advised to follow integrated procedures to manage soil nutrients, water resources, and disease/pest pressures. In case of Integrated Nutrient Management (INM) procedure, crops are provided with nutrients using several different sources such as chemical fertilizers, organic manures, residue recycling, adoption of No tillage that avoids soil erosion, and mulching that restricts emissions and erosion. Split application of nutrients that allow supply of exact quantities of nutrients at various stages of growth is adopted. Cropping sequences and rotations are selected carefully by including legumes and deep rooted crops. Nutrient need of entire cropping sequence is considered, so that deficit/excess of nutrients in the soil profile is avoided. INM helps in avoiding excessive use of inorganic chemicals and amendments. It restricts loss of SOC and maintains soil quality.

1.5.8 PRECISION FARMING

Precision farming technique is a recent introduction to Great Plains. It involves meticulous mapping of soil fertility of a field using large-scale grid sampling and automated analysis of nutrient levels. It adopts computer simulated fertilizer input schedules based on soil nutrient level at each spot and yield goals prescribed. It uses GPS-guided vehicles extensively to mark the soil spots in each field. The computer-guided variable-rate applicators then supply nutrients, water, or insecticides accurately at each point in the field (Lambert and Lowenberg-DeBoer, 2000; Krishna, 2013). In case of soil nutrients, adoption of Precision technique improves soil fertility as prescribed. It renders uniformity to entire field with regard to soil nutrient availability and leads us to uniform crop growth and yield. Precision techniques often need slightly lower levels of nutrients to achieve the same yield target. It avoids excess or deficits of soil nutrients. Soil nutrient accumulation that may lead to emissions, percolation, or seepage loss is reduced to a great extent. Currently,

GPS-guided variable-rate techniques are in vogue on crops such as wheat, corn, cotton, and sorghum in many states such as North Dakota, Oklahoma, and Texas.

1.5.8.1   CROPS, CROPPING PATTERN, AND PRODUCTIVITY IN GREAT PLAINS

Major crops cultivated in the Great Plains of North America can be grouped as follows:

Cereals:—Sorghum for grain, crop residue and fuel; Triticale; Wheat (Durum, Hard Red, Spring); Rye, Corn, Barley (six rowed, two rowed and forage); Buckwheat; Millets (Proso, Foxtail and Pearl); Oats;

Legumes:—Field peas; Lentils; Soybeans; Cowpeas; Chickpeas, Beans (Edible Beans, Navy Beans, Pinto Beans, Dark Red Kidney Beans, Small Red Beans, Light Red Beans, Black Turtle Beans, Cranberry Beans, Pinto Beans, and Great Northern Beans;

Oilseeds:—Soybean, Groundnut, Canola; Sunflower; Flax, Mustard; Safflower;

Sugar and Starch crops: Sugar beet, Potato, Sweet Potato;

Fiber:—Cotton;

Forages and Pasture Species:—Sorghum, Proso millet, *Pennesitum, Paspalum, Panicum, Hordeum,* Napier, Alfalfa, *Trifolium spp, Lens culinaris, Vigna sp,* and so on.

Source: Northern Crops Institute, North Dakota State University, Fargo, North Dakota, USA; Weaver (1927)

Agricultural Prairies cover about 24.2 percent of land in the Great Plains; rest 75.8 percent supports native prairies, shrubs, trees, and wasteland. Following is the list of important crops and percentage area that they occupy in the cultivable regions of Great Plains: wheat—10.2 percent; corn—4.5 percent; hay—3.39 percent; soybeans—2.6 percent, sorghum—1.62; cotton—0.95 percent; barley—0.89 percent; sunflowers—0.57 percent; oats—0.33; beans—0.21 percent; potato—0.04 percent; rye—0.02 percent (Bradford et al., 2005). Obviously, agricultural prairies in the Great Plains are dominated by cereal crops such as wheat, sorghum, corn, barley, rye, and cereal pastures followed by legumes such as soybean, lentil, groundnut, and so on.

1.5.8.2   WHEAT

Wheat is a dominant crop of the North American Prairies. More than 60 percent of the wheat production in United States of America is derived from the prairie agroecosystem that thrives in the Great Plains. Great Plains wheat belt is actually occupied by three major classes of common hexaploid wheat (*Triticum aestivum*), namely hard red spring wheat, hard red winter wheat, hard white, and durum wheat (*Triticum var durum*); (Paulsen and Shroyer, 2008). Currently, these wheat genotypes cover about 16 m ha of the prairies from Alberta in the North to Texas in South. Several wheat researchers, extension service personnel, and private companies have contributed to the conversion of natural prairies into wheat belt in the Great Plains region. Crop specialists such as David Fife who selected and spread Red Fife Spring Wheat; Mark Carleton who introduced Kharkhof Red winter wheat, and Durum production; Bernhard Warkentin who introduced Turkey Hard Red winter wheat; Charles Georgeson who recognized high yield potential of Turkey Red wheat and Elmer Heynes, who initiated Hard White wheat cultivation are important. They generated over 10 elite cultivars that spread into Great Plains and contributed to conversion of natural prairies into a wheat belt. Wheat cultivars introduced by the above crop scientists and American farmer's persistence has transformed several states within Great Plains region into top wheat producers and exporters (Paulsen and Shroyer, 2008). Currently, Hard Red Spring wheat covers about 40 percent and Hard Red winter wheat covers 48 percent of the 16 million ha of wheat-based prairies in the Great Plains. They account for 60 percent of total wheat produce generated annually in United States of America. Durum wheat extends into 7 percent and Hard White wheat into 5 percent of Great Plains wheat belt. The Durum and Hard White types are used to prepare pasta and baked foods. Distribution of wheat classes varies depending on location within the Great Plains. Hard Red Spring wheat dominates Great Plains regions in North Dakota and Montana; Hard Red winter wheat in Kansas, Nebraska, Oklahoma, and North Texas; and Hard white wheat in Colorado and parts of Kansas (National Agricultural Statistics Services, 2005).

Historical facts about introduction and spread of wheat culture in Great Plains of North America—dates and facts:

1812—Wheat production began in small farms of North Dakota. It was meant for settlers in Winnipeg.

1833—Wheat cultivation began in Texas High Plains and Kansas.

1840—Red Fife from Scotland was introduced into Ontario by David Fife.

1853—Durum wheat was introduced into North Dakota Plains

1855—Red Fife Spring wheat introduced into Ontario and Mid-Western USA.

1876—Turkey Red Hard winter wheat introduced into Kansas from Ukraine and Crimea by Mennonite settlers

1900—Kubanka Durum wheat from Kirghiz steppes was introduced into Kansas.

1900–10—Turkey Red Wheat becomes popular and graduates to a commercial crop in the Great Plains.

1900–1920—Disease resistant selections of wheat begin to occupy Great Plains.

1904—Wheat stem rust epidemic affects crop.

1910—States such as North Dakota, Kansas, and Minnesota dominate wheat production in the region. Also, "Durum Wheat" becomes a commercial crop.

1909—1920—Great Plains experiences a boom of dryland wheat production.

1910—1920—Wheat cropping region expands even into dry and arid areas of Plains.

1915—Hard white genotypes were cultivated by early settlers.

1917—Kansas Red wheat distributed into Central Plains.

1961—Semidwarf wheat cultivar "Gaines" distributed into Northern Plains.

1966—"Fortuna" wheat variety spreads into the plains.

1975—"Lancota" wheat is introduced for cultivation.

Source: Paulsen and Shroyer (2008); inventors.about.com/library/inventors/blfarm4.htm;en. wikipedia.org/wiki/ History_of_ agriculture_in_the_United_ States;

Foremost, wheat culture and introduction of improved genotypes with adaptability to environment and greater grain productivity replaced the erstwhile natural prairies and sod grass. Wheat genotypes gained ground in many places to become dominant herbage of the ecosystem. The wheat belt in Great Plains experienced several constraints such as drought and disease pressure, fluctuations in demand from domestic and external consumers, civil war, and of late severe competition from other major wheat producing countries such as Argentina, Brazil, and Australia, and so on. Despite it, wheat belt in Great Plains has progressed from being a subsistence farming enterprise during early 1800s to well-fertilized commercial units currently. This result is attributable to a large extent to wheat genotypes that resist diseases and offer higher grain yield. Regarding grain yield potential, Graybosch and Peterson (2013) believe that genetic gain for grain productivity has been 1.98 percent year^{-1} in most subzones of Great Plains. However, genotype versus environment interaction has shown up some constraints to greater wheat grain yield. Magnitude of abiotic and biotic factors that restrict full expression of grain yield is a matter concern.

1.5.8.3 CORN

Maize, popularly known as "Corn" in North American Plains was initially domesticated in Meso-America, more precisely in Mexican slopes. Archaeological evidences drawn from prehistoric sites of 6300 B.C.–1500 B.C. suggest that maize was cultivated in several agricultural regions of Meso-America during Neolithic Period. Maize was introduced into Great Plains through human migration from Mexico. Maize diffused into Southern and Northern Great Plains and Canadian Prairies during first millennium A.D. It is said that migration of natives into plains and rapid spread of maize in the plains did affect the landscape and food habits of the inhabitants. There were vast expanses of forests that gave way for prairies based on maize to flourish. Spread of maize into Canadian Prairies was prominent during first–fifth century A.D. (See Krishna, 2013). Maize diffused into Great Plains during 700 A.D. through Rio Grande valley. Initially, flint types reached the Northern Plains. However, native Indians grew several varieties of maize that suited their diets and preferences. Later, cultivation spread into vast stretches of Great Plains along the rivers Missouri, Platt, Arkansas, and Rio Grande. Natural genetic crossing, selection by native tribes, and periodic introductions by European settlers have all contributed to genetic diversity of maize grown in North American plains. Maize landraces and composites dominated the Great Plains landscape for several centuries. Their ability to negotiate drought and cold spell decided the productivity to a certain extent. Soil fertility and irrigation sources did affect the crop genotype that was preferred by the natives and European settlers. Maize production techniques got intensified with the advent of fertilizers and hybrids. During 1950s, maize yield that hovered around 3–5 t grain per hectare, got enhanced to 7–9 t grain hectare during 2000–2005. Currently, high-yielding hybrids dominate both the Midwestern Corn Belt region and Great Plains. There are also maize cultivars grown in the plains that cater to forage and biofuel.

1.5.8.4   SORGHUM

Sorghum was introduced into Southern Great Plains during mid-1600s, through Spaniards. Initially, sorghum cultivation was confined to Texas, Oklahoma, Kansas, and Nebraska. Sorghum cultivation spread all through the warmer zones of Southern Plains during nineteenth century. Sorghum was produced both for grains and forage. During 1850s, a forage sorghum called "Black Amber" was introduced from France (Undersander et al., 1990). Sweet sorghums were preferred more and were dominant type of sorghums in the Southern Plains during 1950s. Sorghum belt in USA is filled with different types such as sweet sorghums or "sorghos" that are converted into syrups and molasses; broom corns that yield fiber and Sudan grass that is grown for forage and hay. Several types of grain sorghums are traceable in the Great Plains. Sorghum thrives well in the Great Plains (Plate.1.1). It is C_4 species with potential to yield higher amounts of biomass. Sorghum tolerates drought better than other cereals grown in the Great Plains. Sorghum needs warm day and night temperatures ranging from 75 to 85°F. It tolerates soil pH between 5.5 and 8.0. Sorghum is a short day plant. Hybrids are photosensitive but there are several photo-insensitive types (University of Wisconsin Extension Services, 2012). Most of the sorghums found in Great Plains are hybrids between *Milo* and *Kafir*. Identification of cytoplasmic male sterile lines helped in generating large number of hybrids that suits the environmental parameters of sub regions of Great Plains.

PLATE 1.1 Sorghum based prairie vegetation near Lubbock, Texas in Southern Great Plains of United States of America. *Source:* Lindsay Kennedy, Sorghum *Checkoff*, Lubbock, Texas, USA

1.5.8.5 BARLEY

It seems barley was introduced into the Great Plains region by the Spanish conquerors during seventeenth century. Its cultivation spread rapidly into other regions of the plains. It served as a preferred cereal in parts of Canadian Prairies and Dakotas. Barley withstands relatively cooler and droughty conditions and hence is preferred to fill rotations and fallows. In this region, barley meant for malt predominates. Barley traits such as bright plum kernels, mellow starchy endosperm, white aleurone layer, and high malt extract are important for those meant for malt production. Both, six and three row types are cultivated in the Great Plains. It seems only 25 percent of barley area is utilized for malting, rest offers forage to feed farm animals. Feed barley production zones are common in Dakotas and Central Plains. Commonly, barley varieties that fail to meet malting standards are used as feed barley (Sanderson and Price, 2002). Currently, farmers in the Northern Great Plains grow high-yielding, tall or mediums statured, medium maturity, and white grain types of barley. Cultivars that are resistant to rust, blotch, and smut thrive better in the plains. Barley cultivation spread throughout Canadian Prairies and Western regions during mid-1800s. Barley is sown in spring season, mainly for grains. About 15 percent barley area in Canada is harvested for forage (USDA, 2004).

1.5.8.6   SOYBEAN

Soybean cultivation for grains is wide spread across all states of Great Plains. It is mostly rotated with cereals such as wheat or maize. Soybean was introduced into the Plains during mid-1800s. Since then, its cultivation has extended into all regions in Great Plains and Midwestern states of USA. It is a major legume and protein source for the region. Soybean productivity has increased steadily during the past century. Yet, average grain harvest of even irrigated crop is relatively low at 2.5 t ha^{-1}. During recent years, there has been greater stress on supplying soybean crop with balanced amounts of nutrients such as P, K, and micronutrients. Soybeans are sometimes provided with starter-N to help the roots to grow rapidly at seedling stage. It is legume hence needs relatively smaller quantities of fertilizer-N (Gordon, 2008). Soybean area in the Great Plains has depended on productivity levels possible and demand for its export and pricing (See Shurtleff and Aoyagi, 2007).

Soybean forage is a popular crop in the Great Plains. Its cultivation stretches into soils with low fertility. It is said that initially soybean was introduced into the plains region as a forage crop (Nielsen, 2011) and it flourished so until 1940s when soybean grain crop overtook the forage regions in terms of acreage and productivity. We should note that demand for grain and oil increased rapidly during 1940s and that had to be satisfied. Soybean forage was preferred by the Great Plains farmers because it did not suffer much from water deficits, unlike grain types that mandatorily need irrigation or precipitation, during critical grain fill and maturation stages. This actually helped farmers to tide over risks that otherwise affects grain types. The productivity of forage type soybean ranges widely from 1–2 t biomass per hectare to 12 t biomass per hectare, depending on location, its soil fertility, and water resources. The crude protein contents too vary depending on genotype and inputs. Forage soybeans have spread into vast stretches of Northern plains, because they need less inputs, establish well even in soils that are not fertile, and most importantly it adds to soil N, if included in the rotation during fallows.

1.5.8.7   GROUNDNUTS

Groundnuts were introduced into North American Plains by the Spanish settlers. In the Southern region of USA, peanuts were planted along with maize and beans. Groundnuts were accepted in wider scale during civil war and in the aftermath. At present, groundnuts are grown in large areas in many states of USA. In the Great Plains region, Texas and Oklahoma are dominant groundnut regions. Runners and bunch types are common in Southern Great Plains region. Valencia types occur in some parts of New Mexico. Groundnuts thrive on Alfisols, Mollisols, and Aridisols found in the Southern Great Plains. Groundnuts meant for both oil and confectionary purposes are cultivated in Texas and Oklahoma.

1.5.8.8 CHICKPEA

Chickpea is a high value crop that has gained in popularity in the Canadian Prairie provinces. Its production increased from almost negligible levels of 5,000 t year⁻¹ in 1995–4,50,000 t year⁻¹ in 2010 (MacKay et al., 2002). Chickpea produced in the plains is actually exported to Europe. Chickpea production in Canada experiences certain constraints such as Aschochyta blight, improper weather, and soil fertility loss. For example, during 2003–2005, only 70 percent of chickpea area reached harvest stage. Uncertain market prices too cause reductions in farmer's intent to spread chickpea the Canadian Prairies. Chickpea is a legume that adds to soil-N through symbiotic N fixation. It obviously needs relatively low fertilizer-N supply, if any as primer doze. Chickpeas add to soil organic matter, if incorporated. Otherwise, it serves as nutritious forage to farm animals. The spread of chickpeas in the Canadian Prairies may actually depend on economic advantages, demand for it from Europe and versatile cropping patterns that farmers may adopt in future. Crop genetic potential for biomass and grain yield is much higher than present harvests. There is no dearth for germplasm lines and elite cultivars that can be screened for higher productivity in the Prairies. Introductions from Middle-eastern region, Central Asia, and Indian subcontinent may be worthwhile.

1.5.8.9 LENTIL

Lentil is commonly traced in the Northern Great Plains states. It is being accepted as a substitute for fallows in Dakotas, Montana, and Canadian provinces such as Alberta. Lentils actually serve as a good legume crop and help in recycling residues and nutrients. Lentils add to soil-N via biological nitrogen fixation.

1.5.8.10 CANOLA

Canola (*Brassica species*) was grown in the Canadian Prairies during World War-II, to provide lubricants and oil for lamps. Canola cultivation began in 1970s in the Canadian Prairies. Canadian farmers grew cultivars with low euric acid in over 1.0 m ac during 1970s. Soil moisture, irrigation supplements, and fertilizer application are important aspects that regulate productivity of canola in the Canadian Prairies (Canola Council of Canada, 2001). Canola is actually a derivative of several species of genus *Brassica*. Predominant species are *B. campestris* and *B. napus*). The name canola is derived from Canadian Oil Low Acid (University of Nebraska, 2011). Canola is an important oilseed crop grown in many regions of Great Plains of USA. It is a prominent crop in Northern Plains, Kansas, and Nebraska (Boyles et al., 2003; Nielsen, 2004). It is a good source of cooking oil and provides high protein meal to livestock. Several varieties of canola are grown in the Great Plains. Cultivars resistant to diseases and tolerant to cold temperatures are preferred in the Northern Plains. Canola grown in both spring and winter are common in the Great Plains. In some parts of Southern Great Plains, canola is harvested 1–2 months late due to hot spell and this may reduce grain yield.

1.5.8.11　COTTON

At present, cotton crop thrives in the Central and Southern Great Plains states. It is a dominant crop in Highland region of Texas, parts of Oklahoma, and Kansas. Cotton crop is sown in February and harvested by July in the three Great Plains states Texas, Oklahoma, and Kansas. In some regions, cotton belt gets sown and initiated during April to early June and harvested by late September (Cotton Council International, 2009). The New world cotton (*Gossypium hirsutum*) is superior quality cotton, because of its staple length. It was domesticated in Mexico and Southern Great Plains, a few millennia ago. Evidences indicate that native prairie populations grew new world cotton around 3000 B.C. During past century, cotton belt that spread rapidly in the plains has seen upheavals due to civil war, pests such as boll weevils, beetles, and red bugs. This problem was overcome during 1930s till 1950s, by using insecticide in the vast cotton stretches. During mid-1800s, cotton production zone and its processing improved enormously in the Southern Plains due to introduction cotton boll separator by E. Whitney. During recent years, the cotton filled prairie is being conquered by Bt cotton. The Bt cotton has insect resistance imbibed into its genetic makeup through DNA hybridization techniques. It harbors a potent protein crystal toxin from *Bacillus thuringiensis* in its tissue.

1.5.8.12　FORAGE SPECIES

Forage and pasture is wide spread in the Great Plains. Fallows are often utilized to establish short duration pastures. This is in addition to regular long-term pastures that are prominent in Northern Plains. During recent years, there has been tendency to prefer pastures, because, they demand less investment and maintenance compared to a short duration cover crop. Soils prone to deterioration, or those affected by maladies, such as high pH or saline/alkaline conditions, are often reclaimed using perennial grass species that tolerate soil constraints to a certain extent. Great Plains farmers are known to introduce multiple grass species in mixtures of different ratios to establish a pasture. A mixture of grass species may perform better on degraded and low fertility soils. For example, in Western Plains, farmers use mixtures of Tall fescue forage, Revenue slender Wheat grass, and Salt lander Green Wheat grass. Such a mixture is supposed to aid establishment of pastures even in degraded soils.

They are useful in forage recycling programs and as animal feeds. Since, inputs are meager, forage quality is dependent on inherent soil fertility, precipitation pattern, and forage species. Mixed forages are common in many locations within Great Plains. For example, a mixture of Bermuda Grass (*Cynodon dactylon*) and Grass pea (*Lathyrus sp*) or Lentil (*Lens culinaris*) is supposed to provide better harvest (Rao et al., 2007).

Forage grass species commonly traced in the Great Plains are Tall Fescue, Perennial Rye grass, Kentucky Blue grass, Timothy grass, Red Canary grass, Smooth Brome grass, Big Blue stem grass, Switch grass, Indian grass, Sorghum × Sudan grass, and Sudan grass. Farmers in Great Plains have cultivated several different cool season pasture legumes. Interjecting crop sequences with such leguminous species is useful, since it adds to soil-N fertility, improves SOM content, enhances precipitation-use efficiency, and reduces erosion.

Tall fescue wheat grass, intermediate wheat grass, alfalfa, and sweet clover are common to Canadian Prairie provinces, Dakotas, and Montana. They are preferred in areas receiving 25–40 cm annual rainfall. Grass-legume mixtures are more common in this region. Thickspike wheat grass (*Agropyron sp*), Grama grass (*Bouteloua curtipendula*), and Canary grass (*Phalaris arundinaceae*) are other pasture species common to Northern Great Plains (USGS, 2013).

Annual pasture legumes common to Southern Great Plains regions, especially, Oklahoma and Texas are Burr Medic (*Medicago polymorpha*), Button Medic (*M.orbicularis*), Little Burr (*M. minima*), Arrow leaf clover (*Trifolium vesciculosum*), Ball clover (*T. nigrescens*), Crimson clover (*T. incarnatum*), Rose clover (*T hirsutum*), and Annual Sweet clover (*Mellilotus alba*) (Bailey Seed, 2012; Smith et al., 2012; and Bouton, 2013). Perennial pastures common to the plains are Alfalfa (*Medicago sativa*) and White Clover (*Trifolium repens*). Warm season pasture legumes grown in Great Plains are Cowpea (*Vigna unguiculata*), Lablab (*Lablab purpureus*), and Soybean (*Glycine max*). Warm season perennial legume pasture such as *Desmanthus* is also grown.

Following is list of grass/legume species used to establish pastures in the Great Plains with estimates of risk in establishment, vigor, sustainability, and growth shown in brackets:

Perennial Forages: Bermuda grass (1); Kleingrass (2–4); Bluestem (2–4); Buffelgrass (4–6); Wilman love grass (2–7); Alfalfa (7); Trifolium (7)

Annual Forages: Sorghum (2–4); Ryegrass (1–2); Small grains (2–4); Clovers (3–7)

Source: Ocumpaugh and Rodriguez (1999)
Note: Values in parenthesis suggest estimates of risk in establishment of forage species in the Great Plains.
Scale: Zero = No risk in establishment, good crop stand, and vigorous growth. 10 = Risky, very low, or nil establishment.

1.5.8.13   CROPPING SYSTEMS IN THE GREAT PLAINS

Monocrops, multiple crops, crop rotations, and sequential changes in the crop species that occupy the land scape in the Great Plains have an impact on variety of biotic and physicochemical aspects of soil profile. They have direct influence on natural
resources and their use by farmers. They also affect farmer's economic advantages
depending on the market value of the grain/forage. There is no doubt that dominant
crop species, related flora, and fauna that inhabit Great Plains are dynamic, depending on the cropping systems preferred. Cropping systems that are wide spread,
those needing high inputs, or others that thrive under subsistence farming conditions
with low supply of fertilizers and farm chemicals, may each have specific effect
on general environment and various aspects of soil nutrient dynamics. In general,
several natural and man-made factors affect and restrict farmer's preferences for
different cropping systems. Singer and Bauer (2009) have pointed out that crop
rotations were used prudently even before farm mechanization became prominent.
Later, chemical fertilizers were used to prop up crop yield and synthetic chemical
pesticides/herbicides were applied to control insects, diseases, and weeds.

Farmers in the Canadian Prairies have matched their cropping systems and agronomic procedures to suit the prevailing agroclimate and fluctuations in economic
returns since 1920s. Nyirfa and Harron (2003) believe that abandonment of farms
in Alberta and Saskatchewan during 1920s and 1930s is a response to climate and
economic situation. During recent years, farmers in the same region have been
changing from crop-fallow system to multiple crop sequences that include pulses,
oilseeds, and vegetables. Crop rotation, as an agronomic procedure has improved
crop production. In fact, crop rotation is among the earliest of integrated approaches
tested in the Canadian prairies. Despite a 100-year history of evaluating various
crop combinations and rotations, it continues to interest Canadian farmers to try new
ones. There are several long-term crop rotations that are continuing for over 20 years
(Campbell et al., 2012; Lafond and Harker, 2012). Long-term studies on crop rotations adopted in the Canadian Prairies suggest that fertilizer-based nutrient supply
and crop diversity are aspects that significantly affect soil quality and productivity
of a farm. Bio-diversity in terms of crop species and continuous cropping is essential, if C sequestration has to improve. Currently, wheat-fallow, mixed cereal, and
oilseed cultivation in sequence helps to obtain better grain yield and crop residues
that could be recycled. High returns are dependent on carefully tailored crop rotation, genetic diversity, and input schedules (Lemke et al., 2012b). Further, several
other trials conducted for over 35 years at a stretch at Swift Current, Saskatchewan,
and Canada have shown that conservation tillage may lessen loss of N via nitrate
leaching. However, striking a right combination of tillage system, crop rotation, and
diversity of species seems essential to maximize productivity (Lemke et al., 2012c).

Crop rotations are actually relatively longer-term plans by the farmers that help
them in improving land use efficiency, sustaining soil quality, and productivity.
Farmers in the Great Plains utilize crop rotations for various reasons. They are to

match crop species with soil type and its fertility, season, water resources, and minimize risk of loss. In the drylands that thrive under rain-fed conditions, farmers adopt
wheat-fallow rotations. If the season is still short for major cereals, for example,
in Northern Great Plains of Canada, farmers tend to grow short season small grain
cereal-that is, millets. Crop rotations are used to minimize weeds, insect, and disease
inoculum. For example, in some locations within Southern Plains, peanut cannot
be grown repeatedly, since it induces buildup of insects and diseases. Therefore,
farmers prefer to use longer rotations involving sorghum-maize—peanut and fallow. Carefully, tailored crop rotations in the Great Plains reduce soil maladies such
as erosion, deterioration due to nutrient depletion, and loss of SOM. Crop rotations
that include legumes add to soil-N through biological N fixation. Crop rotations
practiced in the Great Plains have also kept pace with several other natural and
man-made factors. For example, in Northeastern regions, higher density of cattle
ranches has induced farmers to adopt wheat-pasture rotations. Allen et al. (2010)
suggest that since fallows reduce precipitation retention and its use efficiency, farmers in Northern Great Plains are adopting regular cropping instead of fallows. They
are introducing lentils as grain or green manure crop. It adds to soil-N because it is
a legume, restricts loss of topsoil and nutrients, recovers unused nutrients and water from the profile, and improves productivity. It enhances C sequestration in the
soil profile. There are indeed several other suggestions and variations that farmers
in this region adopt. Yet, we may have to realize that natural grass/legume species
that erupts during a fallow gets suppressed to great extent, if continuous cropping
is practiced. Since, fallows are not preferred any more, careful selection of crop
rotations and species/genotypes can effectively thwart weed flora that periodically
appear in the plains (Derksen et al., 2002). In fact, in the Central Plains, summer
cover crops such as Hairy vetch (*Vicia villosa*), Sun hemp (*Crotalaria juncea*), and
late maturing soybean (*Glycine max*) have been effectively used to curb weeds,
improve soil-N, and crop productivity (Blanco-Canqui et al., 2012). Oilseed crops
such as canola, flax, sunflower, and soybean are an important aspect of crop rotations practiced currently in Great Plains. Inferences from computer-based simulations, experimental evaluations, and farmer's experiences suggest that oilseed crops
are compatible with cropping systems adopted in the Great Plains. Interjection of
oilseeds in the rotations improves land-use efficiency. Oilseed cultivation is also
profitable to farmers. Oilseeds cultivation increases net biomass generated and C
sequestered in the fields (Johnston et al., 2002; Nielsen et al., 2012).

Long-term strategies about cropping systems are essential, since they are dynamic and the advantage each system offers too fluctuates within a range. In the
Northern Great Plains, fertile soils and optimum precipitation pattern does allow
modifications to crop sequences practiced. It is an area with wheat-fallow as the
dominant sequence since many decades. However, evaluations of crop sequences
that include a variety of different crops like grain legumes, oilseeds, and vegetables suggest that crop sequence that had continuous cereal, that is, wheat-fallow or
wheat-wheat fallow offered more crop residue that could be used as mulch or animal

feed. At the same time, if cereals were repeated, in other words, a crop was seeded on residues from the same species, then growth rate and productivity was relatively low (Krupinsky, 2006). Inclusion of a legume crop in the sequence induced positive influence on soil-fertility and crop production. Crops like sunflower offered better water-use efficiency for the crop sequence. Generally, avoiding continuous cereal or cereal-fallow system, instead adoption of cereal-legume-fallow system meant gain in soil fertility, crop growth, and grain formation. Durum wheat-oilseed rotations are among several crop sequences that farmers from North Great Plains may opt because it offers cereal grains, oilseeds, and biofuel. Water use of durum wheat was 282 mm and that of oilseed 276 mm (Lenssen et al., 2012). Johnston et al. (2002) believed that including an oilseed or vegetable to suit the short summer fallow period is a good idea, in most parts of the Great Plains. Oilseeds such as canola (*Brassica sp*), mustard (*B.juncea*), or flax (*Linum usitatissimum*) are well suited to cool climate prevalent in the Canadian Prairies and Northern Great Plains of USA. Sunflower (*Helianthus annuus*) and safflower (*Carthamus tinctoriuos*) are better in regions that are relatively warmer and with long fallow season. In addition to matching crop season with species, farmers in the Great Plains have modified various agronomic procedures to enhance productivity of the whole crop rotation, instead of just the spring wheat (Nielsen et al. 2005).

In the Central Plains, it seems there are over 2.7 m ha of land scape that supports wheat-fallow system. This region experiences periodic droughts and erratic precipitation pattern. Hence, usual wheat-fallow has not been replaced by multiple cropping that allows crop production throughout the year. According to Merle et al. (2013), about 58% of agricultural landscape in the Central Plains is still under wheat-fallow because other alternatives seem economically nonviable. There is, however, a need to evaluate several latest crop production systems that avoid soil degradation and loss of soil fertility. Techniques that enhance precipitation use efficiency are of utmost relevance in the prairies. It is hoped that adoption of integrated soil fertility and water management methods, precision techniques, and improved crop rotations may delay soil degradation and yet provide better returns to farmers in the Central Plains. There is a strong reason to believe that rapid spread of Precision techniques will impart uniformity to soil fertility and productivity. It may reduce undue accumulation of soil nutrients in the profile and avoid groundwater contamination.

Retention of precipitation and its efficient use are essential aspects in the drier tracts of Southern Plains. Cropping systems and tillage adopted are among the vital factors that influence precipitation-use efficiency. It seems water storage is actually influenced more by ET of the crops grown in the prairies. No doubt, dominant crop species in a particular season that covers landscape influences variety of aspects related to water use, fertilizer supply, and productivity (Zhang and Garbretch, 2007). Nielsen et al. (2012) found that in the semiarid regions of Southern Plains, mass production of forages offered better precipitation-use efficiency of 14.5 kg ha^{-1} mm^{-1}. In comparison, small grained cereals (2.8 kg ha^{-1} mm^{-1}) or oilseed 4.2 kg ha^{-1} mm^{-1} resulted in lowered precipitation-use efficiency. Economic returns attributable to better precipitation use efficiency was higher for wheat-forage systems (1.2 US$ ha^{-1}) compared to 0.30 US$ for small grained cereals. Generally, enhancing inputs and intensifying crop production meant higher precipitation-use efficiency. Such an option may induce farmers in Southern Plains to try pastures and forage in place of oilseed or legumes.

Factors such as availability of natural resources, ecological considerations, and compatibility of crop species with soil/environment influence crop rotations in the Great Plains. In addition, factors like infrastructure development (for e.g., railways), market price fluctuations, and governmental legislations also affect crop sequences. For example, in the Midwestern regions, known popularly as "Corn Belt," farmers adopt maize-soybean or wheat-soybean, because soybean provides economic advantages and there is a continuous demand for vegetable oil derived from soybean.

There are indeed several studies about the effects of crop rotations on soil environment and profitability. According to Singer and Bauer (2009), corn grown after soybean-a legume, gave 15 percent extra profit to farmers and improved soil-N status. Further, soybean, if grown after corn gave 10–15 percent higher productivity compared to soybean continuous sequences. Duffy (2008) states that corn production after soybean costs 14 percent less bu^{-1} to the farmer compared with corn continuous.

Agroforestry is a time tested practice in the Great Plains. It involves growing small trees/shrubs of utility in terms of firewood, green manure, leafy manures, and horticultural products if the tree is a fruit crop. Strip cropping of agroforestry trees and cereals is a useful proposition to many farmers. Alley cropping of trees such as *Quercus sp* (Oak) or *Faxinus sp* (Ash) with wheat offers wood and grains. In states like Missouri, black walnut *(Juglans sp)* trees are inter-cropped with cereals. Such intercropping procedures offer specific effects in terms of ecosystematic functions, nutrient recovery from soil profile, its recycling as crop residue, fertilizer-use efficiency, and total productivity. Agroforestry offers opportunity to convert monocropping belts into mixed cropping systems and adds to soil fertility, if the tree species is a legume. They have potential to change monotonous grassy prairies and cereal/legume-based agricultural landscapes into parklands dotted with multipurpose and useful trees plus cereal/legume fields.

Cover crops are grown in the Great Plains region to attain several different advantages. They are helpful to farmers in controlling erosion, during fallow period. They conserve water received via precipitation. If the cover crop is a legume, it adds to soil-N content. A cover crop with rapid root growth and ability to accumulate nutrients in roots will help in avoiding loss of nutrients from the root zone. Rapid root proliferation can actually trap nutrients in the upper layers of soil. Let us consider a few examples. Power and Biederbick (1991) state that crop like medic is useful in conserving water and soil-N, so that wheat that fallows grows better than what is possible with fallows without crops. Timing of cover crops is crucial, if farmers in the Great Plains intend to derive greater advantages. Congenial temperature, precipitation, and other weather parameters will induce cover crops to produce greater biomass that could be recycled and higher quantity of grains could be harvested. Soybeans, faba bean, pea, or hairy vetch grown as cover crops during spring generally attained better growth. *Trifolium* with ability to tolerate cold is also used as cover crops. Soil temperature encountered by roots of cover crops is a factor to be considered by farmers in the Great Plains. In Nebraska and Kansas, grass species such as rye, sorghum, or millets are used as cover crop. In the humid regions, lentils are useful as cover crops. Lentils are also common in the Canadian Prairies. They are effective as interjecting legumes in a wheat-lentil-fallow rotation. Lentils are known to reduce N requirements of the crops. In addition to above advantages, cover crops add to diversity of cropping systems and help in sustaining soil fertility and quality of agricultural prairies.

1.6   PRAIRIE CROPS AND ENVIRONMENTAL CONCERNS

According to Cash and Johnson (2003) for crops to flourish in the prairie zones, the primary concerns are soil resources, water, and people and their economic status. Secondary concerns are native prairie vegetation, stable grain harvest, investment on agricultural enterprises, management methods, effective extension services to transmit information and facilities. We should also note that over past 50 years, the ill effects of soil degradation, loss in inherent fertility, and repetitive cropping has been masked by intensification and excessive use of chemicals that results in better harvests (Trautman and Porter, 2012).

Soil fertility depression is a major concern that gets accentuated due to incessant cropping, inappropriate fertilizer in replenishment schedules, and lack of residue recycling. It seems during past five decades, crop belts in the Great Plains have lost over 50 percent of original soil fertility levels. Soil fertility loss has been especially rampant during past 20 years. Major cropping trends such as wheat-summer fallow are insufficient to maintain soil fertility. Tillage systems adopted under conventional farms has resulted in loss of surface soil and nutrients. It has also reduced C-sequestration. Therefore, extension agencies are actively engaged in improving soil conservation using no-tillage and high residue recycling. Long-term measure to store soil-C is essential to preserve soil quality in the Great Plains region.

Agriculture is the main enterprise that occupies most of Great Plains area. It is also the main consumer of water resources available in the Great Plains. Nearly 80 percent of consumptive use of soil water is attributable to crop production. Great Plains has over 23 million acres of irrigated land that is served directly through channels, pipes, sprayers, and drip systems. Impact of reduction or lack of soil moisture at critical period during crop season is immense. Since three decades, there has been sizeable allocation of water resources to urban consumption. Diversion of water from lakes, dams, and reservoirs has also occurred. It affects the cropping ecosystem and its functions. Excessive use of groundwater has led to depletion of water levels in the Ogallala aquifer. Water shortages are becoming common. The climate change related to water-use trends are slowly but surely becoming conspicu-ous (Cash and Johnson, 2003).

1.6.1   SOIL EROSION AND CONTROL MEASURES ADOPTED IN THE GREAT PLAINS

Great Plains has encountered a series of soil maladies and environment-related con-straints to crop production. Farmers and researchers matched them by modifying existing soil management and agronomic procedures. New and effective measures have appeared periodically to prop up crop productivity of prairies. Soil erosion is a perpetual natural phenomenon on the surface of the earth. Historically, it has af-fected crop production at different intensities. According to Skidmore (2000), soil erosion affects quality of soil. In case of wind erosion, dust obscures vision, affects vehicular movement in the fields, pollutes air and water bodies. Large-scale sedi-mentation can clog channels and irrigation devices.

Soil erosion due to wind reached disastrous proportions during 1930s causing dust bowl conditions and induced farmers in many regions of Great Plains to aban-don crop production and migrate. Currently, it is still an important factor that reduc-es crop production efficiency. It also has long-term deleterious effect on soil fertility. Reports indicate that in United States of America, loss of agricultural exchequer due to wind erosion is estimated at US$44 billion annually (Borade, 2012). On an average, 1.3 billion tons of surface soil gets eroded each year in the farming zones of United States of America (McCauley and Jones, 2005). During a 15-year period from 1985 to 2000, soil erosion control measures adopted by farmers in the Great Plains reduced wind erosion from 3.07 t ha^{-1} year^{-1} to 1.90 t ha^{-1} year^{-1} (USDA, 2000).

There are indeed several estimates made about actual loss of surface soil, soil degradation, crop loss, and reduction in grain yield. For example, reports compiled by International Institute for Sustainable Development, Winnipeg in Canada suggest that prairie provinces accrue a loss of CN$155–177 million worth grains each year that is attributable to water erosion and about CN$22 million attributable to wind (IISD, 2012b). In terms of soil, it amounts to loss of 160 million ton surface soil due to water and 117 million ton due to wind annually. We should note that wheat yield showed a mild depression from 1905 to 1932, coinciding with initial settlement spree in Prairie Provinces, but since then grain yield has improved steadily reaching 2.2 t grains ha^{-1} year^{-1}. Wheat yield improvements occurred despite the influence of soil degradative forces, mainly erosion due to water and wind.

Farmers in Northern Plains, especially those in Canada adopting wheat crop pro-duction have effectively used organic farming trends and residue recycling to reduce loss of soil and its fertility. In Southern Canada, soil conservation is needed to thwart erosion caused by wind and/or water. Topsoil loss could be rampant if few of the detrimental factors act in combination. Geographical elements such as topography of fields, soil type, climatic factors, extent and type of tillage, irrigation frequency, and other farm operations, and crop species are aspects that need due consideration, if the aim is to reduce soil erosion. Crop residue application is of course an important procedure that reduces loss of soil fertility. Farmers in Canadian prairies utilize crop residue that includes roots, chaff, stem, and leaves from previously harvested crop. Application of crop residues improves soil traits such as water infiltration, retention, and storage; soil particle aggregation; and nutrient content and soil organic matter. Application of crop residue reduces loss of soil moisture and ill effects of cold climate. The type of crop residue, its stage and crop species are important to note. Crops such as field peas and flax offer half the biomass as residues. Residue from crops such as canola, lentil, and sunflower decay relatively rapidly.

Reports from Government of Saskatchewan suggests that soil erosion control measures depend primarily on factors such as soil texture, field slope, and crop species cultivated (Agriculture Knowledge Centre, 2008). Following is an example of influence of soil texture and slope on quantity of crop residues to be applied:

Soil Texture	Medium	Fine	Coarse	-
Crop Residue required (lb ac^{-1})	900	1,350	1,800	-
Field slope	Gentle (0–6%)	Moderate (10–15%)	Steep (16–30%)	Very Steep (>30%)
Crop Residue Requirement (lb ac^{-1})	700–1,000	1,000–1,500	Continuous grass	Native vegetation

Source: Agricultural Knowledge Centre (2008), Saskatchewan Agriculture, Canada

Several short-term and long-term evaluations about the effects of soil erosion, especially, loss of topsoil and nutrients from the upper horizons of soil have been conducted in the Canadian Prairies. Most, if not all, highlight the relevance of soil conservation measures, no-tillage, and appropriate crop rotations in areas prone to soil erosion. The extent of surface soil loss may actually depend on several other factors too. Wind- and water-induced erosion that is common to Canadian Prairies may remove variable depths of soil. According to Lamey and Janzen (2012), extent of wheat grain yield reduced was 10 percent for a soil loss 5 cm depth, 19 percent for 10 cm depth, 29 percent for 15 cm depth and 38 percent for 20 cm depth. No doubt, surface soil is precious to Canadian farmers. They do adopt control measures such as contouring, application of crop residue and manures, cover crops, and so on.

Long-term studies of 15 years in the arable fields of Lethbridge Agricultural Research Station, in Alberta aimed at understanding the effect of erosion of topsoil of different thickness and manure supply on wheat grain yield. Soil erosion that results in removal of top 15–20 cm soil affects wheat productivity severely. Wheat grain yield decreases by 25–43 percent compared with nil soil erosion plots. Chemical fertilizer supply may reduce effects of soil erosion, but organic manure supply to soil, reduces soil erosion drastically. Soil erosion reduces to almost 2–8 percent of zero loss check, if organic matter is applied (Lamey and Jenzen, 2012). Overall, soil fertility and organic matter loss that occurs due to erosion needs to be refurbished.

In North America, water and tillage may act in combination to enhance soil erosion. Segmentation of fields and mapping of areas that are affected by soil erosion are needed (Li et al., 2008). The type of tillage and its intensity has influence on soil organic matter, crop residue left in the field, and soil erosion. Shallow tillage allows crop residues to localize in the upper surface. It incorporates crop residue, improves water infiltration, and reduces both wind and water-induced soil erosion. Tillage equipment used is an important factor. Farmers in the Canadian prairies use several different types of tillage instruments. Wide blade cultivators usually conserve larger fraction of crop residues. Farmers allow stubbles to continue and conserve both moisture and organic matter in the field. Direct seeding with no-tillage is a popular trend among farmers in most regions of Great Plains. It effectively reduces loss of topsoil and nutrients, but may induce weed growth if not checked timely. Carefully designed crop rotations and genotypes also help in reducing soil erosion. Forage and cover crops are another set of practices that help farmers to reduce loss of soil fertility and control erosion. Wind barriers such as annual crop barriers, summer fallows, perennial grass barriers, shelter belts, strip cropping, and cover crops are among several practices that reduce soil erosion in the farming zones of Great Plains (Agriculture Knowledge Centre, 2008).

Soil erosion due to wind is an important factor to consider during farming in the Central Great Plains. It is wide spread in this region of the plains. Soil erosion due to wind occurs if the wind speed crosses threshold levels and soil surface or crop field is left unprotected. Lyons and Smith (2010) state that wind-induced soil erosion gets initiated if wind speed is more than 6 mph above 1.0 ft of soil surface. We should appreciate that, if wind speed increases from 20 to 30 mph, soil erosion and loss triple over the base levels at 20 mph. At this wind speed, soil particles begin moving along the surface and could be lifted. There are actually two types of wind aided soil erosion. They are "Sweeping drifts" and "Active turbulent drifts." Sweeping drifts are often gentle and cause loss of loose soils. Active drifts are vigorous and remove massive quantities of soil particles from a place (Borade, 2012). Techniques that enhance minimum wind speed required to start erosion or those which reduce wind speed itself are required. Actually, wind erosion becomes conspicuous if soils are sandy, vast stretches of semiarid plains are left unattended, crop fires become frequent, and droughts occur repeatedly. For example, soils are loose in the Loess region. Soil particles from such loess area can be easily lifted and allowed to drift away (Fryrear et al., 2001). However, wind erosion effects could be curbed by regularly monitoring wind speeds and state of crop fields. Computer-based models help farmers in Central Plains and other locations to forecast severe soil erosion events and adopt appropriate measures. Lyons and Smith (2010) suggest that wind breaks, crop residues, cover crops, and strip cropping are useful measures to reduce wind speed. Ridging reduces soil loss by effectively trapping the particle. Soil aggregates could be enhanced and conserved by crop residue and manure application. In the Northern Plains, wheat fields are said to loose surface soil at a rate of 5.1–5.5 t ha^{-1} year^{-1} (USDA, 2000) due to wind erosion. Baumhardt (2003) states that farmers and researchers have tried and tested several ingenious agronomic methods that either completely control or decrease wind caused soil erosion in the Great Plains. A few of them are measures such as wind breaks, emergency tillage with Graham-Hoeme chisel plow, ridging, and mulching. Chisel plowing helped in creating large clods and aggregates that resisted easy disintegration and erosion. Tree wind breaks reduced wind speed to below threshold levels. Application of crop residue gave thorough control of soil erosion due to wind.

Evaluation of several sub regions of North American Great Plains for soil erosion process that afflict farming zone and the extent of soil loss has been carried out since early twentieth century. Forecasts about propensity of soil and fertility loss have varied periodically. Soil loss has been rampant in some part of Great Plains compared with others, wherein soil conservation processes have been in vogue. Reports by Wagner and Dodds (1971) that relate to soil erosion during first half of 1900s, suggest that annually soil loss ranged from 0 to 5 t ha^{-1} year^{-1} in about 16 m ha; 5–10 t ha^{-1} year^{-1} in about 9.5 m ha and 11–20 t ha^{-1} year^{-1} in about 1.2 m ha^{-1}.

1.6.2 GREENHOUSE GAS PRODUCTION BY AGRICULTURAL AND NATURAL PRAIRIES IN GREAT PLAINS

During past two centuries, Great Plains has experienced perceptible changes in natural vegetation and agricultural cropping pattern. According to Hartman et al. (2011), European settlement of Great Plains region has induced land use changes, mainly from natural prairies that were stabilized into agricultural prairies that are supplied with large amounts of fertilizers, chemicals, and water. This has dramatically affected greenhouse gas (GHG) emissions such as CO_2, CH_4, N_2O, NO_2, and N_2. Initially, until 1870s, pastures were more common. Since 1870s, cropping systems gained in popularity. From 1930 until date, intensive cropping supported by high-yielding genotypes, fertilizers, irrigation from Ogallala, and high density planting have dominated the prairies. As stated earlier, Great Plains is a vast expanse of crops covering over 16 m ha. Winter wheat, spring wheat, maize, sorghum, cotton, and soybean are a few dominant crops found in the prairies. Such changes in cropping systems have indeed affected emissions from soils and above-ground portions of ecosystems (Parton et al., 2011). Dryland cropping that depends much on the precipitation is the predominant form of crop production. Drylands generate large quantities of greenhouse gasses such as N_2, N_2O, NO_2, CO_2, and CH_4. Crop production techniques were improved to suit the high grain yield expectations of farmers. This necessitated application of high amounts of fertilizers. Fertilizers, especially inorganic-N applied to soil became vulnerable to loss via emissions. Organic manures too generated large amounts of CO_2 that escaped into atmosphere. In fact, Parton et al. (2011) state that, greenhouse gas emission from agricultural belts of Great Plains is high and is dependent on land use category. About 30–50 percent of soil organic carbon was lost just during past 30 years. Mineralization also led to rampant loss of soil-N as N_2O and NH_4. Cattle farms that are found all around the Great Plains have became an important source of CH_4 emission. CH_4 has high greenhouse effect potential. Following is data accrued through simulations (DAYCENT model) and actual measurements of greenhouse gas fluxes in Great Plains:

Greenhouse Gas Flux (Gg CO_2-Ce)							
Years	Pasture	Dryland	Irrigated Area	Fuel	Fertilizer	Livestock	Total
1870–1879	−1,491	441	0	0	0	824	−225
1900–1909	−4,154	15,597	0	0	0	8,318	19,871
1950–1959	376	9,900	−306	3,839	844	10 645	24,947
1990–1999	−4,109	−2,311	293	3,175	2,720	12,300	9,150

Source: Excerpted from Parton et al. (2011)
Note: Greenhouse emissions are calculated for 10 year intervals. Data for four representative periods are shown. GHG emission from only few major sources are shown, hence, total does not correspond to actual arithmetic sum. For details, please see Parton et al. (2011). Their study has shown that overall production of GHG from 1,870 to 2,000 in the Great Plains amounted to 2.7 million Gg CO_2-Ce. Agricultural emissions amounted to 9,000 Gg CO_2-Ce.

Clearly, drylands with crops such as wheat, sorghum, corn, and legumes have generated large amounts of GHG. Other major sources are fertilizers and livestock. According to Parton et al. (2011), conversion of native prairies into cropland-induced N_2O emissions rather significantly. It was easily attributable to use of high quantities of fertilizer-N and organic manure. Fluxes began to increase significantly during early 1900s. Nitrous oxide emission increased by 30 percent between 1940 and 2000. Nitrogenous fertilizers applied to drylands that were still unused, formed the most vulnerable fraction. Fertilizer-N accumulated in soil got converted into gaseous emissions easily. Gaseous emissions induced loss of soil-N fertility and soil degradation. During past few decades, cattle population has steadily increased and has caused perceptible increase in CH_4 emission in the Great Plains.

Nitrous oxide emissions draw greater attention because it has 270 times, the global warming potential of CO_2. It also affects the ozone layer. Minimizing its emission is of utmost necessity. Nitrous oxide emissions may vary depending on several factors related to cropping systems, fertilizer formulation, its timing and quantity of supply, and weather pattern. Cropping system is among the major factors that influence N-emissions (Engel et al., 2005; Dusenbury et al., 2008). For example, in the Northern Great Plains, where wheat-fallow cropping system is common, conventional tillage produced high N_2O emissions. No-tillage systems generated perceptibly lower levels of N_2O. Greatest fraction of N_2O emission occurred immediately within 10 weeks after fertilizer-N supply. During this period, about 73 percent of total N_2O emissions occurred (Engel et al., 2005; Table 1.4). Nitrous oxide emissions were also influenced by soil moisture during spring season.

TABLE 1.4 Nitrous-Oxide emission from wheat-based cropping systems practiced at Swift Current, Saskatchewan, Canada, in Northern Great Plains

Cropping System	Fallow-Wheat Conventional-Till	Fallow-Wheat No-till	Wheat-Wheat No-till	Winter Pea-Wheat No-till
N_2O emission in 10 weeks after N input (g N ha^{-1})	94	84	49	87
Full Season N_2O Emissions (g N ha^{-1})	133	116	70	155
N_2O-N Emissions from fertilizer-N (kg N ha^{-1})	0.087	0.067	0.022	0.094

Source: Excerpted from Engel et al. (2005)

As stated earlier, nitrogen fertilizers applied to prop up crops are among the major sources of N emissions. For example, in the Canadian Prairies, agricultural cropping accounts for 60 percent of N_2O emissions. About 82 percent of fertilizer-N is consumed to support cropping in three states: Alberta, Manitoba, and Saskatchewan. Therefore, this region induces large amounts of N_2O emissions (Johnston, 2005) for agricultural fields. Farmers who supply fertilizer-N in quantities more than that removed by crops actually accumulate soil-N that could be vulnerable to loss via emissions. Therefore, most fertilizer-N schedules aim at matching the N supply with crop's demand at various stages of growth. In other words, rate of fertilizer-N supplied and chemical formulation are important choices to make, if Great Plains farmers aim at reducing N-emissions. Reduction of N-emissions improves fertilizer-N efficiency. Let us consider an example from Central Plains region that supports corn production. Fertilizer-N rates range from nil to 260 kg N ha^{-1} depending on soil-N fertility status. Most commonly used formulations are urea, ammonium nitrate, and liquid Ammonia (Halvorson and Del Grosso, 2012). They are all prone to rapid loss as emissions. Field investigations by Halvorson et al. (2009, 2011) suggest that N_2O-N emissions increase linearly with quantity of fertilizer-N supplied to corn crop. Nitrous oxide emissions increased from 0.2 kg N_2O-N ha^{-1} to 1.7 kg N_2O-N ha^{-1}. Adoption of No-tillage systems and continuous cropping reduced N-emissions. Fertilizer formulations such as urea or urea-based chemicals induced greater amounts of N-emissions. Fertilizer placement techniques also affect N-emissions from the soil. Overall, farmers in Great Plains must try to reduce accumulation of N in the surface layers of soil, if they intend to reduce N-emissions. Rapid exhaustion of fertilizer-N inputs is essential, if groundwater contamination is intended. Hence, periodic examination of soil and water resources for NO_3-N is essential.

1.6.3   LAND USE CHANGE IN GREAT PLAINS: SOME EXAMPLES

Land use change is an aspect of great concern to naturalists, agricultural specialists, and environmental specialists. It also has bearings on economics and human settlement patterns. Drummond and Auch (2012) point out that knowledge about land use and land cover change and understanding factors that control land use patterns is important. During past five decades, Great Plains, as a complex geographical region has been periodically analyzed in detail for land cover, water resources, cropping pattern, forests, and urban development using Landsat MSS and other satellites. Great Plains region has been demarcated into 17 ecoregions. It helps in assessing effect of various natural and man-made factors more accurately and in detail. Uniform grids of 10 × 10 km are used to collect data at 60 m resolution (Omernik, 1987; Stehman et al., 2005; Table 1.5). The 17 ecoregions identified within Great Plains extend into 2,231,167 km². They are as follows:

Expanse of each ecoregion and average spatial change during past five decades:

Western Great Plains:

Western High Plains—2,88,752 km²; Southwestern Table Land—1,59,938 km²; Central Great Plains—2,73,189 km²; Northwestern Great Plains—3,46,883 km²; Nebraska Sand Hills—60,541 km²

Over all spatial change discerned in 40 years since 1973 = 8.7%

Glaciated Great Plains:

Northwestern Glaciated Plains—1,60,684 km²; Northern Glaciated Plains—1,41,341 km²; Western Corn Belt Plains—2,16,363 km²; Agassiz Plains—40,636 km²

Over all spatial change discerned = 7.1%

East Central Great Plains:

Central Irregular Plains 1,22,589 km²; Flint Hills—27,911 km²; Central Oklahoma/Texas Plains—1,03,412 km²

Over all spatial change discerned = 6.4%

Southern Great Plains:

East Central Texas Plains—44,076 km²; Southern Texas Plains—54,744 km²; Texas Black land Prairies—50,5501 km²; Western gulf Coast Plain—80,972 km²; Edwards Plateau—58,634 km²

Over all spatial change discerned = 10.0%

Great Plains of North America Total area—2,231,167 km²

Over all spatial change discerned = 8.2%

Source: USGS Land Cover Trends Project; Drummond and Auch (2012); Auch (2012).

Over all, Great Plains experienced a land cover change of 8.2 percent during the past five decades beginning in 1973. Maximum change occurred in the Northern Glaciated region (13.6%) that includes areas in Wisconsin and lakes region. Small perceptible change in land cover occurred in the Agassiz Plains (1.4%). Similarly, ecoregions such as Nebraska sandhills and Flint hills that supported grass land and shrubs has a relatively stable vegetation pattern. Fluctuations between grass land and cropland was conspicuous in Northwestern Glaciated Plains (13.6%) and Western Great Plains (11.6%). In case of Southern Plains that includes Texas, it is said that cyclical clearance of junipers, oaks, shrubs, and mesquite enhanced water availability and induced land cover change to crops and pastures. Pastures that required least investment on maintenance and fertilizers were preferred better by farmers during recent years. Climate and population changes too induced land cover changes in the Southern Plains (Drummond and Auch, 2012).

TABLE 1.5 Land cover changes in the 17 ecoregions of the Great Plains during 1973–2010 as detected by satellite

Land Cover Classification	Net Change in Area from 1973 to 2010	
-	(km²)	(%)
Grassland/shrubland	38,922	1.7
Developed land	9,563	0.4
Water bodies	5,811	0.3
Mining zones	1,420	0.1
Mechanically disturbed land	874	0.0
Barren land	−28	0.0
Nonmechanically disturbed land	−1,045	0.0
Wetland	−3,024	−0.1
Forest land	−3,745	−0.2
Agriculture	−48,776	−2.2

Source: Drummond and Auch (2012); USGS Land cover Trends Project, Washington, D.C

Agriculture is the mainstay of Great Plains of North America. Agricultural crop production experienced the most prominent change during past five decades. Agricultural zone in the Great Plains of North America has depreciated in its expanse by 2.2 percent compared to levels known during 1973. Approximately 8.4 percent of crop cover either disappeared or got changed to grasslands and pastures. However, three ecoregions namely, Nebraska sandhills, Southern Texas Plains and Edwards plateau experienced expansion of 0.6–1.4 percent over levels in 1973. It is interesting to note that most ecoregions within Great Plains expanded initially during 1970s. It was attributed to better economic opportunity due to grain export to other countries. Improvement in water use efficiency due to advent of "Pivot Irrigation," further enhanced economic advantages to farmers. Agricultural land depreciated since mid-1980s, because pastures were profitable and easy to maintain. Pastures offered better protection against soil degradation. Surface soil erosion was minimized. Overall, it is said that land cover change from crops to pastures was most frequent (Drummond and Auch, 2012). Other than area, aspects such as crop species, its genotype, cropping sequence, and productivity also experienced marked changes in the Great Plains. In fact, we may realize that total productivity of Great Plains area *per se*, has improved enormously due to use of improved cultivars. This has actually offset the decrease in agricultural area experienced during past five decades.

Great Plains abound with grass lands, pastures, and shrubs. During past five decades, land cover with natural prairies experienced a net change of 38,975 km^2 (1.7%) over 1973 levels. Grass lands depreciated during 1970s but gained in expanse during 1986–2010 s as a consequence of change in cropping pattern. Abandonment of agricultural enterprises also caused gains to pasture land. At the same time, grass lands got converted to developed land forests, which amounted to loss of approximately 6,012 km^2 natural prairies. Grass land decreases were marginal in Central Great Plains, Southern Texas Plains and Nebraska Sand Hills.

Forest area in the Great Plains decreased by 0.2 percent, (equivalent to −3,750 km^2) during past five decades. Most forest cover change was felt in the Southern region, mainly Texas. Forest loss was attributed to clear cutting and construction of water reservoirs.

Wetlands gained in area in the Northern Glaciated ecoregion. Wetter weather during 1980s and 1990s induced improvements in wetlands. Change over from cropland to wetland also occurred for example in Western Gulf region.

During past five decades, developed land increased by 9,545 km^2. New developed land was primarily meant for state highways, industrial infrastructure and metropolitan areas. Developed land gained about 5,180 km^2 from agricultural land cover and 3,117 km^2 from grass lands.

Let us consider dynamics of land cover change and resources (e.g., water) meant for purposes such as agriculture, grassland, wetland, developed land, and forest land water. Overall, water cover in the Great Plains experienced a change of 20,766 km^2, which is equivalent to 4.5 percent (Drummond and Auch, 2012). Wetland development was the chief cause of change in water cover. Erratic climate, drying up of lakes, and channels also induced changes in water cover. Construction of stock dams, stock tanks, and perennial dams also affected water cover, although in a small way.

Great Plains experienced changes in land area allocated to mining. Net gain in mining zone was small 0.1 percent or 1,410 km². Perceptible, yet small changes occurred in land cover classified as "Transitional land" and "Barren land".

1.6.4   TEXAS BLACK LAND PRAIRIES—AN EXAMPLE

This is a small ecoregion within the Southern Great Plains that has been surveyed and examined for changes in vegetation cover, agricultural enterprises, and land development programs. It is located in Central Texas covering areas from Dallas to San Antonio. Black land Prairies extends into 50,501 km². Soils are fine-textured and clayey. Its climate is warm, temperate, and semiarid with regard to precipitation pattern. Rainfall ranges from 710 to 1,015 mm annually. Precipitation decreases from East to West (Natural Resources Conservation Service, 2008). Regarding land cover, it is said that like most areas in Southern Plains, Black land Prairies experienced prominent changes in natural vegetation during early nineteenth century. It supported a tall grass prairie with scattered small patches of forests around lakes and rivulets until then. Human migration and incentive to farm induced rapid changes during late 1800s. Large portion of Black land Prairies was utilized for crop production. It eventually became agricultural prairie with hay, corn, wheat, sorghum, cotton, and soybeans. During the study period from 1970 to 2000, cropping expanses were the dominant form of vegetation. However, substantial changes from agricultural crops to pasture land that required less stringent attention from farmers did occur during past 30 years. Pastures with tame grass were prominent. They were interspersed with vast regions of cropland. It is said during past 30 years, agricultural land experienced net shrinkage by 5.6 percent in area over initial levels. Majority of loss was attributable to conversion of crops to pastures and later to shrubland. Grass lands increased in area by 436 km² (+3.9%) during 30 years. De-intensification of crop production was perceptible (National Agricultural Statistics Service, 2008; Auch, 2012).

1.6.5   RECONVERSION OF ANNUAL CROPLAND TO NATIVE GRASSLANDS: CANADIAN EXPERIENCE

Initially, that is before the advent of settlers in the Canadian Prairies, native grasslands stretched into 61.5 m ha. During past two centuries considerable area of vast Prairies has been tilled, cropped, and converted into regular cropland. At present only 11.4 m ha of native grasslands that produce large quantity of pasture grass/ legumes useful for cattle and other animals survives. In the Great Plains region, Alberta has 49 percent of native grass lands still intact compared to mid-1800s; Saskatchewan has 41 percent and Manitoba only 10 percent. The agricultural expanses have lost sizeable amount of SOC compared to original levels known when native prairies were commissioned for crop production. About 30–50 percent of carbon pools in cropland has been lost to atmosphere (Bailey et al., 2010; Iwaasa et al., 2012). Soil organic matter is well below the minimum levels required in agricultural zones. Further, soil degradation has been rampant in some of parts of Canadian Prairies. Agricultural cropping, if any, has been practiced on marginal land in a few million hectare. Hence, reconversion of marginal cropland back to more productive native grass land has been initiated. It allows better C fixation rates, higher biomass formation, reduces soil erosion, and provides forage useful for farm animals. "Canada Land Conservation Programs" initiated the process of converting cropland into native grass land (Iwaasa et al., 2008). Native grass lands produce 1,100 kg dry matter ha^{-1}. Grass species such as Russian wild rye and wheat grass are used to reclaim marginal croplands. Grass and legume mixtures containing 7–14 species, in different ratios have also been used on the marginal lands. The extent of improvement in SOC due to conversion to native grass land ranged from 0.94 to 3.59 t C ha^{-1} in excess of that derived under regular agricultural cropping on marginal lands.

Wilbert (2012) states that during past 150–200 years, a period coinciding with intensification of agricultural cropping and industrialization, emission of greenhouse gas, and its consequence on climate change has been marked. Emission of CO_2 and CH_4 has increased due to conversion of native forest and prairies to cropland. However, we should take note that greenhouse gas emissions did occur since the time agriculture was invented, cropping zones expanded, and crop production strategies became more intense. Restoring grasslands where ever feasible seems a good proposition. It seems improvising 75 percent of global grasslands will improve atmospheric CO_2 by 330 ppm. Great Plains with large expanses of native prairies that can be repaired or improvised is a good zone to enhance C sequestration. Wilbert (2012), believes that in addition to enhancing species diversity, enriching soil organic carbon, and net biomass productivity, native prairies of Great Plains have direct impact on global climate change.

1.6.6   BIOFUEL PRODUCTION AND LAND USE IN THE GREAT PLAINS-RECENT TRENDS

Estimates during previous decades suggest that Great Plains that serves as a major cereal production zone is also an important forage crop residue generating agroecosystem. The cropland that potentially generates crop residue extends into 187 million hectare covering the 10 Great Plains states of USA. Each year, a certain portion of cropland may get affected by erosion. Therefore, it reduces the effective cropping zone. Great Plains supports several cereals and legume crop that leave a share of biomass to be used as biofuel. The amount of residue generated depends on cereal/legume species, its total above-ground biomass and harvest index. According to Heid (1992), Great Plains states such as Nebraska, Kansas, and Texas offers 75 percent of total crop residue generated in the states. Total crop residue generated accounted to more than 45 million tons during past decade. Crop residue generated in the Great Plains has been used as fuel directly by burning. In many places, it is competitive with coal or other conventional fuels. It has been most effectively used to generate electricity by establishing generating stations close to point of production. Bulky crop residues are not worthwhile, if they have to be transported to long distance for supporting electric companies.

1.6.7   GREAT PLAINS CROP PRODUCTION TRENDS

Cereal grains, especially wheat, maize, and sorghum exports from Great Plains region boomed during 1970s. It was caused by a consistent demand from developing nations. However, it lasted for only a decade and crashed during mid-1980s. Some of the reasons attributed are large-scale wheat and maize supply by Argentina, Brazil, and European nations. The cost of production for cereal grains was higher in the Great Plains. Wheat production in Great Plains needed 112–147 US\$ t^{-1}, but in Argentine Pampas it required only 70.3 US\$ t^{-1}, in United Kingdom, Canada, and Australia cost of wheat production was 117–137 US\$ t^{-1} in mid-1980s. In addition, a large section of Great Plains cereal zones was developed on marginal lands, in response to demand during 1970s. These zones continued yielding relatively low quantity of grains (1.7–2.8 t ha^{-1}) despite higher investments on land conversion. Wheat farmers in Great Plains had to be competitive in the World markets, if they desired to expand cropping zones and harvest more of cereals (Barkema and Drabenstott, 1988; Dhyuvetter et al., 1996).

Mason (2012) opines that Prairie agriculture in Canada is at cross road for various reasons related to soil management, production efficiency, demography and succession of farmers into farms and farm business, competition from grain producers from other continents, and demand for Canadian grain harvests. Number of Prairie farmers who operate individually may also dwindle. One of the suggestions is expanding Canadian exports to as many different crops instead of concentrating on only wheat and barley. Such decisions may change the landscape and several aspects of natural resources and nutrient dynamics. They need to be monitored. A report by Veeman and Gray (2010) suggests that during past 100 years, Canadian Prairies have shown perceptible increase in grain, forage, and livestock production. Introduction of wheat varieties such as "Marquis" into the prairies landscape has actually, tripled grain production. About 40 percent of the farms in Canada are crop based enterprises. During recent years, there is tendency to adopt organic farming. During 1940–2010, average increase in productivity of cereal farms in the prairies has been 1.77 percent annually. During recent years, introduction of crops such as canola and lentils in the prairies has further improved productivity of crop based farms. These crops have changed the landscape predominantly made of wheat-fallow to wheat-oilseed—fallow or wheat-legume—fallow. Maize and sorghum for forage and fuel are other major products offered by farmers in the Canadian Prairies.

During mid-1990s, wheat-fallow system was in vogue in many locations of the Great Plains. This was changed to wheat, summer row crop, and fallow. Cereal crops such as sorghum, maize, or legume intensified the cropping belts. It increased the total cropped area and at the same decreased farmer's economic risks. Dhyuvet-ter et al. (1996) point out that a wheat-sorghum-fallow rotation was less risky to farmer in the Southern Great Plains than just wheat-fallow. Farmer's net returns from the enterprise increased proportionately. Therefore, in this case, less economic risks, stable returns, and better productivity from farms almost changed the land-scape rapidly.

Antle (1997) has clearly pointed out that the farmers in Great Plains are versatile, and many of the farms are highly productive. Farmers do practice integrated systems that allow them to harvest different crops in addition to dominant cereals such as wheat, sorghum, and maize. Yet, they consistently face two economic constraints. They are competition from International markets for cereal grains. Wheat grain productivity has increased steadily since 1950s and real price for their produce has declined. Forecasts suggest that by 2020 price for wheat generated in the Great Plains could be at lowest levels in real terms. Secondly, farmers in Great Plains face production risks due to climate change, vagaries of precipitation pattern, soil loss, and pathogen and pest pressures. These could restrain them from investing and intensi-fying wheat production.

Reports suggest that since early 1900s, Great Plains states such as Dakotas and Kansas have been competing each other for top spot regarding total wheat harvest and export. These two states are closely followed by Oklahoma and Montana in terms of wheat harvest. Hard Red wheat is the predominant type produced and ex-ported by US Great Plains farmers.

During mid-1990s, wheat belt in the Great Plains contributed about 69 million ton grains produced on 79 million acre. Wheat belt in the Great Plains and other states in USA have shown a marginal decline in area since mid-1980s. It dropped sharply due to "Area Reduction Program." About 30 percent of wheat area was left idle voluntarily by the Great Plains farmers. Total wheat acreage in the country dropped from 89 m ac in 1985 to 58.2 m ac during 2009. It has further decreased during recent years. In several locations, fields meant for wheat during yester years have been switched to soybean or corn. As a consequence, wheat grain contribution to food grains *per se* has decreased markedly during past 5 years (USDA Economic Research Service, 2012). In addition to Governmental legislation, disease pressure due to blight has also affected wheat planting in the Great Plains. More recently, ethanol production has gained in importance due to its use in biofuel. Cereal crop meant for biofuel production is corn or at best sweet sorghum and not wheat. Hence, farmers interested in biofuel prospects have consistently replaced wheat zones with corn. Whatever be the causes for fluctuations in area and productivity of wheat belt, we should recognize the fact that Great Plains farmers still offer 73 percent of total wheat grains produced by United States of America. Almost 95 percent of total wheat produced by Canada is derived in the Canadian Prairie provinces. These agricultural prairies feed not only population within Great Plains, but nourish millions in other regions of the world.

Wheat Production in the Great Plains States of USA during 2012

	Great Plain States				USA-Total					
	Ok	Tx	Ne	NM	Ks	Co	Wy	Mo	ND	SD
Area harvested (1,000 ac)	4,300	3,000	1,300	85	9,100	2,182	120	5,585	7,760	2,235	48,991
Production (10^6 bu)	155	96	53	2	382	75	3	195	340	102	2270
Productivity (bu ac^{-1})	36	32	41	55	42	24	25	35	44	46	46

Source: National Agricultural Statistics Service (2012), USDA

Maize has served as one of the most important cereals of the Great Plains, both to natives and European settlers. Maize-based prairies are predominant in certain areas of Great Plains. Maize genotypes cultivated by farmers in this region have been dynamic since ages. Several factors related to soil fertility, its quality and productivity, ambient atmospheric conditions, precipitation pattern, season, human preferences, demand from domestic, and external sources have all affected production trends and area sown to maize. Egli (2008) has pointed out that maize productivity was almost stagnant during six decades from 1866 to 1930. Maize grain yield increased by a marginal 5.3 kg ha^{-1} year^{-1} during this period. Such small or negligible increase of grain productivity was also perceived with many other crops grown in the Great Plains. Absence of chemical fertilizer-N, reduced irrigation facilities, lack of genotypes with high yield potential and low capital were some the reasons for stagnant grain yield in agricultural prairies of the Great Plains. Maize grain/forage yield in the prairies registered a perceptible increase from 1950s, which is termed as the beginning of high yield era for crop production in USA and many other regions of the world. This period coincided with high input era. Chemical fertilizer supply per unit area increased rapidly with concomitant increase in water supply. Maize composites and hybrids with high grain yield changed the land scape of parts of Great Plains situated closer to Corn Belt. It seems, by 1960, more than 90 percent of maize zone was filled with hybrids. Next, maize belt experienced high-intensity planting, use of high dosages of fertilizers, pesticides and herbicides, and enhanced use of labor and mechanization. These changes increased maize grain yield to 5.5 t ha^{-1}. As a caution, Egli (2008) suggests that high inputs, better infrastructure and technology may always be the prime reason for increase or depressions in area and productivity of a crop. However, agroclimatic parameters such as precipitation pattern, large climatic changes, dust bowl situation may also affect corn production trends in the Great Plains. Cassman (2012) has made interesting suggestions regarding the causes for increased corn production in the Great Plains and Corn Belt. During the years 1966–2010, linear gain in corn grain yield is 1.86 bu ac^{-1} year^{-1}. Corn productivity increased from 60 bu ac^{-1} in 1966, to 165 bu ac^{-1} in 2010. The initial gains from 60 bu to 80 bu ac^{-1} has been attributed to application of fertilizer at higher rates, soil testing and conservation tillage practices. This phenomenon occurred during 1965–1978. Next, an increase from 80 bu to 100 bu ac^{-1} during 1978–1988 is mainly due to use of double-cross hybrids and integrated pest management practices. The maize landscape got intensified further during 1980–2000 resulting in higher productivity. Corn yield increased to 140 bu ac^{-1} in 2005 due to factors such as expansion of irrigated areas, transgenic maize and precision planters. At present, in 2010, the grain yield has reached 170 bu ac^{-1}. It has been attributed to higher fertilizer supply at balanced proportions. Cassman (2012) believes that maize yield will further show steep increase, once precision techniques, and auto-steer electronically controlled farm vehicles are introduced in large number. Clearly, agricultural techniques have changed the maize land scape. It has intensified the region's agricultural cropping trends and improved economic advantages of farmers. Intensification of crop production has initially involved higher labor hours and ingredients. Later, techniques and automation of farming procedures has lessened the burden on human labor requirements. Grain yield potential of corn in Great Plains is much higher than that reaped currently. This allows for further improvement in grain yield (see Cassman et al. 2003; Duvick and Cassman, 1999; Duvick, 2005). During 2013, farmers in USA harvested maize at an average productivity of 123.4 bu ac^{-1} (National Agricultural Statistics Service, 2012).

Sorghum production is predominant in Kansas (2.5 m ac) and Texas (2.1 m ac). These two states support 90 percent of sorghum area in the Great Plains. Sorghum belt is expansive in the Southern drylands and extends into 5.5 m ac. Sorghum cultivation is feeble, yet useful quantities of grain/forage are produced in other regions of Great Plains, such as Nebraska (110 thousand ac), New Mexico (50 thousand ac), Oklahoma 220 thousand ac) and Dakotas (180 thousand ac). The productivity of sorghum in the Great Plains fluctuates between 45 and 93 bu ac^{-1}. The United States of America harvested about 7.27 m acres of sorghum belt that yielded 472 million bushels during 2009 (Centre for Sorghum Improvement, 2009). Sorghum belt experienced reduction in expanse during the years 1983–2000. It decreased from 18.5 million acres in 1983 to just 9.4 m ha in 2000. Such drastic effects on the Great Plains cropping landscape was attributed to the "1985 Farm Bill", spread of drought-tolerant corn into areas hitherto occupied by sorghum, and fluctuations in demand by other regions of the world. During 2012, sorghum cultivation zones in USA extended into 6.24 m ac and yielded about 49.8 bu ac^{-1}. Total production of sorghum grains reached 246.9 m bu.

Barley production occurs mostly in the Northern Plains states such as Montana and Dakotas. This region accounts for 45 percent of total barley production zones of the Great Plains. Barley crop extends into other states such as Wyoming, Nebraska, and Colorado. During past two decades, barley cropping expanse has fluctuated within limits from 2.9 to 6.7 m ac annually. However, it is clear that barley belt in the Great Plains, especially in Southern and Central region has steadily declined. The price fetched by a bu of barley has gained from 2.38 US$ in 1996 to 5.37 US$ bu^{-1} in 2010 (USDA, 2010). Barley grain productivity during past decade has ranged from 58 to 73 bu ac^{-1}. Canadian Prairie states and other regions produce over 14 m t barley annually. Great Plains Provinces contribute 90 percent of Canadian barley produce. The average barley grain yield in these states is 2.4–3.8 t ha^{-1} (USDA, 2004). Barley production is feeble and covers <5,000 ha in states such as British Columbia, Quebec, and Ontario. During 2012, Barley belt in USA extended into 3.24 m ac and gave grains at an average productivity of 67.9 bu ac^{-1} (National Agricultural Statistics Service, 2013a).

Oats are popular in Dakotas although its cropping zones are small. It is grown in most of the Great Plains states and area in each of them varies between 30,000 and 1,65,000 acre. Productivity of oats ranges from 49 to 78 but ac^{-1} depending on location. During 2012, US farmers planted oats on 2.7 m ac but harvested grains from only 1.04 m ha at a rate of 61.3 bu ac^{-1} (National Agricultural Statistics Service, 2013a).

Soybean is an important legume derived from the Great Plains, although its production is more pronounced in the Midwestern states such as Indiana, Illinois, and Ohio. Egli (2008a) points out that soybean yield in the Great Plains and Midwestern USA increased perceptibly from 1924 to 1930. Soybean was a new crop introduced into Great Plains. Soybean grain yield was less than 1.0 t ha^{-1}. The gain in soybean productivity continued from 1930 to 1950. The ratio of corn and soybean production changed from 2:1 in 1960s to 3:1 in 2000 (Egli, 2008a). During 2012, farmers in USA harvested soybeans from 76 million acres of fertile lands mostly situated in the midwest regions at an average productivity of 39.6 bu ac^{-1} (National Agricultural Statistics Service, 2012).

Canola is an important oilseed crop in the Northern Plains. It is found grown as a rotational crop with wheat or barley. Canola is also an important oilseed crop in the Canadian Prairie states. Its cultivation became more profitable as years lapsed. During 1997, Saskatchewan grew canola on 5.6 m ac^{-1} and produced 2.6 m t seeds, Manitoba grew 2.3 m ac and produced 1.4 m t, while Alberta grew canola on 4 m ac and produced 2.7 m t seeds (University of Nebraska, 2011). During 2012, canola growing regions in USA extended into 1.73 m ac and provided 1416 lb seeds ac^{-1}. Major canola producing states in the Great Plains are Montana, North and South Dakota, and Wyoming. Oklahoma also supports canola production.

Lentils are gaining ground in the Northern Plains covering regions of Alberta, Saskatchewan, and Manitoba in the Canadian Prairies. Relatively small area of Dakota and Montana also supports lentils as legume cover crop. During 2012, US farmers harvested lentils from 0.46 m ac at an average productivity of 1178 lb ac⁻¹.

Peanut is an important cash crop grown in the Great Plain states such as Oklahoma, Texas, and small regions in New Mexico. A greater share of peanuts is cultivated in eastern states such as Georgia, Florida, North Carolina, and Virginias. During 2012, US farmers harvested about 1.6 m ac of peanut at an average productivity of 4,192 lb ac⁻¹ (National Agricultural Statistics Service, 2013c).

Sunflower is an important oilseed crop in Central and Southern plains, and other parts of USA. During 2012, farmers in entire USA harvested 1.84 m acres of sunflower crop that gave 1513 lb oilseed ac⁻¹. Sunflower cultivation is slightly pronounced in the Southern and Central Plains region. It is a dryland crop with better water use efficiency than other crops (National Agricultural Statistics Service, 2013b).

Cotton cultivation occurs in four different geographic regions of USA termed commonly as Southeast, Mid-South, Southwest, and West. It is grown in the Great Plains as well as outside this region. Within the present context, cotton cultivation in Great Plains states such as Texas, Oklahoma, Kansas, and New Mexico is important. The Great Plains cotton belt supplies, about 38 percent of the annual upland cotton produce of the United States of America. The average staple length of cotton produced in these states is 34.7–36 sec of an inch (Cotton Council International, 2009). Following is the statistics for cotton production in Great Plains during 2009:

State	Area Harvested in 1,000 Acres	Yield in Pounds Ac⁻¹	Total Production in 1,000 Bales
Kansas	32	720	48
New Mexico	29.4	936	57.3
Oklahoma	195	837	340
Texas	3,716	650	5,032

Source: USDA National Agricultural Statistics Service (2009)

The Texas cotton belt is predominant in the Northern Highlands and Rolling plains. It is larger by 100-folds in area compared to other states such as New Mexico or Kansas, although productivity is lower. Texas plains contribute about 40 percent of cotton produce generated in the USA (12,592 × 1,000 bales in 2009). The average farm size in the belt is 1,241 ha with a productivity of 985 kg ha⁻¹ (Osakwe, 2009). During past 3 years, cotton belt was dominated by Upland cotton varieties such as Deltapine and Fibermax. A small portion of highlands produced ELS cotton and Pima cotton. A major share of cotton produced in the Great Plains finds its way to China, which accounts for 32 percent of cotton exports by USA. During 2012, cotton belt in USA extended into 9.42 m ha and yielded 866 lb ac⁻¹ (National Agricultural Statistics Service, 2012).

Forage production, preparation of dry hay and silage is a major preoccupation of farmers in the Great Plains. Farmers fill the landscape with forage grass/legume species wherever possible in the drylands, wastelands, regions with deteriorated soil, and most importantly in the crop fields during fallows. A planted fallow with forage grass/legume mixture is a useful proposition to farmers in Great Plains. Hay and silage productivity has varied depending on soil, water resources, environmental parameters, and forage species sown. Overall, during 2012, US farmers harvested 34.5 m acres of pastures that gave 79.5 m t of hay on dry weight basis (National Agricultural Statistics Service, 2013d). A large part of forage is contributed by Great Plains region especially by Dakotas and Montana in the North and Texas, Oklahoma, and Kansas in the South. Canadian Prairies too contribute large quantities of forage grasses that get exported.

KEYWORDS

- **Agroecoregion**
- **Dryland**
- **Frost-free**
- **Great plains of North America**
- **Migration**
- **Prairies**
- **Semiarid regions**
- **Tillage**
- **Tillage prairies**

REFERENCES

1. Agriculture Knowledge Centre. Organic Crop Production: Soil Conservation Practices. Government of Saskatchewan—Agriculture. **2008,** 1–5 p.
2. Aiken, R. M.; and Lamm, F. R.; Irrigation guidelines for oilseed crops in the US Central Great Plains. In: Technical Conference Proceedings of the 27th Annual Irrigation Association. San Antonio: TX. November 5–7, **2006,** 14–24 p. March 2, **2013,** http://www.reeis.usda.gov/web/crisprojectpages/0185313-managing-transpiration-efficiency-productivity-and-stability-in-semi-arid-crop-systems.html
3. Albrecht, D. E.; "The renewal of population loss in the non-metropolitan Great Plains" *Rural Sociol.* **1993,** *58,* 233–46
4. Allamaras, R. R.; Schonberg, H. H.; Douglas, C. L.; and Dao, T. H.; Soil organic carbon sequestration potential of adopting Conservation Tillage in United States Crop land. *J. Water Soil Conserv.* **2000,** *55,* 365–373.

5. Allen, B.; Pikul, J. L.; Waddell, J. T.; and Cochran, V. L.; Long term lentil green-manure replacement for fallow in the semi-arid Northern Great Plains. *Agron. J.* **2010**, *103,* 1292–1298.
6. Anderson, D.; and Cerkowniak, D.; Soil formation in the Canadian Prairies. *Prairie Soils Crops J.* **2010**, *3,* 57–64.
7. Antle, J.; Climate change and Economic Constraints facing Great Plains Agriculture. **1997**, 1–5, January 17, **2013**, http://www.nrel.colostate. edu/projects/climate_impacts/antle.htm
8. Auch, R. F.; Texas Black land Prairies Ecoregion Summary. United States Department of Geological Survey. **2012**, 1–9, http://landcovertrends.usgs.gov/gp/ecoregion32Report.htm
9. Bailey, A.W.; McCartney, D.; and Schellenberg, M. P.; Management of Canadian Prairie Rangeland. AAFC Publication No. 10144, **2010**, 1–23.
10. Bailey Seed, Legumes. **2012**, 1–8, April 9, **2013**, http://baileyseed.com/infolegumes.asp
11. Barkema, A.; and Drabestott, M.; Can US and Great plains agriculture compete in the world market? Federal Reserve Bank of Kansas City. *Econ. Rev.* **1988**, 1–17.
12. Barker, D. W.; and Sawyer, J. E.; Using Active Canopy sensing to adjust nitrogen application rate in corn. *Agron. J.* **2013**, *104,* 926–933.
13. Baumhardt, R. L.; Wind, water and growing season: Cropping system Selection pressures in the Southern Great Plains. In: Proceedings of Dynamic Cropping Systems Symposium. Colorado, USA; **2003**, 114–123.
14. Baumhardt, R. L.; and Salinas-Garcia, J.; Mexico and United States Southern Great Plains. In: Dry Land Agriculture. Peterson, R. L.; Ed. *Agron. Monogr.* 23, **2005**, 85 p.
15. Bedard-Haughn, A.; Prairie wetland soils: Gleysoilic and organic. Prairie Soils and Crops. **2012**, *3,* 9–15.
16. Borade, G.; Facts about Wind Erosion. Buzzle.com **2012**, 1–2.
17. Bouton, J.; Ecotype-derived White Clover Cultivars and their Place in Southern Great Plains. **2013**, 1–3 p. April 9, **2013**, http://www.noble.org /as/pasture/whiteclovercultivars/
18. Bowman, R. A.; Vigil, M. F.; Andrson, R. L.; Nielsen, D. C.; and Benjamin, J. G.; The soil organic matter dynamics in dry land cropping systems in the Central Great Plains. In: Central Great Plains Research Station. Colorado, USA: USDA-ARS, Akron; Internal Report, **2005**, 1–3.
19. Boyles, M.; Peeper, T.; and Stamm, M.; Canola production. Oklahoma State University Extension Services. USA: Stillwater; **2003**, 1–18.
20. Blanco-Canqui, H.; Claassen, M. M.; and Presley, D. R.; Summer cover crop fix nitrogen, increase crop yield and improve soil-crop relationships. *Agron. J.* **2012**, *104,* 137–147.
21. Bradford, J. B.; Lauenroth, W. K.; and Burke, I. C.; The impact of cropping on primary production in the US. Great Plains. *Ecology.* **2005**, *86,* 1863–72.
22. Bradford, J. B.; Lauenroth, W. K.; and Burke, I. C.; and Pareulo, J. M.; The influence of climate, soils, weather and land use on primary production and biomass seasonality in the Great Plains. *Ecosystem.* **2006**, *9,* 934–950.
23. Campbell, C. A.; Janzen, H. H.; Paustian, K.; Gregorich, E. G., Sherrod, L.; Liang, B. C.; and Zentner, R. P.; Carbon storage in soils of the North American Great Plains: Effect of cropping frequency. *Agron. J.* **2005**, *97,* 349–363.
24. Campbell, C. A.; et al. A Bibliography of Scientific Publications Based on Long Term Crop Rotations in the Canadian Prairies. Winnipeg, Canada: International Institute for Sustainable Development; **2012**, 1–2, January 20, 2013, http://www.prairiesoilandcrops.ca/display_ summary.html?id=92
25. Canola Council of Canada. Effccts of Moisture on Canola Growth. **2001**, 1–22, April 9, 2013, http://www.canolacouncil.org/crop-production/canola-grower's-manual-contents/chapter-4-effects-of-moisture/effects-of-moisture
26. Cash, D.; and Johnson, K.; US National Assessment of the Potential consequences of Climate Variability and change US climate forum. The Great Plains. **2003**, 1–1, February 10, 2013, http://www.usgcrp.gov/usgcrp/nacc/background/meetings/forum/ greatplains_summary.html

27. Cassman, K. G.; Yield gap analysis: Implications for research and policy. Nebraska Centre for Energy Sciences. Nebraska, USA: University of Nebraska-Lincoln; **2012,** 1–23, March 27, **2013,** www.ncesr.unl.edu

28. Cassman, K. G.; Dobermann, A.; Walters, D. T.; and Yang, Y.; Meeting cereal demand while protecting natural resources and improving environmental quality. *Ann. Rev. Environ. Res.* **2003,** *28,* 15–28.

29. Centre for Sorghum Improvement. Great Plains Sorghum Improvement and Utilization Centre: Federal Initiative Accomplishments. **2009,** 1–2, April 1, **2013,** http://www.sorghumcenter.com/annual_report.html

30. Cliff-Notes. The Agricultural Frontier. **2012,** 1–6, January 4, 2013, http://www.cliffsnotes.com/study_guide/The-Agricultural-Frontier.topic ArticleId-25238,articleId-25173.html

31. CNHP. Western Great Plains Short grass Prairies. Colorado Natural Heritage Program. Colorado State University, Fort Collins, Colorado, USA. **2013,** 1–4, February 18, 2013, http://www.cnhp.colostate.edu/projects/eco_systems/pdf/ WGP_ Shortgrass_Prairie.pdf

32. Cochrane, W. W.; The Development of American Agriculture: A Historical Analysis. Minneapolis: University of Minnesota Press; **1993,** 249 p.

33. Cole, C. V.; Burke, I. C.; Parton, W. J.; Schimel, D. S.; Ojima, D. S.; and Stewart, J. W. B.; Analysis of historical changes in the soil fertility and organic matter levels of the North American Great Plains. In: Proceedings of International Conference on Dry Lands. Texas, USA: Amarillo; **1990,** 235–238.

34. Cotton Council International. Regions of US Cotton Production. **2009,** 1, March 27, 2013, http://www.cottonusa.org/directories/ BuyersGuide.cfm?ItemNumber=1267&sn.ItemNumber

35. Cunfer, G. On the Great Plains: Agriculture and Environment. Texas, USA: Texas A and M University Press, College Station; **2005,** 78 p.

36. Derksen, D. A.; Anderson, R. L.; Blackshaw, R. E.; and Maxwell, B.; Weed dynamics and management strategies for cropping systems in the Northern Great Plains. *Agron. J.* **2002,** *94,* 174–185.

37. Dhyuvetter, K. C.; Thompson, C. R.; Norwood, C. A.; and Halvorson, A. D.; Economics of dry land cropping systems in the Great Plains: A review. *J. Prod. Agric.* **1996,** *9,* 216–222.

38. Drummond, M. A.; and Auch, R.; Land Cover Change in the United States Great Plains. Washington DC: USGS Land Cover Trends Project, **2012,** 1–18.

39. Duffy, M.; Estimated Costs of Crop Production in Iowa-2009. Iowa State University Extension Bulletin, FM1712. **2008,** 34 p February 6, 2013, http://www.econ.iastate.edu/faculty/duffy/documents/COP2009.pdf

40. Dusenbury, M. P.; Engel, R. E.; Miller, P. E.; Lemke, R. L.; and Wallander, R.; Nitrous oxide emissions from a Northern Great Plains soil as influenced by Nitrogen management and cropping systems. *J. Environ. Qual.* **2008,** *37,* 542–550.

41. Duvick, D. N.; The contribution of breeding to yield advances in Maize (Zea mays). *Adv. Agron.* **2005,** *86,* 84–145

42. Duvick, D. N.; and Cassman, K. G.; Post-green revolution trends in Yield potential of temperate maize in the North Central United States. Crop Science; **1999,** *37,* 1622–1630 p.

43. Egli, D. B.; Comparison of corn and soybean yields in the United States: Historical trends and future prospects. *Agron. J.* **2008,** *100,* s-79-S-88.

44. Egli, D. B.; Soybean yield trends from 1972–2003 in the Mid-Western USA. *Field Crops Res.* **2008a,** *106,* 53–59.

45. Elgaai, E.; and Garcia, L. A.; Sensitivity of Irrigation Water Supply to Climate Change in the Great Plains Region of Colorado. Fort Collins, Co, USA: Colorado State University; **2005,**

1–8, February 28, **2013,** http://www.hydrologydays.colostate.edu/Abstracts_ 05/Elgaali_abs. pdf

46. Engel, R.; Dusenbury, M.; Miller, P.; and Lemke, R.; A First Check of Nitrous Oxide Emissions Under Cropping Systems Adapted for the Northern Great Plains. Salt Lake City, Utah, USA: Western Nutrient Management Conference; **2005,** *6,* 25–31.

47. Fenneman, N. M.; Physiographic subdivision of the United States. Proceedings of the National Academy of Sciences of the United States of America. **1917,** *3,* 17–22, October 5, 2010, http://www.worldcat.org/oclc/186331193,__http://digitool.library.colostate.edu///exlibris/dtl/ d3_1/apache_media/L2V4bGlcmlzL2R0bC9kM18xL2FwYWNoZV9tZWRpYS8xMTIx- MTA=.pdf

48. Flach, K. W.; Barnwell, T. O.; and Crosson, P.; Impact of agriculture on atmospheric CO_2. In: Soil Organic Matter in Temperate Agroecosystems. Paul, E. A.; Paustian, K.; Elliot, E. T.; and Cole, C. V.; Eds. Florida, USA: CRC Press, Boca Raton; **1997,** 343–352.

49. Fore, Z.; Eroding Your Profits. University of Minnesota Extension Service. **2008,** 1–5, March 13, **2013,** http://www.soybeans.umn.edu/pdfs/ regional/nw/Eroding%20Your%20Profits%20 Soil%20Erosion.pdf

50. Friedmann, H.; "World market, state, and family farm: Social bases of household production in the era of wage labor". *Comp. Stud. Soc. Hist.* **1978,** *20,* 545–86.

51. Fryrear, D. W.; Sutherland, P. L.; Davis, G.; Hardee, G.; and Dollar, M.; Wind erosion estimates with RWEQ and WEQ. In: Sustaining the Global Farm. 10th International Soil Conservation Organization Meeting. Stott, D. E.; Mohtar, R. H.; and Steinhardt, G. C.; Eds. United States Department of Agriculture-ARS National Soil Erosion Research Laboratory. Indiana, USA: Purdue, West Lafayette; **2001,** 760–765.

52. Gordon, W. B.; Maximizing irrigated soybean yields in the Great Plains. *Better Crops.* **2008,** *92,* 6–8.

53. Grant, C. A.; Peterson, G. A.; and Campbell, C. A.; Nutrient considerations for diversified cropping systems in the Northern Great Plains. *Agron. J.* **2002,** *94,* 186–198.

54. Graybosch, R. A.; and Peterson, J.; Specific adaptation and genetic progress for grain yield in the Great Plains Hard Red Winter Wheats from 1987 to 2010. *Crop Sci.* **2013,** *52,* 631–643.

55. Gutmann, M. P.; Parton, W. J.; Cunfer, G.; and Burke, I.; Population and Environment in the Great Plains. Washington, DC: National Academy of Sciences; NCBI Bookshelf.htm **2005,** 1–10, January 20, **2013.**

56. Gutmann, M. P.; and Sample, C. G.; Land, climate and settlement in the Texas frontier. *South-Western Hist. Quar.* **1995,** *99,* 137–172.

57. Haak, D.; Sustainable agricultural land management around Wetlands on the Canadian Prairies. Soil Resources Division. Agriculture and Agri-Food Canada. Ottawa, Canada; **2008,** 1–11.

58. Haas, H. J.; Grunes, D. L.; and Reichman, G. A.; Phosphorus changes in Great Plains soils as influenced by cropping and manure applications. *Soil Sci. Soc. Am. J.* **1960,** *25,* 214–218.

59. Halvorson, A. D.; and Del Grosso, S. J.; Nitrogen source and placement affect soil nitrous oxide emissions from irrigated corn in Colorado. *Better Crops.* **2012,** *96,* 7–9.

60. Halvorson, A. D.; Del Grosso, S. J.; and Alluvione, F.; Nitrogen rate and source effects on nitrous oxide emissions from irrigated cropping systems in Colorado. *Better Crops.* **2009,** *93,* 16–18.

61. Halvorson, A. D.; Del Grosso, S. J.; and Jantalia, C. P.; Nitrous oxide emissions from several nitrogen sources applied to a strip-tilled corn field. Proceedings of Fluid form, Fluid Fertilizer Foundation. Scottsdale, Arizona; **2011,** 1–6.

62. Harding, K. J.; and Snyder, P. K.; Modelling the atmospheric response to irrigation in the Great Plains. Part-1. General impacts on precipitation and energy budgets. *J. Hydrometeorol.* **2012,** *13,* 1667–1686.

63. Hartman, M. D.; Merchant, E. R.; Parton, E. J.; Gutmann, M. P.; Lutz, S. M.; and Williams, S. A.; Impact of historical land use changes in the US Great Plains, 1883 to 2003. *Ecol. Appl.* **2011,** *21,* 1105–1119.

64. Heathcote, R. L.; Perception of Desertification on the Southern Great Plains-a Preliminary Enquiry. Tokyo, Japan: United Nations University; Internal Report, **1980,** 134 p.

65. Heid, W. G.; Turing great plains crop residues and other products into energy. *United States Dep. Agric.-Econ. Res. Ser.* Research Report 623, **1992,** 1–27 p.

66. Helms, D.; The soil conservation service in the Great Plains. Natural Resource Conservation Service. Beltsville, Maryland, USA: United States Department of Agriculture; **1990,** 8, January 16, **2013,** http://www.nrcs.usda.gov/wps/portal/ nrcs/detail/ national/about/ history/?&cid=nrcs143_021395

67. Herridges, D. F.; Peoples, M. B.; and Boddey, R. M.; Global inputs of biological nitrogen fixation in agricultural systems. *Plant Soil.* **2008,** *311,* 1–18.

68. Hubner, B.; Agriculture in the prairie provinces. In: History of Agriculture in the Prairie Provinces. University of Manitoba Libraries. Manitoba, USA: University of Manitoba; **1998,** 1–3 p.

69. Hudson, J. C.; Agriculture. Encyclopaedia of the Great Plains. Wishart, D. J.; ed. **2012,** 1–8, February 24, 2013, http://www.plainshumanities.unl.edu/ encyclopaedia/

70. IISD. On the Great Plains. International Institute for Sustainable Development. Winnipeg, Canada: Manitoba; **2012a,** 1–2, http://www.iisd.org/agri/GPissues.htm

71. IISD. Degradation of prairie soil resources. International Institute for Sustainable Development. Winnipeg, Canada; 2012b, 1–5.

72. IRSNDSU. Northern Great Plains-1880 to 1920. Fargo, North Dakota, USA: Institute for Regional Studies North Dakota State University. **2013,** 1–2, February, 20, 2013, memory.loc. gov/ammem/award97/ndfahtml/paz_ag.html

73. Iwaasa, A. D.; McConkey, B.; and Schellenberg, M.; Effect of grazing and re-establishment of native Grass land on Soil organic matter sequestration for the semi-arid central grasslands of Canada. Joint Meetings: IGC-IRC, Hohhot, China, Jones, O. R.; and Stewart, B. A.; 1979 Basin Tillage. Soil Tillage Research **2008,** *18,* 249–265.

74. Iwaasa, A. D.; Schellenberg, M. P.; and McConkey, B.; Re-establishment of Native Mixed Grassland species into Annual Cropping land. *Prairie Soils Crops J.* **2012,** *5,* 85–95.

75. Jabro. J. D.; Iverson, W. M.; Evans, R. G.; and Stevans, W. B.; Water use and Water Productivity of sugar beet, malt barley, and potato as affected by irrigation. *Agron. J.* **2013,** *104,* 1510–1516.

76. Johnson, S. E.; Irrigation policies and programs in the Northern Great Plains region. *J. Farm Econ.* **1936,** *18,* 543–55.

77. Johnston, A.; Western Canada Research Report. Saskatchewan, Canada: Potash and Phosphate Institute of Canada; **2000,** 1–6.

78. Johnston, A.; Potassium Nutrition in the Northern Great Plains. Norcross, Georgia, USA: Potash and Phosphate Institute (PPI); **2004,** 1–5.

79. Johnston, A.; Nitrous Oxide Emissions from Fertilizer Nitrogen. Saskatoon, Saskatchewan: Potash and Phosphate Institute of Canada; News and Views. **2005,** 1–3.

80. Johnston, A.; Fertilizer Best Management Practices for the Northern Great Plains—How do You Measure up. Norcross, Georgia, USA: Potash and Phosphate Institute (PPI); News and Views. **2006,** 1–8.

81. Johnston, A. M.; Tanaka, D. L.; Miller, P. R.; Brandt, S. A.; Nielsen, D. C.; Lafond, G. A.; and Riveland, N. R.; Oilseed crops for semiarid cropping systems in the Northern Great Plains. *Agron. J.* **2002,** *94,* 231–240.

82. Jones, O. R.; and Stewart, B. A.; Basin tillage. *Soil. Tillage Res.* **1990,** 18(2–3), 249–265.

83. Klingberg, K.; and Weisenbeck, C.; Shallow vertical tillage impact on soil disturbance and crop residue. Proceedings of the 2011 Wisconsin Crop Management Conference. **2011,** *50,* 46–49.

84. Krishna, K. R.; Historical aspects of soil fertility and crop production research. In: Soil Fertility and Crop Production. Krishna, K. R.; ed. New Hampshire, USA: Science Publishers Inc. Enfield; **2002,** 1–43.

85. Krishna, K. R.; Precision Farming: Soil Fertility and Productivity Aspects. Toronto, Canada: Apple Academic Press Inc.; **2013,** 189 p.

86. Krupinsky, J. M.; Tanak, D. L.; Merill, S. D.; Liebig, M. A.; and Hanson, J. D.; Crop sequence effects of 10 crops in the Northern Great Plains. *Agric. Syst.* **2006,** 88-227-254.

87. Lafond, G. P.; and Harker, K. N.; Long-term Cropping Studies on the Canadian Prairies: An Introduction. Winnipeg, Canada: International Institute for Sustainable Development; **2012,** January 20, **2013,** http://www.prairiesoilsandcrops.ca/display_ summay.html?id=88

88. Lambert, D.; and Lowenberg-DeBoer, J.; Precision Agriculture: Profitability Review. Site Specific Management Centre, School of Agriculture. Lafayette, In, USA: Purdue University; **2000,** 154.

89. Lamey, F. J.; and Janzen, H. H.; Long term erosion-productivity relationships: The lethbridge soil scalping studies. Lethbridge, Alberta, Canada: Agriculture-Canada, Eastern Cereal and Oilseed Research Centre; *Prairie Soils Crops J.* **2012,** 5: 139–146, 1–2 p. January 20, **2013,** http://www.prairiesoilsandcrops.ca/display_summary.html?id=82

90. Lamm, F.; Corn Production in the Central Great Plains as Related to Irrigation Capacity. **2005,** 1–11, March 1, **2013,** http://digitool.library. colostate.edu//exlibris/dtl/d3_1/apache_media/ L2V4bGlicmlzL2R0bC9kM18xL2FwYWNoZV9tZWRpYS8xMTIxMTA=.pdf

91. Lange, H. R.; Guide to the Records of the Great Plains Agricultural Council. Fort Collins, Colorado, USA: Colorado State University, Colorado Agricultural Archive; **2008,** 1–20 p.

92. Lemke, R.; Campbell, C.; Zentner, R.; and Wang, H.; Old rotation study-swift current, Saskatchewan. *Prairies Soils Crops J.* **2012a,** *5,* 59–66.

93. Lemke, R.; Campbell, C.; Zentner, R.; and Wang, H.; New Rotation Study-Swift Current, Saskatchewan. **2012b,** 1–2, January 12, 2013, http://www.prairiesoilandcrops.ca/display_summary.html?id=75

94. Lemke, R.; Malhi, S.; Johnson, E.; Brandt, S.; Zentner, R.; and Offert, O.; Alternative Cropping Systems Study-Scott, Saskatchewan. **2012c,** 1–2, January 20, 2013, http://www.prairiesoilsand crops.ca/display_summay.html?id=75

95. Lenssen, A. W.; Iversen, W. M.; Sainju, U. M.; Caesar-Tonthat, T. C.; Blodgett, S. L.; Allen, B. L.; and Evans, R. G.; Yield pests and water use of durum and selected crucifer oilseeds in two-year rotations. *Agron. J.* **2012,** *104,* 1295–1303.

96. Lewis, F.; Causes of the Great Migration: Searching for the Promised Land. African-American History. **2012,** 1–2, January 23, 2013, Africanamericanimmigr.html

97. Li, S.; Lobb, D. S.; Lindstorm, M. J.; and Farehorst, A.; Patterns of water and tillage erosion on topographically complex landscapes in the North American Great Plains. *J. Soil Water Conserv.* **2008,** *63,* 37–46.

98. Library of Congress. Primary Documents of the American History: Homestead Act. United State of America Library of Congress Digital Section. **2012,** 1–12, January 4, **2013,** http:// www.loc.gov/rr/program/bib/ourdocs/Homestead.html

99. Lindstorm, M. J.; Skidmore, E. L.; Gupta, S. C.; and Onstad, C. A.; Soil conservation limitations on removal of crop residues for energy production. *J. Environ. Qual.* **1979,** *8,* 533–537.

100. Loring, K.; Soil Conservation and Allotment Act-1935. Major Acts of Congress. **2004,** 1–2, January 16, **2013,** http://www.encyclopedia.com /topic/soil_erosion.aspx

101. Lyons, D. J.; and Smith, J. A.; Wind Erosion and Its Control. Institute for Agriculture and Natural Resources, University of Nebraska, Neb Guide G1537, **2010,** 1–8 p.

102. Mackay, K.; Miller, P.; Jenks, B.; Riesselman, J.; Neil, Buschena, D.; and Bussan, A. J.; Growing Chickpea in the Northern Great Plains. Fargo, USA: North Dakota Extension Service, North Dakota State University; **2002,** 1–8 p, February 28, 2013, http://www.ag. ndsu. nodak.edu

103. Malhi, S. S.; Gill, K. S.; McCartney, D. H.; and Malmgren, R.; Fertilizer Management of Forage Crops in the Canadian Great Plains. *Recent Res. Dev. Crop Sci.* **2004,** *1,* 237–271.

104. Mason, G.; Prairie Agriculture at the Cross Roads: Global Competition and Success. Canada West Foundation. **2012,** 1–3, January 20, 2013, http:/ /www. cwf.ca

105. McCauley, A.; and Jones, C.; Managing for Soil Erosion. Bozeman, USA: Montana State University; Soil and Water Management Module 3. **2005,** *4481–3,* 1–12.

106. McConkey, B. G.; Campbell, C. A.; Zentner, R. P.; Peru, M.; and VandenBygaart, A. J.; Effect of Tillage and Cropping Frequency on Sustainable Agriculture in the Brown Soil Zone. Alberta, Canada: Agriculture-Canada, Eastern Cereal and Oilseed Research Centre, Lethbridge; **2012,** 1–2, January 20, 2013, http://www.prairiesoilsandcrops.ca/display_summary. html?id=73

107. McGinley, M.; Great Plains Ecoregion. Washington DC: National Council of Science and Environment; **2008,** 1–8, February 22, 2013, http://www.eoearth.org/article/Great_Plains_ ecoregion_(CEC)

108. Merle, V.; Francisco, C.; Maysoon, M.; Joseph, B.; and David, N.; Sustainable Dry Land Cropping System for the Central Great Plains. **2013,** 1, February 24, 2013, http://www.ars. usda.gov/research/projects/projects.htm?ACCN_NO=420253

109. Mogel, K. H.; Interactions Key to Beating Future Droughts. CSA News. **2013,** *58,* 4–9.

110. Mohr, R. M.; Grant, C. A.; and May, W. R.; Nitrogen, phosphorus and potassium fertilizer management for oats. *Better Crops.* **2004,** *88,* 12–14.

111. Moore, N.; and Rojstacer, S.; Irrigation-induced rainfall and the Great Plains. *Am. Meteorol. Soc. AMS J.* **2001,** *40,* 1–2.

112. National Agricultural Statistics Service. Soybean Statistics. **2012,** February, 8, 2013, http:// www.nass.usda.gov

113. National Agricultural Statistics Service. Barley and Oats Statistics. **2013a,** February 20, 2013, http://www.nass.usda.gov

114. National Agricultural Statistics Service. Sunflower Statistics. **2013b,** February 20, 2013, http://www.nass.usda.gov

115. National Agricultural Statistics Service. Groundnuts Statistics. **2013c,** February 20, 2013, http://www.nass.usda.gov

116. National Agricultural Statistics Service. Forage Production Statistics. **2013d,** February 20, 2013, http://www.nass.usda.gov

117. National Agricultural Statistics Service. Wheat: area, Yield, and Production by Major States. 1991–2005. **2005,** February 20, 2013, http://www.nass.usda.gov

118. Natural Resources Conservation Service. The Soil Orders of Texas. **2008,** February, 2013, http//www.tx.nrcs.usda.gov/soil/ soilorder.html

119. NDSU. Fertilizer Requirements of Corn NDSU Extension Service. **2012a,** 1, April 10th, 2013, http://www.ag.ndsu.edu/procrop/crn/crnftr05.htm

120. NDSU. Fertilize Winter Wheat. NDSU Extension Service. **2012b,** 1, April 10, **2013,** http:// www.ag.ndsu.edu/procrop/wwh.ftrwin04.htm

121. Nielsen, D. C.; Oilseed productivity under varying water availability. Akron, CO, USA: Central Great Plains Research Station; **2004,** 30–33.

122. Nielsen, D. C.; Forage soybean yield and quality response to water use. *Field Crops Res.* **2011,** *124,* 400–407.

123. Nielsen, D. C.; Saseendren, S. A.; Ma, L.; and Ahuja, L. R.; Simulating the production potential of dry land spring canola in the Central Great Plains. *Agron. J.* **2012,** *104,* 1182–1188.

124. Nielsen, D. C.; Unger, P. E.; and Miller, P. R.; Efficient water use in dry land cropping systems in the Great Plains. *Agron. J.* **2005,** *97,* 364–372.

125. NRC. Integrated Assessment of Prairies Agricultural Resilience and Adaptation Options. **2009,** http://www.nrcan.gc.ca/earth-sciences/climate-change/landscape-ecosystem/economic-natural-environment/3503

126. Nyirfa, W. N.; and Harron, B.; Assessment of Climate change on the Agricultural Resources of the Canadian Prairies. Winnipeg, Manitoba. Canada: International Institute for Sustainable Development; **2003,** 1–27, March 26, 2013, http://www.parc.ca/pdf/ research_publications/agriculture4.pdf

127. Ocumpaugh, W. R.; and Rodriguez, S.; Pasture Forage Production: Integration of Improved Pasture Species into South Texas Livestock Production Systems. **1999,** 1–11, April 9, 2013, http://www.cnrit.tamu.edu/cgrm/whatzhot/laredo/ocumpaugh.html

128. Omernik, J. M.; Eco-regions of the conterminous United States of America. *Ann. Assoc. Am. Geogr.* **1987,** *77,* 118–125.

129. Onstad, C. A.; and Otterby, M. A.; Crop Residue effects on runoff. *J. Soil Water Conserv.* **1979,** 34, 94–96.

130. Osakwe, E.; Cotton fact sheet. **2009,** 1–3, March 29, 2013, https://www.icac.org/econ_stats/country_facts/e_usa.pdf

131. OSU. Wheat Production and Management Practices Used by Oklahoma Grain and Livestock Producers. Stillwater, OK, USA: Oklahoma State University; **2000,** http://pods.dasnr.okstate.edu/docushare/dsweb/Get/Document-1806/B-818.pdf

132. Padbury, G.; Waltman, S.; Caprio, J.; Goen, G.; McGinn, S.; Mortensen, D.; Nielson, G.; and Sinclair, R.; Agroecosystems and land resources of the Northern Great Plains. *Agron. J.* **2002,** *94,* 251–261.

133. Parton, W. J.; Gutmann, M. P.; Mechant, E. R.; Hartman, M. D.; Adler, P. R.; McNeal, F.; and Lutz, S. M.; Measuring and Mitigating Agricultural Greenhouse Gas Production in the US Great Plains, **2011,** 1870–2000, 1–33, March 15, **2013,** ftp://ftp.icpsr.umich.edu/pub/project//GreatPlains_ Nature.doc

134. Patton, J. J.; and Marston, R. A.; Great Plains. Encyclopaedia of Oklahoma History and Culture. Oklahoma City, USA: Oklahoma Historical Society; **1996,** 1–3.

135. Paulsen, G. M.; and Shroyer, P.; The Early History of Wheat Improvement in the Great Plains. Agronomy Journal **2008,** *100,* S-7—S-78.

136. Peterson, G. A.; Schlegal, A. J.; Tanaka, D. L.; and Jones, O. R.; Great plains cropping systems symposium. *J. Product. Agric.* **1996,** *9,* 179–186.

137. Pieper, R. D.; Grasslands of Central North America. Food and Agricultural Organization of the United Nations. Rome, Italy: FAO Corporate Document Repository. **2002,** 1–35, March 19, 2013, http://www.fao.org/docrep/008/y8344e/y8344e0d.htm

138. Pierson, D. G.; Geology of North Central Texas. **2012,** 1–2, February 20, 2013, http://www.nhnct.org/geology/geohist.html

139. Popper, D. E.; and Popper, F. J.; The Great Plains: From Dust to Dust. **2004,** 1–7, http://www.planning.org/25anniversary/ planning/1987dec.htm

140. Powell, J. W.; Report on the Lands of the Arid Region. 45th Congress 2nd session, United States Government. Washington, D.C. House Executive Document No 73 **1878,** 128 p.

141. Power, F. F.; and Biederbeck, V. O.; Role of cover crops in integrated crop production systems. *In*: Cover Crops for Clean Water. Proc. Int. Conf. Jackson Hargrove, W. L.; ed. TN. April 9–11, 1991. Soil Water Conserv. Soc. Ankeny, IA. P. **1991**, 167–174.

142. Randall, G. W.; and Vetsch, J. A.; Corn production on a subsurface-drained mollisol as affected by fall versus spring application of nitrogen and nitrapyrin. *Agron. J.* **2005**, *97*, 472–478.

143. Rao, S. C.; Northup, B. K.; Phillips, W. A.; and Mayeux, J. S.; Improving forage production in the Bermuda grass paddocks with novel cool-season legumes. *Crop Sci.* **2007**, *47*, 168–173.

144. Rodrigues, M. S.; Repository of North American Migration History: Modern Continental Migration, Citizenship and Community. New York, USA: University of Rochester Press; **2004**, 114–117.

145. Ross, C. I.; and Anderson, M. R.; Spatial and Temporal Precipitation patterns for the Great Plains. Proceedings of the American Meteorological Society, Abstract No P1.80 **2008**, 1–2.

146. Sanderson, E. E.; and Price, P. B.; Barley Production in South Dakota. **2002**, http://agbio-pubs.sdstate.edu/articles/FS$*%.pdf

147. Schlegal, A. J.; Grant, C. A.; and Havlin, J. L.; Challenging approaches to Nitrogen fertilizer recommendation in continuous cropping systems in the Great Plains. *Agron. J.* **2005**, *97*, 391–398.

148. Setiyono, T. D.; et al. A decision tool for nitrogen management in maize. *Agron. J.* **2011**, *103*, 1276–1283.

149. Shurtleff, W.; and Aoyagi, A.; A special report on Soybean Production and Trade around the World. **2007**, 1–8, April 21, 2013, http://www.soyinfocenter.com/HSS/production_and_trade1.php

150. Singer, J.; and Bauer, P.; Soil quality for Environmental health: Crop Rotations for Row Crops. **2009**, 1–3, May 27th, **2013**, http://soilquality.org/ practices/row_crop_rotations.html

151. Skidmore, E. L.; Air, Soil and Water quality as influenced by wind erosion and strategies for mitigation. Proceedings of 2nd International symposium of New Technologies for Environmental Monitoring and Agro-Applications. Agroenviron-2000. Tekrdag, Turkey, **2000**, 216–221.

152. Smika, D. E.; Seed zone water conditions with reduced tillage in the semi-arid Central Great Plains. Proceedings of the 7th Conference of International Soil Tillage Organization. Uppsala, Sweden; **1976**, 75–83.

153. Smith, G. S.; Evers, G. W.; Ocumpaugh, W. R.; and Rouquette, F. M.; Forage legumes for Texas. AgriLife Research. Texas A and M system, College Station. Internal Report. **2012**, 1–11.

154. Smith, E.; Lamey, F. J.; Nakonechny, D. J.; Barbieri, J. M.; and Lindwall, C. W.; Productivity of long term No-till plots at Lethbridge, Alberta, Canada. Agriculture-Canada, Eastern Cereal and Oilseed Research Centre, Lethbridge, Alberta, Canada, **2012**, 1–2, January 20, 2013, http://www.prairiesoilsandcrops.ca/display_summary.html?id=80

155. Snyder, C. S.; Bruulsen, T. W.; Jensen, T. L.; and Fixen, P. E.; Review of Greenhouse Gas production systems and Fertilizer management effects. *Agric, Ecosys. Environ.* **2009**, *133*, 247–266.

156. Soil Survey Staff. Land resource regions and major land resource areas of the United States of America. USDA Soil Conservation Service. Washington, DC. Agricultural Hand Book. **1981**, 296, 156.

157. Stehman, S. V.; Sohl, T. L.; and Loveland, T. R.; an evaluation of sampling strategies to improve precision of estimates of gross changes in land use and land cover. *Int. J. Remote Sensing.* **2005**, *26*, 4941–4957.

158. Stewart, W. M.; Fertilizer contributions to Crop Yield. Potash and Phosphorus Institute of Canada. News and Views. **2002,** 1–2, February 28, 2013, www.ipni.net/ppiweb/ppinews. nsf/0/.../$FILE/Crop%20Yield.pdf

159. Stichler, C.; and McFarland, M.; Crop nutrient needs in South and Southwest Texas. AgriLife Extension Texas A and M system B-6053 04-01. **2012,** 1–8, April 9, **2013,** http://www. texaserc.tamu.edu

160. Trautman, N. M.; and Porter, K. S.; Modern Agriculture: Its effects on the Environment. Cornell University cooperative Extension. **2012,** 1–4, http://psep.cce.cornell.edu/facts-slides-self/facts/mod-ag-grw85.aspx

161. Tremblay, N.; et al. Corn response to nitrogen is influenced by soil texture and weather. *Agron. J.* **2012,** *104,* 1658–1671.

162. Trimble, D. E.; The Geologic story of the Great Plains. United States Geological Survey Bulletin No 1493, United Sates department, Washington D.C. **1980,** 1–19, February 20, 2013, library.ndsu.edu/exhibits/text/greatplains/text.html

163. Triplett, G. B.; and Dick, W. A.; No-tillage crop production: A revolution in agriculture. *Agron. J.* **2008,** *100,* S153–S165.

164. Undersander, D. J.; Smith, L. H.; Kaminski, A. R.; Kelling, K. A.; and Doll, J. D.; Sorghum-Forage. Alternative Field Crops. Madison, USA: University of Wisconsin; **1990,** *1–7,* April 1, **2013,** *http://www.hort.purdue.edu/newcrop/afcm/forage.html*

165. Unger, P. W.; and Musick, J. T.; Ridge tillage for managing irrigation water on the US Southern Great Plains. *Soil Tillage Res.* **1990,** *18,* 267–282.

166. United States Department. The Great Plains and Prairies. Washington DC: United States Department of Geology; **2013,** 1–8, February 20, 2013, http://countrystudies.us/united-states/geography-17.htm

167. University of Nebraska. Canola. **2011,** 1, April 9, **2013,** http://plains humanities.unl.edu/encyclopedia/credits.html

168. University of Wisconsin Extension Services. Sorghum. Agronomy Department, University of Wisconsin, Madison, USA. **2012,** 1–3, April 1, 2013, http://corn.agronomy.wisc.edu/Crops/sorghum/Default.aspx

169. USCCSP. Preliminary review of Adaptation options for Climate-Sensitive Ecosystems and Resources. Report by the US Climate Change Science Program and the Subcommittee on Global Change Research. Julius, S. H.; and West, J. M.; eds. Washington, DC: United States Environmental Protection Agency; **2008,** 675 p.

170. USDA. Summary Report-1997. National Resources Laboratory. Natural Resource Conservation Service, United States Department of Agriculture, Washington D.C. and Statistical laboratory, Ames, Iowa: Iowa State University; **2000,** 89 p.

171. USDA. Canada: Barley. Production Estimates and Crop assessment Division-Foreign Agricultural Service. **2004,** 1–2 p. March 27, 2013, http://www.fas.usda.gov/remote/Canadia/can_bar.htm.

172. USDA. USDA Agricultural Statistics—Feedstocks section. United States Department of Agriculure, Beltsville, Maryland, USA, **2010,** 1, March 26, 2013, http://www.nass.usda.gov/Publictions/Ag_Statistics/index.asp

173. USDA. Harney Silt loam-Kansas State soil. Natural Resources Conservation Service. United States Department of Agriculture, Beltsville Maryland, USA, **2012a,** 1–3, January 18, 2013, http://www.ks.nrcs.usda.gov/soil/stsoil.html

174. USDA. Soil Orders of Texas. Natural Resources Conservation Service. United States Department of Agriculture, Beltsville Maryland, USA; **2012b,** 1–3, January 18, 2013, http://www. Txnrces.usda.gov/soil/soilorder.html

175. USDA Economic Research Service. USDA Wheat Base line-2012-2021. USDA Economic Research Service. **2012,** 1–14 March 27, **2013,** http://www.ers.usda.gov/ topics/cops/wheat/ usda-wheat-baseline-2012-21.aspx#

176. USEPA. Climate Change: Great Plains Impacts and Adaptation. United States Environmental Protection Agency. Washington DC; **2012,** 1–3, January 11, 2013, http://www.epa.gov/climatechange/impacts-adaptation/greatplains.html

177. USGCRP. Great Plains: Global Climate Impacts in the United States'. Washington DC: United States Global Change Program; **2009a,** 1–4.

178. USGCRP. Global Climate change impacts in the United States of America. In: United States Global Change Research Program. Karl, T. R.; Melillo, J. M.; and Peterson, T. C.; eds. New York, USA: Cambridge University Press; **2009b,** 596.

179. USGCRP. Overview: Great Plains. Washington, DC: United States Global Change Program; **2012,** 1–4.

180. USGS. Establishment of seeded grasslands for wildlife habitat in the Prairie Pothole Region. UGSG Northern Prairie Wild Life Research Centre. **2013,** 1–5, April 9, 2013, http://www.npwrc.usgs.gov/resource/habitat/grassland/types.htm

181. Veeman, T. S.; and Gray, R.; The Shifting Patterns of Agricultural Production and Productivity World-Wide. Ames, Iowa: The Midwest Agribusiness Trade Research and Information Centre, Iowa State University; **2010,** 123–148.

182. Wagner, D. F.; and Dodds, D. L.; Soil erosion as a pollution agent. Farm Research **1971,** *28,* 58–59.

183. Weaver, J. E.; Some ecological aspects of agriculture in the Prairies. *Ecology.* **1927,** *7,* 391–523.

184. Webb, W. P.; The Great Plains. Texas Ginn and Company, Dallas, Texas, USA, (reprinted by University of Nebraska, Lincoln). **1931,** 54.

185. White, K. J. C.; Population change and farm dependence: Temporal and spatial variation in the US Great Plains, 1900–2000. *Demography.* **2008,** *45,* 363–386.

186. Wikipedia Foundation. Canadian Prairies. **2013,** 5, February 22, 2013, http://en.wikipedia.org/w/index.php?title=Canadian_Prairies&oldid= 530977224

187. Wilbert. S.; Plows and Carbon—The time line of Global warming. Climate Emergency Institute. **2012,** www.climateemergency institute.com/cei_library.html

188. Woodruff, N. P.; Fenster, C. R.; Harris, W. W.; and Lundquist, M.; Stubble-mulch tillage and planting in crop residue in the Great Plains. *Transact. Am. Soc. Agric. Eng.* **2011,** *9,* 849–853.

189. Wortmann, C. S.; Dobermann Ferguson, R. B.; Hergert, G. W.; Shapiro, C. A.; Tarkalson, D. D.; and Walters. High yielding corn response to applied phosphorus, potassium and sulphur in Nebraska. *Agron. J.* **2009,** *101,* 546–555.

190. Zhang, X. C.; and Garbretch, J. D.; Precipitation retention and soil erosion under varying climate system, land use and tillage and cropping systems. *J. Water Resour. Assoc.* **2007,** *38,* 1241–1253.

CHAPTER 2

THE CERRADOS OF BRAZIL: NATURAL RESOURCES, ENVIRONMENT, AND CROP PRODUCTION

CONTENTS

2.1   CERRADOS: AN INTRODUCTION

Cerrados means "dense or closed" in Portuguese language. The Cerrados region in East-Central part of Brazil is a large expanse of Savannahs supporting natural grasses and shrubs. Cerrados are most pronounced in the states of Goias and Minas Garias of Brazil. The savannahs have developed on weathered acidic Oxisols of low to moderate fertility in terms of nutrients. Cerrados extends into 204 million hectare of land. It is equivalent to 24 percent of total land area of Brazil. Incidentally, the Brazilian Cerrados accounts for 15 percent of total savannahs that occur on the earth (Trigo et al., 2009; Figure 2.1). According to estimates by FAO (2008), total area of Cerrados biome is 204 m ha, out of which 127 m ha is utilized for agricultural crop production. Within the agricultural area, pastures extend into 35 m ha, annual field crops into 10 m ha and perennial crops, plantations, and forestry together constitute 2 m ha. About 80 m ha is still free to be developed for crop production. The total grain production potential of Cerrados is currently 240 m t and fruits harvested annually is about 90 m t. Agricultural prowess of Cerrados got expressed in a big way since 1970s, with the advent of mechanized farming, fertilizer technology, and soil amendments to correct soil pH and Al/Mn toxicity. At present, 45–50 percent of Cerrados is engulfed by crop production enterprises. Soil productivity has been improved enormously since past five decades. Cerrados contribute 59 percent of coffee, 54 percent of soybean, 28 percent of maize, and 18 percent of rice generated in entire Brazil (Yaganiantz et al., 2009).

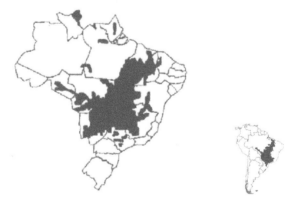

FIGURE 2.1   Cerrados of Brazil.
Source: FAO, 2008
Note: Cerrados of Brazil is a vast expanse of natural grass lands, shrubs, and tropical vegetation. It covers areas in following states of Brazil namely: Mato Grosso du Sul, Sao Paulo, Minas Garias, Mato Grosso, and small isolated zones in Rondanio and Para, Tocantins, Bahia, Piaui, and Maranhao.
(Area marked in dark color represents the biome; smaller inset shows location of biome in the South American continent).

Meaning of the word "Cerrado": In Portuguese it means thick, dense, shut, closed, or broad perhaps referring to thick savannah vegetation and a closed ecosystem. In Spanish, derived from original word carer, it means shut, covered, or closed. In English, meaning suggested are closed, shut, thick, dense, overcast, cloudy, and quiet/reserved. Ecologically, meaning of the word "Cerrado" is understood as vast plains with well characterized prairie vegetation interspersed with shrubs and short trees.

2.1.1 GEOLOGICAL ASPECTS

Cerrados existed as a large biome during Cretaceous period about 145–165 million years ago, even before the geologic separation of South American continent and African main land (Schmidt, 2009). Cerrados is currently a savannah where grassland crops and pastures dominate the landscape. However, it also supports a sizeable posse of shrubs and forests with tall trees. Ledru (2002) suggests that Cerrados has experienced periodical changes in climatic parameters in Pleistocene that alternatively accentuated forests or savannah vegetation. Such alternation of climate and vegetation dynamics has provided Cerrados with a peculiar mixture of savannah, shrubs, and forest species. In certain periods forests contracted, while in other times savannahs shrunk in area.

2.1.2 PHYSIOGRAPHY AND AGROCLIMATE OF BRAZILIAN CERRADOS

The Cerrados region in Brazil extends from the fringes of Amazonian forests in the North-West to the edges of Minas Garias state in the East (6° and 20°S latitudes; 42° and 62°W longitude). It almost touches the Atlantic coast in the southeast and Chaco region in the South-west. Large part of Cerrados occurs on a plateau of sedimentary rocks that is frequently interrupted by depressions. The elevation within Cerrados ranges from 500 to 1,700 m.a.s.l. (DaSilva and Bates, 2002). About 73 percent of Cerrados occurs within the altitude of 300–600 m.a.s.l. (FAO, 2008). Much of the cropland occurs at 500–600 m.a.s.l. As stated earlier, Cerrados occur in the central-west region of Brazil. It is a vast belt of natural and agricultural prairies that experience semiarid climate and thrive on variety of soils that are afflicted with low-P, Al/Mn toxicity, and uncongenial pH for farming.

Annual precipitation pattern in the Western fringes of Cerrados is congenial to crop production and is about 2,500 mm, but decreases to 1,000 mm year^{-1} as we traverse to Eastern tip. Peak precipitation occurs during October–March. Average monthly precipitation during these 6 months is above 130 mm/month. Precipitation during November, December, January, and February ranges from 220 to 300 mm/month (FAO, 2008). About 72–80 percent of annual precipitation is received between October and March. During April to September precipitation is very low and totals only 180–220 mm for these 6 months (see FAO, 2012).

The arable cropping belt in the western Cerrados does not require extraneous irrigation. In the east, rainfall lasts for only 6 months, and hence, the supply of water through irrigation may be needed, depending on cropping system (FAOAQUA-STAT, 2002). Riverine irrigation is utilized in eastern states to produce crops such as soybean, cotton, maize, and wheat. The ambient temperature in the Cerrados is most congenial for cropping enterprises. The yearly averages range from 22°C to 26°C. The overall agroclimate of Cerrados is mostly tropical or subtropical and devoid of frost.

2.1.3   NATURAL VEGETATION AND CROPPING HISTORY

Schmidt (2009) remarks that, naturalists often state about the consequences of human incursions and developmental activities, on the more popular biome in Brazil namely Amazonia. However, Cerrados is unique in its own way and is a repository of large number of well diversified plant species. It has over 160,000 species in its environs that include plants, animals, and birds. Currently, only 20 percent of tropical savannah vegetation of Cerrados is undisturbed. Factors like human settlements, agricultural cropping, commercial pastures, plantations, and infrastructure development have all affected the natural vegetation perceptibly.

The natural habitat and vegetation of Cerrados is easily classifiable into savannah, graminaceous woody savannah, park savannah, wooded savannah, and forest. Earliest descriptions of Cerrados vegetation was made available by Danish Botanist Eugenius Warming (1892). However, Goodland (1971) demarcates Cerrados vegetation and canopy into *Campo Limpo*, which is exclusively a grass land made of short grass/legume species; *Campo Sujo*, which is herbaceous stretch scattered with small trees of 3 m height; *Campo Cerrados* consists of slightly taller trees distributed more frequently; *Cerrado Sensu Stricto* consists of vegetation with savannah grass species, tall trees gives an orchard like appearance; *Cerradao* has a canopy that is endowed with grass and tall trees.

According Klink and Machado (2005), Cerrados of South America is among most important natural preservations with regard to biodiversity and species richness. However, during past 35 years, development of agricultural expanses and pastures has reduced area under natural prairie and scrubland vegetation. About 50 percent of two million square kilometer of natural prairies dotted with shrubs has been transformed into agricultural savannahs, with excessive monotony regarding vegetation. Despite, such large-scale conversions of vegetation, Cerrados, still supports a rich flora of over 7,000 plant species. It seems deforestation of Cerrados has been more rampant than in the dense Amazonia. Plus, the area under legal protection and surveillance is small at just 2.2 percent of Cerrados. It is also true that a large of share of species that need protection does not actually occur in the natural preserves, but are in open expanses that could any time be vulnerable to development activities and conversion to cropland.

A large section of Cerrados has been cleared for cropping, industry, infrastructure, and urbanization. The original vegetation was no doubt a diverse grass/legume filled savannah. It also supported several other genera, small trees, and shrubs. Medium sized forests occur along the rivers, lakes, and ponds. Trees could also be traced in soils with high fertility. According to reports by FAO (2008), natural vegetation of Cerrados as we perceive it today is derived through evolutionary processes spanning over 100 million years, since the Cretaceous period. Cerrados may harbor over 10,000 species of vascular plants. Of course, it also supports large number of highly diversified insects, terrestrial animal species, and birds.

Reports by FAO (2000), suggest that Cerrados of Brazil continues to be a mosaic of natural prairies that predominate the region, plus small shrubs, and woodlands. As a large biome, Cerrados includes semideciduous forest trees in patches. The prairies vegetation includes a large posse of grass/legumes species and several genera of shrubs and trees are found in the trees. Most prominent and frequently occurring species are those of *Bowdichia, Hymenaea, Piptadenia, Incurialis, Machaerium, Curatella americanum, Qualea, Kielmeyera coriacea*, and others. The Cerrados are filled with natural grass prairies, shrubs, and trees at different densities. Most of the trees species are relatively short at 3–8 m height. Leguminosae and Myrtaceae are dominant under the tree/shrub canopies in many locations of Cerrados. Species such as *Caryocar brasiliensis, Curtella americana, Kielmeyera coriacea*, and *Qualea sp* are few examples (FAO, 2000; see Oliviera and Marquis, 2002; Oliviera et al., 2012).

Biodiversity and Natural vegetation in the Cerrados is under constant interaction with several detrimental factors. The detrimental effects may be perceived at different intensities and for various periods at various stages. Klink and Machado (2005) believe that, among them, soil and ecosystem deterioration are wide spread. Poor soil management of cropland, lack of soil conservation in natural expanses, and loss of topsoil, almost severely reduced establishment and sustenance of natural vegetation. Rapidly expanding soybean belt and invasive African grasses are also major causes of vanishing biodiversity of native Cerrados vegetation. Molassa (*Melinus minutiflora*), it seems is highly disruptive to native flora of Cerrados. Natural fires too affect survival and regeneration of several species of plants that are native to Cerrados. In fact, Cerrados are also highly vulnerable to man-made fires initiated to clear the native vegetation prior to development of cropland. It seems over 67 percent of surviving native vegetation is vulnerable to clearing exercises at any time (Tansey et al., 2004).

Conservation of Cerrados landscape and biodiversity has received greater attention during past three decades. It seems states such as Goias, Mato Grosso, and Mato Grosso du Sul have created protected areas and ecological corridors. International conservation agencies too have invested in preserving what remains of Cerrados biodiversity. Klink and Machado (2005) state that during recent years agricultural expansion and intensification has been effected with due care to preserve natural heritage and species richness of Cerrados. In fact, there are several arguments for intensification of crop production that leads to higher grain productivity. It is said, we produce same quantity of grains utilizing less area, if we intensify. It may actually reduce undue expansion and spare Cerrados from being converted entirely into a cropping expanse.

During past two decades, deforestation in Amazonia has been reduced markedly through conservation procedures. However, Cerrados have not received similar attention. Pearce (2011) states that 60 percent of original vegetation that occupied 200 million hectare has got replaced or disappeared under plow. However, there are still 90 million hectare of Cerrados capable of being converted to cropland, without excessive effect on ecosystematic functions and environment perceptibly.

2.1.4   DEVELOPMENT OF AGRICULTURAL FARMING IN CERRADOS

Archaeological studies in the Cerrados suggest that Neolithic culture, there grew small amounts of maize, vegetables, and a few other plant species. Ceramics and grains recorded in the settlements of Una Tradition indicate that beginnings of agriculture are not prior to 3500 B.C. (Barbosa, 2002). Una Tradition normally inhabited rocky zones nearer to river or lake that allowed crop growth. These dwellings were also nearer the trees that allowed fruit picking and hunting grounds. Human migration from Andean regions and elsewhere within the continent might have induced spread of farming in the Cerrados. It is believed that human diffusion composed of nomadics, hunters/gatherers, and those with knowledge of rudiments of agriculture induced settlements and cropping in the Cerrados (Barbosa and Schimiz, 1998; Feltran-Barbieri, 2011). Human populations of Una Tradition that localized in the Central Plateau and Southeast Coastal zones, it seems developed their agricultural skills, developed implements, and learnt seeding from Bolivian migrants (Barbosa, 2008). More recent evidences suggest that human groups such as Aratu/Sapucai and Uru of 1200 B.C. and Tupi-Gurani of 1000 B.C. derived agricultural skills, crop species, and implements from people who diffused into their region from Southern Amazonia as well as Atlantic Coast (Robrahn-Gonzales, 2001). It seems Una, Aratu/ Sapucai, and Pintado people were knowledgeable about use of fire to clear small regions so as to initiate the crops. In fact, Feltran-Barbieri (2011) points out that fire itself and its utility was known to Cerrados human way back in 8000 B.C. Deforestation-burn and tree cutting techniques were mastered by Cerrados dwellers during Neolithic age. There are suggestions that Macro-je, a cultural group of humans from Midwest and Northeast Cerrados and Gorotire tribe's criss-crossed Cerrado plains in search of fruits, and in the process planted/transplanted many useful horticultural species and disseminated farming techniques. They also hunted and gathered during their transit (Feltran-Barbieri, 2011). The Macro-je mainly colonized savannah zones to initiate crops.

Like most other regions, agriculture was a subsistent enterprise in the Cerrados. However, during past five decades, development of infrastructure, crop research centers, railways, roads, and marketing yards has converted the landscape of Cerrados. In addition to coffee, Cerrados offers soybean, maize, wheat, rice, and cotton for export. Monocrop of soybeans or maize is popular in the Cerrados. Cerrados experienced first incursions in 1950s and 1960s due to agricultural expansion. Prior to it, natives stayed in scattered settlements and practiced subsistence crop production. Large-scale expansion of agricultural cropping began in 1960s (The Nature Conservancy, 2013). Although expansive agriculture had taken roots by mid-twentieth century, it was only in 1980s that Cerrados ecosystem was rapidly covered by large farming enterprises known commonly as "Corporate Agriculture firms." This phenomenon relegated individual farms and subsistence farmers. Small farms either folded up or got amalgamated into large ones. A few of them migrated to urban zones in search of jobs. The Cerrados produced large quantities of soybean, maize, coffee, cotton, and sugarcane. It required large well-managed infrastructure (Castillo et al., 2011). Logistics, especially, transport of grains and other farm products became an important constraint to overcome. Since, mid-1980s, Brazilian rail roads that crisscross the Cerrados prairies has been utilized efficiently to transport farm produce to ports for export to other continents. Today, we have well-evolved agribusiness cities and large farms in the Cerrados. The Cerrados has been termed a great success in Brazilian agriculture. However, well evolved and entrenched natural prairies had to make way for economically highly profitable crop production systems.

Cerrados experienced phenomenal increase in soybean cropping zones from a mere 200,000 ha in 1955 – 4.0 million hectare in 2005. The Cerrados, in general, became highly productive during past five decades. Further, application of dolomite or pure gypsum to overcome Al and Mn toxicity and correct soil pH enthused Cerrados farmers and Brazilian National Agencies (EMBRAPA), to enhance crop production. Soybean production in the Cerrados registered a 20-fold increase. The Cerrados region currently provides 54 percent of total Brazilian soybean production, 28 percent of maize, and 59 percent of coffee beans. Cerrados farming belt is well entrenched. Its produce is consumed by over 180 million humans and large number of domestic farm animals. The Cerrados farmers also export a large quantity of grains to Europe and other continents (World Food Prize Foundation, 2013).

Ferreira et al. (2006) have classified the agricultural and commercial development within Cerrados further into sub areas as follows:

(i)   Region with commercial enterprise and livestock – Western and Central Mato Grosso

(ii) Regions with predominance of livestock and agricultural farms – Southern Region of Goias

(iii) Region with predominance of livestock but normal levels of farming – Cerrados zones of Mato Grosso and Mato Grosso du sul

(iv) Region with reduced livestock presence and large agricultural farms, but with presence of many family farms – Western and Northern surroundings of Minas Garias and Piaui states

(v) Regions of new large agricultural enterprises and livestock is also conspicuous – Southern region of Maranhao and Piaui state

(vi) Regions with expansion of livestock but marginal reduction rates of agricultural farming – Northern region of Goias, Central Maranhao and Southwestern Mato Grosso

Charcoal production in Cerrados is linked to development of agriculture and cropping zones. Forest trees and shrub vegetation are cleared to develop fields for soybean or maize. The timber, roots, and twigs from cleared forest vegetation are utilized to produce charcoal. Rampant clearing of native vegetation for the sake of field crops may affect general ecosystematic functions markedly. Hence, during recent years, destruction of forest trees/shrubs to manufacture charcoal is being curbed (Wikepedia, 2013). However, during recent decades, *Eucalyptus* and *Pinus* is grown in Cerrados to serve the paper and other industries.

2.1.5 HUMAN POPULATION, SETTLEMENTS AND MIGRATION

Native South Americans lived undisturbed by migrations from other regions for a long stretch of time. Native settlements supported few crops, grazing land, and cattle. Hunting animals and fishing were other sources of nourishment. The dwellings were prepared using native vegetation and timber. Brazil still has large number of indigenous tribes that confine to Amazonian forests, Cerrados, and similar ecosystems. They stay partially exposed to developed world. Brazil has at least over 200 localized settlements of native tribes (Emmet Duffy, 2008). However, within Cerrados ecosystem, we can easily trace 38 ethnic groups and their population is around 45,000. A few of the examples of native tribes are *Kraho, Xavante, Xerante, Bororo, Kraja, Kayapo, and Canela*. A few of them such as *Tapayu* and *Karaje* are depleting in numbers due to development, migration to cities and intermixing. General opinion suggests that incursions by Europeans in search of gold and precious stones affected the native tribes first. Later, the development of infrastructure such as roads, introduction of automobiles, and industrial enterprises affected the Cerrados native population. Agricultural development then induced migration into developed farming zones within and outside Cerrados. Most conspicuous influence to native communities was due to large-scale clearing and development of farming zones by the Brazilian agency "EMBRAPA" for soybean/sugarcane production. It threatened the native zones perceptibly and induced migrations away from their settlements (Emmet Duffy, 2008).

Portuguese colonized Brazil during sixteenth century. These European settlers introduced several crops and developed suitable agricultural practices (Schmidt, 2009). During past two centuries, Brazilian administration has aided colonization of Cerrados by developing human settlements (villages), agricultural infrastructure, and cities (for e.g., Brasilia). During last century, human population has increased 50-fold in the Cerrados. Following is the trend in human population in the Cerrados:

Year	1872	1900	1940	1960	1980	1991	2000	2005	2012
Population (x1000)	221	373	1259	3007	7545	12,600	18,000	19,500	20,100

Source: Schmidt (2009)

Incentives to form settlements in the Cerrados have induced a cost effective farming zone. Human migration into Cerrados is perceptible. Yet, human population in the Cerrados is relatively feeble. It is 6.76 inhabitants per kilometer. Rangel et al. (2007) studied the influence of spread and density of human population in Cerrados versus its interaction with biodiversity and species richness. They arrived at a conclusion that says species richness (i.e., flora and fauna) were related to expression of modern agriculture and cattle ranching. Human population size or its density did not account much for species richness in Cerrados. Human settlements and size was not a good indicator of biodiversity within Cerrados.

2.2   NATURAL RESOURCE OF CERRADOS OF BRAZIL

2.2.1   SOILS OF THE CERRADOS

Soil is an important ingredient of the Cerrados as a biome. It supports vast patches of natural vegetation and cropland. Estimations in 1970 indicated that out of 204 million hectare that Cerrados occupy in Brazil, following is the proportion of each major soil type encountered:

Soil Type	Area in Cerrados	
	(km²)	(%)
Oxisols (Ferralsols, mainly Orthic, Xanthic, and Rhodic)	935,870	46.0
Ultisols (Red and Yellow Podzols)	307,667	15.1
Psamments (Ferralic Arenosols)	310,715	15.2
Entisols (Lithosols)	148,134	7.3
Alfisols (Luvisols) and Inceptisols (Gleysols)	122,664	6.1
Inceptisols (Gleyic soils)	40,752	2.0
Ultisols (Acric Ferralsols)	57,460	2.8

Source: FAO (2008)
Note: A few other soil types encountered in Cerrados are sporadic in occurrence.

Oxisols are most widely spread in Cerrados. Oxisols have been derived from parent material, which is rich in silicates and carbonates. Tropical climate with warm and moist conditions has induced the weathering and soil formation. Leaching of alkaline material and silica has created soil rich in Al, Fe, and Mn, and acidic in reaction. The concentration Al is perceptibly high in soils of Cerrados. It has deleterious effect on rooting and absorption of soil-P. In many regions within Cerrados, parent material with shale and sandstone poor in minerals has obviously led to soils with low fertility (Ratter et al., 1997; Schmidt, 2009).

Cerrados soils are deep, well weathered, and arable. They are porous and show high infiltration rates. Regarding soil texture, it is said that most of the soil types are endowed with clay minerals such as kaolinite or gibbsite of 1:1 type. They possess relatively low cation exchange capacity (CEC) compared to other clays (e.g., montomorillonite or Illite). Typically, CEC for dystric soils ranges from 5 to 8 m equal per 100 g soil (FAO, 2008). Cerrados soils are deficient in major nutrients required for crop production. They show high P-fixation trends, especially, Oxisols and Lateritic soils. About 20–40 percent of fertilizer-P applied could be fixed chemically into soil matrix. Soil K depletion is also high and needs periodic replenishment. Cerrados soils are deficient for micronutrients such as Zn and B. Soil pH and acidity can be a major constraint to establishment of pastures and crops. Amendment with gypsum becomes compulsory in many areas. Soil organic matter reported ranges from 1.4 to 2.7 percent. (FAO, 2008).

Ruggiero et al. (2002) believe that soil properties such as depth, texture, aggregates, nutrient distribution, ratios, and availability are prime determinants of natural vegetation and crop productivity. Following is an example that depicts physicochemical properties, especially, nutrient content of soil derived from Cerrados in Sao Paulo region:

Physicochemical Characteristics of Soil									
pH	OM	P	K	Ca	Mg	Al	V	m	Clay
(g kg^{-1})	(mg kg^{-1})	--	(mmolc kg^{-1})	--	--	--	(%)	(%)	(%)
Camp Cerrado	3.48	31.10	7.70	0.45	1.50	1.70	10.60	5.20	74.40	7.80
Cerrado Sensu Stricto	3.52	35.30	4.90	0.76	1.20	1.70	11.30	4.70	75.30	9.00
Cerradao	3.35	53.40	6.10	0.63	2.30	1.80	18.40	3.70	81.20	14.20

Source: Ruggeiro et al. (2002)
Note: Soil properties depicted pertain to upper layer 0–10 cm depth; V = base saturation, m = Al saturation. Cerrados soils are acidic and high Al content.

The correlation between soil types, especially, its characteristics with vegetation is pronounced, if we consider soil data from upper horizon. The subsoil seems fairly uniform and not so immediately relevant to establishment of grasslands or even shrubs (Furley, 1996; Ruggiero et al., 2002). Overall, there are reports that soil fertility may be an important factor affecting the pattern of vegetation or crop productivity in Cerrados region. Ruggiero et al. (2002), for example, have reported that establishment and growth of semideciduous forests is dependent on soil fertility.

According to Kluck (2003), in the nutrient poor soils of Cerrados, nutrients derived from atmospheric dusts, precipitation, and wind drifts are fairly important to the farmers. Nutrient budgeting studies showed that nutrient deposited on soil surface is useful in preserving a positive balance of nutrients. In fact, soil analysis of permanent pastures indicate that nutrients applied are exhausted rather rapidly and there is generally very low quantities of nutrient that percolate to subsoil.

Over all, soil maps that depict soil characteristics and fertility variations at different scales are available (Prado et al., 2012). They are helpful to policy makers who deal with larger areas within Cerrados, and while deciding on crops, fertilizer supply, irrigation, cropping systems, and yield goals.

2.2.2 SOIL TILLAGE SYSTEMS AND CROP PRODUCTIVITY IN THE CERRADOS

The natural vegetation in Cerrados has given way for agricultural cropping, may be to grow soybean, sugarcane, or any other species, through a systematic effort that uproots and destroys above-ground greenery. On an average 100–200 shrub/trees may be cut and uprooted per hectare (AgBrazil, 2006b). Like any other region of world, tillage begins in Cerrados too from the moment land clearing begins. Initial land clearing is often accomplished by dragging 100 m steel cables between two tractors. In case, the vegetation is dense, a 75 m cable of greater thickness is dragged over the region. Land clearing and removal of native prairies occur during rainy season. Woodcutting, slash burning, and passing heavy plow then follow, to achieve deep tillage. Tillage is also aimed at removing below ground roots of the erstwhile native vegetation. Once the native flora has been cleared, regular farming exercise that follows is chisel plowing. It offers a deep tilth and disturbs upper horizon of soil plus a layer of subsoil. It allows for good mixing of surface and subsoil. As stated earlier, the Cerrados soils are acidic and need commensurate amendment with gypsum to correct soil pH and Al toxicity. A basal application of 2–3 t gypsum per hectare to the cleared field is common. The cost of clearing Cerrados prairies is high at 100–200 US$ per hectare.

The environmental effect of clearing, plowing, and disturbing the soil ecosystem is indeed marked. Clearing removes large amounts of biomass (i.e., C and nutrients from a region). It affects the natural crop residue recycling phenomena and depletes nutrients. Tillage and exposure of soil profile induces oxidative processes that liberate large quantities of SOC to atmosphere. No doubt, C sequestration trends are affected detrimentally. Removal of natural vegetation also accentuates soil erosion and loss of nutrients in surface layers. In fact, reports by FAO (2003, 2008) suggest that planting crops in the residues of previous crop and without disturbing, which is the essence of conservation agriculture is becoming too popular in Cerrados. It is also a sustainable practice in terms of energy costs and economic advantages.

2.2.3 NO-TILLAGE SYSTEMS

Direct seeding or no-tillage system was adopted in the Brazilian Cerrados about 30 years ago, mainly to thwart soil erosion and reduce loss of soil organic carbon (FAO, 2003; Bernoux et al., 2006). Scopel et al. (2004) have argued in favor of no-tillage systems because conventional tillage has induced soil maladies and reduced the productivity. Fields brought into cropping just 10 years ago by clearing natural vegetation have lost their topsoil fertility and SOM content beyond threshold. The SOC content has in fact declined by 40–45 percent of the original levels in 10 years period. Soil erosion and loss of topsoil nutrients has been catastrophic in some locations within Cerrados (Seguy et al., 1996, 2006). Soybean produced under conventional tillage system has consistently declined in productivity. Crop rotations carefully designed too have not performed well if conventional tillage is continued for long term. During past three decades, conventional tillage systems have been effectively replaced by direct seeding, no-tillage system with application of mulches and use of herbicides.

World-wide surveys indicate that Brazilian Cerrados is among the top areas where no-tillage direct seeding technique has been mastered well and adopted in large tracts. Brazil had 173,000 hectare under no-tillage system. The area under no-tillage system is only 5–10 percent within the intensive farming zones of Cerrados (Moreas Sa et al., 2008). Based on C sequestration rates known for Cerrado soils supplied with FYM, it is said C sequestration potential ranges from 0.34 to 0.81 t C ha^{-1} and the total C sequestered could potentially reach 7.7 to 18.4 Tg year^{-1} (Moreas Sa et al., 2008). In case, C sequestration is further improved by 20–40 percent by adopting appropriate agronomic procedures, then average C-sequestration increases to 2.17 t ha^{-1} and total C sequestered in Cerrados could reach 18.5–30.5 T g year^{-1}. No-tillage systems and FYM supplies seem to hold the key. It is generally accepted that, if residue recycling and FYM supply controls C-sequestration under no-tillage, it is the reduction in loss of C via emissions and leaching that regulated soil C content under conventional tillage systems (Moreas Sa et al., 2008). Reports about no-tillage systems during soybean-corn rotation suggests reduced soil erosion and organic matter depletion compared with conventional deep plow systems using disks and harrows (Kluck, 2003). However, no-tillage systems lost soil moisture and nutrients to subsurface at a faster rate. Hence, it has been suggested that we should provide a break in the long-term no-tillage system. At least superficial tillage with harrows and loosening of surface soil is necessary. Stubble mulching is a useful idea.

Brazilian Cerrados supports quite a large number of small farms, in addition to large corporate agricultural companies. No-tillage systems are equally important even in small farms, because they conserve topsoil fertility and moisture. In fact, often, farmers adopt a relay planting system with no-tillage. For example, maize is sown first followed by relay crop of pigeon pea or Brachiaria. It avoids tillage before planting the second crop and improves land use efficiency (Balde et al., 2011).

Regarding interaction effects of soil acidity and practice of no-tillage system, Martinhao et al. (2011) have opined that it is mandatory to judge both surface and subsurface soil acidity, carefully estimate requirements of gypsum to correct acidity and then opt to follow no-tillage system. Martinhao et al. (2011) have reported that on a long term, Cerrados farmers will benefit from adopting no-tillage systems in several ways. Among them, improvement in SOM content and soil pH are conspicuous advantages. Following is an example related to Oxisols found in Planaltina, in Cerrados:

Soil Depth		SOM (g kg soil)		pH (Cacl2 0.01 M)	
		CT	NT	CT	NT
Surface layer	0–10 cm	30	38	4.5	5.4
Subsurface layer	50–60 cm	18	17	4.3	4.6

Source: Nunes et al. (2008)
Note: CT = Conventional tillage; NT = No tillage. No tillage improves SOM and soil pH in the upper surface (i.e., in the rooting zone crops).

Soybean is the dominant monocrop of Cerrados that has been grown under conventional tillage system. However, since 1980s, conventional deep plowing system has been replaced with direct sowing, no-tillage mulching system. Monocrop is replaced with dual crops and two sequential crops a year. No-tillage system disturbs the surface soil least. No-tillage system allowed higher N accumulation in the surface soil. The C sequestration too was higher in plots under No-tillage. Based on evaluation of N-mineralization rates, it was concluded that long-term adoption of No-tillage say for 10 years may still not affect the N dynamics in the root zone adversely (Maltas et al., 2007). According to Corbeels et al. (2006), enhanced C sequestration in soil is easily attributable to two crops grown under direct seeding with mulch system.

According to CIRAD (2009a), no-tillage systems that involve direct seeding of pasture or cover crop seeds are highly beneficial to even small farms. Cover crops sown mechanically without disturbing soil structure improves soil nutrient content, especially, C and N. Dual crops of maize-grass is also in vogue in Cerrados.

2.2.4 SOIL FERTILITY AND CROP PRODUCTIVITY IN THE CERRADOS

According to Lopes (1996), Cerrados soils were generally considered marginal and less suitable for intensive crop production. Instead, it was thought that low fertility soils with several serious maladies such as acidity, low-P and K, high al/Mn toxicity may after all only allow subsistence crop production or at best, we may achieve low productivity. The Oxisols are highly weathered and low in essential nutrients for crop production. Oxisols occupy about 46 percent of Cerrados land. Ultisols and Entisols occupy 15 percent area each. Again these two soil types too are endowed with low fertility levels. In addition to problems related to soil physicochemical properties, there are some basic constraints that need to be satisfactorily overcome, if Cerrados crop production has to be competitive with other regions of the world. The native grass land converted to cropland experiences typically 5–6 months of dry spell without precipitation from April to October. Short dry spells called "veranicos" occur during main crop season. Hence, due care is needed to supply water. Low water holding capacity of soils is another problem; yet another, but most important problem is excessive Al content that impedes rooting and nutrient acquisition, especially P and K absorption. Therefore, higher yield goals require proportionate soil and crop management techniques.

The Oxisol that supports field crops, developed pastures or natural grasslands needs lime and fertilizer supply, if farmers aim at even moderate grain/forage yield. Fertilizer supply is almost mandatory in all sub regions of Cerrados. Since fertilizer technology is relatively costly, one of the major thrusts is to economize on fertilizer usage, improve fertilizer-use efficiency and adopt alternative programs. According to Benites (2010), soil fertility management is slightly different for crop production, from those meant for pastures developed on wasteland. Integrated soil fertility management procedures are in place in many regions that adopt various sources of soil nutrients, in addition to inorganic fertilizers. It is said that Brazilian administration aims at adopting integrated farms and nutrient management procedures in over 70 million hectare. Crop residue generated is utilized shrewdly. A portion is used to feed livestock and rest is recycled to sustain SOM levels. Nutrient recycling procedures are important considering that large quantities of grains filled with nutrients are transported out of the farms and even the region/nation. Cerrados farmers supply soluble fertilizers that are utilized rapidly by crops. Yet, a sizeable fraction ranging from 30 to 40 percent could remain in the soil profile as residual fertilizers. This fraction of residual nutrients is efficiently utilized by fixing fertilizer schedules based on entire crop sequence and not just one season. The succeeding cover crop may derive benefits from residual nutrients. Soybean adds to soil-N via symbiosis with Bradyrhizobium. Maintaining most appropriate nutrient ratios in soil is also important. Cerrados farmers often prefer to utilize "Best Management Practices" that prescribes fertilizers and FYM to attain high grain/forage yield.

Agronomic procedures for almost all crops grown in the Cerrados involve application of major and micronutrients, soil amendment for correcting pH, and Al toxicity such as gypsum or lime. Micronutrient deficiencies too are conspicuous, hence foliar spray or soil application once in 3–4 years is common. Soybean is among the top priority crops. It is a legume, hence needs relatively less fertilizer-N supply. The BNF process adds soil-N that is perhaps sufficient to satisfy 50 percent of N needs of soybean. Priming fields with small quantities of fertilizer-N stimulates rapid root growth, seedling, and crop establishment. Overall, fertilizer recommendations for soybean culture in Cerrados include 80 kg P, 90 kg K_2O, 20 kg S, 1.0 kg Zn, 1.0 kg Cu, 1.5–2.0 kg Mn, and 400 g B per hectare. Lime stone 1–2 t ha^{-1} is applied during land preparation. During second and third years, farmers are suggested to supply similar levels of fertilizer-based nutrients for a crop that yields about 3 t grains per hectare. Micronutrient supply for the 4 years is based on soil analysis. We may also note that for soybean crop that replaces natural vegetation; fertilizer inputs are often dependent on soil moisture status, water resources, and irrigation schedules. Soybean-based prairies may get stabilized in due course. It is dependent on residue recycling, FYM, and inorganic fertilizer supply patterns.

Farmers in Bahia and adjacent regions tend to produce maize or cotton during 5 years and later, after the initial clearing of natural prairies. By then, soil fertility is supposedly enhanced compared to virgin fields and ecosystematically well stabilized. Yet, grain productivity of a crop such as maize may be low because inherent soil fertility of Oxisols common to Cerrados is low. Productivity of maize may reach 4–6 t ha^{-1}, if maize is supplied with 100 kg N, 60 kg P, and 80 kg k_2O plus micronutrients. Cotton appears on the fields, usually after soybean and maize has been cultivated in sequence for about 7 years. Cotton is a cash crop and fertilizer inputs are high at 100 kg N, 60 kg P, and 100 kg K plus micronutrients (AgBrazil, 2006b). As a thumb rule, cotton grown on Oxisols has to be provided with 125–210 kg N per ton lint (Carvalho et al., 2007; Francisco, 2013; and Francisco and Hoogerheide, 2013). Nutrient management for cotton production often involves series of soil sample analysis prior to fertilizer supply and leaf analysis during the crop season. This way, farmers actually decide nutrient input schedules, of course based on yield goals. There are also general guidelines that help farmers in ascertaining need for nutrient supply to prop up the soil fertility. Following is an example:

Soil Analysis Data and Interpretation for Cotton in Cerrados: Interpretation								
	P	K	Mg	S	B	Cu	Mn	Zn
	(mg dm^{-2})	(mmol dm^{-2})	(mmol dm^{-2})	-	-	(mg dm^{-2})	-	-
Low	6–12	<15	<5	<4	0–0.2	0–0.4	0–1.9	0–1.0
High	>25	>40	>10	>10	>0.5	>0.8	>5.0	1.6

Leaf Analysis Data for Cotton in Cerrados											
	N	P	K	Ca	Mg	S	B	Cu	Fe	Mn	Mo	Zn
	(g kg⁻¹)					-			(mg kg⁻¹)			

	N	P	K	Ca	Mg	S	B	Cu	Fe	Mn	Mo	Zn
Regular Yield	35–43	2.5–4	15–25	20–35	3–8	4–8	30–50	5–25	40–250	25–300	0–5.1	25–200
High Yield	40–45	3–4	20–25	25–35	4–8	4–6	40–80	8–15	70–250	35–80	1–3	30–65

Source: Francisco and Hoogerheide, (2013)

Note: Fifth leaf from top is plucked for chemical analysis. High yield refers to 6.0 t ha⁻¹ and low to 3.0 t ha⁻¹

Lilienfein et al. (2013) compared the soil fertility status of undisturbed native grass land, with pastures that are neglected or not fertilized, but just utilized year after year to feed farm animals with well fertilized ones. Productivity of fertilized pastures was relatively higher due to periodic supply of inorganic and organic fertilizers. They found that if biomass productivity of unfertilized pasture was 2.1 t ha⁻¹, well fertilized pastures gave 4.1 t ha⁻¹. Soil nutrient levels were generally lowest in native grass land < unfertilized pasture < regularly fertilized pasture in that order. The above trend was also noticeable for major and micro nutrients in the topsoil. Despite, lack of fertilizer inputs for 13 years in a stretch, pastures with no fertilizer supply still possessed better soil fertility status than natural grass lands. High fertilizer supply into pastures could lead to leaching losses.

We may note that in many areas of Cerrados, upland paddy is the first crop grown after natural vegetation has been cleared. Upland paddy is supposedly tolerant to low soil fertility conditions. Consequently, grain yields are low at 1–2 t ha⁻¹. Fields are supplied with relatively high quantities of FYM to enhance SOM status. Upland Paddy fields are supplied with major nutrients, micronutrients, and amendment such as lime or gypsum (Ag Brazil, 2006b).

2.2.5 DYNAMICS OF MAJOR NUTRIENTS

Regulation of soil-N dynamics is an important aspect of Cerrados crop production systems. Nitrogen supply to fields has to be carefully decided based on soil characteristics, inherent soil-N levels, mineralization rates, ratio of soil-N:C and P, extent of soil-N leaching, percolation and emission, and most importantly crop demand and fertilizer-N efficiency. Indeed quite a number of soil, crop, and environment-related parameters are to be considered before supplying fertilizer-N. Add to it the traits that are specific to Cerrados soils. Soybean that dominates the Cerrados landscape escapes from stringent management requirements of soil-N, because it adds to soil-N and satisfies its needs partly through biological nitrogen fixation. Martinhao et al. (2011) suggest that any type of fertilizer-N or formulations that are available in market may offer equally high efficiency, if it is managed shrewdly in the soil and crop phase. Application of fertilizer-N in splits is recommended to avoid undue accumulation. It seems N-mineralization rates in the surface soil is relatively low. Hence, N is to be channeled at slow pace. Top-dressing is most useful, since we often match N supply with the period when crop is most efficient and absorbs soil-N in greatest quantities. The period when crop's demand for N is greatest needs to be deciphered and N inputs regulated accordingly. Rapid absorption of fertilize-N avoids undue leaching, percolation, and emission mediated loss form soil. For crops such as corn, Cerrados farmers apply 100 kg N ha^{-1} at seeding and in two splits at vegetative and silking stage. Overall, for each ton of maize grain, Cerrado farmers are advised to supply 20 kg fertilizer-N, for wheat 30 kg N, for rice 20 kg N, for barley 25 kg N, and for sorghum 30 kg N. For crops such as soybean or other legumes, fertilizer-N supply is restricted to small primer dosage of 5–10 kg N ha^{-1} and is aimed to stimulate rapid rooting at initial seedling stage of the crop. In case of cotton, it is said crop response to N input depends to a certain extent on SOM content. Generally, a starter dose of 25 kg N ha^{-1} is followed by two splits of 40 and 60 percent of recommended dosage at fourth leaf stage and flowering respectively (Francisco and Hoogerheide, 2013). Although splitting fertilizer-N into two to four portions is a shrewd way to supply N and match it with the rate at which crops demand and absorb, it is not always needed. For example, in case of wheat, Filho et al. (2011) found that wheat genotypes given fertilizer-N just before sowing did perform as well as those provided in split dosage. Application of 122–126 kg N ha^{-1} was essential and it had to be placed in the soil at beginning of the crop.

Soil N dynamics in the natural prairies and cropland, both are influenced to a certain extent by the legume-nodule bacterial symbiosis and asymbiotic-N fixation in soil. Regarding soybean expanses in Cerrados, their successful establishment and spread were to a large extent dependent on soil-N dynamics. We ought to realize that whenever legumes production zones are expanded, the native Bradyrhizobial flora that is compatible and effective should also flourish. Soil-N input via Biological Nitrogen Fixation is highly dependent on efficiency of legume-rhizobium symbiosis. In this case, it is soybean-Bradyrhizobial symbiosis. Mendes et al. (2003) have recounted the way establishment of both soybean and effective Bradyrhizobial strains were achieved in the Cerrados. It seems initially when soybean cultivation zones were expanded rapidly in the Cerrados, the native Bradyrhizobial flora was not endowed with efficient strains. Hence, lack of efficient nodule bacterial strain became an important constraint. Therefore, aiming high grain yield with low N inputs were not possible. Soil acidity was a major problem that impeded rapid establishment of Bradyrhizobial inoculant stains. Bradyrhizobial strains that were acid soil-tolerant had to be searched, isolated, and multiplied selectively and inoculated. Soybean genotypes that nodulated efficiently in the acid Oxisols were later screened and utilized. During late 1970s, a few Bradyrhizobial strains that were acid-tolerant and efficient in N fixation were isolated and used an inoculum strain for entire Cerrados. They are *Bradyrhizobium elkanii* strain 29W and SEMIA 587. These strains were released as commercial inoculants and spread all across the soybean belt. Next, in

1992 *Bradyrhizbium japnicum* strains SEMIA 5079 and SEMIA 5080 were spread into soybean production zones (Peres et al., 1993; Vargas et al., 1993). Inoculation with *Bradyrhizobium* did reduce stress on use of chemical fertilizers. Yet only a few agronomic procedures had been tailored to improve efficiency of soybean production using legume bacterial inoculation. For example, most farmers still prefer to add a starter dose of 5–10 kg N to stimulate rapid growth of roots in the initial stages of crop. This enhances rooting and nodule formation plus takes care of N needs during early stages when N derived from atmosphere is still not available. In some cases, farmers preferred to add slightly larger dosage of FYM of 26 t ha^{-1}, a couple of weeks earlier. This procedure again helps to induce rapid seedling growth and better nodulation. We should note that excessive N may reduce nodulation.

Build-up of phosphorus in Cerrados soils is a wise concept formulated to overcome simultaneously the phosphate fixation capacity of soil, detrimental effects of high Al saturation, and the crop's requirements for soluble Phosphorus. Lopes (1996) had suggested that among soil fertility aspects, maintaining optimum soil-P in the soluble-P fraction is essential, if one aims at moderate or high yield goals. In many locations within Cerrados, average soluble-P content is very low at <1.0 ppm and the fixation capacity is high at 20–60 percent depending on soil characteristics at that spot. Soil clay content also affects the buildup of soil-P. Generally, 3–5 kg soluble P_2O_5 is added for each one percent of clay content in soil, in order to accumulate soil-P at appropriate levels. Farmers in Cerrados may also opt for a different procedure to maintain soil-P. A common method is to add 20 kg P ha fertilizer, more than what is required in the general course to meet crop's demand. The excess P added supposedly takes care of P fixation capacity. Farmers also apply slightly larger dosages of rock phosphates that release P into soil at a slow rate. This is suitable to pastures sown on newly claimed acid soils. As such agronomic efficiency of rock phosphates found in situ within Cerrados varies. For example, if agronomic efficiency of TSP is considered as 100, then that of Gafsa hydrophosphate is 93, Thermophosphate-Mg is 92, Pirocuoa rock phosphate is 76, Araxa-PO_4 is 27, Patos-PO_4 is 45, Abaete-PO_4 is 21, and Catalo-PO_4 is 8 (Thomas et al., 2000). Further, on clayey Oxisols, it is said powdered and acidulated rock-phosphate is a good bet to build soil-P levels in the Cerrados. Liming needs to be controlled since it affects agronomic efficiency of rock phosphates. Subsurface characteristics are equally important, if farmers aim utilizing stored moisture and nutrients to cultivate the crops. For example, deep-rooted crops like cotton, soybean, and sugarcane may explore and derive large share of their moisture/nutrient requirement from subsurface. Soils in Cerrados often show Ca deficiency in conjunction with Al/Mn toxicity. Generally, farmers are advised to add phosphogypsum on the surface and allow it infiltrate freely to subsurface. This induces soil pH to rise, makes Ca available to crop roots and most importantly corrects Al$^+$. Following is an example that explains improvement in crop yield due to application of Phosphogypsum:

Phosphogypsum	Cotton	Soybean
t ha^{-1}	t ha^{-1}	
0	1.8	3.3
3	2.6	4.6

Source: **Sousa** and Rein (2009)
Note: Phosphogypsum is applied once in 3 years. It is aimed at supplying P and reducing pH as well as Al toxicity.

Yet another concept in vogue in Cerrados, aimed specifically to solve P requirements of crops involves two steps. First one is the corrective step, then second a P fertility maintenance step. As stated above, soil P availability is extremely low in Cerrados soils. Therefore, before adopting no-tillage and sowing crops such as soybean or cotton, farmers are advised firstly to supply fertilizer-P to achieve the critical concentration of P in soil. Next, the fertilizer-P required for crops, mostly between 60 and 100 kg P$_2$O$_5$ is added. This second application of P helps in maintaining and generating a good crop of soybean yielding 3.5 t grain per hectare or 6–10 t grain soybean per hectare of maize (Martinhao et al., 2011). Phosphorus needs could also be calculated for the entire sequence in a year. For example, bean crop responds to fertilizer-P supply. Further, response of bean to fertilizer-P is higher, if the crop is irrigated. Maize crop sown in sequence after the harvest of first crop too responds, utilizing the residual-P left unused by bean crop (Miranda et al., 2000).

Brazil is among the largest importers of Potassic fertilizer and mineral ores. This is done mainly to replenish the vast cropping expanses of Cerrados. During 2012, Brazil imported 7.5 m tons of fertilizer-K. Build-up of K too is recommended in the Cerrados soils. Oxisols with less than 30 ppm exchangeable-K and more than 15 percent clay need careful management procedures aimed at enhancing soil-K availability (Sousa, 1989). Generally, fertilizer-K supply rate ranges from 50 to 80 kg K$_2$O per hectare depending on crop, its genotype, soil characteristics, and irrigation facilities. Sometimes fertilizer-K inputs are calculated based on critical soil-K levels that need to be maintained to achieve yield goals. Potassium buildup in soil is often done along with P. Such a step restores soil nutrient balance. In fact, micronutrient nutrient buildup is also practiced in farms. Micronutrients are broadcast once in 2–3 years to achieve appropriate levels that maintain soil nutrient ratios and provide balanced nutrition to crops. Micronutrient buildup is often confined to nutrients such as Zn and Cu.

Cerrados soils are not satisfactorily endowed with S in surface and subsurface horizons. Crops provided with fertilizer-S or elemental-S is known to respond with enhanced biomass/grain formation. It means soils are deficient. Rein and Sousa (2004) state the Cerrados soils show dearth for S, mainly because, rain fall events may leach this element from the upper layers of soil profile. Lack of crop residue recycling and frequent loss of S via fire and burning trash, reduces S in the ecosystem. Further, during recent years farmers use high efficiency non-sulfur fertilizers such as TSP, DAP, and MAP. Hence, inadvertent S inputs too are minimal in farms. In the Cerrados, farmers prop up soil-S by supplementing it with fertilizers. Sulphur deficiency has been reported in the Cerrados. For example, SPAD readings of soybean, wheat, and pasture grasses such as *Brachiaria* clearly show inadequacy of N and S in the tissue. Sometimes plants get stunted due to S dearth. For example, wheat crop diagnosis with SPAD measurements show S deficiency (Batista and Monteiro, 2007; Gomes de Santos et al., 2010).

Main sources of S are Calcium and Ammonium sulfates. For example, single super phosphate, phosphogypsum, and gypsite. Phosphgypsym that contains 15 percent S is supplied to correct S deficiency if any. Phosphgypsum inputs are dependent on soil type, especially texture, subsoil acidity, and rooting pattern of the crop and yield goal (Gomes de Santos et al., 2010). Elemental S is also added to soil, either in powder or particulate form. However, its usage in Cerrados seems limited by its availability. Sulphur addition to fields also occurs when farmers utilize sulfur-coated fertilizers and slow release fertilizers with S coating. During recent years, farmers apply fertilizer-S based on chemical evaluation of crop residues recycled and its S supply capacity.

2.2.6 SOIL ACIDITY

Soil surveys aimed at estimating area affected with soil acidity have been useful to farmers and policy makers. It is said 70–80 percent of pastures sown after clearing natural vegetation may actually encounter severe soil acidity. Vendrame et al. (2010) surveyed different soil types for soil acidity in the surface (0–0.2 m depth) and subsurface (0.6–0.8 m depth) layers of profile. They have reported that whatever the texture or SOM content, over 90 percent of samples from different locations expressed low soil pH, Al saturation above permissible levels, and low-P availability. Soil types such as Oxisols, Entisols, Latossols, Vermelhos, and Armelhos were all similarly afflicted with acidity. Soil pH often ranged well below 4.5–5.0. Reports by IAEA, Vienna have suggested that since acidity is rampant all across soils of Cerrados, measures to alleviate the situation has to be integrated and we should use as many avenues to thwart ill effects of soil acidity on crop productivity. Major thrusts include careful nutrient management procedures that do not accentuate acidity in soil profile; identification of crops and their genotypes that adapt better to soil acidity and those which tolerate acidity, application of plant residues; phosphate rocks, and green manures along with lime to stabilize soil pH at higher levels; and ecoregional approach that considers entire region while devising large-scale soil pH alleviation methods. There is also need to understand water and nutrient cycles in fields that are afflicted with soil acidity (Thomas et al., 2000). Adoption of no-tillage systems is said to reduce acidity problems. No-tillage system increases soil pH in the upper 0–10 cm layer (Martinhao et al., 2011).

2.2.7 LIMING—AN ESSENTIAL SOIL AMENDMENT IN CERRADOS

Liming is among the most important features of soil management in the Cerrados. It is essential to correct soil acidity that often occurs along with Al toxicity (Lopes, 1996). On an average, 3 t Aglime per hectare is broadcast and incorporated into soils. Liming effects may often get localized at the point of application, say surface layer, if not well mixed into subsoil. Deeper placement is essential if rooting is to be encouraged in subsoil and to a greater depth. Fields under no-tillage or restricted tillage systems need relatively lower quantity of lime. Most of the regions with low pH also suffer low Ca and Mg. Hence, farmers are generally advised to supply dolomitic lime so that it adds Ca and Mg to soil. Lime recommendations are usually based on Al content and Ca + Mg saturation in soil. Amelioration of subsoil acidity is equally important during crop production. As stated earlier, beneficial effects of lime may get localized to surface, if not it is incorporated carefully into subsoil.

2.2.8 SOIL ORGANIC MATTER AND CARBON SEQUESTRATION

Brazilian Cerrados is a large repository of global carbon stock. Cerrados accommodates large quantities of biomass and C, both in the above and below-ground portion of the biome. Coutinho and Balieira (2011) suggest that carbon stock in the Cerrados is comparable to forests in the adjacent Amazonia, despite the fact that vegetation is either savannah or shrubs in most locations. Cerrados possess 190–236 t C ha^{-1} in the upper 0–100 cm layer of soil profile. Generally, conversion of native grasslands to cropland reduces soil C stocks (Bayer et al., 2006; Jantalia et al., 2007). Remote sensing techniques involving satellite observations (LANDSAT) and spectral data of foliage and soil color also help in assessing the C-stocks of Cerrados vegetation (Szakacs et al., 2004). Techniques based on ^{13}C abundance have also been used to estimate C sequestration in the natural vegetation and croplands of Cerrados (Buurman et al., 2004).

The average soil carbon stocks for various types of land use differ. Of course, it is attributable to factors related to soil and crop management procedures adopted and environment. For example, in southern Cerrados, soil C stock in the profile at 0–40 cm depth is 61.83 t ha^{-1} in locations with native vegetation; 59.4 if it has pastures; 54.2 t ha^{-1} if it is cropped and conventional tillage is practiced and 75.1 t ha^{-1}, if it is cropped and No-tillage is adopted (Coutinho and Balieira, 2011; see Table 2.1).

TABLE 2.1 Soil carbon stocks in the different types of soil encountered in the cerrados biome

Soil Type/Vegetation	Mean Soil Organic Carbon Stock (t ha^{-1})
All soil types-average	
Native vegetation	51.26
Pasture	46.24
Agricultural prairie	50.12
Neosols	
Native vegetation	24.68
Pasture	23.96
Agricultural prairie	52.53
Latosols	
Native vegetation	57.49
Pasture	50.10
Agricultural prairie	52.91
Argisols	
Native vegetation	51.20
Pasture	NA
Agricultural prairie	52.53
Cambisols	
Native vegetation	51.26

Source: Coutinho and Balieira (2011)

Soil organic matter status is an important determinant of crop productivity in Cerrados region. Cerrados soils are low in clay content, and their cation exchange capacity (CEC) is also low. Nearly 70 percent of CEC is attributed to organic fraction of soil. According to Lopes (1996), management techniques that include mono-crops, conventional tillage, liming, and inorganic fertilizer usage all induce rapid loss of organic fraction of Cerrados soil. Hence, it has been suggested that integrated management schemes that carefully maintain optimum SOC levels are important. Cerrados farmers have in fact adopted crop rotations, organic manure supply, and adequate residue recycling to preserve SOM levels.

At this juncture, we should realize that agricultural practices can make a soil sequester C and become a sink or source leading to emissions that possess C (e.g., CO_2). There are several factors that literally control sequestration trends in the Cerrados soils used to cultivate crops. Some of the factors listed are crop rotation, mineralization rates regulated by C:N ratios, crop residue supply schedules and its incorporation, soil type and mineralogy, soil moisture and weather patterns, and C emissions rates (Cerri et al., 2007).

Crops and cropping systems adopted are major factors that regulate C-sequestration in Cerrados. Soil organic matter management procedures may also vary depending on crop species. Reports by FAO (2008) indicate that soybean crops may cause rapid decline of SOM. During first 6 years of soybean monocrop, SOM depletes fast from initial level of 3–4 to 1.6 percent. It obviously induces soil deterioration, loss of aggregates, and reduction in moisture holding capacity and an increase in topsoil erosion. Based on long-term studies at EMBRAPA experimental station, Bayer et al. (2006, 2009) suggested that intensive cultivation of soybean with reduced or no-tillage adopted for 11 years invariably increased SOC content.

Marchao et al. (2009) have compared natural Cerrados vegetation, Integrated Crop-Livestock systems (ICLS), and Continuous cropping systems for net carbon storage in the soil profile at 0–30 cm depth. They reported that C storage in Cerrados native vegetation was 60.87 t ha^{-1}, 52.37 t ha^{-1} in ICLS, and 59.89 t ha^{-1} in continuous cropping systems. Continuous cropping for 13 years decreased soil C stocks by 7.5 t C ha^{-1}. No-tillage system is known to enhance C sequestration, but in a long term it may also induce soil compaction that needs to be carefully managed. Computer-based models and simulations suggest that in Cerrados, continuous soybean or maize for 10–15 years without disturbance to soil profile (i.e., in no-tillage systems) improved carbon stocks by 0.83 t ha^{-1} year^{-1} in the topsoil (0–20 cm) layer. Inclusion of a cover crop such as Pearl millet or Brachiaria further improves C sequestration in No-tillage plots (Corbeels et al., 2004).

Sugarcane production is intense in the South-eastern Cerrados. There are several reports on the consequences of land use change from natural prairies/shrubs to sugarcane (Silva et al., 2004, 2007; Resende et al., 2006; and Galdos et al., 2009). Sugarcane, in fact, grows well in slightly wetter terrain and forms tall prairie vegetation within the Cerrados. It is preferred since it adds to exchequer through sugar and ethanol. Ethanol is an important ingredient of Biofuel. Let us consider a few other aspects of sugarcane expanses that occupy a perceptibly large area of soils of low fertility. The consequences of sugarcane production on SOC content, C-emissions, residue recycling, SOM dynamics, soil microbes and their activity as related to SOM accumulation, and nutrient transformation need due attention. According to Santana et al. (2009) and Galdos et al. (2009), sugarcane farming has immediate effects on soil physicochemical characteristics such as structure, aggregate percentage, soil moisture holding capacity, percolation, soil organic matter content, and so on. Burning stubbles and surface weeds/vegetation prior to planting sugarcane sets is a common practice. It affects firstly the C dynamics by emitting CO_2. It also reduces C sequestered in the ecosystem (Balieiro et al., 2008; Pinheiro et al., 2010). It reduces recyclable C but adds to mineral nutrient content when ash is recycled. Pinheiro et al. (2010) reported that in the Cerrados, sugarcane-based prairies showed an increase in soil carbon stocks of 14 t ha^{-1} compared to regions where burning prior to harvest was practiced. Actually maintenance of sugarcane residue and stubble on soil surface itself helps in improving SOC content and reducing weed emergence. It also acts as mulch. Burning residue may enhance soil temperature and reduce soil microbial population. Presence of sugarcane residue on the soil surface affects soil microbial activity to a certain extent. Residue may provide congenial temperature and moisture regime for multiplication and activity of soil microbe. For example, Angelini et al. (2010) state that soil covered with sugarcane residues supported better growth and colonization of a beneficial fungus – Arbuscular Mycorrhizas. Mulches also induce multiplication of several other beneficial bacteria in the soil such as Bradyrhizobium.

Soil quality indicators envisaged for Cerrados soils are many and they vary with regard to priority and importance bestowed within each farm. Yet, in general, soil traits considered and expected values for them are as shown in Table 2.2

TABLE 2.2   Soil quality traits, indicators, and expected optimum values for cerrados soils

Soil Quality Trait	Indicator	Expected Value/Character
Soil color	Topsoil color	Black or dark brown
Organic matter	SOC (%)	>1.3
Acidity	pH	5.6–6.5
Al toxicity	Al saturation (%)	<20
Phosphorus	Extractable P (mg p kg $^{-1}$)	>10 for clayey soils
	(Mehlich-1 extractant)	>20 for loamy soils and >30 for sandy soils
Nutrient balance	Crop appearance/vigor	Dark green and healthy leaves
	Base saturation (V% at pH 7)	40–60%
	Base equilibrium (CEC %)	
	Ca, Mg, K % respectively	60, 15, 5
Compaction	Runoff, infiltration test	Water soaks in slowly, with some runoff after heavy rain
	Soil bulk density	Average value for representative soil
	Penetration test	Cannot get into thin hard pan or plow layer
Erosion	Soil loss	Signs of sheet and rill erosion
	A horizon thickness	Average for representative soils
Ground cover	Permanent ground cover (%)	50%

Source: Santana et al. (1998); Freitas et al. (2013)

2.2.9 PRECISION FARMING AND CERRADOS AGRICULTURAL LANDSCAPE

Precision farming is one of the more recent techniques that is being tried and introduced into the cropping zones of Cerrados landscape. Precision farming involves supply of fertilizers/water based on soil analysis and maps that depict variations in nutrient/moisture distribution. Precision farming could be highly mechanized and GPS guided variable-rate applicators could be used to supply inputs accurately. Precision farming as related to soil fertility and crop productivity, if applied in large area may have tremendous influence on cropping expanse. Firstly, it brings about uniformity to fields with regard to soil fertility and nutrient availability. Since nutrient supply is well staggered and accurate as per crop's needs, it avoids undue accumulation. Therefore, loss to vadose zone/groundwater will reduce. Fertilizer needs could be marginally low. Nutrient loss via emission gets reduced. Most importantly, precision farming brings about uniformity to grain harvest per unit area even within a small field. No, doubt agricultural prairies will experience some modification in soil fertility, moisture, and crop harvest pattern. Consequently, residue recycling and nutrient dynamics will also be affected.

There are reports about feasibility and constraints that have to be overcome while applying Precision techniques in the Cerrados. Lowenberg-DeBoer and Griffin (2006) state that, precision technique could be applied with greater advantage in the Cerrados because it supports several large farms. Precision techniques could also be adopted after due modification to fit the small holders of Cerrados. Soil fertility variation induced due to agronomic procedures and soil-management-related activities is comparatively recent. Since land is available in plenty in Cerrados, Precision farming could also be utilized in newly cleared zones. In the large farms and Cerrados cropping expanse as a whole, record keeping, and decision making could become very accurate and controlled by computer-based models. Precision techniques involve electronically controlled tillage implements, seeders, robotics, and dispensers of water/nutrients. Hence, labor requirement is perceptibly reduced.

Cerrados farmers may play diffident because profits are sometimes just marginal. Much of the profits are to be perceived as gain in uniformity in soil fertility, small reduction in fertilizer usage, and environment-related advantages. High cost of GPS-guided equipment, variable-rate applicators, GPS-guided combine harvesters, soil analysis maps from satellites are other reasons for restricted use of Precision techniques. Right now, Lowenberg-DeBoer and Griffin (2006) opine that adoption of Precision farming in the prairies is slow, since several of the large equipments like combines, variable-rate applicators, computer models, and soil maps may have to be imported. However, in due course, precision techniques will affect the Cerrados landscape.

2.3   WATER RESOURCES AND IRRIGATION

Cerrados is a massive repository of fresh water. It harbors large hydrographic basins namely Sao Fransisco, Prata, and Tocantins (Stefanini, 2012). Cerrados is naturally served by several rivers that flow in either direction from north to south and south to north within the biome. For example, rivers such as Sao Francisco and Tocantins flow south to north. While Parana and Paraguay flow North to Southeast (Figure 2.2).

Reports by FAOAQUASTAT (2002) suggest that irrigation potential of Cerrados is relatively high and needs to be harnessed efficiently. It seems about 3 m ha of low lands and 4.2 m ha of highlands could be easily brought under irrigation. The irrigated area is about 2.8 m ha equating to 5.7 percent of Cerrados region and this could be improved.

Groundwater resource of Brazil (not Cerrados) is distributed well all over the regions. It ranges from <5 $m^3 ha^{-1}$ under metamorphic rock formations of semiarid north east to 1,000 $m^3 ha^{-1}$ in the sedimentary rock regions. It is said that metamorphic rock formations in the northeast allows good storage of groundwater. Groundwater is major source of potable water. It is also extensively used to irrigate crops in the Cerrados. Groundwater attains greater importance in the northeast since precipitation pattern in many locations is only 600 mm $year^{-1}$, but the evapotranspiration of tropics is high at 2,000 mm $year^{-1}$. In addition, rivers flow intermittently and riverine irrigation could get constrained. There are several reports about fluctuations in water table. Maps that delineate the depths of water table within Cerrados helps policy makers in deciding about crops that can be cultivated, irrigation requirements, and yield goals. According to Manzione et al. (2006), precipitation levels and pattern, evapo-transpiration, and vegetation are among important factors that induce fluctuations in water table. Models that assess the influence of above factors and several others are available. For example, "PIRFICT" is a computer aided model that assesses groundwater storage. It also helps in alerting farmers about potential risks of any shortage that may occur. Water table data is also provided periodically to the needy farmers almost every 2 weeks.

The irrigated area in Brazil was negligible prior to 1950s. Much of the Cerrados depended on rainfall pattern. It increased to 500 thousand ha in 1960 and 800 thousand ha in 1970. Later, there was a spurt in irrigated area to 2,600 thousand ha in 2000. Currently about 2.3 m ha are under irrigation.

FIGURE 2.2   Major rivers of Brazil.
Note: Rivers of greater importance to natural prairies and croplands in the Cerrados are Parana, Paraguay, Sao Francisco, Tocantins, Tapajos, and Xingu.

The large commercial crop production enterprises in the Cerrados, especially, those generating soybean, sugar cane, and wheat utilize a wide range of irrigation systems and devices. They are all aimed at improvising on water use efficiency and crop productivity. For example, they use surface irrigation, mobile sprinkle systems, and modern center-pivot system. The water-use efficiencies of these systems range from 40 to 78 percent. Microirrigation methods are more popular in the northeast because of shortage of water resources (FAOAQUASTAT, 2002). During recent years, GPS guided spatial models for sprinkler placements and water requirements in various locations within major cropping belts like soybean in Cerrados have also been adopted (Zeilhofer and Kemp, 2011).

The water requirements of crops grown in the Cerrados varies widely based on species, its genotype, season, soil type, soil fertility, and yield goals envisaged. Yet on an average, crops such as soybean may need 500–600 mm, sugarcane 900–1000, maize 550–600 mm, beans 450 mm, and wetland rice 1,000–1,200 mm, in order to complete life cycle. To quote an example, farmers supply about 450 mm water for a bean crop that produces 4.8 t grain per hectare (Guerra et al., 2000).

Generally, water availability in the soil profile determines the nature of vegetation, crop species, and intensity of cropping (Assis et al., 2012). Yet, we know that *vice versa*, that is vegetation/crops that flourish also have an impact on water in the profile. The natural vegetation that fills the Cerrados region such as savannahs, shrubs, trees, and crop species cultivated by farmers all have a say in the extent of water consumed and recycled. Dynamics of water in the Cerrados is known to change, whenever, natural savannahs are cleared and land use changes to agricultural crops. In case of *Campo denso* with shrubs and trees, soil profile contained about 293–689 mm plant available water in the soil profile at the end of dry season. Savannahs filled with native grass species stored only 155–362 mm available water at the end of dry season. Factors such as evapotranspiration, water absorption rates, biomass formation, and loss via percolation or leaching may all affect the water storage in the ecosystem. For example, evapotranspiration of plant species in *Campo denso* was 1.5–5.8 mm day^{-1}. In case of *Campo sujo*, it was much less at 0.9–4.5 mm day^{-1} during the dry season. The water absorption rates from soil profile and evapotranspiration of cropland depends on crop species, genotypes, biomass accumulation rates, and yield goals envisaged by the farmers. In the natural savannahs and cropland of Cerrados, researchers have often concentrated on studying the fate of soil moisture just within 0–0.5 m depth of the soil profile. This is because fine roots are dense and mostly distributed well within first 0–30 cm depth of soil profile. However, many of the deep-rooted plant species are actually affected by water conserved in the deeper layers of soil. Deep rooting seems to be essential trait for crops that experience drought and periodic dry spells in the Cerrados. It is interesting to note that during dry season, 82 percent of water extracted by species in the *Campo denso* and 67 percent by those in *Campo sujo* were derived from depths greater than 1.0 m in the soil profile.

2.3.1   CROPS, CROPPING SYSTEMS, AND PRODUCTIVITY

Brazilians generate several crop species in the Cerrados. Most prominent among them are soybean, maize, coffee, cotton, and sugarcane. Beans and several cover crops are also produced. Pastures filled with monocrops of cereals or legumes and mixtures, in fact dominate the landscape. Large expanses of soybean, cotton, or sugarcane are a characteristic of Cerrado agricultural prairies (Britto, 2011).

Records suggest that earliest of soybean cultivation occurred in 1900–1901 in the State of Sao Paulo and Rio Grande do Sul (Britto 2011). Today, states of Mato Grosso, Goias, Parana, and Rio Grande together generate 82 percent of Brazilian soybean produce. Soybean zones have also extended into Maranhao, Tocantins, Piaui, and Bahia. In all, about five states in Brazil support soybean cultivation (Britto, 2011).

Soybean production was an unknown occupation to farmers in the Cerrados, until three decades ago. However, Brazilian farmers have conquered Cerrados natural vegetation and in its place have cultivated soybean in large expanses. Britto (2011) opines that topography of Cerrados, the savannahs, and agroclimate suited soybean production rather immaculately and hence it got highly preferred among farmers.

Soybean-based savannah engulfed the Cerrados rather rapidly (Figure 2.3). Simultaneously, its productivity per unit land too registered marked increase (Coutinho and Balieiro, 2011). Soybean production had reached 13 million ha by 2000 A.D. (Kaimowitz and Smith, 2001). Soybean cultivation induced human migration. Several individuals, it seems gave up lucrative jobs to embark on soybean production in large scale and acquire riches. It is generally agreed that, among various factors listed, soybean technology and its incursions into Cerrados effectively replaced natural vegetation markedly. During 1960s, soybean productivity in the Cerrados was relatively low at 1.06 t grains per hectare (Wilkinson and Sorj, 1992). During 1970s, the productivity of soybean planted in Sao Paulo region and elsewhere in Cerrados had increased by 15 percent–1,200 kg ha^{-1}. Crop improvement programs of EMBRAPA introduced several high-yielding genotypes of soybean during 1970s and 1980s. The productivity of about 26 out of 48 soybean genotypes was higher by 36–40 percent over previous cultivars. To match the soybean genotype, better performing Bradyrhizobium strains were also introduced into Cerrados soil (Kaster and Bonata, 1980). Average soybean yield during 1975 reached 1.7 t grains per hectare. In addition, massive demand by European nations and high International price for protein rich grain aided expansion of soybean cultivation zones into entire Cerrados. Overall, causes for spread of soybean farming in Cerrados are improved technology, agronomic procedures, high-yielding soybean cultivars, market conditions and consistent demand, private sector interest, large corporate farms, and most importantly good soil management techniques and congenial tropical environment of Cerrados.

Reports suggest that one of the earliest high-yielding soybean cultivars fit for tropical climate of Cerrados known as "Cristalina" triggered rapid spread of soybean expanses. This variety "Cristalina" was accepted rather rapidly during late 1970s and by mid-1980s, it had spread into 80 percent of soybean belt of Cerrados. Its productivity was enhanced by standardizing fertilizer requirements, especially, lime and micronutrients, in addition to major nutrients. Soybean production became profitable, if such high-yielding varieties were planted after rice and/or Brachiara pasture grass (FAO, 2008).

Maize is an important cereal crop of Brazilian Cerrados. Landraces of maize were cultivated for past few millennia by the natives of Cerrados. During recent decades, improved cultivars and hybrids have dominated the Cerrados maize landscape. Maize production systems have vastly improved in terms of crop genotypes, soil, and water management as well as harvest techniques. Maize-based prairies dominate the landscape in many parts of Cerrados, especially, in areas endowed with low fertility soils and meager irrigation facilities (Figure 2.3). Models that simulate and forecast maize crop growth and yield in the Cerrados are available. Such computer-based models usually consider meteorological parameters like precipitation, evapotranspiration, diurnal pattern, heat units received, and relative humidity; soil characteristics such as fertility, water storage capacity, and several crop characteristics (Affholder et al., 1997). There are also popular models such as CERES-maize that include several submodels (routines) for soil fertility and fertilizer inputs, water requirements, plant growth, and grain formation traits, and so on.

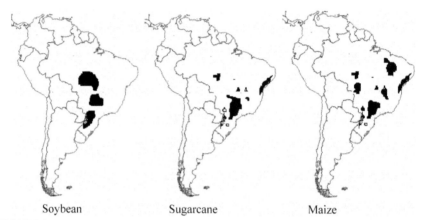

| Soybean | Sugarcane | Maize |

FIGURE 2.3   Soybean, sugarcane and maize cultivation zones of Brazil including Cerrados where they are dominant crop species.
Source: Redrawn based on Baker and Small (2005)

Sugarcane is originally a crop native to Indo-Gangetic Plains in South Asia. Larger patches of sugarcane production began in Cerrados with the advent of Portuguese settlers in sixteenth century. However, currently, Brazilian Cerrados is the top-rated sugarcane producing zone in the world. Literally, Cerrados contributes about 30 percent of global sugarcane production. Large portion of sugarcane produced in Cerrados is exported as finished product like sugar or ethanol. Sugarcane production zones in Cerrados increased mainly due to demand for ethanol and biofuel that replaces fossil fuel. In addition to Cerrados, sugarcane producing zones have also expanded into diverse regions within Brazil, such as Amazonia and Pantanal. Nearly 90 percent of sugarcane in Brazil is localized in the south central region (i.e., Cerrados) (Figure 2.3). Sugarcane has often replaced natural vegetation or pastures in the Cerrados. Its area is supposed to increase markedly during next decade and may double by 2018 (Friends of Earth-Europe, 2012). Interestingly, sugarcane is also re-placing soybean in some parts of Goias. It is also expanding at the expense of forest land and pastures in Mato Grosso. About 8–10 percent of sugarcane expansion has occurred by removal of natural forests in this state. Analysis by Nasser et al. (2012) indicates that during 2005–2008, sugarcane expansion displaced about 1.38 m ha of area that otherwise supported natural grass lands or cropland. Sugar productivity increased and this helped in reducing burden on expansion. It seems sugar cane yield improvement compensated or released an estimated 0.57 m ha in Brazil. Actually, Brazil has about 554 million ha of arable land, out of which 34 million ha is suitable for sugarcane production. At present, sugarcane area covers one percent of total land area in Brazil (Hermele, 2011).

Wheat has been grown predominantly in the temperate regions of Brazil. Since, 1970s, wheat production has been commercially viable and profitable (Cargnin et al., 2008). However, since past four decades, wheat production has become popular in the Cerrados region, especially in Minas Garias and Mato Grosso. Wheat cultivation localizes into areas about 400 m.a.s.l. Wheat genotypes derived from breeding programs of International Maize and Wheat Centre (CIMMYT), National Wheat Research Centre of Brazil and Cerrados Research Centre (CPAC) have been disseminated in this region. The annual progress in yield due to crop improvement program is said to be 48 kg ha^{-1} for the period 1976–2005 (Cargnin et al., 2008). During the same period, refinement in soil fertility and crop management techniques seems to have contributed up to 32 percent grain yield increase. The average productivity of recently developed wheat hybrids is about 3.5–6.3 t grain per hectare depending on location (Silva et al., 2008). Wheat production zone in southern Cerrados and in other regions of Brazil has experienced marked fluctuation, despite constant support from government agencies. It peaked during mid-1980s at 3.5 m ha^{-1} but declined rapidly due to replacement by soybean, to a low at 1.4 ha in 2005. The spread of wheat cropping in Cerrados is also constrained by factors such as high costs incurred on fertilizers and irrigation. Hence, agencies such EMBRAPA are aiming at deriving more of drought-tolerant wheat genotypes from the crop improvement programs. In fact, Wada et al. (1997) have long ago suggested that there is indeed large number of wheat genotypes available within the Brazilian collections that are drought-tolerant.

Coffee was the dominant crop since early 1900s till 1970, until soybean entered the Cerrados region. Brazil is among the top producers and exporters of coffee. Coffee plantations are spread into many states of Brazil. Coffee is among the most important agricultural produce generated in the Cerrados. Arabica coffee is the most dominant species. During recent years, Minas Garias state has become major coffee region within Cerrados. The coffee gardens here are located at 900–1,100 m.a.s.l. and they generate top quality seeds. Factors such as topography, soil type, agroclimate, and general economic conditions are said to be favorable to coffee production (Gale, 2013). Coffee production is highly mechanized. Computer-guided irrigation systems that adopt drip irrigation are used. Several new varieties known for quality seeds and higher productivity have been introduced into Cerrado region. Cerrados generates at least four types of coffee, commonly known as naturally processed, fully washed, semiwashed, and pulp removed before drying. The Café do Cerrados exports 90 percent of its produce mostly to USA, Europe, and Japan.

Rice is grown in parts of Tocantins State of Brazil. Rice genotypes sown, irrigation, and fertilizer supply seem to decide the productivity. Rice genotypes that tolerate drought spells are preferred in this region of Brazil. Among the several genotypes of rice grown, it is said that "Quebra Cacho" that tolerates drought and offers better grain productivity is preferred (Terra et al., 2010).

2.3.2 CROPPING SYSTEMS IN CERRADOS

Agricultural cropping became conspicuous in the Cerrados about 150 years ago, with extensive ranching that replaced native grass lands. Ranches were replaced in many places by rice-pasture system. It was followed by grain crops such as wheat, sorghum, and soybean, then sugar cane. Spehar (2012) points out that for the past 20 years, cropping systems in the Cerrados is lopsided with maize and soybean that dominate the landscape during rainy season, followed by a dry season with least crop production activity. Specialized monocrops of maize, soybean, cotton, sorghum, or millet, seems to threaten the ecosystem with monotony and buildup of insects, diseases, loss of soil fertility, and several other environmental concerns. A recent trend has been to grow cover crops or those that provide forage and mulching material for the succeeding season. This is done immediately after the first major crop during rainy season and it utilizes stored moisture and residual soil fertility. Among the alternatives, introduction of crops from Andean region that may allow Cerrados farmers to fill up the fallow season as break-crops is being popularized. Andean crops such as Quinoa (*Chenopodium quinoa*) and Amaranth (*Amaranthus sp*) are short season crops that fill the fallow excellently and offer vegetables and roughages (Pacheco and Peter, 2010; Spehar et al., 2012). They are also characterized by relatively high protein and mineral content. Grain amaranth (*Amaranthus cruentas; A. hybridus*) could also replace the fallow period in the Cerrados.

Field evaluations in the Cerrados have shown that no-till system for main crops followed by a cover crop is remunerative. Reports by CIRAD (2009b) suggest that cover crops such as *Brachiaria ruziziensis*, sorghum, millet, *Cajanus*, or *Crotalaria* sown as relay crop within maize fields serve the farms beneficially in many ways. Cover crops reduce soil erosion, they add to photosynthetic efficiency and biomass accumulation. Cover crops offer forage and material for mulching.

Relay cropping with cereal such as maize or wheat as first crop followed by a legume crop is a common practice in Cerrados. Planting the grain legume such as pigeonpea or a cover crop like *Brachiaria* helps farmers in many ways. It reduces the cost on tillage and it conserves topsoil fertility. It reduces soil erosion and loss of nutrients. Most importantly, relay planting of crops enhances land-use and precipitation-use efficiency (Balde et al., 2011). Following is an example dealing with N recovery from soil that supports a monocrop or a relay intercrop:

Nitrogen Recovery by Crops and Relay Inter-Crops (in kilogram N per hactare[-1])

	Sole Crop	Maize + Brachiaria	Maize + Pigeonpea
Maize	70	$75 + 56 = 131$	$67 + 167 = 234$
Brachiaria	35	-	-
Pigeonpea	213	-	-

Source: Balde et al. (2011)

Generally, relay intercropped maize with a legume seemed to be more efficient than a cereal and grass cover crop. No doubt, the efficiency of mixed-pasture or natural vegetation with mixed species is generally higher than a monocrop of maize or wheat grown in the Cerrados.

Cover crops grown in no-tillage soybean fields is an important aspect of cropping systems adopted in the Cerrados. Cover crops firstly affect photosynthetic light interception and net biomass generated per unit area. Cover crops, upon recycling improve C sequestration in soil. Cover crops reduce soil erosion. Cover crops improve precipitation-use efficiency. However, during dry season from April to October, cover crops sown after harvest of soybean may encounter paucity of soil water. Nutrient recycling induced by cover crops, especially, leguminous cover crops is highly pertinent. Leguminous cover crops add to soil-N, since they support Biological N fixation by Bradyrhizobial species. Cover crops that grow rapidly in between rows of soybean reduce proliferation of weeds. Cover crops upon harvest at succulent stage serve as excellent green manures and enhance soil-C and N (Giti et al., 2012). The incorporation of biomass generated by cover crops helps in improving soil structure and microbial activity. Cover crop species preferred by farmers in the Cerrados are cereals such as Pearl millet (*Pennesitum glaucum*), Brachiaria (*Brachiaria spp*), and Rye (*Secale cereale*). Legumes preferred are Pigeonpea (*Cajanus cajan*); Crotalarias (*Crotalaria junce, C. spectabilis*); and Stylo (*Stylosanthes sp*). Scopel et al. (2004) suggest that cover crops are useful to farmers in Cerrados in many ways. However, we need to carefully select and accentuate the species and attain net advantage aimed at while planting cover crops. For example, in the cotton growing regions within Cerrados, soil erosion was rampant. This was attributed to consistent use of disk plows. Hence, cover crops with mulch were used to thwart soil erosion. In many areas, it is important to generate green manure. Usually, cover crops that produce succulent leaves and twigs in greater quantities are preferred.

2.3.3   CERRADOS GENERATE BIOFUEL: CROPS REPLACE NATURAL PRAIRIES

Brazil generates a large quantity of ethanol used as biofuel, with sugarcane as the substrate. It seems, currently, ethanol is economically competitive with petroleum. In fact, this Brazilian model could be replicated in other savannah zones. Brazil generates 4.4 billion gallons of ethanol annually using sugarcane. Biofuels are known to reduce greenhouse gas emissions. Biofuel production in the Cerrados is a successful exercise that has been effectively utilized since two to three decades. It has involved conversion of natural grasslands into tall, mildly wet agricultural prairies filled with sugarcane. Yet, there are reports that argue either way suggesting that biofuel is a necessity. On the other hand, there are others who say the cost of land-use change and environmental effects are high. In the following paragraphs discussions that highlight both usefulness and ill effects of biofuel production in Cerrados are listed.

It seems expanding sugarcane zones without proper regulation may affect the environment (Heam, 2007). In fact, there are many farmers and policy makers who believe that Brazil's ethanol from sugarcane is not so sweet, since it affects soil, environment, and farming community in a long run. (Corporate Europe Observatory, 2009). Buckridge et al. (2012) have reviewed the current state of sugarcane-based ethanol production. They have also compared advantages and detriments, if any, to soil and environment. They argue for a midway situation wherein sugarcane cropping expands but in carefully regulated way, so that biodiversity and environmental parameters of the Cerrados are unaffected or at best the ill effects are least. Xavante (2009) suggests that Cerrados is deteriorating. It is really suffering from massive disturbance to ecosystematic functions that were entrenched since long. Cerrados, in fact, is threatened with gradual extinction, if the present trends in farm enterprise establishment and agribusiness continue for many years to come. There are corporate firms that hire either large tracts of native areas or purchase it outright to cultivate crops that yield biofuel. Biofuel is an excellent substitute to petroleum. It is lucrative to farming companies that generate crops such a sugarcane or sweet sorghum or maize. It is believed that initially, soybean export lured native population out of their zones and now it is followed by equally lucrative sugarcane production systems that yield biofuel. At this juncture, we should note that Brazilian government prescribes admixtures of biofuel and this too adds to expansion of sugar cane farming zones meant for ethanol production. According to Nasser et al. (2012), agricultural component, in other words sugarcane-based alcohol production is a key component. It emits greenhouse gas. Land use change from native vegetation to sugarcane production does enhance greenhouse gas generation. However, Leggett (2011) states that 25 percent of car transportation in Brazil is energized by sugarcane-based alcohol mixtures. Sugarcane-based ethanol production is relatively low on greenhouse gas emissions. It affects general environment much less compared to fossil fuel. Further, switching from annual crops or pastures to sugarcane for biofuel may even help cool the ambient temperature. Evaluations by Carnegies' Scientists suggest that if shift from native vegetation to annual crops increased temperature by 2.7°C, then next shift from annuals to sugarcane actually reduced ambient temperature by 1.7°C (See Derra, 2013). Handley (2013) too has made a similar remark that shift from grasslands to sugarcane decreases air temperature by 1.0°C.

Monocropping of sugarcane in the Cerrados was in vogue long ago in 1800s during Portuguese colonization of Brazil (Mendonca, 2011). Brazilian government's initiative to expand and intensify sugarcane cultivation in order to obtain ethanol for use in biofuel has led to significant improvement in biofuel production. It replaces use of fossil fuel, but in addition, it replaces food crops in many locations within Cerrados. In addition to deforestation, sugarcane cultivation displaces the natives who thrived on food crops and by coexisting with biodiversity. Natives have been lured into temporary labor in sugarcane farms and some migrate to towns/cities. Food crop production for local consumption gets reduced since sugarcane meant for alcohol has taken over the place (Mendonca, 2008, 2011). Brazil has embarked on biofuel production using sugarcane as policy. There are several advantages listed that deal with cane production, its processing, conversion to alcohol, and most importantly the economic gains attainable due its export (AgBrazil, 2006a). Biofuels reduce Brazil's burden on importing petroleum from other regions.

Biodiesel is a new idea being touted in the Brazil Cerrados. Biodiesel is again alternative to fossil fuel. It is derived from renewable resources (i.e., crops). The vegetable oils derived from variety of crop species is not directly amenable to be used in IC engines. Its viscosity needs to be reduced and its ignition properties need modification to be suitable for use in vehicles and motors. It seems low clouding point; high octane number and high flash point of biodiesel are its advantages (Lopes and Neto, 2011). Biodiesel is nontoxic, biodegradable, and could be recycled. List of crops that are useful in biodiesel production and are suitable for cultivation in the vast plains of Brazilian Cerrados are as follows: Jatropa (*Jatropa curcas*), Castor (*Castor communis),*African palm (*Elaeis guinensis*), Cotton (*Gossypium spp*), Peanut (*Arachis hypogaea*), Linseed (Linu*m usitatissimum*), etc. Perhaps, in future farmers and corporates will find series of reasons to expand agricultural cropping in Cerrados at the cost of natural vegetation. Depending on demand and economic benefits, these biodiesel crops may find place in the Cerrados and expand.

2.3.4 GENETICALLY MODIFIED (TRANSGENIC) CROPS IN CERRADOS

Forecasts, a decade ago had suggested that constraints to crop production in Cerrados could be partly overcome by producing "Genetically Modified (GM)" crops. They have to be tailored to adapt to environmental constraints and produce higher grains. Resistance to disease, insect tolerance and adaptation to soil fertility constraints were among the major problems that were to be tackled using GM crops. Review by Herrera-Estrella (1997) had suggested that problems that afflict crop production in Cerrados, such as soil acidity, Al toxicity and drought tolerance were good candidates. However, problems such as soil acidity seem to have been overcome more easily using gypsum. In case of Al toxicity, actual problem is the impedance and reduced uptake of nutrients by plant roots. Absorption of major nutrients is affected by high Al content in soil. The idea proposed was to genetically modify roots to tolerate Al toxicity by excreting larger quantities of organic acids and chelating substance. Then, Al toxicity will be reduced and nutrient absorption will be optimum. However, during recent years, greater successes have been achieved in producing GM crops that resist insects that incorporated with Bt—endotoxin genes and herbicide tolerance. Most of the cultivars of Soybean and Maize sown in Cerrados are GM modified with BT genes and herbicide tolerance. The influence of crops with resistance genes on the population of weeds, insects, and other species need to be studied. Insects adapted to soybean may encounter difficulty in surviving, if the Cerrado landscape is entirely filled up with Bt crops.

Most of the recent reports suggest that during 2012, about 18.1 million hectare of soybean sown in Cerrado region were Herbicide-tolerant GM cultivars and it is expected to increase to 23.7 million hectare in the next couple of years. It amounts to 76.2 percent of Cerrados soybean belt being occupied by herbicide resistance genes. Obviously, farmers could be either lured or may inadvertently spray greater quantities of herbicide, into the ecosystem than required. Further, surveys by private seed companies have shown that in case of maize, 45.5 percent of area of its belt in Cerrados is filled with herbicide-tolerant GM cultivars. In case of cotton, much of the belt is sown with Bt Cotton that restricts attack of leaves and boll by insects (Etinger, 2011; GBI, 2011; GMO Compass, 2013a, b).

Genetically modified wheat genotypes are yet to make a mark in the Cerrados or anywhere in different prairies of the world. GM wheat with resistance to Fusarial fungal attack is being tried and tested (GMO Compass, 2013c). We should note that GM wheat resistant to broad spectrum of soil fungi may affect fungal flora in general. Soil fungal flora may get affected. The consequence of broad spectral resistance to fungi, if it occurs needs to be understood in greater detail. Soil fungi are excellent bio-digesters of wood and other organic material in soil.

As stated earlier, sugarcane is an important crop produced in the Brazilian Cerrados. It is commercially useful in many ways such as for sugar production, bagasse, molasses, ethanol, and biofuel. Like any other crop, sugarcane crop is afflicted by several insect pests, disease causing fungi, viruses, weed infestation, drought, and acidity/Al toxicity at different intensities and for different length of time depending on the location within Cerrados. Reports by GMO Compass (2013d, e) suggest that researchers at EMBRAPA and elsewhere in other continents are examining and developing GM sugarcane genotypes with transgenic modifications that help them tide over as many natural and man-made problems. Transgenic sugarcane genotype with 25 percent yield advantage over the present cultivars is also being contemplated (GMO Compass, 2013e). Such a cane helps in generating more biofuel per unit area. It may also reduce stress on land required to expand biofuel production in Cerrados. Whatever the advantage, we have to realize higher cane productivity means greater supply of nutrients and water proportionate to biomass accumulation, higher quantity of nutrients recycled in the ecosystem. Basically, it intensifies agricultural prairies.

2.4 AGRICULTURAL CROPPING IN CERRADOS AND ENVIRONMENTAL CONCERNS

2.4.1 SOIL EROSION AND NUTRIENT LOSS IN CERRADOS

Brazilian Cerrados have indeed undergone rapid change from a well-evolved and established natural savannah grassland, with shrubs and occasional trees into a vast cropland within past four decades. It is fairly drastic effect on the land scape. Vrieling (2012) states that, currently, pastures dominate the land scape along with agricultural crops. This has affected the stability of soil phase of the biome. Soil erosion and with it loss of surface nutrients has become rampant. By definition, soil erosion is movement of soil particles from one location to other through wind, water, or any other means. Erosion may have detrimental effect on soil fertility, seed germination, and seedling establishment. It may also displace well established crop plants from their anchorage. Soil erosion processes such as splash erosion, sheet erosion, rill erosion, gullies, and total degradation of farm land through heavy floods are all common occurrence in the Cerrados. In addition, fluvial processes in the subsurface such as tunneling, seepage, and percolation cause loss of moisture and dissolved nutrients (Vrieling and Rodrigues, 2005). Soil erosion can be easily detected in farms through periodic observation of the fields. However, in large farms and expanses such as Cerrados, monitoring well spread out erosion locations need satellite mediated observation. The resolution of satellite pictures/observations may have to be high, if erosion events are small and details are difficult to decipher. We can study the factors that induce soil erosion in good detail, its magnitude and perhaps get an idea about consequences on field surface and soil profile, plus its consequences on crop productivity. Satellites and remote sensing devices can be effectively used to assess the intensity of factors that induce erosion. For example, Vrieling (2012) suggests that satellites can help providing information on rainfall pattern, its intensity, and extent of erosion caused. Similarly, we can assess the effects of topography, slope, vegetation cover, and land use as agents for erosion. Most often, land cover and its usage is monitored to assess and forecast erosion. Soil and nutrient loss from surface and subsurface horizon actually costs the farmers and the governmental agencies. Conservation measures are not easy to adopt. Telles et al. (2011) estimated the cost of soil erosion in the State of Sao Paulo alone at 212 million US$ each year.

Nitrogen is key element, which is supplied periodically to crop fields. Knowledge about its dynamics in soil is essential while preparing fertilizer-N schedules for major crops such as soybean, maize, or wheat grown in the Cerrados. It is said that generally, fertilizer-N loss through topsoil erosion and leaching/percolation is severe. Hence, matching fertilizer-N inputs with crop's uptake rate and pattern is essential. Otherwise, large portion of accumulated N could be leached from the root zone. Lehman et al. (2004) studied the dynamics of soil-N and fertilizer-N applied to Cerrados soil, with an aim to find out how much N is accumulated in the subsoil, and the way it is lost from the profile. They suggest that much of soil-N leached is derived from upper layer of soil. About 80 percent of soil-N leached within first 30 days of the crop season was derived from topsoil (0–15 cm depth). Subsoil N retention was important, since it helped the crop during later stages of growth. Actually, large amount of applied fertilizer-N was adsorbed in the sub soil. It almost amounted to 100–300 kg N ha^{-1}. This fraction of N is sufficient to support a good crop of maize or wheat. However, subsoil-N too was vulnerable to loss via leaching and percolation whenever intense rainfall events occurred. Loss of N from subsoil to groundwater and aqueducts could be detrimental. Carefully tailored crop sequences and intercrop combinations could reduce loss of subsoil-N. Relay cropping and cover crops are other helpful suggestions to reduce undue accumulation of N in subsoil that may become vulnerable to leaching.

2.4.2 GREEN HOUSE GAS EMISSIONS IN CERRADOS

Major anthropogenic activities that generate perceptible quantities of greenhouse gas are land-use, land use change, agricultural crop production, forestry, industrial development, and urban expansion (Munoz et al., 2010). Cerrados of Brazil has a good share of all of these activities, hence GHG emissions are conspicuous. Cerri et al. (2009) state that Amazonia and Cerrados are important GHG emitting biomes. The major cause for GHG emission during recent period seems to be land use change and farming, rather than, burning of fossil fuel. Agricultural activities initiated in large scale in the Cerrados during past four decades has become a key issue and is mainly related to disruption of normal N and C cycling in the biome. The contribution of agricultural soils to CO_2, N_2O, and CH_4 emissions depends on variety of factors related to soil, ambient weather pattern, and cropping systems adopted in the Cerrados. Soil physic-chemical processes, moisture, redox potential and C:N ratio, N-immobilization, nitrification, de-nitrification, and mineralization of organic matter are among few important parameters that may suggest about emissions.

Cruvinel (2011) states that Brazilian Cerrados with its natural savannahs and cropland is indeed a major source of emissions such as NO, N_2O, and CO_2. During past four decades, aspects such as expansion of cropland, large supply of fertilizers, and organic matter into fields, irrigation, and several other management procedures related to soils have induced high rates of gaseous emissions. In fact, gaseous emissions have caused loss of soil-N and reduced fertilizer-N efficiency. The crops that fill the Cerrados and cropping systems adopted have a major say on the rates of gaseous-N emissions. In this regard, estimations by Cruvinel et al. (2011) suggest that, in the maize field NO-N fluxes were higher, if fertilizer-N sources were broadcast on the surface. Cultivation of cover crops such as *Brachiaria* induced marginally lower levels of NO-N fluxes than an undisturbed natural vegetation of Cerrados. In case fields with beans were irrigated and fertilizer-N supply was optimum, then N emissions were perceptibly higher than field not provided with fertilizer-N. Regarding pattern of NO emissions during a crop season, Cruvinel et al. (2011) have reported that a pulse of NO efflux (8.4 ng NO-N cm^{-1} h^{-1}) occurred after first rains. The NO emission after rains are however variable. The NO emission actually depended on soil moisture status. Any transition from dry to wet soil may induce NO emission. Lowest NO emission measured in soybean fields were of the order 0.3 ng NO-N cm^{-1} h^{-1}.

Generally, N_2O emissions were of lower order, irrespective land use. The N_2O emissions were 0.5–0.6 ng N_2O-N cm^{-1} h^{-1} in both natural vegetation and cropland. Adoption of no-tillage systems increased N_2O emissions, if the soil was poorly aerated (Rochette, 2008). This has been attributed to de-nitrification rates that are higher if fields are under no-tillage systems. No-tillage systems induced greater N_2O

emissions mostly during growing stages of the crop. Investigations spread across 25 fields in the Cerrados, during past two decades suggest that soil aeration controlled the extent of de-nitrification and consequent N_2O emission. In poorly aerated soils, No-tillage plots lost 2 kg N ha^{-1}. It is said that availability of NO_3-N in the soil profile was a prime factor inducing N_2O emission in cultivated fields. Rainfall or ir-rigation events also induced N_2O-N emission. In case of legume such as soybean or field bean, N_2O-N emissions increased during senescence. Fertilizer-N broadcasted as urea also induced N_2O-N emission. For example, Cruvinel et al. (2011) reported a loss of 2.3–10.6 ng N_2O-N cm^{-1} h^{-1}. Despite series of studies on GHG emissions in Cerrados that covers various locations, seasons, crops, soil types, fertilizer sched-ules, and soil moisture regimes, Bustamante et al. (2009) suggest that large uncer-tainties about GHG emission exist. Further detailed studies and data accrual about N-emission and CO_2 evolution is required.

The natural grass lands in Cerrados emanated CO_2 at a rate of 0.4 µmol cm^{-1} s^{-1} in dry seasons and it increased upto12 µmol cm^{-1} s^{-1} in wet period. Soil moisture derived from precipitation was crucial factor inducing soil microbial respiration and CO_2 emission (Cruvinel et al., 2011). Regarding soil-C loss through respiration, it was found that soybean fields and comparable natural vegetation in the vicinity, both emanated similar levels of CO_2 per unit time and area. In case of soybean fields, the CO_2 emission ranged from 0.2 to 1.0 µm mol cm^{-1} s^{-1} (see Table 2.3). However, in case of a maize field, soil respiration was generally higher and CO_2 emission was 0.8–1.8 µmol cm^{-1} s^{-1}. Soil respiration measured in maize fields were three times higher than in native vegetation. Availability of NH_4-N in the soil profile and micro-bial metabolism was important factor. However, CO_2 emissions from fields sown to maize and irrigated bean showed high CO_2 evolution. It seemed that fertilizer supply and irrigation induced greater levels of CO_2 emission. Agronomic procedures and irrigation may increase CO_2 emission significantly. In a bean field, soil respiration and CO_2 emission continued almost until harvest time at the same rates.

TABLE 2.3 Loss of soil nitrogen due to nitrous oxide and nitric oxide emissions from native savannahs, and fields grown with maize, bean, soybean, and cotton in Central Brazil

Nitrogen Emission	Native Cerrados	Maize	Irrigated Bean	Soybean	Cotton
	kg ha^{-1} Crop Season^{-1}				
NO-N	0.1	0.3	0.3	0.2	0.8
N_2O-N	0.01	0.2	0.2	0.1	0.1

Source: Cruvinel et al. (2011)

Pastures developed after clearing the native savannahs is among major land use patterns in the Brazilian Cerrados. Pastures and grass land may extend into 50 ×

10^6 ha^{-1}. Knowledge about nutrient dynamics, especially loss of essential elements through gaseous emissions and its consequence to pastures and environment is no doubt very useful. Pastures developed using grass/legume mixture such as *Brachiaria* and *Stylosanthes* is said to emanate 2–8.3 µmol CO2 m^{-1} s^{-1}. Rainfall events and irrigation, if any may induce higher levels of CO2 evolution. The transition from dry period to wet season was an important factor inducing NO-N and N$_2$O emissions in the pastures. However, rates of N-emissions varied even within small marked plots (Pinto et al., 2006). Bustamante et al. (2009) mention that in some areas, N-oxide emissions progressively decreased as the pastures aged. The NO flux from a 20 year old *Brachiaria* ranged from 0.1 to 0.15 kg N ha^{-1} annually. The N$_2$O emissions were of very low order around 0.5–0.6 kg N ha^{-1} annually. Almost similar level of 0.5 kg N$_2$O-N emission was noticed in a field with *Andropogon sp.*

2.4.3 FIRES IN CERRADOS

Paleo-ecological studies have shown that fire has been endemic occurrence in the Cerrado vegetation. Fires that occurred in Central Cerrados, way back in 32000 B.C. have been discovered (Roitman et al., 2008). Schmidt (2009) suggests that fire that occurs in Cerrado bear large number of similarities to those noticed in savannahs of different continents. Fires often destroy herbaceous vegetation quite rapidly. Woody regions may resist damage for longer period because they have thick, corky, and fluted barks. Fires increase soil temperature in the region and therefore are detrimental to soil organisms and microbial activity. Microbe mediated transformation in the below-ground ecosystem too is affected. Basically, it destroys flora and fauna within the region of its influence.

Fire in open savannahs such as Cerrado affects herbaceous vegetation and causes soil heating. Soil temperatures may range from 74°C to 350°C depending on the intensity of the fire and vegetation (Pivello et al., 2010). However, fire has also been extensively used to clear natural vegetation of Cerrados and in its place establish pastures or cropland. Klink and Machao (2005) point out that fire as an environmental detriment has affected even the so called fire-tolerant Cerrados trees and shrubs. Fire, no doubt is an important factor that affects Cerrados vegetation. It may occur periodically, either through natural or man-made causes. Soil temperature may increase to 32°C–42°C, but temperature may not be affected significantly below 10 cm soil depth. The ambient air temperatures may fluctuate around 85°C. Air temperature above flames may reach a high of 800°C though for short period during active burning (Miranda et al., 1993). It affects regeneration and resurgence of prairie grasses and establishment of replants of woody species. In the process, it induces small to massive loss of C from the Cerrados ecosystem. Carbon emission due to fire is significant. For example, Castro and Kauffman (1998) studied the influence of fire in different types of Cerrados vegetation. They actually tested effect of fire on a gradient of vegetation such as *Cerrados limpo* that is filled with grasses. Here, total above-ground biomass (TAGB) was 5 t ha^{-1}. The TAGB in *Cerrados denso* was 29.4 t ha^{-1}. The savannah vegetation in the *Cerrado limpo* was prone to excessive loss of green matter due to fire. It lost about 84–92 percent if a fire broke out. In the *Cerrados aberto* and *denso* regions, fire engulfed less and destroyed only 54 percent of TAGB. The loss in TAGB, in other words C held in the ecosystem was entire 5 t ha^{-1} in *Cerrados limpo* and ranged upto13.5 t ha^{-1} in *Cerrado aberto*. Castro and Kauffman (1998), opine that as such the TAGB in *Cerrados limpo* and even in *aberto* is relatively small. The TAGB of *Cerrados denso*, which is highest is actually <9 percent of TAGB held by Amazonia in the neighboring regions. The total below-ground biomass (TBGB) also occurs in a gradient across different regions of Cerrados. It ranges from 16.3 t ha^{-1} TBGB in *Cerrados limpo* to 30.4 t ha^{-1} in *Cerrados sujo*, 46.5 t ha^{-1} in *Cerrados aberto* and 52.9 t ha^{-1} in *Cerrados denso* (Castro and Kauffman, 1998). It is believed that fire is major cause of loss of C to atmosphere from Cerrados region. It is based on several measurements of rate at which C is lost to atmosphere due to fire. Along with deforestation trends noticed in Cerrados, which is a very large biome, it could cause large disturbance to global C dynamics and sink/source relations. For example, Klink et al. (1993) reported that 600 thousand km^2 of forests were cleared in 1990s decade at a rate of not less than 20 thousand km^2 year^{-1}.

Sugar cane production in Cerrados involves a series of soil and crop management procedures, many of which may have positive or detrimental results on soil/ ecosystem. Sugarcane residue management, in particular, the tendency to burn the leaves prior to harvest, incorporation of stubbles into soil and removal may have severe effect on soil physicochemical properties. Sugarcane burning may affect soil moisture, structure, greenhouse gas emission pattern, soil carbon stock, and enzyme activity. Fire in sugarcane field firstly raises the soil temperature and incinerates surface flora residing on leaves and canopy. It affects soil aggregation, soil organic matter and its composition. Caio et al. (2012, 2013) state that burning foliage affects labile organic matter, particulate organic matter, and microbial biomass in upper layers of soil. It reduces this fraction since it gets emanated as CO_2. Sugarcane does occupy a significant portion of Cerrados region, and hence burning will affect the ecosystem, at large, especially its soil phase. Microbial transformations may get severely altered due to heat generated by burning the foliage. Further, Caio et al. (2013, 2013) have found that sugarcane cultivation in a long term reduces soil bacterial population due to repeated burning cycles. Population of bacterial species belonging to *Proteobaceria, Acidobacteria, and Acinetobacteria* may get affected adversely. Overall, spread of sugarcane technology has affected the bacterial diversity of Cerrados soil in that zone. We need to understand the consequences, restore microbial biodiversity, if it is possible or change only to a point where alterations in microbial activity do not affect ecosystematic functions.

Soil organic matter is an integral part of Cerrados soil profile that has immense influence on soil biotic component, especially, microbes that mediate many of the important nutrient transformations. In many regions within Cerrados, SOM literally regulates soil structure, aggregate stability, and soil microbial flora and its activity.

Hence, it is imperative that any change of land use or agronomic procedures that impoverish SOM content will automatically affect soil microbes and their activity. Conversion of Cerrados land from natural grass land to pastures or crops, loss of residue due to burning, excessive tillage may all affect SOM and biological activity (Bayer et al., 2006, 2009). Further, reports by Pivello (2010) suggest that fire may affect availability of nutrients to crops. Long-term evaluation in Cerrados suggests that fire affects soil pH, SOC content, and mineral nutrient availability to crops.

2.5   LAND USE CHANGE IN THE CERRADOS

Natural vegetation and land cover that has evolved in the Cerrados has been under threat of destruction. Soil erosion has induced loss of topsoil and made it difficult for natural grasses to establish without fail. Natural fire too affects vegetation cover. Frequent use of fire and chemicals to clear natural grasses to later plant crops has induced changes to ecosystem. The spread of exotic pastures, crops, and other activities related to urbanization are among the major factors causing land use change in the Cerrados (Klink and Machados, 2005). In fact, economic trends indicate that land conversion from natural prairies to agricultural cropping is right now lucrative. Hence, it might induce further development of infrastructure and enterprises related to farming. Large-scale land conversion to farming is no doubt a threat to perpetuation of natural vegetation. Human migration into Cerrados and development of farming practices and other activities that generate livelihood are generally counted as factors that induce land use change.

Zoretto (2008) believes that periodical evaluation of satellite images of ecosystems such as Cerrados can help us trace the dynamics of shrinkage/expansion. Fluctuations in composition of vegetation and its density to a certain extent may help us in devising control measures, if there is deterioration in the Cerrados. Agricultural factors that have induced land use change vary with locations within Cerrados and purpose. Coutinho and Balieiro (2011) identify at least four different agriculture-related factors, each operating at different intensity and length of period. They have listed cultivation of international commodities such as soybean, maize, sunflower, and wheat; forestry, cattle ranching and pastures, and agroenergy as important factors that have caused land use change in the Cerrados. Soybean and pastures have affected land use change in the Southern Cerrados; Forestry has also affected natural vegetation in Cerrados. Biofuel production, mainly through sugarcane has affected Mato Grosso do sul and Goias. Further, they suggest that one of the earliest signs of impact of land use change is loss of remarkable biodiversity due to fragmentation of habitat.

Cerrado agriculture is valuable since it satisfies indigenous demand for food grains plus offers exchequer through large-scale export of soybean, cereals, and forage. The extension of agricultural area is expected to continue despite environmental concerns. In addition, infrastructure needed for agriculture and ancillary industries too will keep pace. Large-scale conversion in land use from Cerrados natural vegetation to farms seems to continue but in a regulated pace, along with several measures that protect the biodiversity and ecosystem in general (Klink and Machado, 2005). The impact of agricultural enterprises have been so dominant that Cerrados is now mostly known for its large soybean, sugarcane, and cereal expanses and less as a store house of native vegetation and naturally evolved biome (Emmet Duffy, 2008).

According to Kaimowitz and Smith (2001), Cerrados was feebly populated. It had large expanse of unattended natural pastures and rugged shrub vegetation. Soybean technology, public roads, rail roads, subsidized farming enterprises, large private corporate farming units, fuel, and highly mechanized crop production programs have converted Cerrados, irreversibly. Highly mechanized and large farms of over 1,000 ha each replaced natural vegetation at a fast pace. Farmers gained in richness due to soybean technology (Taylor, 1998). Initially, poor soils, acidity, Al toxicity and lack of improved soybean genotype that resist disease and yield high were major constraints. However, within a period of 10 years soil amendments and high-yielding soybean genotypes were available and productivity increased to 2.5–3 t grains per hectare. Deforestation rates in 1990s ranged from 26 to 42 thousand ha annually (Morales, 1996). During recent years, Governmental regulations allow farmers to replace natural vegetation with soybean only in area that are classified as wasteland and those with no value as natural repository of plant genetic diversity. Also, areas replaced usually sequester less C compared to other locations (Kaimowitz and Smith, 2001). Areas with value as good repositories of genetic diversity of floral/fauna are not converted.

Xavante (2009) states that despite rapid changes to Cerrados landscape, through introduction of soybean, sugarcane, and large-scale coffee that continues for some time now and various agro-industries, native Cerrados tribes such as Wadera, Xavantes, and others continue to produce crops and vegetables in small farms. Some of them partly hunt for meals and grow native roots, vegetables, and cereals. The Cerrados flora and fauna is still intact in these areas. The native Cerrados tribes continue to learn through folklore and avoid any change to natural vegetation. Land use change is not observable in these preserves. Cerrados prairies continue to thrive on natural precipitation and inherent soil fertility. The native tribes may need to devise procedures of cropping and pasture management to confront the climate changes forecasted to occur in Cerrados as a result of land use change.

Regular surveillance of Cerrados vegetation/cropping zones using satellites is already in vogue. Information on fluctuations of boundaries of reserves, cropland, and ranches are provided by satellite groups such as Landsat, MODIS, and ALOS. They are specially focused to study areas that support soybean, sugarcane, and pastures (Coe, 2012). Computer-based models that analyze carefully the consequences of expansion of agricultural frontiers, especially, with regard to deforestation trends, C emissions, loss of soil-N as NO_2, N_2O, and NH_4 are used to arrive at appropriate decisions. Forecasts about future trends in farming and deforestation are considered while arriving at decisions about expansion/intensification of soybean or sugarcane areas (Coe, 2012). Right now, cattle ranching, it seems is the major cause of deforestation, clearing, and land use change in the Cerrados.

Afforestation is an important aspect of Cerrados, if not right now at least in future, since it restores and sequesters carbon that was lost during field crop production. In case, afforestation takes place on wasteland or deteriorated Cerrados vegetation, it again adds to stability of ecosystem by improving biomass accumulation, turnover of nutrients and providing useful quantities of wood for fuel and paper pulp. In the Brazilian Cerrados, currently, tree species such as Eucalyptus, Pine, and few legume species are being planted regularly to recover deteriorated fields. Demand for wood, cellulose material, and charcoal decides the extent of afforestation (Coutinho and Balieiro, 2011). Factors such as soil water, fertility, especially subsoil characteristics, and management strategies that preserve soil quality need due attention, while achieving afforestation. The tree species planted affects the rate at which biomass is accumulated, the decomposition of tree leaves/twigs, and improvement in SOC and quality. For example, Eucalyptus twigs/leaves are slow to decompose. Therefore, soil-C increases are slow compared to leguminous trees whose succulent tissues if judiciously incorporated, rapidly add to soil fertility. In addition, leguminous trees improve soil-N status. Recent reports suggest that farmers in Cerrados are planting two or three different tree species that include Eucalyptus and legume species. These procedures add to soil-N, soil-C and quality of Cerrados soil (Forrester et al., 2006; Laclau et al., 2008). Afforestation temporarily induces changes in natural Cerrados vegetation or if the replanting zone is an agricultural zone, then it replaces cropping ecosystem. However, sooner or later farmers may bring back savannah grass/shrubs or crops in place of afforested land. It is a cyclical phenomenon in the Cerrados biome. Kluck (2003) opines that despite conversion of natural Cerrado vegetation into intensive cropping and continuing for over 15–20 years under no-tillage stubble mulch, soil properties fluctuated marginally. The organic matter stocks of the Ferralsols stabilized at optimum levels. Yet, a change in land-use such as to pastures or forestry and providing a break in no-till systems were often opted by the farmers.

2.5.1 *CROP PRODUCTION TRENDS IN THE CERRADOS*

The agricultural area and grain harvests per unit land in the Cerrados increased perceptibly during 1980s and 1990s. There is still a vast expanse of natural pastures, shrubs, and wasteland in Cerrados that could be encroached into and crops produced luxuriantly. During mid-1990s, the grain harvest in the Cerrados surpassed the rapidly growing human population (FAOAQUASTAT, 2002). Cerrados supports several different grain crops, cotton seed/fiber, sugarcane, and coffee. Following is list of crops that make up the agricultural prairies within Cerredos, their expanse, total production, and productivity during 2010:

	Area (M ha)	Production (m t)	Productivity (ha⁻¹)
Soybean	10.6	30.94	2.92
Beans	0.76	1.25	1.63
Maize	4.35	21.51	4.94
Rice	0.76	1.61	2.13
Sugar cane	8.43	662	8.27
Cotton	0.74	2.74	3.67
Coffee	0.40	0.55	1.40

Source: IBGE (2009); Franca et al. (2010); USDA Foreign Agricultural Service (2008); and Nasser et al. (2012)

Note: Soybean expanse, its productivity and total grain production has increased markedly since late 1970s.

At present, Brazil is the second largest producer of soybean and its products. It trails USA in terms of area and productivity of soybean. Initial surges for soybean in the Cerrados was created during 1960s and 1970s. By late 1960, increase in domestic consumption of soybean enthused the Brazilian government, to improve soybean production. International demand for soybean added to reason for expansion of soybean in Cerrados. Next, soybean and its products are an important source of protein and nourishment to low and middle income populations. Hence, government policies were aimed to induce soybean farmers to enhance area/productivity of soybean. Soybean production helped large processing units to thrive on the farm produce. Brazilian farm enterprises and small farmers were largely localized in south, east, and northeast. Therefore, to integrate farm population, Brazilian government had policies in place to encourage soybean production in the Southwest of Cerrados. Self-sufficiency in oilseeds was among the major reasons for massive expansion production in the interior Cerrados (Schnepf et al., 2001). Simultaneously, during 1950 till 1970 there was trend to close down coffee gardens in the southern Cerrados. Coffee eradication peaked in the Sao Paulo region due to freeze and high taxes. Soybean became the best crop to replace coffee.

International events too became reasons to expand soybean farming zones in the interior Cerrados. A surge in demand for soybeans in Europe and Fareast, failure of proteineceous feed material in Peru; embargo on export of soybean, oilseeds, and vegetable oil by USA were other reasons for a booming soybean cropping system in Cerrados as well as northeast Pampas (Schnepf et al., 2001). Most importantly, international price for soybean grains was much higher than in domestic Cerrados region. Hence, export oriented soybean farms became common.

Soybean cultivation became remunerative with the introduction of high-yielding tropical varieties. Grain productivity steadily increased from 1.7 t ha⁻¹ in 1990 to 2.4 t ha⁻¹ in 2000 and then on to 3.1 t ha⁻¹ in 2010. Goias state in Cerrados, which is the main soybean and maize producing zone in Cerrados too experienced steady increase of grain yield from 1.3 t ha^{-1} in 1990 to 2.5 t ha^{-1} in 2000 and then to 3.2 t ha^{-1} in 2010 (FAO, 2008). The total soybean grain output increased dramatically between 1960 and 1980. Cerrados produces most of soybeans. Following is the soybean production trend in entire Brazil (not Cerrados):

Year	1960	1970	1970	1980	1990	2000	2005	2010
Grain Production (million t year^{-1})	0.20	1.50	15.15	24.07	32.82	49.54	51.2	58.2

Source: Wikipedia (2012)

During 1970–1990 soybean grain harvest increased by 10 percent each year. It was attributable to 8.3 percent yearly expansion of soybean area in the Cerrados. Soybean belt in the Cerrados increased from 2.0 million ha in 1970 to 18 million ha in 2002. We should note that expansion of soybean in traditional areas of Rio Grande du Sul, Santa Catarina and Sao Paulo increased from 1.5 to 7.3 million ha during the period from 1969 to 1975. Later, from 1975 till date, it has stagnated at the same level of 7–8 million ha. However, much of rapid expansion in response to international market demand occurred mainly in Mato Grosso, Mat Grosso du sul, Minnas Garias, and Maranhao. Further analyses on soybeans suggest that, continued expansion of area in the Cerrados, aided by good soil amendment programs was the main reason for high soybean output. Britto (2011) points out that huge increase in total soybean production was predominantly attributable enhanced grain yield per hactare^{-1}. The soybean production was over 24 million t in the Cerrados during 2009–2010, while it stayed at 9–14 million t in the traditional zones during the same period. The productivity of soybean aided by high fertilizer input, especially P, K, and Gypsum reached 3–3.5 t grains per hectare in the Cerrados but remained at 2.4 t grains per hectare in the Traditional areas (Schnepf et al., 2001). In fact, since 1970s, the Cerrados *per se* has been amended with fertilizer-based nutrients, gypsum, and FYM just to reach high grain/forage yield targets. It is a massive disturbance of soil conditions. Currently, Cerrados contributes 58 percent of total soybean grains generated in Brazil (Emmet Duffy, 2008). Cerrados, with its vast expanse of soybean adds large quantities of atmospheric-N to soil via symbiosis with Bradyrhizobium. It is indeed massive supply of N to soil at a rate 20–40 kg N ha^{-1}. This natural advantage derived by farmers often goes unnoticed. It has a value in terms of soil wealth of the Cerrados region. It saves on fertilizer-N proportionately.

Regarding infrastructure that support soybean production in the Cerrados, it is said, roads, rail roads, storage structures, and export terminals were all propped up during late 1970s. The three major Atlantic seaports, namely Paranagua, Rio Grande, and Santos export 80 percent of soybean produced in the Cerrados and traditional soybean belt of Brazil.

Production efficiency and costs incurred in generating each ton of soybean grain in an area is a crucial factor in many ways. During 2000, cost of producing a bushel of soybean grains was 3.89 US$ that in Argentinian Pampas is 3.92 US$ per bushel. This is 23 percent lower than cost incurred by farmers in Midwestern USA, which is 5.11 US$ per bushel. Soybean produced in the Cerrados has to be economically competitive in the international market. The area planted to soybean, inputs, and productivity goals are all highly dependent on demand, price line in the market and cost effectiveness. Literally, economic gains dictate the soybean expanse whether it is Cerrados or Pampas. This is a case, wherein economic considerations show stringent influence on prairie vegetation and biomass/grain productivity. In a nut shell, profits, and market competitiveness affects agroecosystem functions.

Maize cultivation zone has experienced a marginal reduction from 13.5 million ha in 1970–12.5 million ha in 2000 and then on to 11.2 million ha in 2010. Maize production in Brazil is predominant in the South and Southeast that possess about half the maize area at 5.2 million ha. Central West (2.51 m ha) and Northeast (3.52 m ha) regions, each contribute about 25 percent of maize area. Total maize production too has shown a marginal fluctuation between 30 and 32 m t year^{-1}. The productivity of maize in the Cerrados has however increased from 2.2 t grain per hectare to 2.8 t ha^{-1}. It seems maize production has depreciated in the Northeast mainly due to subsistence agricultural practices. In general, current trend to produce maize with relatively low fertilizer supply seems to curtail total production in the Cerrados. However, there has been a marginal increase in grain out-put due to cultivation of a second crop of maize crop, during post rainy period. This crop seems to add about 0.5–1 t grain per hectare. Cerrados exports about 0.5–0.6 m t maize grains annually (Schnepf et al., 2001). Brazil exports a sizeable portion of its maize produce to nations in Middle-East Asia. During 2010, it seems Brazilian maize export to Middle-East was worth over US$ 2.3 billion (CropSite, 2010).

Wheat is among the most important cereal diets of Brazilian population. It is consumed in all regions and by all tribes and modern settlers. During first half of 1900s, Brazil remained an endemic importer of wheat grains from USA or Europe. During 1970s and 1980s, Brazilian government initiated several programs to support expansion of wheat region in southern Cerrados. High-yielding cultivars and fertilizers were also utilized to prop up wheat production. The wheat belt within Brazil peaked at 3.8 m ha in area and productivity of 5.4 m t ha^{-1} during 1986–1990. Then, it experienced marked decline in area due to drastic shifts by Cerrados farmers to soybean. During 2001–2010, average area declined to 1.4 m ha and total production remained very low at 2.8 m t annually (Schnepf et al., 2001). In Brazil, wheat is consumed predominantly in the form of bread loaf (55%), as pasta (15%) and pastries (15%). A sizeable amount of low quality grains is utilized as animal feed. Brazil imports wheat from other regions of the world. Wheat has been actually expelled from the ecosystem in preference to soybean that is remunerative. Soybean also adds to soil-N fertility through BNF. Perhaps it is a good decision transitorily, in light of economic realities. Otherwise, in any individual region, it is prudent to maintain a naturally required ratio of cereals to legumes, so that indigenous population is self-sufficient. In most areas, cereals are needed at least in quantities more than legumes, perhaps by 3 to 5-folds. Brazil, imported 7.1 m t wheat grains during 2012, mainly to augment the net demand of 11.3 m t wheat grains.

Wheat production in Brazil has increased steadily from 0.71 million t annually to 2.7 million t in 1980 and then to 4.65 million t in 2005. It is interesting to note that Brazil produced about 5.8 m t in 2010/2011, but it suffices to answer just half the demand for wheat. Brazil imports the rest to satisfy local demand of 10.4 m t. Cerrados is not the major wheat producing region. It contributes only about 5–10 percent of wheat grains. About 85 percent of wheat production in Brazil occurs in the southern states of Parana and Santa Catarina. Argentina produces about 90 percent of wheat imported by Brazil. Hence, there is a lot of interest to expand wheat area and productivity in Cerrados, wherever possible (Cortes, 2011). In fact, Brazil intends to double wheat output by increasing area as well as improving on productivity of wheat. Agronomic procedures and wheat genotypes that economize on production costs seems important. Profitability of no-tillage wheat that is supplied with relatively less amount of nitrogen is being examined in many locations. Reduction in inorganic supply by using a green manure crop in the sequence is also being evaluated within Cerrados (Gitti, 2012).

Sugarcane production is localized in the South-Central and North-eastern region of Brazil. Sao Paulo is the most important sugarcane farming state. It supports 56 percent of sugarcane area of Brazil. Sugarcane area also extends into Minas Garias, Parana, Mato Grosso, Mato Grosso du Sul, and Goaias (Nasser et al., 2012). At present, sugarcane area in entire Brazil is about 8.7 m ha. Sugarcane culture for ethanol and biofuel production gained in importance during past two decades. For example, in a single state, since 2000 A.D., its cultivation zones in the Mato Grasso du Sul of Cerrados, increased from 100 thousand ha in 2005 to 220 thousand ha in 2008. Sugar cane zone spread rapidly between 2008 and 2012 and increased from 220 thousand ha to 500 thousand ha in 2011 (Coutinho and Balieira, 2011). The marked improvement in sugarcane area in the Cerrados biome has been attributed mainly to global demand for biofuel and to a less extent on sugar. The sugarcane-based wet prairies expanded by over 3 million ha within a decade from 2000 to 2010. Currently, it has reached over 8.4 million ha. Interestingly, about 95 percent of expansion in sugarcane has occurred in Cerrados (Nasser et al., 2012). Sugarcane is predominantly derived from Cerrados and its production is more localized to within 3–4 states in Cerrados. Following is the Sugarcane production trend in entire Brazil (not Cerrados):

Year	1960	1970	1970	1980	1990	2000	2005	2007	2012
Cane Production (million t year^{-1})	57	80	149	263	326	415	463	558	622

Source: Wikipedia (2012)

Note: Cerrados produces 11 percent of total coffee seeds produce of Brazil.

Brazil is among the top coffee producers and exporters of the world. During 2010, it generated about 49 million bags of coffee seeds that equates to 40 percent of global coffee produce. Coffee seed productivity in Brazil has increased steadily. During last decade, coffee production increased by 9 percent. Coffee is produced in the southern region of Cerrados. Coffee seed produced in Cerrados is almost exclusively Arabica type. Cerrados contributes about 11 percent of total coffee produced in Brazil (Neves et al., 2011).

Rice production has been an important farm enterprise in the Brazilian Cerrados. It most often follows as first crop after clearing natural grass land and shrubs. Rice is grown for two to three seasons and it is followed by soybean or a dryland crop. Hence, expansion of rice culture in Cerrados is mostly linked to rate at which natural prairie locations are cleared. Rice cultivation zone in Cerrados has been experiencing a decline since 1970s (Schnepf et al., 2001). It peaked at 6.9 m ha in 1979/80 and currently, it has declined to 3.3 m ha (Schnepf et al., 2001) suggest that rice is likely to be a stand by crop after soybean and maize in Cerrados.

During the past three decades, beginning in 1980, overall Brazilian balance of trade was positive for both agricultural (8.3 billion US$) and nonagricultural (4.8 billion US$) commodities. The balance of trade fluctuated between 6 and 9 billion US$ for the period from 1986 to 1998, later it experienced rapid increase to 17 billion US$ owing to soybean and sugarcane exports. However, balance of trade for nonagricultural commodities took an abrupt dip to—16 billion US$ (Baker and Small, 2005).

KEYWORDS

- **Cerrados of Brazil**
- **Cropland**
- **Ecosystem**
- **Gypsum**
- **Micronutrient**
- **Monocrop**
- **No-tillage**
- **Water resource**

REFERENCES

1. Affholder, F.; Rodrigues, G. C.; and Assad, E. D.; Agroclimatic model for evaluation of maize behaviour in the Cerrado region. *Pesquisa Agropecuaria Braziliera.* **1997**, *32*, 993–1002.
2. AgBrazil, Biofuel Investment options. **2006a**, 1–2, June 22, 2013, http://www.agbrazil.com/invest_biofuels.htm

3. AgBrazil, Opening Cerrado Land. **2006b,** 1–3, June 10, 2013, http://www.agbrazil.com/opening_virgin_cerrado_land.htm

4. Angelini, G. A. R.; Balieiro, F. C.; Saggin, O. J.; Zannat, J. A.; Coutinho, H. L. C.; Salton, J. C.; and Franco, A. A.; Impacto da retirada da palhad de areas com Cana-de-Acuar sobre os Fungos Micorrizcos Arbsculares em solo de Cerrados, em Dourados-MS. In: XXIX Reuniao Brasileira de Fertilidade do solo e Nutrica de Planatas. **2010,** 1–12.

5. Assis, A. C. C.; Coelho, R. M.; Pinheiro, E. D. S.; and Durigan, G.; Water availability determines physiognomic gradient in an area of low-fertility soils under Cerrado vegetation. *Plant Ecol.* **2012,** *212,* 1135–1147.

6. Baker, M. M.; and Small, D.; Brazil's agricultural "success" in the Cerrados is a disaster. *Execut. Intell. Rev.* **2005,** *32,* 20–25.

7. Balde, A. B.; Scopel, E.; Affholder, E.; Corbeels, M.; Da Silva, F. A. M.; Xavier, J. H. V.; and Wery, J.; Agronomic performance of No-tillage relay inter cropping with maize under small holder conditions in Central Brazil. *Field Crops Res.* **2011,** *124,* 240–251.

8. Balieirao, F. C.; Pereira, M. G.; Alves, B. J. R.; Resende, A. S.; and Franco, A. A.; Soil carbon and Nitrogen in pasture soil reforested with Eucalyptus and Guachapele. *Revista Brasileirade Cienca do Solo.* **2008,** *32,* 1253–1260.

9. Barbosa, A. S.; Andarhilos da claridade: os primeiros habitants do cerrado. Goiania: Univsidade de Goias Instituto do Tropico Umido. **2002,** 1–45.

10. Barbosa, A. S.; A ocupacao humana no Cerrado. In: O universe do Cerrado. Goianaia. Gomes, H.; ed. Editora Universidade Catolica de Goias. **2008,** *1,* 23–87.

11. Barbosa, A. S.; and Schimiz, P. I.; Ocupacao indigena do cerrado:esboco de uma historia. In: Cerrado: ambiente e flora. Sano, S. M.; and Almeida, S. P.; eds. EMBRAPA-CPAC, Planaltina, Brazil. **1998,** 89.

12. Batista, K.; and Monteiro, F.; Nitrogen and sulphur in Marudundu grass: relationship between supply and concentration in leaf tissues. *Scientia Agricola.* **2007,** *64,* 44–51.

13. Bayer, C.; Maratin-Neto, L.; Mielnizcuk, J.; Pavinato, A.; and Diekow, J.; Carbon sequestration in two Brazilian Cerrado soils under No-till. *Soil Tillage Res.* **2006,** *86,* 237–245.

14. Bayer, L. B.; Batjes, N. H.; and Bindraban, P. S.; Soil organic carbon dynamics in soybean-based cropping systems in the Brazilian Cerrados. **2009,** 1–2, June 20, 2013, http://www.isric.org/isric/webdocs/docs/SOC_dynamic_%20in_soybean-based_cropping_ systems_in_the_ Brazilian_Cerrados_(Nov2009).pdf

15. Bernoux, M.; et al. Cropping systems, carbon sequestration and erosion in Brazil, a review. *Agron. Sustain. Dev.* **2006,** *26,* 1–8.

16. Benites, V. D. M.; Sustainable soil fertility management of crop systems in the Brazilian Cerrado. Proceedings of a Workshop "Better Soils for Better Life: Protecting Our Future through Soil Conservation". Jacobs University Brehman and Julius Kuhn Institute; **2010,** 1–3, June 10, 2013, http://www.jacobs-univefrsity.de/SES/better_soils_better_life.htm

17. Britto, M. C.; Production and exportation of Brazilian soybean and the Cerrado-2001–2010 **2011,** 1–27, May 27, 2013, http://www. dznehcbyl9q204.cloudfront.net/downloads/wwf-soy-cerrado-english.pdf

18. Buckridge, M. S.; De souza, A. P.; Arundale, R. A.; Anderson-Teixeira, K. J.; and Delucia, E.; Ethanol from sugarcane in Brazil: A midwaystategy for increasing ethanol production while maximizing environmental benefits. *GCB Bioenergy.* **2012,** *4,* 119–126.

19. Buurman, P.; Roscoe, R.; and Alcantara, F. E.; Carbon sequestration research in Brazilian savannah systems: Problems and results. *Edafologia.* **2004,** *11,* 157–170.

20. Bustamante, M. M. C.; Keller, M.; and Silva, D. A.; Sources and Sinks of trace gases in Amazonia and the Cerrado. *Amazonia Global Change: Geophys. Monograph Ser.* **2009,** *186,* 337–354.

21. Caio, T. C.; et al. Effect of Sugarcane burning or Green harvest methods on the Brazilian Cerrado soil Bacterial community structure. **2013**, 1–7, May 25, 2013, http://www.plosone.org article/info%3Aclci%2F10.1371%2Fjournal.pone.0059342.htm

22. Caio, T. C.; et al. Physical, Chemical and Microbiological changes in Cerrados soil under differing sugarcane harvest management systems. *BMC Microbiol.* **2012**, *12*, 170–178.

23. Cargnin, A.; Alves De Souzo; and Fronza, V.; Progress in Breeding of irrigated wheat for the Cerrado region of Brazil. *Crop Breed. Appl. Biotechnol.* **2008**, *8*, 39–46.

24. Carvalho, M. C. S.; Ferreira, G. B.; and Staut, L. A.; Nutricao calagem e adubacao do algodoiera. Abrapa, Brasilia, Brazil; **2007**, 581–648.

25. Castillo, R.; Vencovsky, V. P.; Brags, V.; Logistics and intensive agriculture in the Cerrados: The new Brazilian rail road system. *Terrae.* **2011**, *8*, 19–25.

26. Castro, E. A. D.; and Kauffman, J. B.; Ecosystem structure in the Brazilian Cerrado: A vegetation gradient of above ground biomass. Root mass and consumption fire. *J. Tropical Ecol.* **1998**, *14*, 263–283.

27. Cerri, C. C.; Maia, S. M. F.; Galdos, M. V.; Cerri, E. P. G.; Feigi, B. G.; and Bernoux, M.; Brazilian greenhouse gas emissions: The importance of agriculture and livestock. *Sci. Agric.* **2009**, *66*, 831–843.

28. Cerri, C. E. P.; Sparovek, G. M.; Bernoux, W. E.; Easterling, W. E.; Melillo, J. M.; and Cerri, C. C.; Tropical agriculture and Global warming. *Sci. Agric.* **2007**, *64*, 83–99.

29. CIRAD. No-tillage with cover crops for the Brazilian Cerrados. CIRAD Agricultural Research for Development. **2009a**, 1–2, April 23, 2013, http://cirad.fr/enresearch-operations/research-results/2009/no-tillage-with-cover-crops-for-the-brazilian-cerrados. htm

30. CIRAD. Building sustainable production systems in partnership in Brazil. Annual Report-2007 of CIRAD. CIRAD Agricultural Research and Development. Brazil; **2009b**, 1–2, May 25, 2013, http://www.cirad.fr/en/publications-resources/sciences-for-all/rapports-annuels/cirad-2007/inventing/building-sustainable-production-systems-in-partnership-in

31. Coe, M. T.; Ecosystem Studies and Management: Linking Land Use Change to Economic Drivers and Biophysical Limitations of Agricultural Expansion in the Brazilian Cerrado. Massachusetts, USA: Woods Hole Research Centre; **2012**, 1–2, May 25, 2012, http://www.whrc.org/ecosystem/amazon/cerrado_expansion.html

32. Corbeels, M.; Scopel, E.; Cardoso, A.; Bernoux, M.; Douzet, J. M.; and Neto, M. S.; Soil carbon storage potential of direct seeding mulch-based cropping systems in the Cerrados of Brazil. *Global Change Biol.* **2006**, *12*, 1773–1787.

33. Corbeels, M.; Scopel, E.; Cardoso, A.; Douzet, J. M.; Neto, M. S.; and Bernoux, M.; Soil Carbon Sequestration and Mulch-based Cropping in the Cerrados Region of Brazil. **2004**, 1–4, http://www.cropscience.org.au /icsc2004/ poster/2/2/ 1179_ corbeelsm.htm

34. Corporate Europe Observatory. Sugar ethanol not so sweet. **2009**, 1–5, May 25, 2013, http://wwwcorporateeurope.org/news/sugar-cane-ethanol-not-so-sweet

35. Cortes, K.; Brazil to encourage wheat output in Cerrado region. *Embrapa Says.* **2011**, 1–3, June 18, 2013, http://www.bloomberg.com/news/2011-03-18brazil-to-encourage-wheat-output-in-cerrado-region-embrapa-says.html

36. Coutinho, H. L. C.; and Balieiro, F.; Land use change, biofuels and impacts on soil Carbon dynamics in the La Plata Basin. EMBRAPA Solos, Rio de Janeiro, Brazil. Final Report of IDRC Grant No 104783-001. **2011**, 25.

37. Crop Site. Brazil Maize Exports to ME Grow. **2010**, 1–2, May 25, 2013, http://www.thecropsite.com/news/7128/brazil-maize-exports-to-me-grow

38. Cruvinel, E. B.; Bustamente, M. M.; Koovitz, A. R.; and Zepp, R. G.; Soil emissions of NO, NO_2 and CO_2 from croplands in the Savannah region of Central Brazil. *Agric. Ecosyst. Environ.* **2011**, *144*, 29–40.

39. Da Silva, J. M. C.; and Bates, J. M.; Biogeographic patterns and conservation in the South American Cerrado: A tropical Savannah hot spot. *BioScience*. **2002**, *52*, 225–233.

40. Derra, S.; As Brazil ramps up sugarcane production researchers forsee regional climate effects. **2013**, 1–2, May 25, 2013, http://www. eurekalert.org/pub_releases/2013-03/asu-abr030713. php

41. Emmet Duffy, J.; Biological diversity in the Cerrado. *Conserv. Int.* **2008**, 1–12, June 5, 2008, http://www.eoearth.org/article/ Biological_diversity_in_the-Cerrado#gen10.htm

42. Etinger, J.; Brails GMO Crop Growth Outpacing Conventional. **2011**, 1–2, May 25 2013, http://www.organicauthority.com

43. FAO. Global ecological zones: South America. Food and Agricultural Organization of the United Nations. Rome, Italy; **2000**, 1–15, April 15, 2013, http://www.fao.org/docrep/006/ad652e/ad652e18.htm

44. FAO. Conservation agriculture. Food and Agriculture Organization of the United Nations. Rome, Italy; 2003, 1–19, May 25, 2013, http://www.fao.org/docrep/006/y4690E/y4690e0a. htm

45. FAO. The Cerrados. Food and Agricultural Organization of the United Nations. Rome, Italy; **2008**, 1–6, May 23, 2013, http://www.fao.org/ docrep/004/Y2638E/y2638e08.htm

46. FAO. Water profile of Brazil. Food and Agricultural Organization of the United Nations. Rome, Italy; 2012, 1–16, May 24, 2013, http://www. eoearth.org/article/Water_profile_of_ Brazil

47. FAOAQUASTAT. Regional Report—Water Report: Brazil. **2002**, 1–11 May 16, 2013, http://www.fao.org/nr/water/aquastat/countries_regions/ BRA/.htm

48. Feltran, R.; The Other Side of the Agricultural Frontier: A Brief History on Origin and Decline of Indian Agriculture in Cerrado. **2011**, 1–16, May 25, 2013, http://www.socialsciences.scielo. org/pdf/s_assoc/v5rise/sasoc_aO2,pdf

49. Ferreira, C. M.; Villar, P. M.; and Wander, A. E.; Socio-economic contrasts of colonization and utilization of the Brazilian Cerrado. Proceedings of Conference on Prosperity and Poverty in a Globalized World-Challenges for Agricultural Research. Germany: University of Bonn; **2006**, 1–4.

50. Filhos, M. C. M. T.; Buzetti, S.; Andreotti, M.; Arf, O.; and De Sa, M. E.; Application times, sources and doses of nitrogen on wheat cultivars under no-till in the Cerrados region. *Cienca Rural*. **2011**, *41*, 1–8.

51. Forrester, D.; Bauhas, J.; Cowie, A. L.; and Vanclay, J. K.; Mixed species plantations of Eucalyptus and nitrogen fixing trees: A review. *Forest Ecol. Manag.* 2006, 233, 211–230.

52. Franca, J. G.; Peres, J. R.; Anjos, J. R.; The Brazilian Cerrados and its implications for investment to produce surplus food. 2010, 1–37, June 15, 2013, www.rsis.edu.sg/nts/events/docs/ ICAFS-PPT-Jose_Geraldo.pdf

53. Francisco, F. A. H.; How important are phosphorus and potassium for soybean production in the Cerrados of Brazil. *Better Crops*. **2013**, *97*, 25–27.

54. Francisco, E.; and Hoogerheide, H.; Nutrient management for high yield cotton in Brazil. *Better Crops*. **2013**, *97*, 15–17.

55. Freitas, D. A. F.; Silva, M. L. N.; Curi, N.; Silva, M. A.; Oliviera, A. F.; and Silva, S. H. G.; Physical indicators of Soil quality in Oxisols under Brazilian Cerrados. **2013**, 88–110, June 10, 2013, http://dx.doi.org/10.5772/54440

56. Friends of Earth-Europe. Sugarcane. **2012**, 1–8, May 25, 2013, http://www.foe.co.uk/resource/briefings/sugar_cane_and_land_use_ch.pdf

57. Furley, P. A.; The influence of slope on the nature and distribution of soils and plant communities in the Central Brazilian Cerrados. *Adv. Hill Slope Processes*. 1996, *1*, 327–345.

58. Galdos, M. V.; Cerri, C. C.; Cerri, C. V. P.; Soil Carbon stocks under burned and unburned sugarcane in Brazil. *Geoderma*. **2009**, *153*, 347–352.

59. Gale, M.; Cerrado emerges as producer of high quality coffee. **2013**, 1–4, May 24, 2013, http://www.thefreelibrary.com/ Cerrado+emerges+ as+producer+of+high+quality+coffee-a021070457

60. GBI. Brazil sets record in adoption of GM crops. **2011**, 1–2, June 10, 2013, http://www.the-bioenergysite.com/news/ 9313/brazil-sets-record-in-adoption-of-gm-crops.htm

61. Gitti, D. C.; Arf, O.; Melero, M.; Rodrigues, R. A. F.; Tarsitano, M. A. A.; Influence of nitrogen and green manure on the economic feasibility on no-tillage wheat in the Cerrado. *Rev. Ceres.* **2012**, *59*, 1–7.

62. Gomes de Santos, J. D.; Rein, T. A.; Martinhao, D.; Sousa, G.; Muraoka, T.; and Blair, G. J.; Evaluation of sulphur enhanced fertilizers in a soybean-wheat rotation grown in a Brazilian Cerrados Oxisol. Proceedings of World Congress of Soil Science Solutions for a Changing World. Brisbane, Australia; **2010**, 90–93.

63. Goodland. A physiognomic analysis of the Cerrados vegetation of Central Brazil. *J. Ecol.* **1971**, *59*, 411–419.

64. GMO Compass. Brazil utilizes more GM crops than ever before. **2013a**, 1–2, May 25, 2013, http://www.gmo-compass.org/eng/news/ 550.brazil_ utilize_more_gm_crops_ever_before. html

65. GMO Compass. Brazil GMO planting to increase by 14 percent this season. Reuters-UK Edition. **2013b**, 1–3, May 25, 2013, http://www.ukreuters.com/article/2012/12/17/brazil-biotech-idUKL1E8NH1LW20121217.htm

66. GMO Compass. Wheat. **2013c**, 1–2, May 25, 2013, http://www.gmo-compass.org/eng/grocery_shopping/crops/22genetically_modfied_ wheat. html

67. GMO Compass. Sugarcane. **2013d**, 1–2, May 25, 2013, http://www.gmo-compass.org/eng/database/plants/76.sugar_cane.html

68. GMO Compass. Brazil: GM Sugarcane with a 25 percent yield increase expected in 10 years. **2013e**, 1–2, May 25, 2013, http://www.gmo-compass.org/eng/news/456.brazil_gm_sugar_cane_yield_increase_expected.html

69. Guerra, A. F.; Da Silva, D. B.; and Rodrigues, G. C.; Irrigation timing and nitrogen fertilization for common bean crop in the Cerrados region. Pesquisa Agropecuaria Brasiliera. **2000**, *6*, 1229–1236.

70. Handley, T.; Sugarcane production impacting local climate in Brazil. Environmental News. **2013**, 1–4, June 22, 2013, http://www. news.mongabay.com/2013/30501-handley-sugarcane-brazil.html

71. Heam, K.; Ethanol production could be an economic disaster, Brazil's critics say. National Geographic News. **2007**, 1–2, May 24, 2013, http://www.newsnationalgeographicco.in/news/2007/02/070208-ethanol.html

72. Herrera-Estrella, L.; Transgenic Plants for Tropical Regions: Some Considerations about their Development and their Transfer to the Small Farmer Proceedings of National Academy of Sciences. **1997**, *96*, 5978–5981.

73. Hermele, K.; Regulating Sugarcane Cultivation in Brazil. Department of Ecology. University of Lund, Sweden. PhD Thesis. **2011**, 166.

74. IBGE. Crop statistics in Brazil and Cerrados. Instituto Brasiliera de Geograpfiiae Estatistica, Brasilia. **2010**, June 15, 2013, ibge.gov.br/

75. Jantalia, C. P.; Resck, D. V. S.; Alves, B. J. R.; Zotarelli, L.; Urquiga, S.; and Boddey, M.; Tillage effect on C stocks of a clayey Oxisol under a soybean-based crop rotation in the Brazilian Cerrado region. Soil and Tillage Research. Amsterdam; **2007**, *95*, 97–109.

76. Kaimowitz, D.; and Smith, J.; Soybean technology and the loss of natural vegetation in Brazil and Bolivia. In: Agricultural Technologies and Tropical Deforestation. Angelson, A.; and Kaimowitz, D.; eds. United Kingdom: CAB International; **2001**, 195–211.

77. Kaster, M.; and Bonato, E. R.; Contribucao das Scienca Agrarias para o desenvol-vimento: A pesquisa en soja. Revista de Economia Rural (Brasilia). **1980**, *18*, 415–434.

78. Klink, C. A.; Moreira, A. G.; and Solbrig, O. T.; Ecological impacts of agricultural development in the Brazilian Cerrados. In: Young, M. D.; and Solbrig, O. T.; eds. The World's Savannahs—Economic Driving Forces, Ecological Constraints, and Policy Options for Sustainable Land Use. Man and Biosphere Series. , Paris: UNESCO; **1993**, *12*, 259–282.

79. Klink, C. A.; and Machado, R. B.; Conservation of Brazilian Cerrados. *Conserv. Biol.* **2005**, *19*, 707–713.

80. Kluck, J.; Managing Acid Soils. Geography Institute, Guttenberg University. **2003**, 1–4, May 25, 2013, http://www.stadd.uni.mainz.de/ wilckew/CerradoEnglisch.htm

81. Laclau, J. P.; et al. Mixed species plantations of *Acacia mangium* and *Eucalyptus grandis* in Brazil. *Forest Ecol. Manag.* **2008**, *255*, 3905–3917.

82. Ledru, M. P.; Late Quarternary history and evolution of the Cerrados as revealed by palynological evidences. In: The Cerrados of Brazil. Columbia University Press; **2002**, 398.

83. Leggett, M.; Shifting from pasture to sugarcane cools Brazil. *Earth Times.* **2011**, 1–3, June 22, 2013, http://www.earthtimes.org/climates/shifting-pasture-sugarcane-cools-brazilian-cerrado/731

84. Lehman, J. M.; Lilienfien, J.; Rebel, K.; Lima, S. C.; and Wilcke, W.; Subsoil retention of organic and inorganic nitrogen in a Brazilian savannah oxisol. *Soil Use Manag.* **2004**, *20*, 163–172.

85. Lilienfein, J., Wilcke, W., Vilela, L., Ayarza, M.A., Lima, S.C. and Zech, W. 2013 Soil fertility under native Cerrados and pasture in the Brazilian Savannah. *Soil Sci. Soc. Am. J.* **2013**, *67*, 1195–120.

86. Lopes, A. S.; Soils under Cerrado: A success story in soil management. *Better Crops Int.* **1996**, *10*, 9–15.

87. Lopes, D. C.; and Neto, A. J. S.; Potential crops for biodiesel production in Brazil: A review. *World J. Agric. Sci.* **2011**, *7*, 206–217.

88. Lowenberg-Deboer, J.; and Griffin, T. W.; Potential for precision agriculture in Brazil. Site Specific Management Centre Newsletter. West Lafayette, Indiana, USA: Purdue University; **2006**, 1–3.

89. Maltas, A.; Corbeels, M.; Scopel, E.; Oliver, R.; Douzer, J.; Da Silva, F. A. M.; and Wary, J.; Long term effects of continuous direct seeding mulch-based cropping systems on soil nitrogen supply in the Cerrados region of Brazil. *Plant and Soil.* **2007**, *298*, 161–173.

90. Manzione, R. L.; Knotters, M.; and Heuvelink, G. B. M.; Mapping trends in water table depths in a Brazilian Cerrados. In: Proceedings of the 7th International Symposium on Spatial Accuracy Assessment in natural Resources and Environmental Sciences. Lisbao, Portugal: Instituto Geografico Portugues; **2006**, 1–5, May 24, 2013, http://www.spatial-accuracy.org/ Manzione-2006accuracy.htm

91. Marchao, R. L.; Becquer, T.; Brunet, D.; Balbino, L. C.; Vilela, L.; and Brossard, M.; Carbon and nitrogen stocks in a Brazilian clayey oxisol: 13-years effects of integrated crop-livestock management systems. *Soil Tillage Res.* **2009**, *103*, 442–450.

92. Martinhao, D.; Sousa, G.; and Rein, T. A.; Soil fertility evaluation and control for annual crops in the Cerrados. *Better Crops.* **2011**, *95*, 12–15.

93. Mendes, I. C.; Hungria, M.; and Vargas, M. A. T.; Soybean response to starter nitrogen and Bradyrhizobium inoculation on a cerrado oxisol under no-tillage and conventional systems. *Review Brasiliera Cienca Solo.* **2003**, *27*, 81–87.

94. Mendonca, M. L.; Sugarcane plantations destroy Cerrados in Brazil. **2008**, 1–3, June 22, 2013, http://www.social.org.br/indexphp? option=com_ content&view=article&id=126:sugarcane-plata

95. Mendonca, M. L.; Monocropping for agrofuels: The case of Brazil. *Development.* **2011**, *54*, 98–103.

96. Miranda, L. N.; Azevedo, J. A.; Miranda, J. C.; and Gomes, A. C.; Productivity of common bean in relation to phosphate fertilizer and irrigation in a Cerrados soil. *Pesquisa Agropecuaria Brasiliera.* **2000,** *35,* 703–710.

97. Miranda, A. C.; Miranda, H. S.; Oliviera Dias, I. F.; and D'Souza, B. F.; Soil and air temperature during prescribed during fires in the Central Brail. *J. Tropical Ecol.* **1993,** *9,* 313–320.

98. Moreas S. A. J. C.; Seguy, L.; Goze, E.; Bouzinac, S.; Husson, O.; Boulike, S.; Tiver, F.; Forest, F.; and Santos, J. B.; Carbon sequestration rates in no-tillage soils under intensive cropping systems in tropical agroecozones. **2008,** 1–13, June 19, 2013, http:// www.fao.org/ag/ca/ Carbon%20Offset%20Consultation/CARBONMEETING/3FULLPAPERSBYCONSULTAT IONSPEAKERS/PAPERSA.pdf

99. Morales, I.; Monitoreo del Basque en el departments de Santa Cruz. period 1988–1992. Plan de Uso del Suelo. *Santa Cruz.* **1996,** 1–24.

100. Munoz, C.; Paulion, L.; Monreal, C.; and Zagal, E.; Greenhouse gas (CO_2 and N_2O) Emissions from Soils: A review. *Chilean J. Res.* **2010,** *70,* 485–497.

101. Nasser, A. M.; Rudorf, B. F. T.; Antioniazzi, L. B.; Aguilar, D. A.; Bacchi, M. R. P.; and Adami, M.; Prospects of the Sugarcane expansion in Brazil: Impacts on direct and indirect land use changes. **2012,** 1–26, June, 2013, http://www.sugarcane.org/ resource-library/studies/wageningen%20-%20chapter%203.pdf

102. Neves, M. F.; Kalaki, R. F.; and Trombin, V. G.; The future competitiveness of the Café do Cerrado: Key elements for success for success to the first Brazilian Geographical indiction. **2011,** 1–12, June 26, 2013, http://www.ifama.org/events/conferences/2011/ cmsdocs/2011symposiumdocs/427-symposium%20paper.pdf

103. Nunes, R. S.; Sousa, D. M. G.; Goedert, W. J.; and Soares, J. R. R.; Proceedings of Symposia nacional do Cerrados, Brasiliera, Brazil CD-ROM. **2008,** May 25, 2013.

104. Oliviera, R. S.; Bezerra, L.; Davidson, E. A.; Pinto, F.; Klink, C. A.; Nepstead and Moreira, A.; Deep root function in Soil water dynamics in Cerrados savannahs of Central Brazil. *Funct. Ecol.* **2012,** *19,* 574–581.

105. Oliviera, P. S.; and Marquis, R. J.; The Cerrados of Brazil: Ecology and Natural History of a Neotropical Savannah. New York, USA: Columbia University Press; **2002,** 400.

106. Pacheco, L. P.; and Petter, F. A.; Benefits of Cover crops in Soybean Plantation in Brazilian Cerrados. 2010, 1–35, June 23, 2013, http://www.intechopen.com

107. Pearce, F.; The Cerrados: Brazils other Biodiverse Region loses ground. Environment 360. **2011,** 1–5, May 22, 2013, http://www.e360.yale.edu /feature/the_brazils_other_biodiversity_ hotspot_loses_ground/2393/

108. Peres, J. R. R.; Mendes, I. C.; Suhet, A. R.; and Vergas, M. A. T.; Efficiencies a competitivedade de stripes de rizhibia para soja em solos de Cerrados. *Review Brasiliera Cienca Solo.* **1993,** *17,* 357–363.

109. Pinheiro, E. F. M.; Lima, E.; Ceddia, M. B.; Urquiga, S.; and Boddey, R. M.; Impact of preharvest burning versus trash conservation on soil carbon and nitrogen stocks on sugarcane plantation in the Brazilian Atlantic Forest region. *Plant and Soil.* **2010,** *333,* 71–80.

110. Pinto, A. S.; Bustamante, M. C.; and ds Silva, R. S. S.; Effects of different treatments of pasture Restoration on soil trace Gas Emissions in the Cerrados of Central Brazil. *Earth Interactions.* **2006,** *10,* 1–12.

111. Pivello, V. R.; Oliveras, I.; Miranda, S.; Haridasan, M.; Sato, M. N.; Effect of fires on soil nutrient availability in an open savannah in Central Brazil. *Plant and Soil.* **2010,** 1–5, June 20, 2013, DOI 10.1007/s11104-010-0508-x

112. Prado, R. B.; Benites, V. M.; Polidoro, J. C.; and Alexy, G. C. E.; Mapping soil fertility at different scales to support sustainable Brazilian Agriculture. World Academy of Science. *Eng. Technol.* **2012,** 69, 844–852.

113. Rangel, T. F. L. V. B.; Bini, L. M.; Diniz-Filho, J. A. F.; Pinto, M. P.; Carvalo, P.; and Bastos, R. P.; Human development and Biodiversity conservation in Brazilian Cerrado. *Appl. Geography.* **2007,** *27,* 14–27.

114. Ratter, J. A.; Riberio, J. F.; and Bridgewater, S.; The Brazilian Cerrado vegetation and threats to its biodiversity. *Ann. Botany.* **1997,** *80,* 223–230.

115. Rein, T. A.; and Sousa, D. M. G.; Adubacau com enxofre. In: Cerrado-Correcau do Solo e Adubacao. Sousa, D. M. G.; and Lobata, E.; ed. Embrapa Informacio Tecnologica, Brasilia, Brazil; **2004,** 227–334.

116. Resende, A. S.; Xavier, R. P.; Oliviera, O. C.; Urquiga, S.; Alves, B. J. R.; and Boddey, R. M.; Long term effects of pre-harvest burning and nitrogen and vinnasse applications on yield of sugarcane and soil carbon and nitrogen stocks on a plantation in Pernambuco, N.E. Brazil. *Plant and Soil.* **2006,** *281,* 339–351.

117. Robrahn-Gonzalez, E. M.; As aldeias circularis do Brasil central. In: Sao Paulo: Museu de Arqueologia e Etnologia. Brazil 50 mil anos: uma viagem ao passado pre-colonial. Sao Paulo: Edusp; MAE-USP, Sao Paulo Brazil. **2001,** 54.

118. Rochette, P.; No-till only increases N2O emissions in poorly aerated soils. *Soil Tillage Res.* **2008,** *101,* 97–100.

119. Roitmann, I.; Felfili, J. M.; and Rezende, A. V.; Tree dynamics of a fire-protected *Cerrado sensu stricto* surrounded by forest plantations, over a 13 period (1991–2004) in Bahia, Brazil. *Plant Ecol.* **2008,** *197,* 255–267.

120. Ruggiero, P. G. C.; Batalha, M. A.; Pivello, V. R.; and Mierelles, S. T.; Soil-vegetation relationships in Cerrados (Brazilian Savanna) and Semi-deciduous forest, South-eastern Brazil. *Plant Ecol.* **2002,** *160,* 1–16.

121. Santana, D. P.; Filho, B.; and Antonio, F. C.; Soil quality and Agricultural sustainability in the Brazilian Cerrados. **1998,** Symposium no. 37, Paper no. 2281, 1–7, natres.psu.ac.th/Link/Soil Congress/bdd/symp37/2281-t.pdf

122. Santana, S. A. C.; Fernandes, M. F.; Ivo, W. M.; and Costa, J. L. S.; Evaluation of soil quality indictors in sugarcane management in sandy loam soils. *Pedosphere.* **2009,** *19,* 312–322

123. Schmidt, N.; The Cerrados biome in Central Brazil-natural Ecology and threats to its diversity. **2009,** 12, June 5, 2013, http://www.goek.tu-freiberg.de/oberseminar/OS_09/Nadja_Schmidt.pdf

124. Schnepf, R. D.; Dohlman, E.; and Bolling, C.; Agriculture in Brazil and Argentina: Development and prospects for major field crops. United States Department of Agriculture. Washington DC: Foreign Service; **2001,** 1–78.

125. Scopel, E.; Triomphe, B.; Ribeiro, M.; Seguy, L.; Denardin, J.; and Kchahann, A.; Direct seeding mulch-based cropping systems in Latin America. **2004,** 1–8, http://www.cropsscience.org/au/icsc2004/symposia/2/2/1406_scopele.htm

126. Seguy. L.; Bouzinac, S.; Trentini, A.; and Cortez, N. A.; Brazilian agriculture in new immigration zones. *Agricu. et Dev.* **1996,** *12,* 2–61.

127. Seguy, L.; Bouzinac, S.; and Husson, O.; Direct seeded tropical soil systems with permanent soil cover: Learning from Brazilian experience. In: Biological Approaches to Sustainable Soil Systems. Boca Raton, Florida: Taylor and Francis Group, CRC Press; **2006,** 3–14.

128. Silva, A. J. N.; Ribeiro, M. R.; Carvelho, F. G.; Silva, V. N.; Silva, L. E. S. F.; Impact of sugarcane cultivation on soil carbon fractions consistence limits and aggregate stability of a yellow lactose in Northeast Brazil. *Soil Tillage Res.* **2007,** *94,* 420–424.

129. Silva, J. E.; Resck, D. V. S.; Coraza, E. J.; and Vivald, L.; Carbon storage in clayey Oxisol cultivated pasture in the Cerrados region, Brazil. *Agric. Ecosyst. Environ.* **2004,** *103,* 357–363.

130. Silva, M. S.; et al. BRS 254-Wheat cultivar for irrigated conditions. *Crop Breed. Appl. Biotechnol.* **2008,** *8,* 96–98.

131. Sousa, D. M. G.; Calagem e adubacao da soja no cerrado. Porto Alegre DEAGRO/ADUBOS TREVO, S/A; **1989**, 17,

132. Sousa, D. M. G. de.; and Rein, T. A.; Manejo da fertilidade do solo para culturas anuais: experiências no cerrado. Informações Agronômicas, No 126, Junho 2009. IPNI, Piracicaba, SP; **2009**, 1–7.

133. Spehar, C. R.; de Santos, R. L.; and Jacobsen, S.; Andean crop introduction to the Brazilian Savannah. **2012**, 1–22, May 25, 2013, http://www. condesan.org/memoria/AGRO6298.pdf

134. Stefanini, A.; Know your Brazilian Biome, the new agricultural frontier. The Riotimes issue CCXXI. **2012**, 3, June 6, 2013, http:// riotimesonline.com/brazil-news/opinion-editorial/ opinion/know-your-brazilian-biome-the-cerrado-the-new-agricultural-frontier/#

135. Szakacs, G. G.; Eschenbrenner, V.; Cerri, C. C.; and Bernoux, M.; Soil carbon stocks under pastures in the Brazilian Cerrados region their assessment by orbital remote sensing. *Int. Soc. Photogram. Remote Sensing ISPRS Arch.* **2004**, *35,* 790–795.

136. Tansey, K.; et al. Vegetation burning in the year 2000: Global burned area estimates from Spotvegetation data. *J. Geophys. Res.* **2004**, *109,* D14.

137. Taylor, M. Z.; Farming the last Frontier. *Farm J. Today.* **1998**, *16,* 17–23.

138. The Nature Conservancy. Brazil: Cerrado. **2013**, 1–3, June 3, 2013, http://my.nature.org

139. Telles, T. S.; Guimaraes, M. F.; Carmela, S. O.; and Dechen, F.; The costs of Soil erosion. *Revista Brasiliera de Cienca do Solo.* **2011**, *35,* 1–13.

140. Terra, T. G. R.; Leal, T. C. A.; Rengel, P. H. N.; Barros, H. B.; and Santos, A. C.; Tolerance to drought in rice cultivars in Southern Cerrados area from Tocantins state, Brazil. *Acta Sci-Agron.* **2010**, *32,* 715–719.

141. Thomas, R. J.; Ayarza, M.; and Lopes, A. S.; Management and conservation of Acid soils in the Savannahs of Latin America: lessons from the Agricultural development of the Brazilian Cerrados. Proceedings of Consultants Meeting on Management and Conservation of Tropical Acid Soils for Sustainable Crop Production. Vienna: International Atomic Energy Agency; **2000**, 1–73.

142. Trigo, E.; Cap, E.; Malch, V.; and Villareal, F.; The case of Zero-tillage technology in Argentina. International Food Policy Research Institute. Discussion Paper. **2009**, 56–64.

143. USDA Foreign Agricultural Service. Commodity intelligence report: Brazil expected to produce three record 2007/2008 crops. **2008**, 1–7, May 25, 2013, http://www.pecad.fas.usda. gov/hights/2008/04/Brazil.htm

144. Vargas, M. A. T.; Mendes, I. C.; Suhet, A. R.; and Peres, J. R. R.; In: Cultura da Soja nos Cerrados. Arantes, N. E.; and Souza, P. I. M.; eds. Brazil: Sao Paulo; **1993**, 159–182.

145. Vendrame, P. R.; Brito, O. R.; Guimareas, M. F.; Martins, E. S.; and Bacquer, T.; Fertility and acidity status of latossols (Oxisols) under pastures in the Brazilian Cerrados. *Ann. Acad. Brasiliera Cienca.* **2010**, *82,* 1085–94.

146. Vrieling, A.; Erosion assessment by satellite remote sensing in the Brazilian Cerrados. Environmental Sciences Group-Internal Report. The Netherlands: Wageningen; **2012**, 1.

147. Vrieling, A.; and Rodrigues, S. C.; Erosion assessment in the Brazilian Cerrados using multi-temporal SAR imagery. Proceedings of Envisat and ERS symposium. Salzburg, Austria; 2004, 1–2.

148. Wada, M.; Carvelho, J. C. B.; Rodrigues, G. C.; and Ishi, R.; Yield response of spring wheat cultivars at different irrigation rates. *Nippon Sakamotsu Gakkai Kiji.* **1997**, *66,* 92–99.

149. Warming, E.; Lagao Santa: A contribution to the biological plant geography with on the list of Lagao Santos vertebrates. *Royal Danish Sci. Company Writings Sci. Math. Depart.* **1892**, *6(6),* 153-488

150. Wikipedia. Agriculture in Brazil. **2012**, 1–4, June 22, 2013, http://www.wikipedia.org/wiki/ Agriculture_in_Brazil.htm

151. Wikipedia. Cerrados. **2013**, 1–8, May 15, 2013, En.wikipedia.org/wiki/Cerrado

152. Wilkinson, J.; and Sorj, B.; Structural Adjustment and the Institutional Dimensions of Agricultural Research and Development in Brazil: Soybeans, Wheat, and Sugar Cane. OECD/GD (92). Paris: OECD Development Centre; **1992,** 1–22.

153. World Food Prize Foundation. Edson Lobatao, Alysson Paolinelli and Colin McClung. 2013, 1–4, May 22, 2013, http://www.worldfoodprize.org/ enlaureates/200020009_laureates/2006_ lobato_mcclund_paolinelli.htm

154. Xavante, C.; Saving the Cerrado. Cultural survival: 40 years. **2009,** 1–3, June 5, 2013, http:// www.cultural survival.org/publications/cultural-survival-quartery/brazil/saving-cerrado

155. Yaganiantz, L. E.; Alves, A.; and Avila, F. D. A.; Cerrados development in Brazil: A case study. Excerpts of World Food Prize Lecture. International Symposium of World Food Prize Foundation. Des Moines, Iowa, USA; **2009,** 28.

156. Zeilhofer, P.; and Kemp, S. M.; Spatial modelling of sprinkler irrigation suitability in a Central Brazilian Cerrado region. *Geocarto Int.* **2011,** *26,* 227–248.

157. Zoretto, R.; Greener than we thought. *Pesquisa FAPESP.* **2008,** *145,* 1–5.

CHAPTER 3

THE PAMPAS OF SOUTH AMERICA: NATURAL RESOURCES, ENVIRONMENT, AND CROP PRODUCTION

CONTENTS

3.1 PAMPAS: AN INTRODUCTION

Pampas means plains in the native Quechua dialect. Pampas indeed is a vast fertile low land plain. Pampas spreads into 7,50,000 km² (289,577 miles²) zone between latitudes 28°S and 39°S, and longitudes 50–65°W. It encompasses parts of southern tip of Brazil, entire Uruguay and Central to North-eastern parts of Argentina covering provinces such as Buenos Aires, La Pampas, Santa Fe and Cordoba (Figure 3.1). Majority of the plains region occurs in Argentina and Uruguay. Pampas are interrupted by low mountain ranges such as Sierra de laVentana north of Bahia Blanca. These hill ranges are about 1,300 m.a.s.l. River Platta is the major water drainage system of the Plains region. It irrigates expanses in the Cardoba, La Pampa, and Buenos Aires provinces. Rivers Salado, Parana, and Uruguay too supply irrigation water to natural prairies and cropland. Rivers such as Colorado and Negro are excellent water resources to famers in southern Pampas. Pampas region is drier in the west and it becomes progressively wetter and humid as we traverse east to the Atlantic coast.

FIGURE 3.1   Pampas of South America.

Note: Pampas Plains region is shown in gray color. The Pampas region includes provinces such as Rio Negro, small area in Mendoza, La Pampas, San Luis, Cordoba, Buenos Aires, parts of Santa Fe, Entre Rios, entire Uruguay and small region in the southern tip of Brazil.

3.1.1 GEOLOGICAL ASPECTS OF PAMPAS

Following are some descriptions about geology and topography of Pampas region located off the river Salado. It is derived from Charles Darwin's travels into Pampas during 1833, as narrated by Zarate and Folguera (2009). Charles Darwin identifies two major geologic regions—*Pampas interserrana (or intermontane Pampas) and Pampas deprimeda (or depressed Pampas)*. This grouping was based on lithology, large deposits of dust and alluvium and sediments. Some of the estuarine derivatives and marine deposits were attributed to more recent geological phenomenon of recent epochs. Paleantologically, Pampas is said to be home to large number of vertebrates, invertebrates, and molluscs. The sedimentary succession for 11–12 million years was traceable in the dry western and humid Pampas. It is said that vast areas of *Pampas interserrana* experienced relatively stable geologic era during Miocene and Plieocene. These inferences were drawn after studying calcrete deposits. Sedimentation process in the Pampas was more active during later period of Pliocene–Pleistocene especially in the Salado tectonic region and Sierra de Tandil. *Pampas interserrana* is geologically complex in feature. It separates the Salado and Colorado tectonic basins. Fluvial terraces are encountered in the *Pampas interserrana*, Tandila, and Ventania. Steppes are actually created by the flow of rivers/streams such as Quequen Grande, Quequen Salado, and Sauce Grande. Currently, the Pampean sediments are characterized as light brown to reddish sandy silts. The mineral composition of the sediments suggest that it is volcanic, derived from Andean region. Andean rocks have weathered and caused the sediments that have moved into Pampas plains. Pre-Cambrian metamorphic rocks and Paleozoic quartzites in regions such as Tandil and Ventana have also generated sediments (Table 3.1; Taboada, 2006). Sediments from Northern regions of Cardoba and San Luis are also discernible. Darwin believed that all over the Pampas, sediments were derived during the same epoch, based on the uniformity of morphological traits.

More recently, updates on description about the geologic history of Pampas by Taboada (2006) suggest that Pampas is a plain with unconsolidated sediments from quaternary period. Sediments were deposited both through eolian and hydrologic mechanisms. In some places, especially near estuary, sediments were drawn from marine deposits. It is said the eolian deposits were placed on a solid crystalline surface created by igneous and metamorphic rock. In the depressions, deposits are thick (up to 6,000 m), especially in the basins of river Salado and Colorado. It is believed that glacial shifts might have occurred in the Quaternary period (Taboada, 2006). The oldest geologic formation is said to be the *Puelches* Formation, which is composed of sands of different hues and occur in the ravines of Parana river. *Ensenadense* is a formation that has dark greenish sandy silts, with clayey intercalations. They occur in the ravines of Rio de Plata. Bonaerense is composed of sediments rich in clay and silt. They are volcanic in origin and reddish brown in color. Pampean Loess is called so because it has resemblance to such loess soils in other continents. The loess has volcanic ashes and igneous material which has given rise to soils of the Pampas. The *Tandelia* and *Ventania* are sierra systems found in northeast province of Pampas. The *Querandinense* is a Holocene formation caused by ingression of deposits.

TABLE 3.1   Stratigraphical information about Pampas

Geologic Age (in Million Years)	Era	Period	Time/Epoch	Lithology
1	Cenozoic	Quaternary	Holocene	Sands, Clays, Sandy and silty
70	Cenozoic	Tertiary	Sandy Limonites and clays	–
250	Mesozoic	Cretaceous	Limnites, Tobas, anArensics	–
600	Paleozoic	–	–	–
Precambrian	Crystalline basement derived from Granites, migmatites, and gneiss			

Source: Taboada (2006)

Regarding soil formation, Taboada (2006) suggests that Hapludolls or Haplic Phaezems that are predominant in the west of Pampas is derived from invading dunes; Arguidolls or Luvic Phaezems occurring in the east were derived from *Bonaerense* formation; Haplustolls common to inland Pampas were derived from material similar to *Bonaerense*.

3.1.2   PHYSIOGRAPHY AND AGROCLIMATE OF PAMPAS

The geographical classification of Pampas includes regions/features such as Sierras, Continental Plains, Marine Plains, Fluvial Plains, Delta and Lacunar depressions (See Taboada, 2006). We can further reclassify the geomorphological units into subdivisions based on soil types encountered. Then, they are termed as Edaphic domains. It has greater relevance to agricultural cropping pattern possible in different sub regions of Pampas. Based on geographic and ecological considerations together, Viglizzo et al. (2011) classify sub regions of Pampas into Rolling Pampas, Subhumid Pampas, Southern Pampas, Semiarid Pampas, Flooding Pampas, and Mesopotamian Pampas. The Pampas region is also classified into subdivisions such as Rolling Pampas, Inland Pampa (Flat Pampa, Western Pampa), southern Pampas, Flooding Pampas, and Mesopotamic Pampas (Hall et al., 1992). For simplicity, there are others who just identify two physiographic regions—the Dry Pampas that occurs to the west and Humid Pampas in the east.

3.1.2.1   AGROCLIMATE OF PAMPAS

Pampas of South America is a vast stretch of dry or humid cropping zones. The region is classified as semiarid considering precipitation pattern. Koppen's climate classification for the Pampas is Dry Mid-latitude Climates (**Bs)** with dry summers. Winds from Atlantic immensely influence the weather patterns in the Plains. The annual rainfall is 1,200 mm in the North-eastern region. Much of the plains receive 750–850 mm precipitation annually, but it declines as we traverse southwest to 600 mm annually. Thunderstorms that occur during afternoons bring in intense amount of water. Precipitation season begins around late October in the Pampas region. Precipitation peaks at 80–100 mm per month during December till March end. Precipitation is lowest at 2–40 mm month^{-1}, during May/June to September. As stated above, total precipitation hovers at 750 mm year^{-1}in Pampas.

Pampas enjoys long stretches of warm summer with temperature ranging from 29°C to 32°C. Night time temperatures dip to 18°C in the north and to 14°C in the south. Temperature in the coastal plains ranges from 25°C to 28°C. Heat waves are common during summer when the temperature may soar to 40°C for short duration in the mid afternoon. The period of heat waves is usually followed by "Pampero wind" that blows from southern region. It brings cool breeze for 2–3 days at a stretch, resulting in drop in ambient temperatures to 14°C–18°C from 28°C. During fall season that begins approximately in March, rainy weather predominates with mildly cold and dry interruption. Temperature ranges from 24°C to 28°C during day and from 14°C to 18°C during night. Rainfall occurs throughout the fall season in the east, but it becomes sporadic and light as we move to west. Frost may occur on several days during winter. Frosty period stretches from early April to July. Last of the frosty period occurs by mid-September. During frost, temperatures are low at 1°C–5°C and it is uncongenial for standing crops if any. Germination and seedlings survival is difficult. Snowfall is uncommon to plains but sporadic reports do occur. Spring in Pampas is a windy period with hot and cold breezes. Violent thunderstorms may occur on occasions, but equally so, dry spells without rainfall are experienced (Podesta et al., 2007, 2009; Wikipedia, 2013a, b).

The agroclimate of Pampas is influenced by a range of natural factors. The El Nino southern Oscillation (ENSO) is a major factor that influences seasonal to inter-annual climate pattern (Podesta, 2006). ENSO influences precipitation in the Pampas, especially during critical stages of crop that occurs during November/December. The El Ninos events are connected with higher precipitation, but El Nina events are associated with lower levels of precipitation. Timely forecast about El Nino Southern Oscillation has been useful. In fact, Bert et al. (2006) have stated clearly that computer aided predictions using crop models has allowed farmers to make more knowledgeable decisions about crops, genotypes, and inputs. Interestingly, in addition to inter-annual fluctuations, prairies in Pampas experiences inter-decadal fluctuations in precipitation and weather related parameters (Podesta, 2006; Podesta et al., 2007, 2009). Podesta et al. (2007) state that Pampas, which is an important agricultural belt of the world experiences inter-decadal variation in the climate. During past 4–5 decades there has been a gradual increase in the precipitation received by the plains. The increased precipitation has enhanced water flow in the streams and aquifer. In fact, inter-decadal variations in the climate has impacted aspects such as crops species grown, cropping systems, crop production trends, intensity of cropping, and yield goals. For example, since 1970s, soybean production zones expanded in response to high precipitation pattern. Magrin et al. (2005) reported that during 1970–1999, yield increase attributable to climate change was 38 percent for soybean, 18 percent for maize 13 percent for wheat and 12 percent for sunflower. Podesta et al. (2007, 2009) have cautioned that such high productivity of soybean cropping zones may after all experience moisture dearth in the next 25–30 years and hence farmers may have to alter cropping systems suitably. Reports by IPCC (2007) suggest that in the Pampas, the forecasted negative impacts of climate change on maize and soybean productivity could be overcome, to a certain extent. Modifications in planting dates, supplementary irrigation and appropriate genotypes could itself reduce the problem perceptibly.

3.1.2.2   NATURAL VEGETATION AND CROPPING HISTORY OF PAMPAS

The Humid Pampas in its natural state is a tall grass prairie. This region has no tree species except for few imported ones such as American Sycomore or Eucalyptus. "Ombu" is said to be a native tree-like bushy plant that thrives along high ways and in towns. The Western Pampas is relatively a dry belt that supports large set of short prairie grasses (Wikipedia, 2013a, c). Pampas grass (*Gynerium argenteum*), whose feathery panicles reach 8–9 ft is distinctive of Pampas's tall prairies. It dominates the ecosystem and may often affect establishment and spread of other botanical species, especially several short statured herbs. The Pampas natural grass species could be classified into *Pasta durra* (hard pasture species) and *pasta molle* (soft pasturage). *Pasta durra* supports grasses that form tussocks or clumps. *Pasta molle* is filled mostly with succulent, small herbs that flourish as undergrowth in the Pampas prairies. It is said that steppes of Pampas are home to large number of grasses. Among them, graminaceous species belonging to the tribe Stipae, for example Genera *Stipa* and *Piptochaetium*, are common (Smithsonian Institute, 2012).

European settlers have introduced several herbage species such as medicago, trifolium, and grasses. Most of them have got naturalized into the prairies. Native trees are sparse or nil in most of the plains region. A few of the tree species that has established were brought into the region by the European settlers. They are Eucalyptus, Melia, Azadiracta, peach willow, ombu, and few others.

During past four centuries, extensive arable farming and grazing of cattle have definitely altered the flora of Pampas. Natural prairies have given way for domesticated crops, large-scale commercial cropping and pastures (Argentina Autentica, 2012). It is said the least altered area of Pampas, in terms of vegetation, are the coastal lowlands, Parana delta, and Southern Sierras. The marshy coastal plains and sands still support large posse of Pampas grass (*Cortadera selloana)* also known as Silver Pampas grass or Uruguayan Pampas grass. *Cortaderia* is a very common grass all across the plains. It produces over a million seeds per plant during the life time. Purple Pampas grass (*Cortaderia jubata*) is also frequently encountered in the plains (Peterson and Russo, 2012). Many of the species of genus *Cortaderia* are native to Andes mountain regions in the west of Argentina. In the drier west Pampas, also called the Sterile Pampas, brackish and saline water supports only feeble vegetation. Hardy plants such as "chanar" (*Gurliaca decorticans*) are common. Parana delta supports tall-grass prairie which is interspersed with *Prosopis* trees.

Pampas Prairie supports several animal species adapted to the region. A few examples are Pampas hare (*Dolichotis patagonica*), coypu (*Myopotamus coypu*), cui (*Cavia australia*), tucotuco (*Ctenomys magellanica*), wood cat (*Felix geoffroyi*), a small cat (*Leopaldus pajeros*), several types of weasels, skunks, deer, armadillos and opossum. It is said that fauna that had naturalized to natural prairies has been affected because of large-scale farming, replacement of natural grasses with crops and massive use of pesticides and chemicals (Connior, 2008). However, Solbrig (2005) states that despite expansion and intensification of agricultural cropping, faunal diversity, and population too improved in the prairies.

Agricultural crop production in the Pampas involved low input, rotational cropping systems since 1960s. Until 1980s, agricultural prairies expanded replacing natural grass land. This process helped the region to satisfy grain/forage demand created by domestic consumers and those from other continents. Once, option for expansion became minimal, farmers in Pampas resorted to intensification of crop production systems using higher amounts of fertilizers and water (Viglizzo et al., 2011). Land use change to agricultural cropping was most pronounced in the Rolling Pampas. During the period from 1960, annual crops increased in acreage from 30 percent of Pampas in 1960 to 70 percent in 2005. Natural grasslands suffered reduction in area rather drastically. Factors such as fertile Mollisols, excellent crop responses to fertilizer and irrigation supplements and consistent demand for export of grains induced this change in vegetation of the prairies. In addition to change in botanical species, crops, and cropping procedures induced marked changes in soil profile, its nutrient distribution pattern, emissions of gasses such as CO_2, N_2O, NO_2, and so on. Monoculture of wheat, soybean, or maize became common in place of natural grass/legumes. There were perceptible increases in cropping zones in the Subhumid and Southern Pampas. Agricultural cropping belts expanded least in the Semiarid and Flooding Pampas during 1960–2005. Natural grasslands and pastures that require less input, and are adaptable to harsher climates were preferred over risky high-intensity crop production. Mesopotamic Pampas has experienced marked changes to farming enterprises during recent years from 2,000 A.D. Following is the distribution of annual field crops and natural grasslands in the Pampas during past 50 years since 1960.

Land Use (% Area Covered in the Pampas)						
Pampas/Region	**Annual Field Crops**			**Natural Prairies**		
	1960	1987	2005	1960	1987	2005
Rolling	37	57	70	63	43	30
Sub humid	44	40	51	56	60	49
Southern	39	39	53	61	61	47
Semiarid	39	42	43	61	58	57
Flooding	17	9	1	83	90	89
Mesopotamian	18	16	8	82	84	61
Pampas—total area	34	34	45	66	66	55

Source: Viglizzo (2005)

Note: There is a progressive decrease in the area under natural prairies. However, proportionate enhancement in agricultural field crops has occurred during past 5 decades. Since forest with trees is negligible in Pampas, land use change to agriculture mostly affects grasslands. Total area of Pampas studied in the above study was 4,26,160 km².

A long-term study on rates of conversion of natural grass land to cropland was conducted in Pampas by Baldi and Paruelo (2008). They found that area under natural grasses gradually decreased from 67 to 61 percent in the study area around Rio de Platta. Annual crops such as soybean, wheat, maize, and sunflower replaced the natural grass. However, in some patches within Flooding Pampas, conversion of natural grasslands to cropland was not easy. Clearly, natural factors as well as economic gains from land use conversion affect farmer's decisions to change the flora.

3.1.2.3   DEVELOPMENT OF AGRICULTURE IN PAMPAS

Agricultural activity in Pampas was restricted to subsistence levels during colonial era. Although farmers felt the need to intensify crop production as early as 1775, it was constrained due to conflicts, wars, and lack of technology. Earliest effort to improve agricultural skills and import machinery was made through initiation of Institutions in 1870s. Relatively large-scale immigration of Europeans induced intensified farming systems. From 1920 till early 1960s, Pampas farming zone was gradually intensified by the flux of Europeans, machinery and capital. During 1970s, Pampas experienced a boom in soybean production that lasted until mid-1990s and then plateaued. Viglizzo (2005) states that since the early days of colonization by Spaniards in 1879, Pampas as an ecosystem promised to offer high crop productivity and excellent nutrient turnover. The development of prairies was actually gradual and currently it contributes a large share of total grain/forage produced by South America. Further, according to him, current productivity of both natural prairies and cropland is much lower than the genetic/environmental potential. There are still large areas of natural prairies and wasteland that could be transformed into productive cropland. There is also enough possibility to increase fertilizer and water supply and intensify the crop belt.

According to Solbrig (2005), agricultural landscape in Pampas has experienced marked transformation during last 10 years. Grain production has doubled and soybean harvests have tripled. Cereal and oilseed production for export have increased dramatically. Most of the farms aiming at export have been intensified using high fertilizer and water supply. In fact, agricultural cropping in the Pampas is so crucial that, it is said consistently rich harvests have almost saved a state that has failed due to economic mismanagement. Agricultural prairies are still the major transforming agents of the Pampas landscapes. Economic activities like mining, industries and urbanization together occupy only 3 percent of the landscape. Farms meant for grazing occupy 40 percent area and rest is apportioned for grain production.

3.1.2.4   HUMAN MIGRATION, SETTLEMENT AND POPULATION TRENDS IN THE PAMPAS

Human migration into Pampas with an intent to settle and procure food either through hunting and/or crops might have occurred during 13,000 B.C. Archaeological evidences show that initially, hunting and gathering were in vogue in the natural prairies of Pampas and other regions of present day Argentina (Dillehay, 2000). Native tribes of Argentinean Pampas thrived well until medieval times. Migrants from Guarani and Paraguay settled in parts of Pampas and practiced a kind of "slash and burn" farming. They lived transitorily in a place as long as soil fertility supported good crop harvest (Lewis, 2001). It is believed that at the time of Spanish conquest of Pampas, the population of native tribes hovered around 7,50,000. Many of them were engaged in regular crop production (Pyle, 1976). In the upper reaches of Rio Uruguay and Rio Parana, Jesuits missionaries from Europe made their farms and home during early to mid-1600s. European settlers brought crops and several other items into Pampas. Settlements by Europeans that began with Pedro Mendoza in 1536 and Juan de Gray in 1580 affected human population dynamics. It also induced changes in vegetation. Crop species and weeds introduced by Europeans began establishing and naturalizing with the native species. Cattle, it seems were first introduced by Portuguese and Spaniards during 1550s. The La Pampas region, in particular, was explored by voyagers such as Hernandarias and Jeronimo Luis de Cabrera during mid-seventeenth century. However, permanent settlements and farming got initiated in this region of the plains only during eighteenth century. Inhabitation of La Pampa, Rio Negros, and adjoining areas increased after 1880s. During 1915s there were 11,000 inhabitants. Many of them were engaged in farming crops such as wheat and corn (Wikipedia, 2013c). Pastoral production was in vogue during nineteenth century. They established ranches and cattle farms (Adelman, 1994).

During past 100 years, Pampas has experienced rapid development of urban dwellings and industries. This has induced migration out of farms. During 1950s, rural population engaged in crop production was 33 percent. Migration during 1960s changed the scenario. Urban population increased to 80 percent in 1970 and by 2003 urban dwellers increased to 90 percent. Proportionately, farmers residing in the rural dwellings reduced to 30 percent and then to 10 percent by 2003. Mechanization, large-scale farming techniques and large farms made several small farming communities nonviable, inducing them to move to cities. Droller (2013) has made analysis of the effects of European migration into fertile plains of Pampas on settlement pattern, occupation involving farming, ranches or industries and their influence on economic status of people inhabiting the area. He has reported that skilled European communities depending on industries had higher gross domestic product (GDP). Migration occurred for long periods during the past, but it peaked around 1880–1891 prior to World War 1. There are other inferences about economic status of Pampas. Fertile plains that need low inputs are highly profitable to settle and practice farming/cattle ranching. It seems over 5.5 million European settlers moved into Pampas to establish farms during the period from 1857 to 1920. It resulted in a social upheaval to the native tribes of Pampas who practiced subsistence farming techniques (Droller, 2013). It seems calmer and more peaceful pastures in the Pampas induced European to move to Pampas. The indigenous tribes of Pampas did not recognize government until mid-nineteenth century. Major effort to convert natural prairies into farmland occurred with annihilation of native Indians by Spaniards in 1816. It is said a new economic era dawned in the Pampas with advent of farm/industrial workers drawn predominantly from countries such as Italy, Germany, Spain, and Portugal.

Gauchos are native herdsmen who had extensive cattle and sheep ranches. *Chacreras* maintained small farms known as *chacras*. During past decade, there has been a marked shift from ranching and cattle farming to soybean production. Shift from ranch to cropland by the Gauchos is engineered actually by the high export demand and profitability of soybean production (Bolsa, 2011). They say many of the hitherto Gauchos of Pampas have left nomadic ways and cattle ranching due to change in occupation.

The Pampas and most of Argentina are endowed excellently to expand and/intensify agricultural regions. Yet, we should try to know more about future population trends versus the need to sustain or improve food crop generation. Kingsland and Hamilton (2009) have opined that demographic trends show rapidly enhanced population in the Pampas. We ought to give weightage to periodic occurrence of drought spells and reduction in crop production. Add to it, administrative lapses and civilian disturbance. All of the above parameters may induce farms, firstly to get diversified with regard to crops and then resort to intensification of crop production systems. Population migration into or out of Pampas farms will of course depend on agrarian jobs, exchequer and expectation on living standards. We should also note that population influx or increase may induce conversion of natural grasslands to cropland. This aspect needs greater attention in terms of soil deterioration, environmental effects, and productivity of Pampas as an ecosystem.

During past 3 decades, several aspects related to agricultural cropping trends and economics have affected the human dwellings and their sustainability in the Pampas. Factors such as commercialization of farms, excessive involvement of foreign farm enterprises, preference to large farms, exclusion of small and medium sized farms, use of large amounts of capital in farming and marketing, and lack of small time farm jobs has resulted in human movement to urban zones and other regions (Pengue, 2001).

3.2 NATURAL RESOURCES OF THE PRAIRIES OF THE PAMPAS

3.2.1 SOIL RESOURCES OF PAMPAS WITH SPECIAL REFERENCE TO CROP PRODUCTIVITY

There are at least 10 different soil orders that can be traced in Argentina. However, soil types that are dominant in the Pampas region are Mollisols, Entisols, Alfisols, and Inceptisols (Hall, 1992; Hillock Capital Management, 2012). According to reports by INTA (1990), major soil types found in the Pampas and adjoining agricultural belts are Mollisols (0.87 m ha), Entisols (0.62 m ha), Aridisol (0.56 m ha), and Vertisols (0.02 m ha). Soil type dominant in each of the subregions of Pampas differs (Table 3.2)

TABLE 3.2 Major soil orders of Pampas of Argentina

Region in Pampas	Soil Types
Rolling Pampas	Arguidolls, Aquic Arguidolls, Chromousterts
Mesopotamian Pampas	Paleusterts, Ochraqualfs, Calcareous, and Silty clay soils
Southern Pampas	
Plains and table lands	Petroclacic Paleustolls, Haplustolls
Terraces	Aridic Haplustolls, Entic Haplustolls, Udic Haplustolls, Natraqualfs, Argidolls
Parana flood plains	Alluvial complex

Sources: Glatzle et al. (1999); Hall et al. (1992); Hillock Capitol Management (2012); IDRC (2011); Krishna (2003); Mostacelli and Pazos (2000); http://www.isric.org/projects/soil-and-terrain-database-soter-argentina

Mollisols are dominant soil types found in the agricultural stretches of Pampas. They are among the best in terms of fertility and quality that support high crop yield. We should note that the Pampean farmers relied on inherent soil fertility and its productivity so much that they did not refurbish soils using inorganic fertilizers for a long time. Inorganic fertilizers have been used only since 1970s, mainly N and P (Mostacelli and Pazos, 2000). Mollisols traced in humid and Semiarid Pampas are easily sub classified into Udolls and Ustolls. They are derived from repeated deposition of sediments in layers at different periods in the geologic history. Mollisol found in Pampas are richer in $CaCO_3$ that forms a petrocalcic horizon. Mollisols traced in high-intensity production zones are often richer in SOM ranging from 3 to 7 percent. They possess good moisture holding capacity. Mollisols of Pampas are fertile and most suited to raise cereals like wheat and maize.

Alfisols are also common to Pampas. They support diversified natural grass/legume communities and crops. Productivity of Alfisols is moderate. They need replenishment with N and P, if the region supports intensified farming practices. Alfisols possess a surface horizon that is not very rich in organic matter and are usually light colored. They have a natric horizon. Soil moisture holding capacity is moderate. Alfisols that are relatively less fertile are preferred to raise pastures or grow cereals of moderate grain/forage productivity (Mostacelli and Pazos, 2000; Krishna 2003; and Krishna, 2013).

Vertisols are common to Northeast Pampas, the Buenos Aires region and eastern half of Entre Rios region (Mostacelli and Pazos, 2000). Vertisols are used for cropping extensively in the humid and semiarid belts of Pampas. Vertisols are used to produce, wheat, soybean, flax, and sunflower.

Following is an example of mineralogical composition of soil profile from sample drawn around Buenos Aires and de Platta region in the Pampas of Argentina. Here, Mollisols are dominant. Overall, light minerals constitute 95 percent of soil and heavy minerals rest 5 percent.

Light Minerals: Quartz 20 percent; Ca-Na Feldspars 18 percent; K-feldspars 5 percent; volcanic minerals and silicates 15 percent; mineral aggregates 3–5 percent and rock fragments/gravel 20 percent.

Heavy Minerals: RZTA-5 percent; Muscvite-3 percent; Apatite-2 percent; Epidotes-23 percent; Hyperstene-18 percent; Hornblende-20 percent.

Source: Mostacelli and Pazos (2000)

Following is the soil chemical characteristics of A horizon of a typical Agricultural soil classified as Arguidoll and commonly traced in Pampas:

Depth	Organic Matter	Bray P-1	Exchangeable-K	pH	EC
(cm)	(%)	(mg kg^{-1})	(mg kg^{-1})		(ohms cm^{-1})
0–5	3.5	2.4	1,059	6.2	0.17
5–18	2.9	11	795	6.3	0.10

Source: Ghio et al. (2010)

3.2.2 SOIL FERTILITY AND NUTRIENT DYNAMICS IN PAMPAS

The inherent soil fertility of natural prairies of Pampas has supported grass/legume and mixtures of other species. Nutrient cycles that operate in the natural settings have played a key role in sustaining the prairies. Regarding agricultural prairies, it is believed that subsistence farming trends, residue recycling and organic manure supply have determined the crop productivity. However, during 1950s, demand created for grains by other continents, has induced farmers in Pampas to alter the nutrient dynamics rather drastically. The need to reach higher yield goals and intensify the agricultural prairies has meant enhanced supply of fertilizers. In a nutshell, Pampas' farmers had to intensify production of wheat, maize soybean to cater the needs of export demand.

During 1980s and the following decade, Pampas experienced modification with regard to tillage systems adopted and soil fertility programs. No-tillage systems became more common. Soil fertility that had depreciated due to incessant cropping and improper nutrient supply had to be corrected using fertilizer. Since 2002/2003, almost all cereal farms do apply N and P fertilizers in proper ratios. This helps to remove nutrient deficiency and correct nutrient imbalances, if any. Reports by FAO (2007) suggest that since 2003, farmers located in the Pampas plains have consistently applied 40 kg N ha^{-1} and 20 kg P to wheat crop, 30 kg N and 19 kg P to maize, 2–3 kg N and 8 kg P to soybean, 10 kg N and 7 kg P to sunflower and pastures have received 2 kg N and 9 kg P. On a percentage basis, 85 percent of wheat and maize is supplied with extraneous nutrients as fertilizers. About 30 percent of soybean, 38 percent pastures found in Pampas are also fertilized. Interestingly, only 15 percent of cotton fields are fertilized. Rest of cotton belt depends on inherent soil fertility. On an area basis, 23,747,000 ha in Pampas is supplied with fertilizers. Soils in Pampas are often considered as sufficiently endowed with K, hence its supply is more often ignored by Pampas farmers. Yet, we should realize that the vast prairie of Pampas has been impinged with large doses of nutrients (Table 3.3). Clearly, since the advent of fertilizer technology, supply of nutrients to the agricultural prairies has affected the productivity of the region. Fertilizer input has also influenced the nutrient dynamics and ecosystematic functions of the prairies. During recent years, farmers have noticed exhaustion of sulfur in the Pampas cropping zones. Right now, use of crop growth models, fertilizer supply based on soil test and crop yield goals seems more pertinent.

Hall et al. (1992) opine that Mollisols traced in the large tracts of Pampas is among the most fertile agricultural soils. Soil quality in Pampas is relatively superior and organic matter content is optimum at over 3–5 percent. In addition, during past 3–4 decades, farmers in the Pampas have grown soybean during at least one season. It has led to improvement in soil-N status and efficient recycling of residues and C—sequestration in soil. As a proof, it has been argued that farmers in Pampas do not repeatedly add chemical fertilizers to their soils. Instead, nutrient inputs are curtailed and spaced at once in 2–3 years. Studies by Madonni et al. (1999) have clearly shown that soil quality of Pampas is superior enough to provide nutrients in sufficient quantities, allow good rooting, and crop growth of maize, wheat, or soybean. Among the various soil parameters assessed, soil aggregation, soil-N and SOC were found to be important soil quality indicators that have placed Mollisols of Pampas as superior. Knowledge about critical NO_3–N level in soil is important. For example, maize crop grown in Pampas needs at least 3.4 percent NO_3–N in soil to reach its potential yield of 11 t grain ha^{-1} (Bianchini et al., 2003).

TABLE 3.3 Nutrients supplied as inorganic fertilizers to the major crops of Pampas during 2008

Crops Grown in Pampas	Fertilizer Supply (in 000 tons)			
	Nitrogen	Phosphorus	Potassium	Total
Wheat	247	163	Nil	410
Maize	87	58	Nil	145
Soybeans	30	73	Nil	102
Sunflower	23	16	Nil	39
Other crops	8	5	Nil	14

Source: FAO (2007, 2011); Garcia (2013); Alvarez (2007)
Note: Fertilizer products such as Urea, Di-Ammonium Phosphate, and mixtures are popular in the Pampas. Use of Potassic fertilizers is negligible. Current estimates suggest that 95 percent of wheat belt in Pampas is supplied with fertilizer-N, 85 percent with fertilizer-P and about 2 percent at best receives fertilizer-K. On an average, a crop like wheat receives 49 kg N, 30 kg P_2O_5 and 2 kg K_2O ha^{-1}. According to Schnepf et al. (2001) fertilizer usage has been minimal with most crops during the years 1970–2000. In the Pampas, fertilizer usage was nil in 1970–1977, and almost negligible until 1992. Its use began in 1990s (28 kg N, P ha^{-1}). Currently, its use is still feeble excepting in fields intended for intensified production.

Soybean cultivation in the plains of Pampas is a major preoccupation for farmers. It is a source of protein, oil, and exchequer. Soybean and its products enjoy a consistent demand for export. In terms of soil fertility, soybean cultivation and recycling its residues is perhaps one of the most important agronomic procedures. Soybean is a legume that adds to soil-N through its symbiotic association with Bradyrhizobium. Extent of N benefits derived from soybean varies depending on genotype, soil type and its N status, environment, and Bradyrhizobial strain. Pampas is a vast plain and soybean cultivation zones too are large. Hence, even small increase in soil-N ha^{-1} and N derived via atmospheric N fixation is important. According to Di Ciocco et al. (2013), on an average, Mollisols with 3–4 percent SOM that produces 2.2 t grains, benefits from BNF to the tune of 48–129 kg N ha^{-1}. Further, their analysis indicates that N derived from BNF has statistically direct relationship with above ground-N accumulation pattern, which in turn has direct influence on dry matter and grain formation by soybean crop. Soybean cultivation, no doubt, affects the soil-N balance positively. Therefore, fertilizer-N input to soil is perceptibly lower in Pampas compared to other regions of the world. Farmers tend to replenish fields with inorganic fertilizer once in 2–3 years. Crops tend to draw large quantities of soil-N to support luxuriant growth/grain formation. This leads to soil-N deficits. Quite often, Pampean prairies may experience deficit and negative N balance of 20–40 kg ha^{-1} annually. So, soil-N has to be replenished. Based on results from 10 field trials conducted on Arguidolls during the years 1995–2010, it was concluded that BNF adds to soil-N, but cultivation of soybean in the prairies may still create a net negative balance of—8 K N ha^{-1} annually. Clearly, soybean affects not only just exchequer, but also soil N dynamics perceptibly, yet a small amount of priming with fertilizer-N is required. Gutierrez-Boem et al. (1999) have shown that starter-N (18 kg N ha^{-1}) increases above-ground biomass, leaf area, and photosynthetic light interception. Such advantages due to starter-N may not immediately get translated to higher grain yield. Next, late season N management in the soybean dominated prairies is also important. It is believed a small dose of late-season N may delay leaf senescence and support continued carbon-fixation by leaves. In case of crops such as sunflower, fertilizer-N supply has to be as accurate as possible. Its N requirement ranges from 37–42 kg N t grain^{-1}. Over application improves vegetative growth but may reduce oil content of seeds by 2–5 percent (Scheiner et al., 2002). It may also have an impact on NO$_3$ in groundwater, if excess N gets percolated.

Lehman and Pengue (2000) point out that intensification of soybean belt in Pampas resulted in rapid depletion of soil nutrients. Lack of appropriate replenishment created patches of low fertility soils that did not provide nutrients enough for high productivity. Hence, fertilizer consumption was stepped up from 0.3 million ton in 1990 to 2.5 million ton in 1999. Where do we locate the Pampas's agricultural prairies in terms of scale of intensification? As stated earlier, inherent fertility of soils is relatively high. It allows intensification of cropping systems. Fertilizer-based nutrient supply is still very low compared to other intensified ecosystems such as the Corn Belt of America or Northeast China or European Cereal farms. There is still a very large yield and intensification gap in the Pampas that could be gradually covered by the farmers. It may affect the nutrient dynamics, at times rather drastically. Interaction between water and nutrient dynamics needs careful attention. Simple fact to note is that nutrient movement in the soil profile, absorption, translocation, loss and transformations are all mediated in dissolved state and with water as the medium. We may also have to understand the effects of intensification on rotational crops such as maize and wheat, the natural grasslands adjacent to soybean fields, and several ecosystematic functions.

Wheat productivity depends to a great extent on soil-N status and fertilizer-N applied to the crop. Generally, soil NO$_3$-N and other fractions are assessed prior to fertilizer-N inputs. Calvo et al. (2013) have stated that soil-N fractions and precipitation pattern interact to determine wheat grain yield in the Pampas. Precipitation pattern during July–December may influence the crop productivity. Next, soil N dynamics in Pampas is marginally, yet positively improved by soil microbes classified as biological N fixers. Extensive field evaluation in over 300 locations in the Pampas has shown that inoculation with *Azospirillum brasilense* improves productivity of wheat. Cereals such as wheat drive N-benefits in various soil types such as Argiudolls, Petrocalcic Ustolls, and Haplustolls traced in Pampas (Diaz-Zorita et al., 2008). Factors such as crop species, season, inherent soil-N, previous crop, tillage system adopted and fertilizer schedules seem to affect N derived from soil microbes.

Phosphorus budgets need greater attention in the Pampas. A study by Riskin et al. (2013) suggests that intensive production of crops like soybean or maize may need high P inputs. At the same time, high P supply can lead to eutrophication and other environmental consequences such as groundwater contamination. Pampas supports dryland farming in large patches where in crops may encounter soil with high P fixation capacity. The fertilizer-P requirement of such areas is higher, since farmers have to first satisfy the P fixation capacity of soils, then answer the crop's need. Viglizzo et al. (2011) found that as cultivation of crops intensifies, depletion of available-P becomes greater than replenishments. The stock of available-P was 0.87–1.1 Mg ha^{-1} in 1960s when crop production trends were still subsistent. Between 1986 and 1990, soil analysis suggested that available-P stock ranged from 0.25 to 0.48 Mg ha^{-1}. Between 2000 and 2008, available-P stock had depleted markedly to 0.20 Mg ha^{-1}–0.40 Mg ha^{-1}. Crop productivity and response to fertilizer-P input depends to a great extent on available-P stocks in the soil. Further, to quote an example, wheat crop's response to extraneous inputs of fertilizer-P depends much on SOM content, P fixation capacity of soils and most importantly, the available P pool in the soil. Crops grown on soils with low available-P, for example 5 ppm, respond better than those grown on soils with 10 or 20 ppm P (FAO, 2007). Bianchini et al. (2003) found that soil P concentration affects maize grain productivity. The available-P pool in soil has to reach 0.36 percent, for maize crop to reach its high grain potential of 11 t ha^{-1}. Inherent soil-P level is considered moderate if it is 15 ppm Bray-1 P. It is considered high if it reaches beyond 20 ppm Bray-1 P. Berardo et al. (2001) have reported that fertilizer-P inputs to soils in Pampas are profitable, if the field soil has less than 17 mg P kg soil^{-1}. In case of soybean, P supply has consequences on crop growth, grain yield formation, plus the BNF process that occurs in the nodules. Phosphorus supply to soybean expanses adds 300 kg grain ha^{-1} to the harvest. If fertilizers that are combination of P and S are supplied, it affects dynamics of both the elements P and S positively. Grain productivity is enhanced by 670 kg ha^{-1}, if both elements are supplied (Gutierrez-Boem et al., 1999). Maize supplied with P and S fertilizers also responds with better grain/forage productivity (Prystupa et al., 2006).

Sulphur dynamics in the soils of Pampas is gaining in importance, since its deficiency is getting manifested in patches. Further, supply of fertilizer-S does affect its dynamics and productivity of croplands positively (Duggan, 2010). Major crops of Pampas such as wheat, soybean, and maize have shown responses to S inputs, if the crops are grown on degraded soils and those with low organic matter. In regions with long cropping history, fields adopting no-tillage system and where only N and P fertilizers have been supplemented for long period, S supply is beneficial. The ratio of S replenishment versus that removed seems crucial to productivity of prairies. Currently, farmers in Pampas supply S using sources such as gypsum from Catamarca, Entre Rios, and Mendoza. The S content of these sources ranges from 13.7 to 16.9 percent. It seems mobility of S in soil is optimum, hence it allows flexibility in terms of placement of fertilizer-S sources in the soil. Broadcasting or incorporation of bulk blends (particulate or powder) suffices. Liquid sources of S have also been used to prop the crops in Pampas. Evaluation of agronomic procedure related to soil S dynamics and fertilizer supply suggest that crop rotations do respond with better grain/forage production. Following is an example:

Crops/Rotation	Sulphur Source	Rate of Supply	Grain Response
-	-	(kg S ha^{-1})	(kg ha^{-1})
Wheat	Gypsum	15	625
Elemental-S	24–40	208–465	
Soybean	Gypsum	15	160–500
Wheat/Soybean Rotation	Gypsum	20 sequence^{-1}	Soybean-217 / Wheat-495

Source: Duggan (2010)

As stated earlier, generally, farmers in the Pampas produced crops using the inherent soil fertility efficiently by devising appropriate cropping systems. They recycled nutrients in situ and derived stable grain/forage yield. However, incessant cropping, erratic and uneven depletion of soil nutrients and lack uniform recycling procedures has created soil nutrient imbalance. Garcia (2001, 2013) states that, in general, soils of the Pampas, now a days, are deficient for nutrients such as N, P K, Ca, S, and Mg. They need to be replenished periodically. Soil nutrient deficiency and imbalances, if any, are detrimental to achieving high productivity. Major crops produced in the prairies such as wheat, maize, or soybean all require nutrients in tune with soil tests and yield goals. Following is an example that depicts the effect of supplying various essential nutrients to maize and soybean grown on Hapludolls found in Pampas:

Crop	Check	NP	NPS	NPSK	NPSKMg	Complete
Maize	5,620	9,000	10,100	10,100	10,200	10,200
Soybean	3,860	4,020	4,240	4,140	4,330	4,310

Source: Excerpted from Bianchini et al. (2003)
Check = unfertilized check, complete = N, P, S, K, Mg plus micronutrient mix that includes Zn, Fe, Cu, and B.

If the grain yield derived from check plot is equated to 100, then relative yield improved as we supply each and every essential nutrient deficient in the soil is perceptible. A complete set of nutrients (N, P, S, K, Mg plus micronutrients) resulted in balanced nutrition and the relative yield was 144 (Bianchini et al., 2003). Soil tests for all essential nutrients and adoption of "Best Management Practices" or balanced supply of nutrients is essential to achieve high productivity. We should note that dearth for nutrients, uneven distribution in the profile or inappropriate ratios may all affect natural prairies and cropland by restricting optimum productivity.

A section of farmers in Pampas are still prone to follow the traditional systems of fertilizer/manure supply. Fertilizer inputs in such farms may range from 70 to 83 N, 11 to 30 P, and 12 to 24 S plus micronutrients if any through FYM. Nutrient removals by moderately yielding crops such as maize or wheat are higher at 70–108 N, 27–64 P, and 30 S (Ghio et al., 2010). It creates a negative balance each season. There are other recently devised agronomic procedures termed as "Best Management Practices", Maximum Yield Technology, and GPS-guided Precision techniques. These provide extra allowance of nutrients and do not normally create negative balance of nutrients. Nutrient accumulation may occur in the prairie soils. Let us consider an example from Cordoba in the Pampas region. Ghio et al. (2010) measured nutrient balances as difference between net removal of nutrients by grains and that applied. For crops such as soybean about 50 percent of N need was satisfied through Biological N fixation and this value was considered while devising fertilizer schedules. Generally, N and P balances were positive. Highest negative balances noted ranged from 28 to 33 kg N ha^{-1} and 3–18 kg P ha^{-1}. Sulphur accumulation indicates that its needs are being over estimated in many farms. As stated earlier, nutrient balance is a key concept that influences fertilizer inputs, crops grown, and grain/forage harvest levels. It may also affect several other soil processes and eco-systematic functions. We should be alert to appearance of micronutrient deficiencies in the soils of Pampas. Micronutrients recycled using crop residues or via FYM may not suffice as the prairies get intensified. In fact, micronutrient imbalances may depreciate grain productivity by affecting crop's response to major nutrients—N, P, and K. This aspect is easily explained by Liebig's law of minimum. Nutrient most deficient will dictate crop's or grassland's response to fertilizer supply.

3.2.3 SOIL ORGANIC CARBON AND ITS SEQUESTRATION

The Pampas region is an important repository of C in soil, above-ground portions of natural prairie grasses, pastures and crops that fill this biome. In recent years, importance of soils and their C sequestration ability has been highlighted by many experts who have studied the vast cropping expanse of Pampas (Civeira, 2011). Soils that support cereal/soybean in Pampas are part of terrestrial systems that sequester large quantities of C in soil. Crop productivity in Pampas depends on satisfactory C fluxes and sequestration. Generally, SOC levels decrease as crop production progresses. Accurate estimate of C loss/sequestration trends in crop fields is highly useful while devising FYM input schedules and fixing crop yield goals. Let us consider a few useful facts about wheat belt in Pampas with regard to C dynamics. Civeira (2001) points out that wheat productivity is marginally higher in Humid Pampas (2.7 t grain ha^{-1}) compared to Semiarid Pampas (2.1 t ha^{-1}). It is attributed to climate, especially precipitation pattern, soil fertility and C recycling trends during past decades. About 0.9 t C ha^{-1} year^{-1} is recycled in wheat growing regions of Humid Pampas, but it is only 0.75 t C ha^{-1} year^{-1} in Semiarid Pampas. Long-term assessments by Kong et al. (2005) suggest that average organic-C input by wheat, throughout 1993–2000, in Humid Pampa was 8.1 t ha^{-1}, but it was only 6.75 t C ha^{-1} in Semiarid Pampas. Fertilizer supply, irrigation, and intensity of cropping have a say in C accumulation pattern. Fertilizer-N supply that improves vegetative production of wheat increases C sequestration in soil. Crop species and rotations also affect C sequestration in the soils of Pampas (Alvarez et al., 2002; Alvarez 2012). Typical wheat/corn or soybean rotation practiced in Pampas adds 3.3 t C ha^{-1} year^{-1} into soil (Miglierina et al., 2000; Alvarez et al., 2002). Wheat, individually accounted for 1.05 t C ha^{-1} year^{-1} (i.e., 32%). Prior knowledge about soil productivity and C input derived from different crop species seems very useful to farmers.

Soil quality depends on SOM levels to a great extent. Soils in Pampas, especially Mollisols are well endowed with organic matter. Yet, repeated tillage and reductions in residue recycling may lead to loss of SOC as CO_2, or through erosion of topsoil. Deep tillage practiced under conventional tillage systems is detrimental to SOC content. Repeated tillage actually makes soil highly oxidative, increases soil microbial activity, and induces CO_2 evolution. For example, in the Arguidolls of Uruguayan Pampas, no-tillage systems conserved SOC, but conventional tillage reduced SOC by 20 percent of initial levels. Inclusion of pasture in the rotation alleviated the situation. A crop-pasture system induced only 14 percent SOC loss (See Garcia-Prechac et al., 2004).

Diaz et al. (2002) believe that in Hapludolls of Western Pampas, SOM content and soil quality depend both on soil texture and management procedures. Examination of previous data and repeated experimental trials indicate that SOC decreases as crop season becomes lengthier. Pastures included in the rotation may enhance SOC. Further, Diaz et al. (2002) suggest that crop productivity in Pampas is related to SOC content in the top 0–20 cm layer of soil profile. Adoption of no-tillage systems seems efficient in preserving soil-C in the prairies. Bianchini et al. (2003) report that SOC content that ranges from 2.4 to 5.7 percent in large tracts of Pampas is sufficient to support a moderately high-yielding crop of maize/wheat. Finally, we ought to recognize that soil C status, C recycling potential, and accurate estimates of C inputs from different crops may affect global climate, considering that Pampas is a vast expanse of natural vegetation and crops.

3.2.4   ORGANIC FARMING

Agricultural zones within the Pampas of Argentina are endowed with naturally fertile Mollisols, congenial agroclimate, and moderate disease/pest pressure. Hence, farmers have usually reaped good harvests without applying extraneous inorganic fertilizers. Soil nutrient deficiencies have appeared only recently after 2003/2004. For a long stretch of period, since mid-1800s, Pampas has been producing crops with only farmyard manure and residue recycling as major sources of nutrient. Inorganic nutrient replenishment was never common in Pampas. Therefore, in practice, large farms in Pampas have been managed as organic farms, though inadvertently. Since 1985, organic farming has been a well-directed farming concept in the Pampas (FAO, 2004). CANECOS (Centro de Estudios de Cultivos Organicos) is an institution that concentrates on expanding organic farming in Argentina. Organic farming was in rudimentary stages with just 5–10 farms adopting it in 1987. Organic farming became more professional in 1990s. Further expansion of organic farming in Pampas depended much on demand for organically produced grains/forage and food items by European nations. About 80 percent of organic farm products are currently exported to Europe. Nature of farming enterprises, say those depending on inorganic fertilizers or those professing organic farming depends much on economic and external demand. Rapid intensification using inorganic fertilizes, obviously out competes organic farming zones. Organic farming in Pampas is not restricted to field crops such as wheat, maize, or soybean. Horticultural crops such as vegetables, flowers, and fruits are also grown with only organic manures (FAO, 2004; Organic Bouquet, 2013). During past 5 years, organic farming area in Argentina seems to have platued. The certified organic farm land has already expanded to 6.9 million acres. It offers US$ 40 million each season to nation's exchequer. Currently, organic soybean and maize produced in the Pampas is consumed in Europe. The main advantage of organic farming is that it does not disturb the ecoystematic functions to any great extent. Loss of nutrients to atmosphere and via percolation to groundwater is also negligible. Organic farming does not affect soil quality; rather it enhances C sequestration and quality (Non-GMO Report, 2006).

3.2.5   NO-TILLAGE SYSTEMS IN PAMPAS

Earliest efforts to introduce No-tillage system in the Argentina began at Anguil, in La Pampa province in 1960 (Garcia et al., 2000). At that time, several advantages in terms of ease of operation and economics were counted, but at the same time problems such as weeds, residues, and appropriate seeding equipment were also considered. Actually, development of potent herbicides, modified seed drills and planters in 1990s helped in popularizing no-tillage system. The zero-tillage or no-tillage system was introduced into cropping belts of Pampas and popularized vehemently during early 1990s. After an initial lag until 1993, spread of no-tillage system was rapid and reached over 10 m ha in 2000. At present, 22 m ha of cropland in Pampas and other regions of Argentina adopts no-tillage system (Peiretti, 1998; Trigo et al., 2009). The above trend in adoption of no-till indeed matches the forecasts made in 1998 about the potential and spread of no-till acreage in Pampas. Solbrig (2005) has remarked that rapid spread of conservation tillage or no-tillage systems into over

60 percent of cropping belt in Pampas is an achievement, considering its impact on soil fertility.

Some of the earliest experimental evaluations in Pampas suggest that, in case of no-tillage system, parameters such as soil pH in the top 15 cm layer of soil, organic matter content in the topsoil, amount of total and NO_3–N accumulated, $NaHCO_3$ extractable-P, soil evapotranspiration and water efficiency are all affected differently compared to conventional tillage (Peiretti, 1998). No-tillage reduces loss of topsoil and nutrients due to erosion. Reports from INTA suggest that in the sub-humid Pampas, no-tillage and crop residue application together restricted loss of soil moisture. Soil moisture got buffered. Hence, farmers benefitted to the tune of 4 inches of water due to adoption of no-tillage. Pieretti (1998) believes that improved water use efficiency is an important criterion for farmers in Pampas that enthuses them to adopt no-tillage system.

In the Uruguayan region of Pampas, farmers have practiced conventional tillage and continuous cropping for many years since 1970s. This has led to soil degradation, rampant erosion, and loss of SOM. It has also reduced soil productivity. Hence, during recent years, farmers have shifted to no-tillage with crop-pasture system. No-tillage reduces loss of SOC. Inclusion of pasture especially mixed pasture with legumes increases soil-N fertility and sequesters C (Garcia-Prechac et al., 2004).

In Uruguay, the Pampas ecoregion experiences subhumid temperate climate. It has about 80 percent of 16 m ha under natural prairies or agricultural field crops. Grasslands support variety of grass species/legumes and others. They are not exposed to regular tillage operations. Agricultural prairies or cropland is under conventional tillage systems and continuous cropping. According to Garcia-Prechac (2004) conventional tillage seems unsuitable because it has reduced soil productivity. Conventional tillage has also reduced SOM content. Continuous cropping has led to depletion of soil fertility. Hence, farmers in Uruguayan prairie region are suggested with idea of adopting crop-pasture rotation with no-tillage (i.e., direct seeding). Transition from conventional to no-tillage was initially not easy. However, since no-tillage reduces soil erosion, minimizes emissions, and avoids loss of SOC, it was accepted. In a crop-pasture system, no-tillage systems were better in terms of total biomass production. The crop-pasture sequence with no-till system actually adds to diversity of crops and buffers against economic risks.

There are several other reasons that support a shift from conventional tillage or disk plowing to no-tillage systems. We know that no-tillage reduces loss of SOC and erosion. There are still many other soil properties of great relevance to crop production that no-tillage system affects. For example, examination of different soil types that occur in the Pampas has shown that no-tillage improves soil texture, structure, and aggregate percentage. No-tillage causes slow infiltration of water. Soil penetration resistance was higher in no-tillage compared to conventional tillage. We should note that properties of topsoil (0–20 cm depth) of the prairies are important aspects that affect crop establishment, nutrient and water absorption by crops/natural grasses (Alvarez et al., 2009). Laura (2007) opines that no-tillage system that disturbs the soil profile least may yet affect a range of physicochemical traits of prairies. Soil properties such as water/ethanol—stable aggregates, total organic carbon (TOC), bulk density, soil aeration and infiltration are among the properties affected by tillage. In the soils from northern Pampas, SOC content had direct influence on aggregate stability and moisture holding capacity. Soil-N depreciated rapidly with depth under conventional tillage, but only gradually under no-tillage system. For example, at 18 cm depth, no-till plots still had 1.4 g N t^{-1} soil but only 1.02 g N t^{-1} soil under conventional system Again, reduction in NaHCO$_3$ extractable P due to soil depth was marked in conventional tillage plots compared to no-tillage plots (See Pieretti, 1998).

Let us consider an example from INTA, Marcos Juurez in Cordoba, where corn/soybean intercrops and rotations are common. Following is the effect of adopting conventional tillage, reduced tillage, and no-tillage systems on corn/maize rotation:

Grain Yield (kg ha^{-1})

Crops	Conventional Tillage	Reduced Tillage	No-Tillage
Corn/soybean rotation	8,467	8,988	9,334

Source: Excerpted from Sanchez et al. (1998)

Overall, no-tillage is an important agronomic procedure of great utility to farming stretches of Pampas. Along with crop residue/mulches, it has reduced loss of soil and its fertility to a great extent. It has also offered economic advantages. Natural resources such as soil, water, and nutrient in the prairies are conserved better if we adopt no-tillage system.

3.3   SOIL FERTILITY AND ENVIRONMENTAL CONCERNS

Historically, Pampas plains have supported evolution of ecologically well stabilized natural grasslands. Natives grew crops in dispersed locations with perhaps least disturbance to natural vegetation. However, with the advent of European settlers, development of pastures initiated soil degradation processes. Later, expansion and gradual intensification of agriculture became major disturbance to ecosystem. It induced deterioration of soil and other natural resources. Several phenomena such as soil erosion, loss of soil fertility, emission of greenhouse gas, salinity, and change in climatic parameters started appearing as major constraints to crop production in Pampas. Viglizzo (2002, 2005) have remarked that since 1890s, conversion of land for crop production has induced soil deterioration and lessened C sequestration potential of Pampas. Major reasons quoted are extensive conversion of natural grasslands, use of fire to manage range lands, excessive grazing, continuation of conventional tillage, and lack appropriate soil conservation procedures for over 80 years since 1890. Carbon emission has detrimental effects on atmosphere. Next, during past 2 decades, intensification of crop production has clearly induced rapid depletion of soil nutrients. Nutrient recycling trends too have been badly affected in some areas of Pampas.

Solbrig (2005) opines that agricultural cropping has been the may stay of Pampas during past 3–4 decades. At the same, it has also induced drastic changes in natural resources. It has caused soil degradation in vast areas. Fertile soil of Pampas is now wasteland unfit for crops or even natural prairie establishment, in many areas. Awareness about loss of fertility and soil quality is essential. Simplest concept to understand is that incessant cropping depletes nutrients and organic matter; hence they have to be replenished. Traditional fallows take 3 years to refresh a field. It means only one-third of the area in farm will be good enough for farming, at any given time. Not a good proposition considering the economic necessities. Fertilizers applied and inherent soil fertility is vulnerable to loss via several avenues. Now, let us consider in greater detail a few aspects of soil deterioration (e.g., erosion, leaching, percolation) and nutrient loss from Pampas landscapes.

Soil deterioration in Pampas has been generally attributed to incessant cropping necessitated by greater demand for food crops, better grain price, and economic advantages. During 1970s, higher demand for export further accentuated crop production trends. Inadequate measure to conserve soil fertility and quality and neglecting sustainable practices has led to certain degree of soil degradation. According to Mostacelli and Pazos (2000) main causes of soil degradation in the Pampas is incessant cropping and intensification of agricultural enterprises. The introduction of double cropping of wheat-soybean or other variations of crop sequence that includes sorghum or maize has caused rapid depletion of nutrients. Excessive tillage connected with incessant cropping tendencies has further accentuated loss of topsoil and soil organic carbon. Of course, no-tillage systems have alleviated the situation to a certain extent. According to Connior (2008), main cause of rampant soil erosion that leads to desertification is adopting improper farming techniques and neglecting soil. Overgrazing has reduced herbage cover, exposed soil and destabilized natural prairies. Such factors have also led to soil erosion and dust formation. Mining for silver and other elements has also induced soil erosion in Argentina. Soil erosion is a malady that afflicts over 1.3 m ha in the Pampas plains. Soil loss due to erosion is estimated at 20 t ha^{-1} in the Pampas. Wind induced erosion of topsoil is common in entire Pampa. Water erosion is a major detriment to cropping since 1950s (Soriano et al., 1992; Krapovickas and Di Giacomo, 1998). Baldi et al. (2006) have argued that fragmentation of native grassland and cropland has been driven by natural causes such as inadequate drainage and soil erosion processes. Crop production practices have been constrained. Adoption of large-scale soil alleviation programs are almost difficult.

Mostacelli and Pazos (2000) have further reported that soil deterioration is also rampant in many parts of Semiarid Pampas. Over 60,000 ha are exposed to environmental vagaries and damage of topsoil. Wind erosion of topsoil is experienced more in the semiarid regions of the plains. Water induced erosion is more common in the Humid Pampas and Mesopotamic region. Water erosion has direct influence on the cereal grain yield. Reports suggest that even moderate soil erosion due to surface flow of water reduces wheat grain productivity by 12–17 percent, compared to unaffected regions (Irrutia, 1995). If extrapolated to entire Pampas, soil erosion due to water itself reduces profitability of farms and natural pastures, and it results in loss of few billion US$ each year to the exchequer.

Fertilizer supply to crops in Pampas has been increasing gradually yet perceptibly due to intensification of cereal/soybean production. Fertilizer recovery rates are not high. In case of fertilizer-N, crop recovers about 33–39 percent and about 29 percent may get immobilized into SOM. The fraction of fertilizer-N unaccounted is generally presumed to have been lost through leaching, percolation, seepage as dissolved salts and via emissions as NO_2, N_2O, or N_2 (Silvine et al., 2006). Mollisols of Pampas possess relatively higher SOC content. Hence, they may immobilize N in the organic pool. Yet, N traced in drainage channel indicates that its leaching could become a problem, resulting in reduction of fertilizer-N use efficiency. Soil NO_3 is highly mobile. Its percolation into groundwater and aquifers causes serious environmental problems. Hence, fertilizer-N supply should be timed appropriately and applied to match the demand as exactly as possible. Crops should rapidly absorb fertilizer-N without allowing it to accumulate and become vulnerable to natural processes. Techniques that reduce loss of fertilizer-N need priority. Experimental evaluations during 1990s have shown that, if precipitation is high, and soil erosion too is rampant, N loss through leaching gets accentuated. This leads to high N stress index with a detrimental effect on maize grain yield (Paruelo and Sala, 1993). Following is an example:

Precipitation	N Stress Index	N loss	Grain Yield
(mm)	(vegetative)	(kg ha^{-1} d^{-1})	(t ha^{-1})
488	0.31	0.12	3.92
733	0.37	0.22	3.35

Source: excerpted from Paruelo and Sala (1993)

Intensification of cereal production zones in the Pampas has necessitated supply of higher quantities of fertilizer; especially N. Portela et al. (2009) believe that in Rolling Pampas, where water table is variable, NO_3–N levels in topsoil, groundwater and aquifers need to be monitored regularly. It is important to ascertain extent of NO_3–N exchange between aquifers and soil that support crops. Further, they suggest that, landscape in Rolling Pampas that has either natural grass or crops may actually alternate from being a natural sink to excess soil-N to NO_3–N source to crops, based on precipitation trends and direction of flow of nutrients.

According to Peiretti (1998), within the prairies of Pampas, topsoil loss due to erosion caused by wind and/or water is a major problem. Adoption of no-tillage system and application of crop residue on the surface reduces soil erosion. The loss of topsoil is regulated by extent of crop residue and mulches applied. For example, if only 20 percent of soil surface is covered by crop residue, relative loss of soil from top layer would be 48 percent. Portela et al. (2006) believe that loss of nutrients, especially N, through leaching and erosion is relatively less compared to similarly intensified wheat/maize farming systems in Europe.

Field studies during 1990s indicate that wind mediated erosion of topsoil, loss of fertility and quality is indeed a major problem in the semiarid regions of Pampas (Silvia et al., 1999). Wind erosion is particularly severe in Loess soils of La Pampa province (Michelena and Irrutia, 1995; Buschiazzo et al., 1998). Production of cereals and pasture may be impeded due to topsoil loss. Michelena and Irrutia (1995) have reported that frequent intensive winds have blown out topsoil, caused loss of organic matter and loss of soil aggregates. Field soil with greater proportion of small aggregates is prone more to loss due to wind. The actual rate of wind erosion measured in the La Pampa is about 9.4–27.1 t ha^{-1} year^{-1}. Wind erosion is also rampant in the South-western La Pampa province where it reaches >51–53 t ha^{-1} year^{-1}. If unchecked, the potential loss of topsoil due to wind in the South-Western Pampa reaches 176 t ha^{-1} year^{-1}. Buschiazzo et al. (1998) believe that susceptibility of soil to wind erosion is dependent on soil clods and loamy structure. In a different study within la Pampa, they found that topsoil loss due to a wind storm (mean wind speed = 21.4 km ha^{-1}) is dependent on soil type. On Ustipsamments, soil loss was 1.82 t ha^{-1} and on Haplustolls it was 0.29 t ha^{-1}. Shorter wind storms with high wind velocity created more erosion and hastened soil deterioration compared with long duration storms of low wind velocity. Wind erosion has direct consequences on establishment and productivity of natural grass lands, pastures, and cropland. Within limits, there is another aspect of wind mediated erosion of topsoil and nutrient loss. We may realize that dust and soil particles that carry a certain amount of essential nutrients may actually get deposited at distant place, where they actually add to soil fertility. Hence, soil nutrient loss measured is actually the difference between that eroded and inputs derived through deposits drifted from other locations. Ramsperger et al. (1998) found that dust/particulate deposit rate in Pampas is about 370–770 kg ha^{-1} year^{-1} depending on location. In comparison, it is about 200–900 kg ha^{-1} year^{-1} in Kansas, 200–400 kg ha^{-1} year^{-1} in Israel and 140–1,560 kg ha^{-1} year^{-1} in Sahelian West Africa. The mineralogy and nutrient carrying capacity of soil particles deposited is important. Following is an example of nutrients transported and deposited in the La Pampas:

Location	Total Nutrients Deposited (kg ha^{-1} year^{-1})						
	Na	K	Ca	Mg	Cl	N	PO$_4$-P
Maize field in Bahia Blanca	43.3	33.6	55.8	13.5	92.8	29.1	2.5

Source: Ramsperger et al. (1998)
Note: Typically, dust generated in La Pampa may carry 566 mg Na, 273 mg K, 664 mg Ca, 145 mg P, and 125 mg Mg kg^{-1} dust.

Loss of soil-N through volatilization is rampant in the humid regions of Pampas. Fertilizer-N or that inherently found in soil profile is vulnerable to loss via emission as NO_2, N_2O, or N_2. Palma et al. (1998) found that loss of N via ammonia volatilization is influenced by factors such as agroclimate, soil type, soil-N status, agronomic procedures, especially tillage procedures. On Arguidolls common to Pampas, ammonia volatilization becomes marked in 3 days after application of Urea. Ammonia volatilization process could be uneven during the various stages of crop phenology (Ciampitti et al., 2012). Ammonia volatilization was higher at 11.5 percent, if urea was surface applied and no-tillage systems were adopted. In plots under conventional tillage, ammonia volatilization accounted for 5.4 percent of fertilizer-N applied to crops. Steinbach and Alvarez (2006) have also reported that emission of N as NO_2 or other forms increases under no-tillage systems. They have argued that no-tillage systems have indeed improved C sequestration and reduced soil erosion in the Pampas. It seems conversion of whole Pampean cropping zones into a no-till system adds 74 Tg C annually. This perceptible advantage may after all get offset due to enhanced loss of GHG such as NO_2 and N_2. Hence, use of nitrification inhibitor and other measures that reduce N-emissions are important. According to Alverez (2012), N_2O fluxes from natural grasslands or agricultural prairies both are influenced by seasonal variations in weather parameters. Generally, N-emissions were low during winter but peaked during spring. Measured values for N-emissions were lower than simulated values for cropped fields/fallow regions. Further, Gomes et al. (2009) state that, long-term effects of crop rotations, cover crops and fallows on N-emissions need to be understood. Pampas is a vast stretch; hence even small reductions in N-loss may actually be very useful. In order to reduce N-emissions, they suggest better synchronization of N inputs, N-mineralization and its release into root zone with crop growth stage. Mean fluxes of N-emissions are related to total soil-N stocks in the prairies. Hence, undue soil-N accumulation has to be avoided.

3.3.1   WATER RESOURCES, IRRIGATION AND CROP PRODUCTIVITY

Pampas is endowed with several different water sources, in addition to natural precipitation. Major rivers that supply water to crops are Salado, Parana, Uruguay, Rio De Platta, Colorado, and Rio Negros. Pampas is also served by several water bodies such as dams and lakes. Farmers in Pampas use different irrigation systems such as surface (canal), sprinkler, and localized water supply. Irrigation system is relatively not well developed, since only <6.0 percent cropping zones are irrigated. Most farmers rely on annual rainfall and stored soil moisture.

Groundwater resources and water table could gain in importance based on location, crop and other resources, such as precipitation pattern or riverine/canal irrigation. Pampas are endowed with groundwater resources. However, it has not been efficiently utilized. Nosetto et al. (2009) state that both groundwater and crop interact to influence each other. When groundwater level is at the bottom of root system, it serves as valuable source of water. It can replace irrigation water requirements to a certain extent. In humid region, water table above certain limit creates waterlogging and stagnation. It affects crops and natural vegetation detrimentally. Water logging induces root anoxia and salination. Natural grasslands and crops do not establish well, if the groundwater is high and close to surface. It leads to swampy terrain. Jobbagy and Jackson (2004) point out that vegetation changes, especially from tree dominated systems to natural grassland or to cropland has immediate consequences on groundwater storage.

The relevance of groundwater to crop production in Pampas actually depends on crop's need and the extent to which it is satisfied through natural precipitation pattern. Several other factors such as soil/crop evapo-transpiration, crop duration, salinity, water loss via percolation and seepage need attention. Positive effect of groundwater on Pampas is more visible in the drier regions and summer. According to Nosetto et al. (2009), an optimum groundwater depth for maize is 1.4–2.45 m, for wheat it is 1.2–2.0 m and for soybean it is 0.70–1.65 m. Locations with optimum groundwater levels often produce grain yield 1.8–3.7 times greater than in location where groundwater level is more than 4 m deep from surface. We should note that, reciprocally, crops too influence groundwater. Presence of crops on a continuous basis reduces recharge of profile. Crops rapidly deplete water during critical stages and do not allow groundwater level to rise.

In the Pampean region, soil moisture deficiency and lack of supplemental irrigation are highly correlated to wheat crop productivity and total grain generated each season. Water supply level/precipitation pattern explains about 30 percent of grain yield variation in the provinces of Buenos Aires, Santa Fe, Cordoba, La Pampa, and Entre Rios (Hurtado et al., 2009). In fact, water required at critical stages of wheat growth needs most attention. According to Mon (2007), supplementary irrigation is necessary when natural precipitation is deficient or if its distribution pattern does not coincide with crop's need. Supplemental irrigation gains in importance when soil moisture is deficient at critical stages of the crop. In the Rolling Pampas, farmers have consistently adopted supplemental irrigation for the past 15 years. Such irrigation has affected soil quality, salinity levels, Electrical conductivity, ESP (Na%), pH, and SOM content. Irrigation may also have an impact on runoff, soil percolation index and erosion. Hence, prairies need to be supplied with more shrewdly, by timing it exactly and in quantities that match crop's need.

During the past 50 years since 1960s, water use efficiency of Pampas as a region has declined marginally from 65 to 59 percent. In contrast, there are other ecosystems/regions that show higher water use efficiency ranging from 75 to 88 percent (Viglizzo et al., 2011).

3.3.2 DROUGHTS IN THE AGRARIAN ZONES OF PAMPAS

Reports by National Aeronautic and Space Agency of USA (NASA) suggest that droughts have been frequent in the Pampas during past 2 decades. Satellite pictures, ground reports on weather and crop production trends during February 2009 suggest that drought affected zones showed dusty conditions and stunted crop growth. The period during November–March is crucial for various crops such as maize, wheat, sorghum, soybean, and cotton. Shortages in soil moisture and lack of irrigation resources severely affect crop productivity (NASA, 2009). Dust bowl conditions experienced during 2009 resulted in drying up of lakes and rivers. Water table too got depreciated. For example, the Salado river dried up due to longer stretch of drought. Rapid expansion of crops in the Pampas has led to proportionate depletion of water resources. In addition, factors such as climate change, a flat topography with unrestricted possibility for erosion of topsoil and frequent drought cycles have resulted in hydrological shifts in the Pampas. Farmers and ranchers have responded to dust bowl conditions in grass lands of Pampas by moving into areas where at least a low yielding crop could be raised. In 2012, drought conditions were experienced on all major crops grown in the plains. It was attributed to La Nina that struck the region and delayed precipitation events. The net crop loss forecast due to drought that occurred in 2012 was estimated at 5–7 m t corn. Grain productivity of soybean, sorghum, and wheat also reduced due to water scarcity in large farms.

3.4 CROPS AND CROPPING SYSTEMS PRACTICED IN THE PAMPAS

Major crops cultivated in the Pampas plains are:

Cereals: Wheat, Maize, Soybean, Sorghum, Barley, Rice

Legumes and Oilseeds: Soybean, Sunflower, Groundnut

Other crops: Sugarcane, Cotton, Tobacco, Pastures and Horticultural crops (e.g., vegetables).

Wheat culture was introduced into the Pampas plains during medieval conquests by Spaniards. European settlers initially grew wheat in the Buenos Aires region. Wheat crop expanded rapidly in the Pampas during 1880–1920. In fact, major wheat culture zone in the Pampas plains occur in the Buenos Aires province, especially southern portions. Studies adopting computer models/simulations suggest that wheat production in Pampas is affected by precipitation and temperature fluctuations during the crop season. In the Western dry Pampas, productivity of winter wheat is affected by temperature. Periodic droughts and low soil moisture status too reduces productivity. Simulation actually suggests that storing water and irrigating can increase productivity of wheat. A situation easier said than done. According to Menendez and Satorre (2007), Argentinean Pampas are considered most productive with regard to wheat culture. However, there are constraints to higher grain yield of wheat grown in different parts of Pampas. Models developed using information about cropping systems, precipitation pattern, and fertilizer supply during past 30 years are in place to refine wheat production. Plant factors such as photo-thermal unit, grain number, grain weight, and length of grain-filling period seem to affect the grain yield potential.

According to Bono et al. (2004), 40 percent of Argentinean wheat is generated in Semiarid Pampas. Here, the average grain yield is low at 2,000 kg ha^{-1}. Soil fertility, precipitation pattern and crop management decide the grain yield attainable. Wheat grain yield without fertilizer supply is low at 3,500–4,000 kg ha^{-1}. Mean yield potential of an irrigated crop is over 5,500 kg in the Northern Pampas and over 7,200 kg ha^{-1} in the Southern Pampas. Timing of anthesis is crucial. Alvarez et al. (2009) too believes that soil water in the upper 100 cm of soil profile, SOM content and weather parameters such as photo-radiation, evapotranspiration, and anthesis period influence wheat grain productivity in the Pampas. In almost all subregions of Pampas that support wheat, it is said precipitation pattern and anthesis period plays vital role in deciding grain yield potential (USDA Foreign Agricultural Service, 2006). Clearly, developing irrigation potential to regulate soil moisture status is an important aspect of wheat crop production system in Pampas. Frank (2011) states that wheat production is regulated more by the following factors such as two major nutrients N and P, and environmental variable such as rainfall and temperature. During past 4 decades, winter wheat genotypes released for cultivation in Pampas are meant for dual purpose (i.e., forage and grains). Regulation of planting dates, clipping and irrigation schedules are necessary to derive best advantage (Arzadun et al., 2006). Drought tolerance traits of wheat genotypes are being accentuated, since wheat belt in Rolling and Semiarid Pampas often encounters drought spells. Wheat genotypes with long grain fill period and high harvest index were preferred over those with extra rapid vegetative growth (Brisson et al., 2001).

Maize cultivation in the Pampas can be classified into four stages. Coscia (1980) identifies them as: (a) the introductory period (1875–1900); (b) the boom period (1900–1930); (c) the maize crisis (1930–1950); and (d) era of new technology (1950–till date). During past decades, maize cultivation in the Pampas has been concentrated in Northern Buenos Aires, Southern Santa Fe and Central Cordoba provinces.

Schnepf et al. (2001) have stated that during 1950s and 1960s, Pampas was already a major wheat and maize producing biome. However, soybean had not invaded the prairies to any great extent until 1970. The soybean crop in Pampas extended into mere 36,000 ha in 1970. In comparison, during the same period, US soybean cropping belt was 17 million ha and that in Brazil was 1.8 million ha. Grain productivity in USA was 50 percent high than in Pampas. It seems oilseed export embargo in USA created an upsurge in demand for soybean from other regions. This induced rapid expansion and intensification of soybean production in Pampas. The incentive to farmers in Pampas was great. By 1979, about 2.5 million ha of Pampas was covered by soybean crop. Many of the maize producing units shifted to soybean. As a consequence, maize area depreciated appreciably from 4.1 million ha to 2.5 million ha in 1980. Clearly, the cereal/legume ratio in the prairies got affected due to shifts in crop preferences. Overall, soybean area has steadily increased from almost nil to over 7.5 million ha (planted) during 2001 (Schnepf et al., 2001).Soybean cultivation is predominant in southern Santa Fe, eastern Cordoba, and Northern Buenos Aires. Since, its early introduction as an important export oriented crop in 1970s, soybean expanses in the Pampas have undergone considerable changes with regard to fertilizer/irrigation inputs, crop genotype and intensity of production. The total soybean grain production in Argentina (not Pampas) during 2010 had crossed 38 m tons, compared to 5–8 m t during 1980s. Clearly, soybean production practices have improved vastly resulting in high productivity. Currently, soybean belt in Pampas is filled predominantly with genetically engineered (GE) genotypes. Introduction of GE soybean during mid-1990s has again altered the Pampas farming region in terms of crop genotype. Herbicide-tolerant genotypes capable of producing higher amounts of grains per unit area have taken over from previously preferred genotypes. The GE soybeans tolerate higher levels of herbicide application, especially glyphosate. They allow repeated application of herbicides to control weeds. The high-yielding soybean genotypes require proportionately larger dosages of fertilizers. In many farms, primer-N and other essential nutrients are channeled at relatively higher rates to achieve grain yield goal of 3.8–4.2 t ha^{-1} (Kingsland and Hamilton, 2009). Soybean is a legume that adds to soil-N through its ability to form symbiotic association with Bradyrhizobium. The soil-N gain may range from 20 to 80 kg N ha^{-1} or even higher depending on factors related to inherent soil-N fertility, fertilizer input, irrigation, genotype, and productivity. We should note that soybean cultivation zone has expanded rapidly in Pampas. Therefore it has changed soil N dynamics to a perceptible level. Farmers do not supply fertilizer-N. At best, they apply only 5–10 kg N ha as primer-N to soybean compared with 40–60 kg N to wheat or maize. Natural grass lands adjacent to soybean fields and rotation pastures too might have perceived the change in soil-N dynamics. It is a glaring fact in terms of natural phenomena, yet not talked about or accounted for is that, introduction of soybean and its expansion has offered Argentina with large amounts of N at no cost, no extra toiling. The BNF process has added 20–40 g N ha^{-1} multiplied by 7 million ha annually to Pampas. Incidentally, this 20–40 kg N ha^{-1} assumption made above is among the lowest ranges reported for soybean's ability to contribute N via BNF. It has saved exchequer on industrial production of fertilizer-N, distribution and use in fields. It is an intelligent decision by Argentineans in selectively accentuating a legume such as soybean and harvesting atmospheric-N in massive quantities.

Sunflower production zones are well established in the western part of Pampas. The sunflower belt here thrives on Typic and Entic Haplustolls. Soils are coarse textured and well drained. Sunflower cropping zones also occur in southern and western Buenos Aires and La Pampa provinces. They are most preferred in drylands and may often replace a natural grass prairie or fallow field (Scheiner et al., 2002). Sunflower produced in this region is processed to derive crude vegetable oil. Much of the crude oil, about 70 percent is exported and rest is refined *in situ* to produce cooking medium. Sunflower genotypes grown in the Pampas possess about 41 percent oil.

The vast plains support a range of natural grass and legume species. Natural prairie of Pampas is actually a repository of several genera/species of C_3 and C_4 grasses, legumes, and surface turf type flora. The mixture of plant species actually negotiates fluctuations in the environment and rainfall pattern exceedingly well. Grass species that thrive better appear periodically and keep the surface green. The temperate grass species traced in Pampas are of good quality since they offer >20 percent protein with 70–80 percent digestibility (Garbulsky and Derigibus, 2003). In the Humid Pampas, Flooding Pampas and Entre Rios region, pasture grass species are well adapted to warm climate. Several C_4 grass species belonging to Panicoidea, Chlorideae, Andropoganeae, and Oryzae tribes occur. They alternate seasonally with C_3 species of Agrostae, Avenae, Festucae, Phalardeae, and Stipae tribes. Herbaceous legumes such as *Cassia sp, Crotolaria sp, Desmonthes sp, Vicia sp,* and *Phaseolus sp* are common Rio de Platta region. The Semiarid Pampas extends into small regions within Buenos Aires, La Pampa, San Luis, and Cordoba provinces. The climate is dry and it supports drought-tolerant species. Major grass/legumes traced in the natural prairies of Semiarid Pampas are *Poa ligularis, Stipa tricotoma, Stipa filiculmis, Panicum urvilleanum, Elionorus muticus, Sorghastrum pellitum, Eragrostis lugens, Bromus brevis,* and *Chloris retus* (Cleveland, 2008). Pampas supports several species of shrubs and small trees that are scattered all through the vast plains. Of course, like natural grasses, legumes, and pasture species, there is a clear need to try and establish as many shrub and tree species in the Pampas.

In addition to natural flora, Pampas supports regular commercial pasture species that stretch into many regions. Most commonly cultivated grass species that form the commercial pastures are rye grass (*Lollium perenne, L multiflorum*), tall fescue (*Festuca arundinaceae*), canary grass (*Phalaris aquatica*), brome grass (*Bromus catharticus*), cocksfoot (*Dactylis glomerata*). Mixed pastures contain a range of legumes such as *Trifolium repens, Medicago sativa, T. pretense,* and birds foot trefoil (*Lotus corniculatus*). When pastures are well developed and adequately supplied with fertilizer-based nutrients, it is said the productivity ranges from 12 to 15 t herbage ha^{-1} (Garbulsky and Deregibus, 2003). In comparison, natural grassland that flourishes in the Pampas may also supply 8–10 t herbage ha^{-1}. Natural grasslands and commercial pastures, both offer a certain degree of risk free and relatively low investment enterprises to the farmers. Forage productivity in cool temperate grasslands of Pampas has depreciated to 5 t ha^{-1}. This attributed gradual loss of cold-tolerant grass species and over grazing. Soil degradation processes and loss of soil-N are other reasons for reduction in herbage production in temperate regions of Pampas. During recent years, prairies are supplied with quality seeds of grasses/legumes that are used to develop commercial pastures.

3.4.1 CROPPING SYSTEMS OF PAMPAS

Several variations in crop mixtures and rotations are practiced in the Pampas. Their impact on soil fertility and productivity of Pampas needs to be understood in greater detail. Garcia (2001) opines that given the yield goals in Pampas, about 1.2 m t of fertilizers (e.g., Di Ammonium Phosphate) needs to be replenished if zero balance is to be reached in the Argentinean Pampas. In the Northern Pampas, wheat/soybean/maize/soybean is a 2–3 year long rotation that may encounter nutrient deficiencies, if fertilizer inputs are not appropriate. For example, Garcia (2001) reports that P recovered by soybean is 8 kg P t^{-1} grain and that by cereals is 4–5 kg t grain. The total P removed into grains by the rotation is 67.9 kg P ha^{-1}, but replenishment is only 32 kg P ha^{-1}. It creates a deficit of 35.7 kg P ha per sequence. Popular rotation in southern Pampas is wheat/maize/soybean. Again, total P supply to the cropping system is 40 kg P ha^{-1}, which creates a deficit of 15 kg P ha^{-1} per each cycle of rotation. Such deficits may have long-term impact on nutrient absorption pattern, and productivity of entire cropland.

During the past two centuries, natural prairies of Pampas has given way for variety of cropping systems, that include monocrops, dual crops, and crop mixtures. Wheat-fallow or wheat-summer crop has been the main stay of the region for many decades. During recent period, dual crop of soybean—wheat has dominated the ecosystem. Wheat-soybean dual crop has been adopted on a large scale to intensify the Pampas ecosystem, in terms of crop culture and productivity (Caviglia and Sadras, 2004). Coupled with better farming skills, capital and fertilizer inputs, farmers in Pampas have enhanced crop productivity. Annual yield of dual crops are high. Computer simulations and previous data suggest that wheat/maize and wheat/soybean double crop rotated with maize are the best suited options to further intensify Pampas farming zones.

During past 2 decades, the cropping system practiced in Pampas is predominantly soybean-based or to a certain extent it is based on wheat (e.g., Buenos Aires). Soybean-based cropping sequences have expanded from 32 to 65 percent of Pampas in the past 15 years (Van Opstal et al., 2011). The trends in cropping systems practiced in the Uruguayan and Brazilian Pampas is similar to one traced in Argentinean Pampas. The soybean/maize ratio in Argentina is six to seven compared with two for Brazilian Cerrados. Sole crops of soybean are known to capture 24–31 percent solar radiation and 51 percent of soil moisture (Caviglia et al., 2004, 2011). Caviglia and Andrade (2010) opine that Pampas currently has a sort of lopsided cropping system. It depends excessively on summer crop of soybean. They suggest that we should strive to find more efficient crops keeping in mind growing season. Van Opstal et al. (2011) believe that intensification of Pampas using sequential double cropping helps in enhancing water and solar radiation capture. They suggest wheat/soybean or maize/soybean double cropping as better option in terms of production efficiency. The total grain yield derived from double crops has almost always been higher by 40 percent over sole crops. No doubt, wheat-soybean sequence is very important to Pampas. This sequence has been subjected to further evaluation using variety of soil and crop management strategies. Models such as CERES-wheat and CERES-soybean have been employed to test, analyze, and arrive at best options for wheat-soybean production. For example, a few of the options acceptable were short season wheat genotypes, use of different herbicides, relay double cropping of wheat and soybean (Monzon et al., 2007).

Farmers in the Uruguayan and Brazilian Pampas have adopted several variations of cropping systems. They have of course aimed at maximizing photosynthetic light interception, biomass accumulation, and grain productivity. Cereals such as wheat, maize, sorghum or legumes such as soybean have dominated the landscape as either monocrop or dual-crop. Mixed cropping plus livestock combination has served the Uruguayan farmers better during past 3 decades. Farmers in this region have allowed natural prairies to flourish, developed pastures and grown dual crops to make their enterprise more stable in terms of ecosystematic functions and economic advantages. Crop sequences in particular have included 3–4 years of annual crops and 3–4 years of mixed pastures. Crop/pasture for livestock is also common in other regions of Pampas (Gimenez, 2006).

There is a great opportunity in the Pampas to overcome the monotony of soybean expanses being sown season after season each year. Martin et al. (2005) and Martinez et al. (2013) state that soybean is relatively a versatile legume and adapts to vagaries of environment and precipitation pattern both in winter and summer. Hence, it flourishes in the Pampas, covering almost over 50 percent of the fertile belt. They suggest that cover crops sown in the fallow period may alleviate the situation. Cover crops could add to soil-N and C sequestration. On Arguidolls, cover crops added about 6.6 t biomass ha^{-1} yr^{-1}. A legume cover crop may also add to soil-N. In addition, cover crops reduce soil erosion. Since they utilize residual soil nutrients efficiently, loss of nutrients via percolation or seepage and away from root zone is also minimized. Nutrients can be recycled effectively if cover crop is recycled.

3.4.2 GM CROPS IN THE PAMPAS

Crop genotypes developed to produce more grain/forage, resist major diseases, insects and tolerate environmental constraints are being increasingly sought in most of the intensive farming belts and equally so in subsistence farming zones. In the present context, soybean tolerant to herbicides such as glyphosate has been accepted by farms in Pampas, rather rapidly. The genetically engineered, glyphosate-tolerant soybean (i.e., Roundup Ready[RR] Soybean) helps farmers to apply larger dosages of herbicide and control the weed effectively (Bindraban et al., 2009). Lehman and Pengue (2000) state that soybean tolerant to Monsanto's glyphosate herbicide was commercially grown in just 80,000 ha in 1996. The "[RR]Soybean" was so effective that in 5 years, by 2,000 A.D., over 6.8 million ha, equivalent to 80 percent of soybean belt in Pampas was engulfed by this soybean genotype. It is a real upheaval if we consider agricultural flora of Pampas. It allowed excessive use of glyphosate chemicals. The usage of glyphosate chemicals increased from 1.3 million liters in 1990 to 59.2 million L in 2000. They further state that along with no-tillage system that got popularized equally rapidly and got accepted in Pampas, it led to a revolution in soil and weed management in the vast agricultural belt. According to Surman (2007), over 900 field experimental evaluations have been made with GM crops during 1998–2006. The above field trials aimed to know the advantages and ill effects, if any, to the ecosystem and crop productivity. Currently, 98 percent of soybean fields in entire Pampas are sown with herbicide-tolerant GE soybean that is also known as [RR]Soybean. About 70 percent of maize and 60 percent cotton sown in Pampas are genetically engineered genotypes. Among the consequences of introduction of GM crop, Trigo and Cap (2003), believe that it has contributed, though partially, to the improvement of total grain production from 26 million ton in 1988/89 to 75 m t in 2002/2003. During 2012, Argentina harvested 98 million ton grains. Development of resistance to glyphosate by weeds that cohabit the soybean fields is a point to ponder. Binnemalis (2009) has reported that natural prairie vegetation too may acquire resistance to glyphosate (e.g., Johnson grass). It may lead us to develop soybean genotypes that could tolerate incrementally higher dosages of glyphosate than what is resisted by weed flora. Later, it may induce farmers to amply high dosages of herbicide. This is not a situation sought by either farmers or ecologists.

3.4.3   BIOFUEL PRODUCTION AND ITS INFLUENCE ON NATURAL AND AGRICULTURAL PRAIRIES

Biofuel production is a recent trend in the Pampas that may engulf larger areas of natural grasslands, pastures and cropland. Biofuel production may induce changes in cropping pattern and environmental parameters such as C sequestration, N-emissions, soil erosion, and nutrient recycling *in situ*. Forage allocation to farm animals may also get altered. Like many other nations (e.g., Brazil), Argentina too is stipulating on admixtures and inclusion of biodiesels in the fuel. Bioenergy seems a promising alternative to petroleum usage in the Pampas. Pampas is a major agricultural belt in South America. It caters grains and forage to both domestic and international markets. About 80 percent of cereals, soybean, and pastures of Argentina are generated in the Pampas region (Hilbert, 2009; Tomei and Upham, 2011). Right now, Argentina utilizes large amounts of fossil fuel, which in due course may get partially replaced with biodiesel. The demand for biodiesel from both domestic and international consumers is high and promises to last forever. Hilbert (2009) states that, as a consequence of massive global demand for biofuel, International trade for raw material for biofuel, and the encouragement to generate biofuel crops is reasonably high. Agricultural prairies of Pampas have been intensified using GM crops (e.g., soybean, maize), irrigation and fertilizer-based nutrient supply. Pampas supports large area of soybean. Soybean was primarily produced for oil and meal in these prairies. Since soybean is a very good raw material for biodiesel, recent trend with farmers in Pampas is to allocate a certain area for biodiesel generation. During 2009/10 soybean belt extended into 19–20 m ha in the Pampas. About 20 percent of produce was used for oil extraction and 43 percent for making meal. Rest was used for forage and biodiesel production.

One of the major concerns expressed relates to biofuel crops and their localization within the already intensified Pampas prairies. Major biofuel generating crops such as maize and soybean, pastures and natural grasslands are all clustered in Pampas (Tomei and Upham, 2011). It may have environmental consequences in future. As such, it immediately affects the biodiversity and cropping pattern in the Pampas. In due course, more area in Pampas may be occupied by biofuel crops. A large section of natural grass lands too may give way for bio-diesel production. Van Dam et al. (2009) have reported that it is naturally more feasible and profitable to grow crops such as soybean and switch grass in Pampas for export to European destinations such as Netherlands. These two crops are good raw materials for biodiesel generation. Switch grass has high lignin and cellulose content. The high heating value makes switch grass a good crop for biofuel production. Incidentally, Pampas supports vast expanses of such grasses. Switch grass pellets are cost effective. Overall, trend in biofuel generation may actually affect crop diversity, dominant crop species, area occupied by each crop species, permanent pastures and natural grass lands that occur in Pampas. We need to monitor and regulate cropping zones where feasible, so that the GHG emissions are lessened and soil deterioration is avoided.

3.5   CROP PRODUCTION TRENDS IN ARGENTINA

Pampas is an important agricultural belt of the world. During past four–five centuries, it has been converted from natural grassland of cool temperate regions of South America, into a major crop production zone. Major crops produced in the prairies are wheat, maize, sorghum, soybean, groundnuts, sunflower; sugarcane and cotton. Agriculture is a major economic enterprise of Argentina. It is predominant in the fertile plains of Pampas. Crop production strategies are relatively intensified. They are also expansive with large farms. Agriculture provides employment to 7 percent of human population in Argentina. Interestingly, about 10–15 percent of Argentinean farms are owned by foreign-based private companies that support really large farms. It generates food for all in that nation plus exports a large quantity of grains/forage to other region in different continents. Indeed the agricultural prairie of Pampas is an important biome of the world, in terms of food generation. The Pampas agricultural belt is filled with crops such as soybean, maize, wheat and sorghum in addition to few others. Allocation of cropping zone within this vast biome has gradually increased. In 1960, only 23.7 percent area of Pampas was cropped. It increased to 30.3 percent in mid-1980s, 40 percent in mid-1990s and to 47 percent in 2007. Rolling Pampas supports crop production in 65 percent of its area, followed Subhumid Pampas 45 percent area and 39 percent in Semiarid Pampas (Viglizzo, 2002).

During 2011, Argentina exported 86 billion US$ worth raw agricultural product generated mostly in Pampas. It was mainly constituted by three major grains namely soybeans, maize and wheat. Other major products exported were flour, oilseeds and animal feed (INDEC, 2011). Forecast by USDA, suggests that during 2013, Argentina may harvest 55 m t soybean grains, 11.5 m t wheat grains and 28 m t corn (Mercopress, 2013).

Pampas is famous for its wheat-based agricultural prairies. It has gained special attention as a major exporter of grains during past 3 decades. Wheat production has been a major preoccupation in the plains of Pampas. Major wheat producing province in Pampas is Buenos Aires. It is termed as the "Bread basket". Forecasts suggested that wheat crop may extend into 5.1 m ha and offer about 11.1–11.5 m t grain during 2013 (MercoPress, 2013). This region contributes about 2.5 percent of global wheat harvest, but much of it is exported. In fact, wheat belt in Pampas in Argentina is regulated by demand for wheat by other nations situated in Russia, Asia and Fareast. Pampas has generated 2.5–2.7 percent of global wheat grain production during past decade. The wheat belt in Pampas has fluctuated markedly on a yearly basis between 1970 and 2010. It extended into 4.1 m ha in 1970 and peaked at 6.5 m ha in 1982–1985. Later, there was slump in area to 4.3 m ha in 1988 and peaked again in 1996–98 at 6.8 m ha (Schnepf et al., 2001). Currently, wheat belt of Pampas is about 5.2 m ha in expanse.

Soybean belt is large and dominates an area of 16.2 m ha and generates 47.6 m t grains annually. Much of the soybean acreage is in Pampas, especially in Cordoba. In 2007, about 22 percent of global soybean production occurred in Pampas and adjoining areas in Argentina. Agricultural agencies in Argentina state that soybean production may reach 50–55 m t year^{-1}. It is derived from an area of 16.8 to 17.8 m ha. Future increase in soybean production seems to depend on adoption of latest techniques, improved genotypes, enhanced infrastructure and economic returns for the produce (Pengue, 2013). Pampas has supported a vast soybean-based cropping system during past 3 decades. It must have perceptibly altered the soil-N dynamics and fertilizer-N supply to the biome. Soybean's demand for fertilizer-N is least compared to cereal crops that it must have replaced. On an average, soybean adds 20–40 kg N ha^{-1} through BNF. When extrapolated to entire 17.8 million ha of soybean expanse, it becomes clear that Pampas, as a biome has added large quantity of atmospheric N, because of soybean cultivation.

Maize producing zone in Pampas is concentrated in Cordoba, Entre Rios, San Luis, and La Pampa. It is estimated to extend into 4.7 m ha and offer 4.97 m t grains during 2013 (Mercopress, 2013). Argentina offered about 2.8–3.0 percent of global maize grain harvests during 2007–2012. Maize production in Pampas was well entrenched into over 4.2 m ha during early 1970s. However, maize belt has experienced fluctuation in area and has depreciated gradually. Corn belt peaked in 1980 at 3.2 m ha, but reduced markedly in expanse to <1.8 m ha in 1988/89. Currently maize belt extends into 3.1 m ha. It is mostly rotated or intercropped with soybean or wheat.

During the past decade, sunflower production zones expanded in area. It reached 2.4 m ha and produced about 3.6 m t grains annually. Sunflower belt in Argentina accounted for 11 percent of global sunflower grain harvests during 2007–2012 (FAO, 2012).

Sorghum belt within Pampas stretched into 0.6 million ha and produced 3.0 million t grains during 2011. It constitutes about 5 percent of global sorghum grain production.

KEYWORDS

- **Agroclimate**
- **Agronomic**
- **Organic farming**
- **Pampas of South America**
- **Recycling**
- **Sequestration**
- **Topsoil**

REFERENCES

1. Adelman. Frontier Development: Land, Labour and Capital on the Wheat lands of Argentina and Canada 1890–1914. **1994b**, 1–2, April 23, 2013, http://www.oxfordscholarship.com/veiw/10.1093/acprof:oso/9780198204411.001.0001/acprof-9780198204411 .htm
2. Arzadun, M. J.; Arroqquy, J. I.; Laborde, H. E.; and Brevedon, R. E.; Effect of Planting date, Clipping height and cultivar on forage and grain yield of winter wheat in Argentina Pampas. *Agron. J.* **2006**, *98*, 1274–1279.
3. Argentina Authentic. Vegetation. **2012**, 1–3, April 20, 2013, http://www.argentinaautentica.com/vegetation.php
4. Alvarez, R.; Predicting average regional yield and production of wheat in the Argentine Pampas by an artificial neural network approach. European Journal of Agronomy **2007**, *30*, 70–77.
5. Alvarez, R.; Soil Organic Carbon stock in Pampean soils: Changes associated to Rotation and Tillage. **2012**, 1–2, April 21, 2013, http:/www istro2012.congressos-rohr.info/programa/_4_1_Roberto_Alvarez. Htm
6. Alvarez, R.; Alvarez, C. R.; and Steinbach, H. A.; Association between soil organic matter and wheat yield in Humid Pampa of Argentina. Communications in Soil Science and Plant Analysis. **2002**, *33*, 749–757.
7. Alvarez, C. R.; Taboada, M. A.; Gutlerrez Boem, F. H.; Bono, A.; Fernandez, P. L.; and Prystupa, P.; Top soil properties as affected by Tillage systems in the Rolling Pampa region of Argentina. *Soil Sci. Soc. Am. J.* **2009**, *73*, 1242–1250.

8. Baldi, G.; Guerschman, J. P.; and Paruelo, J. S.; Characterizing fragmentation in Temperate South America grasslands. *Agricu. Ecosyst. Environ.* **2006,** *116,* 197–208.

9. Baldi, G.; and Paruelo, J. M.; Land use and land cover dynamics in South American Temperate Grasslands. *Ecol Soc.* **2008,** *13,* 1–10.

10. Binnemalis, R.; Transgenic trademill: Response to the emergence and spread of glyphosate-resistant johnsongrass in Argentina. *Geoforum.* **2009,** 1–4, April 19, 2009, http://dx.doi.org/10.1016/j.geoforum.2009.03.009

11. Berardo, A.; Ehrt, S.; Grattone, F. D.; and Garcia, F. O.; Corn yield response to Phosphorus fertilization in the South-eastern Pampas. Better Crops International **2001,** *15,* 3–5.

12. Bert, F. E.; Satorre, E. H.; Toranzo, F. R.; and Podesta, G. P.; Climatic information and de-cision-making in Maize crop Production systems of the Argentinean Pampas. *Agricu. Syst.* **2006,** *88,* 180–204.

13. Bianchini, A.; Ambrogio, M.; Lorenzatti, S.; and Garcia, F.; Nutrient management for a no-till rotation in the Pampas: Three years of field trials. *Better Crop Int.* **2003,** *17,* 6–10.

14. Bindraban, P. S.; et al. GM-related sustainability: Agro-ecological impacts, risks and opportu-nities of Soy production in Argentina and Brazil. Plant Research International B.V. Wagenin-gen, Netherlands, Report, **2009,** *259,* 1–45.

15. Bolsa, D.; From cattlemen to Crops: Can Argentina's Gauchos cash in on the commodities boom. **2011,** 1–4, April 19, 2013, http://www.drbolsa.com/?p=1654

16. Bono, A.; Paepe, D. J.; and Alvarez, E. A.; In-season wheat yield prediction in the Semi-arid Pampa of Argentina using Artificial Neural Networks. **2004,** 133–150, April 20, 2013, http://www.novapublishers.com/catalog/product_info.php?products_id=27234

17. Brisson, N.; Guevara, E.; Meira, S.; Maturano, M.; and Coca, G.; Response of five wheat cultivars to early drought in the Pampas. *Agronomie.* **2001,** *21,* 483–495.

18. Buschiazzo, D.; Teddy, Z.; and Silvia, A.; Wind erosion in Loess soils of the semi-arid Ar-gentinean Pampas. **1998,** 1, April 20, 2013, http://are.usda.gov/research/publications/publica-tions.htm?seq_no_115=86320

19. Calvo, N. I. R.; Rozas, H. S.; Echevarria, H.; and Berardo, A.; Contribution of Anaerobically incubated Nitrogen to the Diagnosis of Nitrogen Status in spring wheat. *Agron. J.* **2013,** *105,* 321–328.

20. Caviglia, P. O.; and Andrade, F. H.; Sustainable intensification of agriculture in the Argentin-ean Pampas: Capture and use efficiency of environmental resources. *Am. J. Plant Sci. Biotech-nol.* **2010,** *3,* 1–8.

21. Caviglia, P. O.; and Sadras, V. O.; Long-term simulations of productivity in crop sequences differing in intensification in the Argentina Pampas. Proceedings of International Crops Sci-ence Congress, Brisbane. **2004,** 1–4, April 16, 2013, http://www.icsc2004.htm

22. Caviglia, O. P.; Sadras, V. O.; and Andrade, F. H.; Intensification of agriculture in the South-eastern Pampas I. Capture and efficiency in the use of water and radiation in the double cropped wheat-soybean. *Field Crops Res.* **2004,** *87,* 117–129.

23. Caviglia, O. P.; Sadras, V. O.; and Andrade, F. H.; Grain yield and quality of wheat and soy-bean in sole and double cropping. *Agron. J.* **2011,** *103,* 1081–1089.

24. Ciampitti, I. A.; Ciarlo, E. A.; and Conti, M. E.; Nitrous oxide emissions from soil during soy-bean (glycine max) crop phenological stages and stubbles decomposition period. *Biol. Fertil. Soils.* **2012,** *44,* 581–588.

25. Civeira, G.; Estimation of Carbon inputs to soils from what in the pampas region, Argentina. *Czech J. Genet. Plant Breed.* **2011,** *47,* S39–S42.

26. Cleveland, C. J.; Semi-arid Pampas. In: Encyclopaedia of Earth. National Council for Science and Environment. Washington DC; **2008,** 1–7, April 20, 2013, http://www.eoearth.org/article/Semi-arid_Pampas.htm

27. Connior, M. B.; Argentina: A state of the Environment report. *Arkansas State Univ. Elect. J. Integrat. Biosci.* **2008,** *2,* 21–60.

28. Coscia, A. A.; Maize development in Argentina: A hundred years of Maize in the Argentina. Hemisferio Sur. Argentina: Buenos Aires; **1980**, 120.
29. Diaz, Z.; Duarte, G. A.; and Grove, J. H.; A review of no-till systems and soil management for sustainable crop production in the sub-humid and semi-arid Pampas of Argentina. *Soil Tillage Res.* **2002**, *65*, 1–8.
30. Diaz-Zorita, M.; Virgina, M.; and Canigia, F.; Field performance of a liquid performance of *Azospirillum brasilense* on dry land wheat productivity. *Euro. J. Soil Biol.* **2008**, *10*, 1–9.
31. Di Ciocco, C.; Penon, E.; Coviella, C.; Lopez, S.; Diaz-Zoritta, M.; Momo, F.; and Alvarez, R.; Nitrogen fixation by soybean in the Pampas: Relationship between Yield and soil Nitrogen balance. **2013**, 1–12, April 22, 2013, http://www. congresodeseulos.org. ar/site/wp-content/media/PDFS/Miecoles/Columnas_Boulevard?08.30-ps_-_Di_ciocco.pdf.
32. Dillehay, T. D.; The Settlement of Americas: A New Prehistory. New York: Basic Books Inc.; **2000**, 275.
33. Droller, F.; Migration and Long-run Economic Development: Evidence from Settlements in the Pampas. **2013**, http://blogs.brown.edu/fdroller/files/2013/01/DROLLER_JMP.pdf 1-19
34. Duggan, M. T.; Rodriguez, M. B.; Lavado, R. S.; and Melgar, R.; A review of sulphur fertilizer use and technology management in Pampas region of Argentina. Proceedings of 19th World Congress of Soil Science, Brisbane, Australia; **2010**, 14–17 (published on DVD).
35. FAO, Argentina: World markets for organic fruit and vegetable. Food and Agricultural Organization of the United Nations. Rome, Italy; **2004**, 1–14, May 17, 2013, http://www.fao.org/docrep/004/Y1669E/Y1669e0h.htm
36. FAO, Use of Fertilizers by Crop and Region. Rome, Italy: FAO of the United Nations; **2007**, 1–7, April 29, 2013, http://www.fao.org/docrep/ 007/y5210e/y5210e09.htm
37. FAO. FAO Statistics. **2011**, April 16, 2013, http://www..faostat.fao.org/site/567/
38. FAO. FAO Statistics. **2012**, April 16, 2013, http://www.faostat.org/site/567/
39. Frank, L.; The wheat production function in the Pampean region (Argentina). *Ciencas Agronomicas.* **2011**, *18*, 33–41.
40. Garbulsky, M. F.; and Deregibus, V. A.; Argentina. **2003**, 23, April 15, 2013, http://www.fao.org/ag/AGP/AGPC/doc/Counprof /Argentina/ argentina.htm
41. Garcia, F. O.; Phosphorus balance in the argentinean Pampas. *Better Crops Int.* **2001**, *15*, 22–24.
42. Garcia, F. O.; Nutrient Best Management Practices for Wheat fertilization practices for Intensive wheat production in Southern Latin America. **2013**, 1–3, April 20, 2013, http://www.ipni.net/ppiweb/ltams.nst/$webindex/48E840CAD00D2E5903257529007 10F96?opendocument&print=1.htm
43. Garcia, F. O.; Ambroggio, M.; and Trucco, V.; No-tillage in the Pampas of argentina: Success story. *Better Crops Int.* **2000**, *14*, 24–27.
44. Garcia-Prechac, F.; Ernst, O.; Siri-Prieto, G.; and Terra, J. A.; Integrating no-till into crop-pasture rotations in Uruguay. *Soil Tillage Res.* **2004**, *77*, 1–13.
45. Ghio, H.; Gudelj, V.; Espotumo, G.; Boll, M.; Bencardini, J.; and Garcia, F.; Long-term on-farm demonstrations in the central Pampas of Argentina: A case study. *Better Crops.* **2010**, *94*, 28–31.
46. Gimenez, A.; Climate change and variability in the Mixed crop/livestock production systems of the Argentinean, Brazilian and Uruguayan Pampas. Final report on Impacts and Adaptations to Climate Change (AIACC) Project no. LA 27. Washington, DC: The International START Secretariat; **2006**, 138.
47. Glatzle, A.; Compendio para el manejo de pastures en el Chaco. Proyecto Estacion Experimental Chaco Central (MAG-GTZ) GTZ, Ellector. **1999**, 188.
48. Gomes, J.; Bayer, C.; Costa, F. S.; Piccolo, M. C.; Soil nitrous oxide emissions in long-term cover crops-based rotations under sub-tropical climate. *Soil Tillage Res.* **2009**, *106*, 36–44.

49. Gutierrez-Boem, F. H.; Ferraris, G.; Prystupa, P.; and Salvagitti, F.; Corn Response to Phosphorus, Sulphur and Potassium Fertilization. Georgia, USA: International Plant Nutrition Institute, Norcross; **1999,** 1–3.

50. Hall, A. J.; Rebella, C. M.; Ghersa, C. M.; and Cullot, J. H.; Field crop systems of the Pampas. In: Pearson, C. J.; ed. Ecosystems of the World. Field Crop Ecosystems. Amsterdam, Netherlands: Elsevier Scientific; **1992,** 413–450.

51. Hilbert, J. A.; Argentina, an alternative to produce biofuels. **2009,** 39–41, May 8, 2013, http://www.biofuelskeyplayer.com.ar/Download/Cap10.pdf

52. Hillock Capitol Management. Agricultural Soils of Argentina. **2012,** 1–19, April 23, 2013, http://www.agribenchmark.org/fileadmin/freefile/ ccc_2012/ccc12-GF_Otero.pdf

53. Hurtado, R.; Faroni, A.; Murphy, G.; Serio, L.; and Fernandez Long, M. E.; Critical watr supply deficiency for wheat's yield in Pampas region of Argentina. Revista de La Facultad de Agronomia, Universidad de Buenos Aires. **2009,** *29,* 1–12.

54. INDEC. Foreign Trade Export complexes. **2011,** 1–5, April 16, 2013, http://www.indec.gov.ar/principal.asp?id_tema=5187.htm

55. IDRC **2011** Floods Droughts and Farming on the Plains of Argentina and Paraguay, Pampas and Chaco Regions. Consejo Nacional de Investigaciones Clentificas y Tecnicas. 1–3, April 22 2013, http://www.idrc.ca/EN/Regions/Latin_America-and-the-Caribbean/Pages/Project-Details.aspx?/ProjectNumber=106601

56. INTA. Atlas de Suelos de la Republica Argentina., Instituto de Nacional Tropicale Agricultra, Buenos Aires, Argentina, 2 Tomos, 677 paginas, **1990.**

57. IPCC, Agriculture and Forestry. Climate Change 2007: Working Group II: Impacts, Adaptation and Vulnerability. Washington. DC: International Panel on Climate Change; **2007,** 1–2, April 19, 2013, http://.ipcc.ch/publications_and _data/ar4/wg2/en/ ch13s13-5-1-2.html

58. Irrutia, C. B.; Influencia de los procesos de degradacion en la produvidad de suelo.Informe de annual plan de trabajo. Buenos Aires. Argentina: Instuto de suelos, INTA, Castelar; **1995,** 1–22.

59. Jobbagy, E. G.; and Jackson, R. B.; Groundwater use and salinization with grassland afforestation. *Global Change Biol.* **2004,** *10,* 1299–1312.

60. Kingsland, M.; and Hamilton, M.; Population and Natural Resources: How Can Food be Produced Sustainably to Feed Growing Populations-Argentina. **2009,** 1–6, April 18, 2013, http://www.cgga.aag.org/PopulationandNaturalResources1e/ CS_Argentina_ July09/CS_Argentina_July09_print.html

61. Kong, A. Y. Y.; Six, J.; Bryant, D. C.; and Denison, F.; The relationship between carbon input, aggregation, and soil organic carbon stabilization in sustainable cropping systems. *Soil Sci. Soc. Am. J.* **2005,** *69,* 1078–1085.

62. Krapovickas, S.; and Di Giacomo, A. S.; Conservation of Pampas and Campos in Argentina. *Parks.* **1998,** *8,* 47–53.

63. Krishna, K. R.; Agrosphere: Nutrient Dynamics, Ecology and Productivity. Enfield, New Hampshire, USA: Science Publishers Inc.; **2003.**

64. Krishna, K. R.; Maize Agroecosystem: Nutrient Dynamics and Productivity. Waretown, New Jersey, USA: Apple Academic Press Inc.; **2013,** 348.

65. Laura, F.; Physical quality indicators in soils from Northern Pampa region of Argentina under no-till management. *Cienca Seoulo.* **2007,** *25,* 159–172.

66. Lehman, V.; and Pengue, W. A.; Herbicide tolerant soybean: Just another step in a technology trade mill. *Biotechnol. Dev. Monitor.* **2000,** 43, 11–14.

67. Lewis, D. K.; The History of Argentina. Westport, Connecticut, USA: Greenwood Press; **2001,** 278.

68. Maddoni, G. A.; Urricariet, S.; Ghersa, C. M.; and Lavado, R. S.; Assessing soil quality in the Rolling Pampa, using soil properties and Maize characteristics. *Agron. J.* **1999,** *91,* 280–287.

69. Magrín G. O.; Travasso M. I.; and Rodríguez, G. R.; Changes in climate and crop production during the twentieth century in Argentina. *Climate Change.* **2005,** 72, 229–249.

70. Martin, G. O.; Travasso, M. I.; and Rodriguez, G. R.; Changes in Climate and Crop Production during the twentieth century in Argentina. *Climate Change.* **2005,** *72,* 229–249.

71. Martinez, J. P.; Barbieri, P. A.; Sainz Rosas, H. R.; and Echevarria, H. E.; Inclusion of Cover crops in cropping sequences with soybean predominance in the Southeast of the Humid Argentine Pampa. *Open Agricu. J.* **2013,** 7, 3–10.

72. Menendez, F. J.; and Satorre, E. H.; Evaluating wheat yield potential in the Argentine Pampas. *Agricu. Syst.* **2007,** 95, 1–10.

73. MercoPress, Argentine Wheat Forecast Down Because of Flooding in Key Parts of Farm Belt. Duenos Aires, Argentina: South Atlantic News Agency; **2013,** 1–2, April 20, 2013, En.mercopress.com/2012/11/23/argentina-wheat-forecast-down-because-of-flooding-in-key-parts-of-farm-belt

74. Michelena, R. O.; and Irrutia, C. B.; Susceptibility of Soil to wind erosion in La Pampa province, Argentina. *Arid Soil Res. Rehab.* **1995,** *9,* 227–234.

75. Miglierina, A. M.; Iglesias, J. O.; Landriscini, M. R.; Galantini, J. A.; and Rosell, R. A.; The effects of crop rotation and fertilization on wheat productivity in the Pampas semi-arid region of Argentina: Soil physical and chemical properties. *Soil Tillage Res.* **2000,** 53, 129–135.

76. Mon, R.; Irrutia, C.; Bota, G. R.; Pozolo, O.; Melcon, F. M.; Rivero, D.; and Bomben, M.; Effects of supplementary irrigation on chemical and physical soil properties in the Rolling Pampa Region of Argentina. *Cienca e Investigacion Agraria.* **2007,** *34,* 187–194.

77. Monzon, J. P.; Sadras, V. O.; Abbate, P. A.; and Caviglia, O. P.; Modelling management for wheat-soybean crops in the South-eastern Pampas. **2007,** 1–2, April 19, 2013, http://www.academic.research.microsoft.com/Paper/6399166.aspx

78. Mostacelli, G.; and Pazos, M. S.; Soils of Argentina-Nature and Use. **2000,** 1–8, April 19, 2013, http://www.elsitioagricola.com/articulos/ mostacelli/ soilsofargentina-natureanduse.asp

79. NASA. Drought in Southern South America. NASA Goddard Space Centre. **2009,** 1–2, April 21, 2013, http://earthobservatorynasa.gov/ naturalHazards/View.php?id=37239

80. Non-GMO Report. Organic Farming Increasing in Brazil, Peaks in Argentina. **2006,** 1–2, May 17, 2013, http://www.non-gmoreport.com/articles/ mar06/organic_brazil_argentina.php

81. Nosetto, M. D.; Jobbagy, E. G.; Jackson, R. B.; and Sznaider, G. A.; Reciprocal influence of crops and shallow ground water in sandy landscapes of the Inland Pampas. *Field Crops Res.* **2009,** *113,* 138–148.

82. Organic Bouquet. BioGarden La Pampa. **2013,** 1–2, May 17, 2013, http://www.organicbouquet.com/info.aspx?pid=138

83. Palma, R. M.; Saubidet, M. I.; Rimolo, M.; and Utsumi, J.; Nitrogen losses by volatilization in a Corn crop with two tillage systems in the Argentine Pampa. *Commun. Soil Sci. Plant Anal.* **1998,** *29,* 2285–2879.

84. Paruelo, J. M.; and Sala, O. E.; Effect of global change on maize production in the Argentinean Pampas. *Climate Res.* **1993,** *3,* 161–167.

85. Peiretti, R. A.; The development and future of direct seed cropping systems in Argentina. In: Sustaining the Global Farm. Selected Papers from the 10th International Soil Conservation Organization Meeting Held at Purdue University. Stott, D. E.; Mohtar, R. H.; and Steinhardt, G. C.; eds. West Lafayette, Indiana, USA: Purdue University; **1998,** 234–247.

86. Pengue, W. A.; The Impact of Soybean Expansion in Argentina. Grain Publications: Seedling. **2001,** *3,* 1–3, May 15, 2013, http://grain.org/ publications/seed-01-9-3-en.cfm

87. Pengue, W. A.; Environmental and Socio-economic Impacts of Transgenic Crops Releasing in Argentina and South America: An Ecological Economics Approach. **2013,** 1–14, April 20, 2013, http://www.gepama.com.ar/pengue/pdf/ENVIROM-1 PDF pp 1–14

88. Peterson, D. L.; and Russo, M. J.; *Cortaderia jubata.* **2012,** 1–7, April 19, 2013, http://www.wiki.bugwood.org/Cortaderia_jubata.htm

89. Podesta, G.; Project Background—The Argentine Pampas. Climate, Agriculture and Complexity in the Argentine Pampas. **2006,** *1–3,* April 19, 2013, http://www.rsmas.miami.edu/groups/agriculture/science/Background.htm

90. Podesta, G.; et al. Climate complexity in agricultural production systems of the Argentine Pampas. Proceedings of an Expert meeting on Regional Impacts, Adaptation, Vulnerability and Mitigation. Nadi, Fiji; **2007,** 1–13, W.

91. Podesta, G.; et al. Decadal climate variability in the Argentine Pampas: Regional impacts of plausible climate scenerios on agricultural systems. *Climate Res.* **2009,** *40,* 199–210

92. Portela, S. I.; Andruilo, A. E.; Jobbagy, E. G.; and Sasal, M. E.; Water and Nitrate exchange between cultivated ecosystem and groundwater in the Rolling Pampas. *Agric. Ecosyst. Environ.* **2009,** *134,* 277–286.

93. Portela, S. I.; Andruilo, A. E.; Sasal, M. C.; Mary, B.; and Jobaggy, E. G.; Fertilizer versus organic matter contributions to Nitrogen leaching in cropping systems of the Pampas: 15N application in field lysimeters. *Plant Soil.* **2006,** *289,* 265–277.

94. Prystupa, P.; Gutierrez-Boem, F. H.; Salvogiotti, F.; Ferraris, G.; and Couretot, L.; Measuring Corn response to fertilization in the Northern Pampas. *Better Crops.* **2006,** *90,* 25–27.

95. Pyle, J.; A re-examination of aboriginal population estimates for Argentina. In: The Native Population of Americas. Denevan, W. M.; ed. Madison, Wi, USA: University of Wisconsin Press; **1976,** 181–204.

96. Ramsperger, B.; Peinemann, N.; and Stahr, K.; Input and characteristics of Aeolian dust in the Argentinean Pampa. *J. Arid Environ.* **1998,** *39,* 467–476.

97. Riskin, S. H.; Porder, S.; Schipinski, M. E.; Bennet, E. M.; and Neill, C.; Regional differences in phosphorus budgets in intensive soybean agriculture. *Bioscience.* **2013,** *63,* 49–54.

98. Sanchez, H.; Garcia, J.; Caceres, M.; and Corbella, Y. R.; Labranzas en la region Chaco-Pampeana sub-humeda de Tucuman. In: Pannigatti, J.; ed. Siembra Directa Hemisferio Sur-INTA. Argentina: Buenos Aires; **1998,** 27.

99. Scheiner, J. D.; Gutierrez-Boem, F. H.; and Lavado, R. S.; Sunflower Nitrogen requirement and 15N fertilizer recovery in Western Pampas, Argentina. *Euro. J. Agron.* **2002,** *17,* 73–79.

100. Schnepf, R. D.; Dohlman, E.; and Bolling, C.; Agriculture in Brazil and Argentina: Developments and Prospects for Major Field Crops. Washington DC: United States Department of Agriculture, Foreign Service; **2001,** 1–78.

101. Silvia, A.; Buschiazzo, D. E.; and Zobeck, T. M.; Wind erosion in Loess soils of the Semi-arid Argentinean Pampas X. **1999,** 1–5, April 24, 2013, *natres.psu.ac.th/Link/SoilCongress/bdd/symp31/823-t.pdf*

102. Silvine, P.; Adrian, A.; Morton, B.; and Bruno, M.; Fertilizer versus organic matter contributions to Nitrogen leaching in cropping systems of the Pampas: 15N application in field lysimeters. *Plant Soil.* **2006,** *289,* 265.

103. Smithsonian Institute. Regional Overviews: Argentina and Chile. **2012,** 1–6, April 23, 2013, http://www.botany.si.edu/projects/cpd/sa-viii.htm pp. 1–6

104. Solbrig, O. T.; The Dilemma of Biodiversity Conservation: Agricultural Expansion in the Pampas. Revista. **2005,** 1–3, April 20, 2013, http://www.drclas.harvard.edu/publications/revistaonline/fall-2004-winter-2005/dilemma-biodivrsity-conservation. htm

105. Soriano, A.; et al. Rio de Platta grasslands. In: Natural Grasslands: Introduction and Western Hemisphere. Coupland, R. T.; ed. New York, NY, USA: Elsevier Science Publishing Company Inc.; **1992,** 284.

106. Steinbach, H. S.; and Alvarez, R.; Changes in soil organic carbon contents and nitrous oxide emissions after introduction of no-till in Pampean Agroecosystem. *J. Environ. Qual.* **2006,** *35,* 3–13.

107. Surman, W.; GM crops in Argentina. New Agriculturist. **2007**, 1–2, May 15, 2007, http://www.new-ag.info/en/developments/devitem.php?a=32
108. Taboada, M. A.; Soil structural behaviour in Flooded and Agricultural soils of the Argentine Pampas. Doctoral theses submitted to Devant L'institut National Polytechnique De Toulouse, France; **2006**, 135.
109. Tomei, J.; and Upham, P.; Argentine clustering of soy biodiesel production: The role of international networks and the global soy oil and meal markets. *Open Geogr. J.* **2011**, *4*, 45–54.
110. Trigo, E. J.; and Cap, E. J.; The impact of the introduction of Transgenic crops in Argentinean agriculture. *Ag. Bio. Forum.* **2003**, *6*, 1–3.
111. Trigo, E.; Cap, E.; Malach, V.; and Villarreal, F.; Innovating in the Pampas: Zero-tillage soybean cultivation in Argentina. In: The Case of Zero-tillage Technology in Argentina. Washington, DC USA: International Food Policy Research Institute; **2009**, 59–65
112. USDA Foreign Agricultural Service. Argentine Wheat Update: Rainfall Just in Time. **2006**, 1–2, April 29, 2013, http://www.pecad.fas.usda.gov /highlights/2006/10/Argentinea_27Oct2006/
113. Van Dam, J.; Faaij, A. P. C.; Hilbert, J.; Petruzzi, H.; and Tuckenburg, W. C.; Large-scale bioenergy production from soybeans and switch grass in Argentina. Part A. Potential and economic feasibility for national and international markets. *Renew. Sust. Energ. Rev.* **2009**, *13*, 1710–1733.
114. Van Opstal, N. V.; Caviglia, O. P.; and Melchiori, R. J. M.; Water and solar radiation of double cropping in a humid temperate area. 2004 Water and solar radiation productivity of double cropping in a humid temperate area. *Aust. J. Crop Sci.* **2011**, *5*, 1760–1766.
115. Viglizzo, La substentabilidad ambiental del agro pampeano. Ediciones. Institut Nacional Technolgie Agriculture, Buenos Aires, Argentina. **2002**, 123.
116. Viglizzo, E. F.; Argentina: The provision of Ecosystem services and Human well-being in the Pampas of Argentina. **2005**, 1–4, April 4 2013, http://www.milleniumassessment.org/en/SGA.ArgentinePampas.aspx
117. Viglizzo, E. F.; et al. Ecological and environmental foot print of agricultural expansion in Argentina. *Global Change Biol.* **2011**, *17*, 959–973.
118. Wikipedia. Environment of Argentina. **2013a**, 1–3, April 2013, http://www.en.wikipedia.org/wiki/Envronment_of_Argentina
119. Wikipedia. Climate of Argentina. **2013b**, 1–5, April 23, 2013, http://www.en.wikipedia.org/wiki/Climate_of_Argentina
120. Wikipedia. La Pampa province. **2013c**, 1–4, April 22, 2013, En.wikipedia.org/wiki/La_Pampa_Province
121. Zarate, M.; and Folgeura, A.; On the formations of the Pampas in the foot steps of Darwin: South of Salado. *Revista de la Association Geologicas Argentina.* **2009**, *64*, 124–136.

CHAPTER 4

EUROPEAN PLAINS AND RUSSIAN STEPPES: NATURAL RESOURCES, ENVIRONMENT, AND CROP PRODUCTIVITY

CONTENTS

4.1 INTRODUCTION

Northern European Plains and Steppes in the east are among the most prosperous agricultural regions of the world. The Agricultural prairies of this region support high-intensity cropping on fertile Chernozems. These prairies generate several different cereals, legumes, oilseeds, fiber crops and vine yards. Kusters (2000) states that prairies of Europe have offered one of the most stable agricultural belts for human kind to establish, survive, and perpetuate.

Natural prairie vegetation and agricultural cropping areas, mainly those of cereals, legumes, and oilseed are included in this chapter. The main areas included are those in European nations such as France and Germany in Northern European plains, Poland, Czeck Republic, Hungary in Central European Plains and Ukraine and Russian Steppes in Eastern Europe (Figure 4.1). Prairies and meadows of Portugal, Spain, England, Netherlands, Italy, Yugoslavia, and so on, have not been included for broader discussion, mainly to keep brevity of chapters.

FIGURE 4.1   European agricultural prairies discussed in this chapter.
Note: Agricultural Prairies occur in different nations of Europe. Areas shown in black depict the regions discussed in this chapter. Agricultural cropping zone in countries such as Portugal, Spain, England, Italy, and Yugoslavia has not been discussed. Agrarian regions of France, Germany, Poland, Czeck Republic, Hungary, Ukraine, and Southern Russian Steppes have been highlighted throughout the discussions.

4.1.1 GEOLOGICAL ASPECTS OF EUROPEAN PLAINS AND STEPPES

European plains experienced drastic changes to its environment during Pliestocene. It seems glacial activity during Ice age and movement of glaciers from Scandinavia and Alps has affected the geological aspects of plains (Kusters, 2000). As a consequence, forests retreated and they were replaced by scrub tundra vegetation leading to steppe like landscape. During late Glacial age, tundra and steppes gave way for slow but definite expansion of forests.

Geologically, northern European plains are complex stretch of land scape. It encompasses flat and undulating regions influenced by glaciation during Pleistocene (2.6 million years ago) that continued up to 12,000 years ago. The landscape is therefore typically postglacial. "Moraines" are glacial deposits that are well spread in the Northern European plains. Pleistocene ice sheets are traceable in parts of Germany and France. Hilly terminal moraines occur in Belarus and Western Russia. Spill ways for melted ice occur in between moraines. The spill ways got covered by sand and gravel by the gushing melted glacial streams. At present, these regions are associated with poorly drained wetlands (Encyclopaedia Britannica, 2010).

Regarding geology of French plains, it is said that during Pleistocene Ice age, it supported an open grass land. Forests developed as glaciers retreated. Then, during Neolithic period, forests were cleared and lumbering extended appreciably in the region leading to development of agricultural expanses (Wikipedia, 2013f). The central part of plains, that is, Paris Basin, shows well-layered sedimentary rocks and fertile soils derived from parent material.

German plain was created through glacial advances from Scandinavian ice sheets during Pleistocene. They are termed young or old regions based on glacial drifts. German plains are slightly undulating in some places and totally flat in other parts. Low lands occur west of Schleswig-Holstein. These are below the sea level. Plains occur at a slightly higher altitudes in areas derived from glaciation and they are often referred as terminal moraines depending on period during which glaciation was prominent. Coastal plains of Germany are related to Holocene lakes, River marshes, Lagoons, and Old and Yung drifts that occurred during Pleistocene (Wikipedia, 2013d). The North-eastern part of German plains is also known as Young drift. It shows up lakes formed during ice age, some 16,000–13,000 years ago. The dry plains of Saxony and Schleswig-Holstein are heavily weathered and glaciated.

The Upper Rhine valley in the European plains extends for 350 km with an average width of 50 km. The southern section occurs at the borders between Germany and France. The Rhine valley was formed during Cenozoic era and late Eocene epoch. During this period it seems formation of Alps was rudimentary. The core of the Rhine valley shows up crustal gneiss. This European Cenozoic rift system extends across the Central Plains. While the upper Rhine Graben was formed during Oligocene. River Rhine flows in the trough created in the plains (Wikipedia, 2013a; Dezes et al., 2004).

Polish plains that support intense agricultural activity currently it seems were shaped 60 million years ago by the continental collisions and quaternary glaciations of Northern Europe. The soils of North Poland are made of loam and sand, but soils in south are made of loess. The Cracow upland, Pieniny, and Tatars are made of limestone. The high Tatars and Beskids are made of granite and basalts (Wikipedia, 2013c).

Regarding geology of Ukrainian steppes zone, we can distinguish Ukrainian Shield, Volyno-Podiliska slab, Volynska and Dniprovsko-Donetska cavities, and other formations. The largest structural unit is the crystalline massif, which takes up the central part of Ukraine and is composed of Archean and Proterozoic rocks dated 3,500–1,200 million years old. It seems Carpathian region is still in the process of development with regard to geological evolution. The Donetsk Aulakogen is composed of thick carboniferous coal bearing formations and parmian salts. In the northwest of the basin, Triassic and Jurassic deposits are traceable. Ukrainian plain is among major sources of minerals, iron, and coal (Kosivart, 2013c).

A generalized and concise time line for geology of Europe is as follows:

2,300–2,700 m years ago—Baltic shield was formed;

1,900–2,300 m years ago—Volga-Ural shield was formed; also the East European Craton was formed;

330–375 m years ago—Eurasian land mass was formed;

270 m years ago—Cimmerian plate and Laurasia got separated from Gondwana.

50 m years to present—The continental land mass in Europe and adjoining regions in Asia approached present configuration. Europe experienced land connection with American continent via Green land. It allowed migration of animals and dispersal of flora between the regions. European zones have fragmented to small islands in some locations. The sea level retreated allowing plains in Europe and Russian regions to get vegetated (Wikipeida, 2013b).

4.1.2   GEOGRAPHY OF EUROPEAN PLAINS

European Plains, also called as the "Great European plains" is a prominent geographical feature of European continent. It is largely a mountain free flat expanse of different soil types and prairies filled with either natural vegetation or agricultural crops. Much of the Great European Plains is flat land that lies below 153 m.a.s.l. (Evers and West, 2013; Shahagedanova, 2008). European plains are among the most productive agricultural zones of the world. The European plains stretch from Pyrenees in Spain and Bay of Biscay in France in the west to Ural mountain of Russia in the East. European plains are not wide but extend only for 200 km breadth in the west. However, the plains are wide in the Steppe region of Russia and may reach 1,800–2,000 km in width. The Northern European plains include parts of France, Germany, Belgium, and Netherlands. In the Northern Plains terrain is mildly undulating. These are glaciated zones. Northern Plains includes alluvial regions and delta of river Rhine that drains into North Sea. Northern Plains also show up wind-blown deposits that has given rise to Loess soils. North European plains are congenial for agricultural crop production. Central Plains of Europe include agricultural areas in France, Germany, Poland, Czeck Republic and Hungary. Central Plains are occupied by large tracts of fertile Chernozems. Central plains are served by rivers such as Vistula, Dvina, Niemen, Danube, Drava, Oder, Mures, and Tisza. Eastern part of European plains is termed as Russian plains or Steppes. The steppes are vast stretch of fertile soils that supports natural vegetation and crops. The steppes are served by many rivers such as Volga, Don, Dnieper, Dniester, Donets, and few others.

The French plains are among the most intensively cultivated prairie regions of Europe. It experiences a temperate climate and supports large arable cropping. French plains extend into entire nation. Topography is either flat or undulating with meadows, pastures, and cereal fields. Total land area of France is about 551,000 km^2, out of which 59 percent is amenable for agricultural crop production. Arable cropping zones account for half of the agricultural belt (Wicherek and Bossier, 1995).

German plains currently support both natural and agricultural prairies. Entire German plain lies below 100 m.a.s.l. Glacial action has generated three major land forms, namely alluvial region of lower Rhineland, glacial sand, gravel of lower Saxony and Morainic uplands extending into Schleswig-Holstein and Baltic Coastal Plains are lowest in the Moorland and dry northwest of Saxony and Schleswig-Holstein. Undulating topography is also prominent in the cropping zones. About 37 percent of plains support arable copping belt, 17 percent consists of meadows and pastures, 30 percent is filled with forests and 16 percent is utilized for urbanization and other purposes. The topography of Germany can be categorized into North German low land, Central uplands, southern German Loess region and Alpine region. Federal agencies of Germany identify several natural zones such as Meckleburg coastal lowlands, Middle Elbe plains, Flaming heath, Stad Geeste, Weser-Aller plain, Westphalian lowlands, lower Rhine plain, Cologne low land, and so on (Dickenson, 1964). German plains are served by rivers such as Rhine, Ems, Weser, Elbe, and Havel. These rivers drain the plains into North Sea. Their flood plains have supported forests and large-scale intensive cereal cropping zones. North German plains are also served by rivers such as Oder and Neibe, which flow into Baltic Sea (Wikipedia, 2013d).

Geographically, Polish cropping zones occur in the Central Europe, between 49°N and 55°N latitude and 14°E–25°E longitude. Poland includes parts of the North European Plains in the North and Central part of the nation. Polish plains support production of cereals such as wheat, barley, and maize. Rape seed and pastures are also common. Cropping zones also occur in the coastal and undulating zones of Poland. Poland is served by rivers such as Vistula, Oder, Warts, and the Bug. There are other rivers such as Lyna, Agrapa, Pregolya, Czarna, and Danube. Poland is known to possess large number of lakes. They do support farming on the fringes. More than 50 percent of Polish land resource is used for agricultural cropping. Agricultural expanses are decreasing in area, but it has been compensated well by intensification and higher productivity of crops. Poland generates wheat, barley, rye, triticale, sugar beet, and potato. Polish plains are also termed as "Bread Basket in Europe."

The Czeck plains are actually vast stretches of undulating meadows that support large-scale production of wheat, barley, and other crops. Agricultural area occupies 50 percent of the Czeck region. Valleys of river Morava and Labe in Moravia are fertile and support intensive crop production. Czeck highland and Bohemia are other regions of importance agriculturally. Arable land has stagnated at around 30–32 m ha in the entire Czeck Republic (Kralovec, 2006). The altitude ranges from 244 to 675 m.a.s.l. Topography is said to affect productivity of fields. Nutrient distribution is affected by the slopes and undulating land. As a consequence crop productivity is also affected by topography (Kumhalova et al., 2008). Digital models of land surface and nutrient distribution are used in some advanced farms adopting Precision techniques and GPS-guided variable rate application.

Hungary is a country with large plains, undulating terrain and few mountains. Much of the topography lies below 200 m.a.s.l. There are only a few mountains that reach above 300 m.a.s.l. The Great Alfold in the east of Danube River; Trans-Danubia that lies to the west of Danube and North Hungarian mountains that form the upper most border for the Great Hungarian plain are important geographic aspects (Wikipedia, 2013e). Fertile soils classified as Chernozems occupy much of the agricultural zones in Hungary. Danube and Tisza are most important rivers with regard to farming in the Hungarian Plains. Hungary is also served by several fresh water lakes. Transdanubia is primarily an agricultural region with rolling plains. It supports large expanses of cereal crops and viticulture. It is bounded by Danube and Drava rivers. The Tisza river belt encompasses fertile terrain good enough for crop production, swampy zones and sandy waste lands.

The Great Steppes is a very large region on Earth, encompassing prairie vegetation that occur between Ukraine in the West and covering areas in Russia, Kazakhstan, Turkmenistan, Uzbekistan, and until Tian shan ranges. In the deep southern Europe, Steppes occur in parts of Anatolia, Armania, and Iran. However, within the context of this chapter we are concerned only with Steppes that occur in Ukraine and southern Russia. According to FAO (2005), Ukraine is demarcated into three regions based on geomorphology. They are mixed-forest zone, a forest-steppe and a steppe zone that is predominantly cropland. There are mountain regions such as Crimea and Carpathian. Moon (2012, 2013) describes Russian steppes as vast stretch of rolling plains and extremely flat plains, whose monotony is broken by few rivers such as Dnieper, Don, Volga, and Ural and their tributaries. These temperate steppes are characterized by vast stretches of grasslands, scattered shrubs and forest species near water bodies. Russian steppes considered here are mostly filled with short and mixed prairies. They also support large-scale production of cereals such as wheat, barley, and oats; a few legumes; oilseeds; vegetables; and cotton (Wikipedia, 2013b). Steppes encompass large tracts of "Flood Plain Meadows." Several rivers that traverse within the steppes are prone to floods. They inundate the plains. Melting of snow, both *in situ* and that found in the upper reaches of the river as the spring season sets in, causes floods. The natural vegetation and crop production trends are highly dependent on the flooding frequency and intensity (Boonman and Mikhalev, 2008). Prolonged flooding is a feature with rivers such as Dnieper, Don, and Volga. The steppe region is the main agricultural region of Eurasia. Steppes are endowed with richly fertile Chernozems that support arable cropping. The arable cropping belt extends into 1,189 km^2, out of which Chernozems occur on 787 km^2, equivalent to 57 percent (Kovda, 1974; Chibiliyov, 1984; Dudkin, 2013; Kosivart, 2013b).

4.1.3   AGROCLIMATE OF EUROPEAN PRAIRIES

The North European plains experiences very low temperatures of 3°C–10°C between November and March. Cropping season becomes warmer beginning from April and lasts till October, when temperature ranges from 14°C to 25°C. Peak rainy period occurs during May to July (Wikipedia, 2013a).

The agroclimate of French plains is congenial for cereal production. The agrarian zone in Northern France experiences relatively colder climate. The southern region experiences subtropical climate and supports intense cropping. The average annual temperature in north is 15°C. In south the summer temperature may reach 23°C–25°C. The oceanic region is cooler in north and the temperature ranges from 6°C–16°C. Westerly winds influence the precipitation pattern. Rainfall ranges from 850 to 1,270 mm. French plains are exposed to periodic droughts. During past 5 decades, French plains experienced droughts of perceptible intensity during 1976, 1989–1990, and in 2003.

North German plains experience a relatively cold climate in winter and modestly warm summer period. In the southern regions, a continental climate with warmer temperature is common. Warmer climate is found in the rain shadow regions of Flaming, Harz, and Drawehn. Lower temperature ranges occur near coasts and in the lower Elbe river region. It allows fruit production. Ambient temperature regimes and riverine effects decide the vegetation pattern in many locations within the German plains (Wikipedia, 2013d). The annual mean temperature of plains is 9°C. Cold front affects the plains severely during December to March, when the temperature dips to −2°C. Warmer period occurs from May to October. The annual precipitation ranges from 600 to 800 mm in the plains, but reach up to 2000 mm in the mountainous regions of south. Central European plain that includes Poland, Czeck Republic and Hungary experience both maritime, cold climate and continental warmer seasons. Mean temperature during January is low at −1°C to 3°C in the west and −4°C to −6°in the east. Warmest months are July and August, when temperature reaches 17°C–19°C. Annual precipitation ranges from 500 to 800 mm in the Central European plains. Frost and snow cover is sporadic, but may affect crops (Tarasov, 1979). Agricultural zones in Czeck Republic are situated mostly at 300–375 m.a.s.l. altitude (Kumhalova et al., 2008). The temperature in the cropping zones of Czeck plains ranges from −2°C to 23°C, but annual average is 7.9°C. Annual precipitation is 525 mm, but ranges from 400 to 700 mm based on location. According to Kralovec (2006), we can identify 10 different agroclimatic regions within Czeck Republic based on heat units received above 10°C, soil moisture retention, annual precipitation and ambient temperature. In the plains of Czeck Republic, crop yield potential in response to climate change has been studied using models. Actually, a system of crop models is used to assess effects of various weather parameters and expected alterations on crop yield (Dubrovsky et al., 2004).

Hungarian part of European Plains enjoys a continental climate with slightly cold winters and warm summers. Average annual temperature is 10°C, but it ranges from 27°C to 35°C in summers to 0 to −12°C in the peak cold winters. Annual precipitation is about 600 mm, but it is unevenly distributed during the crop season. Drought spells are common in the plains. Yet, study of precipitation pattern suggests that it is higher at 787 mm in the western part of Hungary and it progressively decreases as we travel to east reaching as low as 508 mm.

Ukrainian Plains enjoy a comfortable agricultural weather pattern that suits natural steppe grasses and other flora as well as field crop production. Lowest temperature occurs during January and February (−7°C to −8°C). July and August are warm with temperature ranging from +17°C to +19°C. In the Carpathian region, warmer temperature up to +13°C is experienced. Sunshine hours are more than 12 h day^{-1} during main crop season. Rainfall pattern is also congenial for cereal crop production. It ranges from 600–650 mm in the plains and 300–450 mm in the coastal belt. Mountainous zones and Carpathians receive higher precipitation sometimes reaching 1,200 mm year^{-1}. Ukrainian region experiences clear seasons. Winter is slightly longer, but relatively warmer. Spring sets in when ambient temperatures reach above zero. Summer season arrives as ambient temperature touches +15°C. Frost and severe winter curtail the length of cropping season (Kosivart, 2013a).

The Russian steppes experience semiarid and relatively cold climate (Moon, 2012). Temperature ranges from 35°C during summer and depreciates to as low as −40°C during winter months. Frosty subzero conditions are common in the Russian steppes. The average precipitation recorded ranges from 250 to 500 mm year^{-1}. In the Ukrainian region, Central Chernozems, Volga territory, and Southern Russian prairies, the precipitation ranges from 300 to 400 mm year^{-1}. Potential evapotranspiration is relatively low during winter in the steppes. Precipitation patterns are unreliable and often create droughts that have catastrophic effects on crop fields (Moon, 2012).

The temperature in the Russian/Ukrainian steppes ranges from 0 to 30°C. Generally, average temperature is relatively low in the steppes. Steppes experience coldest weather during winter months of November to March. It is frosty, snowy, and many of the plant species either die or hibernate till spring in the April. Fauna not acclimatized to winter may perish. Lowest temperatures during January may range from −12°C to −16°C in the Volga territory and other parts of southern Russian steppes, and −5°C to 12°C in the Ukrainian Chernozem regions. The total heat unit received by the natural vegetation or crops is important. The vegetation has to complete its life cycle and reproduce within the short span before frosty conditions set in. The average heat units received (>10°C) by crops/grass land species is 2,600–3,200 in the Chernozem belt and Volga territory (Boonman and Mikhalev, 2008). In the Southern Steppe region of Russia, climate is relatively warmer and dry. The P/E coefficient is 0.47 leading to water deficits. Mean summer temperatures are 20°C–24°C in the Western part of Steppes and 17°C–21°C in the Eastern Steppes. Annual heat units received in the Dry steppes is 2,300°C–3,500°C. The crop season is about 180 days, but decreases progressively as we traverse west to east (Blagoveschchenskii et al., 2007).

4.1.4 CLIMATE CHANGE AND CROPPING IN THE PLAINS

Reports by Kjellstrom et al. (2007) and Alcamo et al. (2007) state that European plains and other regions in the continent has experienced increase in ambient air temperature by 0.9°C between 1901 and 2005. However, during past 5 decades, from 1977 to 2001, increase in temperature was 0.4°C. The increase in temperature has been felt more during winter than in summer (Jones and Moberg, 2003). There are also clear indications that frequency of drought events in early spring and summer has increased during past 3 decades. In some regions, such as Northern European plains, annual precipitation has increased (Oleson et al., 2011). High-intensity rainfall events have often led to floods in the plains. Droughts have been frequent in the Central European region. It has been attributed to higher ambient temperature. Air temperature during cropping season has increased by 3°C–5°C in the Central plains during past 3 decades. The heat wave situations that occur have at times caused a deficit of 300 mm precipitation. Droughts, in fact, are the major grain yield retarders during the past 2 decades in the European Plains. In general, it reduces primary production of both agricultural and natural vegetation. Farmers in the European plains have adapted themselves to climate change by altering the planting dates, modifying agronomic procedures to a certain extent, especially with regard to cropping systems. Global weather models and forecasts have also been utilized to advice farmers in the European plains. Oleson et al. (2011) evaluated different models. According to them, Eastern European plains and Russian steppes may face greatest effect of climate change. Expected change in seasonal temperature from June to July may affect crop yield. Also, warmer winters may cause uncertainties in precipitation pattern and general weather pattern (Giorgi et al., 2004; Christensen and Christensen, 2007). In the Southwest European plains and parts of steppes, expected increase in temperature during next decade is 6°C. Agricultural cropping trends and productivity may show net change in France, Germany, Poland, and Ukraine. According to Oleson et al. (2011), detrimental effects from climate change may be most pronounced in the Pannonian plains that include Hungary, Serbia, Bulgaria, and Romania. It may also be severely felt in the Steppes of southern Russia. Increased heat waves may force farmers to adopt shifting agricultural practices and changes in cropping systems. Altering sowing time based on forecasts may become a rule. Farmers may have to alter cropping systems to enhance their earning. Olesen et al. (2011), state that climate change may lead farmers in European Plains to adapt through variety of modifications in the crop management practices. Farmers may alter several aspects such as seeding time, crop season, tillage practices, fertilizer supply trends, residue recycling procedures, cultivars, crop protection procedures, and harvest timing. In general, climate-change-related increase in forage/ grain productivity is expected in the North European plains. Russian steppes may not reap higher grains than at present (Olesen and Bindi, 2002; Marachi et al., 2005; Alcamo et al., 2007). However, we may still overcome climate change effects by generating crop genotypes that adapt better to latest weather patterns, disease, and pestilence in the area. Modifications to agronomic procedures and their schedules may after all answer many of the climate change effects. It needs to be rapidly researched so that problems are fixed rapidly. At the bottom line, several different agricultural and related measures that subside soil erosion, reduce water loss from profile to groundwater, and lessen emissions from crop fields are essential. Several of these measures may have to become mandatory in order to preserve productivity of agricultural enterprises.

4.1.5 NATURAL VEGETATION OF EUROPEAN PLAINS AND RUSSIAN STEPPES

Natural vegetation in the European plains is exposed to consequences of climate change scenarios equally along with cropping expanses. Oleson et al. (2011) have made a few forecasts based on weather models, about the effect of climate change on weeds, natural flora and their ability to put forth biomass and perpetuate. Factors such as frost, drought, high rainfall periods, hails, and biotic stresses may affect natural vegetation and biodiversity to certain extent. Natural selection process will operate, but there is every possibility to preserve and shift the species that may not flourish to regions more congenial and secure its perpetuation.

Natural vegetation in the European cropping belts has undergone periodic changes based on several natural and man-made factors. Factors, such as invasion, extinction, natural succession, dominance, and abiotic aspects have affected vegetation through the ages. According to Romermann et al. (2008), intensification of European cereal belt has been marked. It has affected the flora of the region in many ways. Several plant species have gone extinct or occur only in few individual patches and show scattered distribution. This situation has been attributed to abandonment of cereal fields not used, not utilizing it for fallows to support either natural vegetation or cover crops, and total lack of upkeep of fields. Long term deep plowing has often reduced natural weed flora and left the fallows blank, for it to get degraded. However, application of fertilizers, organic manures and residual moisture in the unused fields allows ruderal species to sprout. A few species still persist irrespective of onslaught of agricultural techniques and abandonment. Field evaluations at Crau in Southern France suggest that fields not utilized to produce high-intensity wheat crop still support a few natural species. Unused fields possess a seed bank of naturally distributed flora. Previous crop grown in those fields seems important with regard density and diversity of flora still retained. Dominant species of the natural flora that establishes after cereal production may differ. For example, species such as *Dactylis glomerata, Sagina apetala* and *Verbena officinalis* was predominant after Coussous. Species such as *Calamintha album, Geranium molle* and *Seneco vulgaris* was more in fields that had supported cereals. The dominant species too fluctuated in the fields periodically. There is a particular need to maintain an inventory of natural vegetation, population, and intensity of flora in all the major crop production regions within European Plains and Russian steppes. We have to monitor and identify the species in danger of extinction from a place and perhaps preserve small seed banks or germplasm in waste lands.

Natural vegetation of the Russian Steppes includes a large number of herbs, shrubs, and tree species. However, progressive clearing of native species due to spread of crop production has affected the biodiversity of both flora and fauna. Boonman and Mikhalev (2008) classify the steppes vegetation based on maturity groups into very early, early, medium, and late maturing plants. Let us consider a few examples:

Very early maturing species: *Poa bulbosa* is a prominent species in the semiarid, Chernozem and Chestnut soil belts of Steppes. It dries in summer and regenerates with fresh rains.

Early maturing species: Festuca sulcata is a tufted species with grayish leaves. It is common in virgin steppes and old fallows. It is also traceable as understory shrub lands and forested zones with in steppes. *Stipa lessingiana,* and *Koeleria gracilis* are again common in virgin steppes and old fallows.

Medium maturity species: *Agropyron pectiniforme* is a slightly tall plant that grows up to 90 cm height. It is spread densely in the clayey and loamy soils of semi-arid belt. It is common in the river banks and near water bodies. *A. sibiricum* and *A racemosum* are common in the virgin steppes and old fallows. These are hardy species that tolerate drought and salty soils. *Bromus inermis* is frequent in the flood plains of rivers within steppes. They are often found as pure stands of over 100 cm tall and in large patches. They are well adapted to Chernozems and fallow soils.

Late maturing species: *Agropyron repens* is rhizmatous species which grows up to <80 cm height. It is predominant in the flood plains and old fallows in the Chernozem belts. *A. repens* tolerates salinity to a certain extent. *Stipa capillata* is a tufted species of 60 cm height. It is common in steppes but less frequent compared to *S. lessingiana*. It is easily traced in semiarid area and forested steppes.

The natural flora of steppes may alter or follow a systematic succession, based on several factors related to weather pattern, soil fertility levels, grazing levels, and adaptability of plant species to a particular region within the steppes. For example, in fertile Chernozems and Chestnut soils, swards with *Stipa lessingiana* is more common than *Festuca sulcata*. After a few years of dominance by *S. lessingiana,*

Festuca species take over and dominate the ecosystem. In areas with intense grazing, *Poa bulbosa* dominates the flora. Locations prone to excessive grazing usually support only *Polygonum aviculare*. In the flood plains region within steppes, natural vegetation differs based on flooding pattern and ability of species to withstand inundation. For example, if flooding is experienced for more than 40 days, species such as *Bromus inermis, Stipa pretense,* and *Phalaris* survive and thrive after the water recedes. The flood meadow vegetation, in general, may include species such as *Festuca ovina, Nardus stricta, Phalaris sp, Agrosits sp, Deschampsia, Alopecurus,* and tall graminaceous species. Shrubs and conifers may also occur in the flood plains.

Typical succession sequences traced in the steppes are as follows; 1st year annuals, 2nd and 3rd years biennials and perennial herbs; third and 4th years *Agropyron racemossum*; 5–8 years, *A racemosum* continues but small patches of *Festuca sulcata* begins to appear in the Chernozems. During later years, *Stipa lessingiana* and *Festuca sulcatta* form the dominant species of the flora within the natural prairies. Plant species such as *Phleum pratense, Agrostis alba,* and *Geranium pratense* are common in the flood plains within steppes.

Fallows that occur naturally or those farmed during cropping sequences may harbor different species of plants. Plant species that colonize the fallows may follow a particular pattern. According to Boonman and Mikhalev (2008), young fallows are often filled with annual herbaceous species. Then, the green fallows may support variety of plant species and become dense with thick canopy. Rhizomatous perennials are common in the fallows beyond 4–5 years.

4.1.6 DEVELOPMENT OF AGRICULTURE IN EUROPEAN PLAINS AND RUSSIA

Earliest recordings of human inhabitation in the present day North European plains and Eastern Steppes pertain to humans practicing hunting and gathering. They moved several times from south to northern regions of the plains and back, during the course of history. Prehistoric dwellers located themselves mostly in the vantage hilltops and mountains scattered in the European terrain (Kusters, 2000). Excavations in Paleolithic camps of the plains suggest that early Europeans gathered several wild plants and tubers while on their hunting expeditions.

It seems Mesolithic Europeans who dwelled in forests and plains alike, found hunting not so easy and hence supplemented their food with hazelnuts. Archeological evidences suggest use of hazelnuts by the hunter/gatherer. Evidence of "Hazel nut meals" in the trails used by migrating Europeans, suggests that gathering was in vogue. Other source of nourishment was from wild apples, strawberries, and pears (Kusters, 1986). Transition from Mesolithic to Neolithic age that marked the beginning of sedentary cropping and group settlements, it seems occurred gradually in the European plains. Kusters (2000) further suggests that much of the "crops package" grown by Neolithic Europeans populations found in the Central Plains and Steppes actually were derived from Middle-east. Incidentally, domestication of crops and their cultivation had already taken root in the Middle-east much prior to European plains. Therefore, spread of agriculture in European Plains included crops that supported the Neolithic settlements with cereals (carbohydrates), pulses (proteins), and vegetable fats (oilseeds). The Einkorn wheat (*Triticum monoccocum*) and Emmer wheat (*T. dicoccum*) were the earliest domesticated cereals that induced agricultural settlements in the Plains. These wheat types were grown along with lentil (*Lens esculentus*) and peas (*Pisum sativum*) in the Balkans and Steppes of Ukraine and Russia. Barley (*Hordeum vulgare*), Hexaploid wheat (*T. aestivum*), and Poppy (*Papaver somniferum*) were other crops easily traceable in the archeological excavations of Mediterranean and Western European Neolithic settlements. The cluster of domesticated crops utilized during establishment of agriculture in European plains included still many species. A few examples from German plains are Parsley (*Petroselium crispum*), Dill (*Anethum graveolens*), and Celery (*Apium sp*). Sedentary agriculture began in the Steppes of Ukraine during early Neolithic period (Pashkevich, 2000). Cereals such as hulled wheat (*T. dicoccum*), barley (*Hordeum vulgare var coeltese*), and *Panicum milleaceum* were introduced by the tribes from southwest Asia and Balkans. Several other crops were introduced into steppes by the nomadic tribes, colonizers, and conquerors from Greece. It has been pointed out that crop genotypes derived from the West Asian assemblages got naturalized rather rapidly in the steppes, because of similar weather pattern. The dry environment, coldness, and low precipitation levels coincided between center of origin of the crops and steppes in Ukraine.

During Bronze age and Iron age, agricultural intensification was conspicuous in the European plains. Migrant settlers and natives found the plains to be congenial in terms of soil fertility and agroenvironment for consolidation of agricultural practices and food crops generation. Relatively better biomass and grain producers such as Emmer wheat, barley, and spelt (*T. spelta*) were preferred to substitute natural vegetation. Einkorn wheat that provided grains sparsely gave way for better crops during Bronze age. Agricultural cropping expanded with introduction of draught horses. The "crop package" included many legumes such as lentils, vicia, vetch and cereals such as setaria and pearl millet (Wyss, 1971). During Iron age, agricultural implements were improved and agronomic procedures got refined. Agricultural cropping got stabilized further and encompassed more crop species during ancient period (Roman age). Many of the exotic crop species such as *Oryza sativa* (rice), *Secale cereale* (rye), *Avena sativa* (oats), variety of beans, vines, and vegetable entered into cropping belts of Northern Europe especially in the Danube region and Steppes. During Medieval period, crop production techniques were intensified around citadels. In the modern period, agricultural products were used in industries as raw material. Crops such as Potato (*Solanum tuberosum*), Cotton (*Gossypium sp*), and oilseed Brassicas were cultivated more frequently in the Russian plains (Kusters 1992).

The Russian steppes supported nomadic pastoralism and very few agricultural settlements for more than millennium prior to eighteenth century. It seems, nomads had been burning the vegetation and destroying trees to develop almost a treeless expanse. "Steppe Nomads" were Turkish and Mongolian who preferred to inhabit the area for hunting, gathering, and adopted farming settlements only feebly. The *Slavs* too were nomadic for greater part, but during nineteenth century, farming and larger agricultural settlements were initiated by them and perpetuated since it helped them economically (Moon, 2012). The population of the steppes increased markedly during nineteenth and twentieth centuries. Arable cropping replaced nomadic pastoralism almost rapidly. Chernozem (Virgin Black Earth) was used to develop cereal fields, meadows, and pastures. These were the earliest stages of conversion of natural steppe and into cropland (Moon, 2008). Agricultural development in the Russian Steppes was rapid during the period from 1725 to 1887. Again, during 1954–1965, large-scale conversion of Russian steppes occurred due to state programs of Soviet Union that aimed at enlarging agricultural area and total production.

During 1900s, agricultural cropping trends have improved enormously, owing to improvements in roads, railways, irrigation through dams and channels, improved disease-tolerant and high-yielding crop varieties, postharvest techniques and marketing facilities. In addition to pastures, production of maize for fodder is a practice that has become popular since mid-1950s.Today, farmers in European plains produced a variety of food crops, firstly to feed the native population and then to export to different regions. For example, Ukrainian Cereal belt is called "Bread Basket for Russians." Production of cereals such as wheat, oats, barley, and rye, and pulses such as lentils and fiber cotton has become highly intensive. Farmers supply high dosages of organic and inorganic manures and use hybrids. Currently, agricultural prairies in the Northern European plains, Poland, Ukraine, Hungary, and Russian steppes are highly developed with intricate irrigation channels, fertilizer supply depots, and postharvest marketing yards. They are among world's most intensive wheat, barley, and oilseed producing centers. Development of farming communities and regular crop production in the Russian steppes has long history. According to Kovda (1974), three major phases of agricultural development can be identified in the Russian steppes. As usual, during early period, perhaps for several centuries, the land is still fertile and crop production is at subsistence level or higher, but without any addition of amendments. Next, consistent farming depletes soil fertility level, nutrients are removed from the profile and crop productivity decreases progressively. Reports suggest loss of fertility of Chernozems. Loss in N and P from the profile accentuates and organic matter too declines. The third phase termed as "cultivation phase" refers to sustainable crop production using fertilizers, organic manures and irrigation. Soil nutrient balance is carefully maintained and cropping practices are well stabilized and continued for long stretch of time. Preservation of fertility of Chernozems and crop productivity is the key concept during this phase in the European plains.

During past five decades, European plains has supported a highly mechanized agricultural expanse and supported production of wheat, barley, rapeseed, and grape wines in relatively large quantities. It has also generated other crops such as maize, sorghum, legumes, peas, and so on. Agronomic procedures including deep tillage with disk harrows once in few years, mechanized and GPS-guided seeding vehicles, fertilizer supply based on soil test, computer-aided decision making models, precision techniques and combine harvesting have been in use in many regions within European Plains and Steppes.

4.1.7 HUMAN MIGRATION, POPULATION, SETTLEMENTS, AND AGRICULTURAL CROPPING

Human inhabitation and migratory trend has been immensely affected since the invention of agricultural cropping. Most believe that crop production and steady availability of food grains did induce sedentary tendencies. However, it also induced small and large-scale migration to more fertile zones. Migration of farming communities from Middle-east to European plains is an example. Actually, several natural and man-made factors have played crucial role in deciding agricultural settlements and migratory habits of humans relying more on farming. Crop species and its ability to adapt to a weather pattern is among major reasons. Soil fertility and productivity levels possible, irrigation, postharvest procedures, and consumption centers have also affected human migration.

According to Bogucki (1992, 2000), low mobility hunter-gathers appeared in the North Central plains around the present day Germany, Poland, and Ukraine as early as 5–6 millennium B.C. The partly sedentary tendencies were localized at wet lands, lake shores and riverine areas. Evidences are available from Rhine, Weser, and Danube regions (Cappers and Raemaekers, 2008). Further, during Neolithic period, both sedentary farming communities and agropastoral populations inhabited the European plains and Steppes. Moore (1985), believes that the European region was a mosaic of farming settlements and pastoral nomadics during 5000–3000 B.C. Bogucki (1992, 2000) identifies different types of human groups in the European plains. They are; foragers who thrived by picking tubers, fruits, and fauna; farmers who are partly sedentary, yet continue to hunt and agriculturalists/pastoralists who do not depend on hunting, but practice cropping and cattle grazing. Movement of above groups all across the European plains has been a major factor in dispersal of crop species, introductions, and replacement of natural vegetation surrounding the settlements (Ingold, 1985).

Prior to World War II, agricultural farms in France were small and about 35 percent of nations population were engaged in producing crops in the prairies region. However, rapid mechanization, improved crop varieties and irrigation facilities has intensified crop production systems. France currently employs only 3.5 percent of its active population on farms. Large farms in the Aquitine basin in the Southwest

France adopt highly specialized agronomic procedures and farm machinery. Wheat and grape vines are most dominant in the landscape. However, the French plains do support production of several other crop species such as barley, maize, legumes, vegetable, and so on. Valleys of Rhine, Loire, and Garonne in the Northern France support large wheat farms, cattle ranching, pastures, and dairy.

Germany, it seems was an agricultural country during 1800s. Farming units in German plains supported a relatively higher share of 62 percent of nation's population. However, population employed on farming reduced to 42 percent in the early 1900s and it further dropped to 31 percent during the period prior to world war-II. This led to food insufficiency. German population that shifted to industries had generated goods that could be bartered for grains and other food products from countries like United States, Hungary, and Argentina. German agricultural productivity got enhanced with advent of inorganic fertilizers. The productivity of cereals, rapeseed, and legumes increased markedly due to high fertilizer supply and irrigation. Some of the best agricultural settlements and corporate farming companies are currently situated in the plains and valleys of Rhine and its tributaries. Currently, Germany employs only 10 percent of its population in farming and related activities.

Agricultural farming in the Czeck plains and uplands employs large section of human population. It has ranged from 20 to 24 percent of the total work force of the nation (Kralovec, 2006). Yet, there has been a steady decline in total number of farms and farm workers. Farm workers employed has declined from 507,000 in 1980 to 190,000 in 2010. It is easily attributable to agricultural intensification, highly mechanized procedures and automation. Agricultural output per person employed has increased. Hence, it has led to less number of workers being engaged in farming.

Hungary is largely an agricultural country, known more frequently as "Bread Basket" for Eastern European people. Farmers and human inhabitation thrive on Arable land (4.5 m ha), Grassland (0.76 m ha), Forest (1.9 m ha), orchards and grapevines (0.17 m ha) and uncultivated land (1.9 m ha) (Birkas, 2012). The per capita agricultural area at present is 0.55 ha for a total population of 99,72,000. Soil degradation due to rampant use for crop production without conservation measures, reduced refurbishment of soil nutrients and low productivity has often induced local migration and abandonment of farms. Severe erosion due to wind and water induces loss of top soil in many parts. It creates dust bowl conditions and drought. As a consequence, human migration to cities/towns in search of lively hood from industries and other sectors occurred. However, a sizeable portion of Hungarian population equivalent to 53 percent of total population thrives on agricultural cropping and related occupation (Wikipedia, 2013e). We should note that native Hungarians have thrived on these Great Plains for over a millennium, mainly depending on agriculture and allied professions.

Agricultural soils in Russian Steppes have been in use since the Neolithic period when native population and migrants from West Asia both cultivated crops in this region. Soil degradation was not experienced yet, because farming was feeble. Earliest of the symptoms of soil degradation and depreciation in crop yield, it seems was noticed in medieval period. Chronicles of fourteenth century that make clear suggestions about soil erosion, dust storms and other soil degradative processes that induced migration of farmers from their erstwhile settlements are available (Chibiliyov, 1984). Russian steppes were colonized rapidly during sixteenth and seventeenth century, mostly by the migrants form Middle-East. In many regions of the forested steppes, fields were converted to arable cropland. Agricultural cropping became slightly extensive since farming communities spread out well into all parts of Steppes. Migration into Steppes was once again rapid and marked during mid-nineteenth century. Nearly 66 percent of steppes were explored by migrants. They had been settled where ever agricultural cropping was feasible. During this period, rural population too had grown and the human population in the steppes had doubled from 6 to 12.8 million (Stebelsky, 1974, 1983). The population increase had generated pressure on land as a resource. Intensification of crop production did occur but feebly. Black soil region has been thickly colonized by the farmers. Eventually, as soil degradation and erosion of top soil became rampant in some areas, farmers started movement in different directions within steppes. During late 1800s, eroded regions and gullies in the black soil zone were conspicuous. Gullies sometimes covered about 25 percent of a farm land forcing the farmer to find a new location and migrate (Masalsky, 1897). Farmers, in many other areas depended on soil conservation processes to continue in the same place within the steppes. Therefore, initially highly productive Chernozems were the main attraction for human migration into steppes, latter soil degradation, dust bowl conditions, gullies in the fields and loss of productivity initiated human movements in the Russian steppes. During very recent period, spanning past two decades since the formation of Russian Federation, there have been changes to agricultural enterprises and their management. Large agricultural farms that were state supported have been dismantled into smaller farms and individual farms. Farm workers have shifted to their own small farms. Some workers have migrated to other regions, to get occupied in different professions (Blagoveschchenskii et al., 2007).

Dobrovolsky (2009) states that drought and dust bowl conditions in the Russian Plains and Southern Steppes affected human settlements and cropping zones. Initially, prescriptions made by the famous Soil Scientist Dokuchev were adopted to rescue agricultural expanses from soil degradation and loss in fertility. It seems during 1861–1885 over 3,00,000 farmers moved out of their native lands. Then, the Socialist Revolution induced changes in human dwellings and farm enterprises. The "Stolypin reform" created large changes in farm holdings, individual farms and farmer populations. It did induce human migrations to other regions in the Steppes and cities. During World War-I, large portion of fertile plains in Ukraine was wasted without being plowed. The Kovkhoz and Sovkhoz during socialist period created large farms. Extensive soil analysis and fertilizer supply was carried out (Dobrovolsky, 2009). Farm workers often stayed close to large agricultural units that were spread all across Ukrainian Plains. World War-II, again created large-scale migrations away from farm. Soil degradation and lack of upkeep of farms was prominent.

Then, large-scale conversion and colonization of steppes occurred due to conversion of natural grasslands to cropland during mid-1950s. This again created migration to regions within the Ukrainian and Russian Steppe zones that was hitherto not inhabited. During past two decades, dismantling of state run farms and development of individual farms that were small and easily manageable has once again affected human dwelling patterns in the Ukrainian plains. However, it is said, even today, migration of population intending to settle in Ukrainian farm belt is continuing. The major spree of migration is from cold Siberia, parts of Germany, Russian steppes and Middle Eastern regions. Private ownership of land in Ukraine that began with the end of Soviet era is also enthusing human migration and settlement. The arable land meant to be used for agricultural cropping is still in plenty (FAO, 2005). About 0.68 ha cropland is available per person compared to 0.25 ha in other parts of Europe, for example, Western Europe. Report by IEASSA (2013) suggests that for a while Ukrainians adopted extensive farming with large section of its population engaged in farming enterprises, about 65 percent. However, with intensification of crop production and adoption of modern techniques, population thriving on farming is about 16 percent during past few years.

4.2   NATURAL RESOURCES WITH REFERENCE TO AGRICULTURAL CROP PRODUCTION

4.2.1   SOIL RESOURCES OF EUROPEAN PLAINS AND STEPPES

4.2.1.1   SOILS OF EUROPEAN PLAINS

European prairie is supported by a few different soil types that are highly weathered, deep, and fertile. They are (a) Chernozems or Black earths that are most widely spread across the plains and most pronounced in Poland, Hungary, Ukraine, and southern Russian Steppes; and (b) Luvisols are common in the Central European plains of France, Germany, Czeck Republic and parts of Poland; Albeluvisols are traced in the North-eastern European plains (see Brady, 1975).

Chernozems are mostly widely spread and dominant soil group found in the European plains region. Chernozems are dark-colored soils rich in humus (7–15%) and minerals. They are often loamy, arable, and highly congenial for crop production. No doubt, they support good growth of pastures or natural prairie vegetation. Chernozems contain optimum levels of Ca deposits in the profile, hence are not afflicted by high acidity. Chernozems are highly aerated soils with excellent buffering capacity for moisture and minerals (Brady, 1975; Beyer 2002; and Devaney, 2013). Chernozems vary in thickness from 0.2 to 1.5 m in depth. Productivity of Chernozems is among the highest. Hence, in the cereal belt of Ukraine, fields with Chernozems are costlier than others in the region. In Ukraine, about 74 percent of Chernozem region is flat and occupied by arable cropping systems. The remaining

20 percent of land is under natural grasslands and small portion is retained as forested steppes (Chibiliyov, 1984).

European Loess soils are among the most fertile substrates for agricultural cropping. In fact, German soil specialists at Helmholtz Centre provide a value 100 to Loess soils on a scale depicting fertility and crop productivity (Helmholtz Association of German Research Centre, 2007). Loess soil extends into one fifth of the entire European plains. It spans region from foothills of Alps to deltas of Danube in the central Europe and it is scattered well in the Eastern European Steppes region. The European Loess is powdery product of glaciation during the Ice ages. The fine, lightweight silty dust from glaciated zone has swept into the plains in Germany, Poland, and into Russian Steppes. Loess is largely composed of quartz, silicate, and lime material. The fine mineral grains provide excellent aeration and percolation of water. It supports arable crop production. Digital map for entire Loess region of Europe including the Steppes of Russia is available (Haase, 2007; Haase et al., 2007). It helps in judging soil nature and fertility versus productivity of natural prairie vegetation and crops.

Polish soils are generally considered low in fertility, light textured and mostly sandy. Nearly 60 percent of the plains has light or very light soil of relatively low productivity. Low productivity is attributable to both sandy nature and acidity. Further, soils are low in organic matter which is often 1–2 percent (Rutkowska and Pikula, 2013). Soil with high or very high content of organic matter (>3%) is not common. Acidity affects over 55 percent of soils, wherein pH is 4–4.5. Low productivity of soils in the Polish plains has also been attributed to incessant cropping that depletes nutrients, intensification of crop production tactics without much care about its effects on soil fertility, especially nutrient imbalance.

Total agricultural land in Czeck Republic has stagnated at 4.3 m ha for the past 25 years (Kralovec, 2006). Agricultural expansion is perhaps difficult owing to lack of land available for conversion, soil fertility traits, and availability of easier methods, such as intensification to enhance crop productivity. Soil types encountered and its fertility are therefore important aspects in the Czeck plains. Soil traced in the cropping zones of Czeck plains and uplands can be grouped into five categories. They are Cambisols that occupies 40 percent agricultural land in Czeck Republic; Stagno-gleyic Luvisols that occupy 20 percent area; Luvisols 19 percent; Chernozems 11 percent, and Fluvisols 10 percent. Based on texture, it is said soils in Czeck cropping belt are categorized as heavy (15%) medium heavy (60%), light (20%), and gravely (5%).

Several soil types are detectable in the Hungarian plains. Among them, Chernozems are predominant and they are highly fertile and congenial to produce arable field crops. In addition to field crops, Hungarian soils support natural forests, grasslands, agroforestry species, meadows, and pastures. Birkas et al. (2012) identifies at least 6 different soil types. They are (a) Chernozems that are used for cereal production; (b) Medium–heavey textured soils used to develop forest plantations; (c) Medium textured soils used for meadows and pastures; (d) Solonetz and Solonchecks are alkaline soils afflicted with salinity/alkalinity. They possess excessive Na ion. They invariably need amendments and drainage before utilizing the field for crop production; (e) Humic and quick sand soils are classified as Arenosols or Cambisosls. These soils are relatively rich in nutrients and they allow crop production. Quick sand soils are easy to plow and are low in organic fraction; and (f) Eroded and shallow soils more commonly classified as Laptosols and Entisols are well distributed in slopes and undulating plains. Soil erosion induces loss of fertility and variability in crop production. Overall, soils found in the Hungarian plains can be characterized as sandy (15%), sandy loam (12%), loams (47%) and loamy clay or clay (26%). According to Birkas et al. (2012), in Hungary, soil degradation threatens over 1.9 m ha; water logging occurs on 0.3 m ha; soil erosion due to water/wind is felt as a problem on 3.7 m ha; acidification affects 0.65 m ha; and salinization reduces crop yield on 1.1 m ha.

Several orders of soils are traced in the agricultural plains of Ukraine and Southern Russian Steppes. Agriculturally, most prominent soil types are Chernozems, Fluvisols, Greyzems, Phaeozems, and Haplic Kastanozems. Chernozems are either gray-black or black in color. However, brown Chernozems too occur in some areas. These soils are rich in organic matter and possess calcareous deposits. They are highly suited for crop production. A few different versions of Chernozems that are easily traced in the Russian Steppes are Luvic Chernozems (leached), Podzolized Chernozems (Luvic Phaezems), Ordinary Chernozems, and Calcic Southern Chernozems (Blagoveschchenskii et al., 2007). Calcic Chernozems are more commonly encountered in the Steppes. Gypsum and soluble salts are accumulated at about 30 cm depth in these soil types. They possess high CEC of 35–40 cmol kg^{-1}. In the northern regions of Steppes, Brown Chestnut soils also known as Haplic Kastanozems are predominant. Gypsum and soluble salts occur at a greater depth of 2 m from surface. Meadow Chestnut soils are also referred as Haplic Phaeozems. Phaeozems are sometimes connected with high groundwater level. Hence they show in up wetter regions. Phaeozems support permanent growth of herbaceous vegetation. Soil types encountered in the dry steppes are exposed to drought. They are sandy in nature and possess. They are often highly heterogeneous with regard to fertility. Hence, crop productivity could be erratic unless supplied with nutrients and water.

Major soil types found in Ukraine and Russian Steppes and their fertility classification:

Fertility Status	Soil Types
Low	Arenosols, Planoslols, Podzols
Moderate	Regosols, Andosols, Greyzems, Histosols
High	Fluvisols, Gleysols, Vertisols, Kastanozems, Chernozems, Phaeozems, Cambisols

Source: Stolbolov (2000); Rode (1975); FAO (2005); Krupennikov, I. A. (1983); World Soil Resources Reports, 66th Review1, FAO Rome, pp. 66
Note: Fertility capability of soil is defined as its ability to store, retain in the profile and release plant nutrients into the root zone in quantities required by the crop species cultivated.

According to FAO (2005), Soddy podzolic and nongleyed sandy loams are common in Polissya region of Ukraine. They possess thick humus layer of 15–150 cm. Soils in the Forested steppes are mainly Chernozems. In the Southern dry Steppes, soils are either Chernozems or Haplic Kastanozems also known as chestnut soils. Again, they possess a thick humus layer. Soil pH of agricultural landscapes ranges from 6.0 to 7.5. Soil P and K content is classified as average to high. The exchangeable-K content ranges from 11 to 13 mg K_2O 100 g^{-1} soil in the steppes.

4.2.1.2   SOIL FERTILITY ASPECTS

During eighteenth century, fertilizer use on crop fields and pastures was confined to FYM, animal manures and residue recycling. Farmers were destined to harvest depending on crop's response to organic matter recycling and natural atmospheric inputs if any. Low crop yield and famine together did affect the agrarian populations of European plains/steppes. During 1850s, it became gradually clear that appropriate crop rotations and propped up soil fertility is essential for higher crop productivity. Fertilizer consumption, either in organic or inorganic form became the main indictor of crop production. For example, wheat grain production increased from 1,000 kg ha^{-1} to 1,600 kg ha^{-1} due to manure use, but it took about 100 years from 1850 to 1950 to register this increase. Fertilizer supply in wheat belt of France was enhanced from 1.2 million t $year^{-1}$ in 1950 to 6 million t in 1973. As a consequence, wheat yield jumped from 1,600 kg ha^{-1} in 1950 to 3,000 kg ha^{-1} in 1973. Next, wheat productivity jumped from 3,500 kg ha^{-1} in 1973 to 7,150 kg ha^{-1} in 2005, but fertilizer use and its supply was not well timed accurately. Hence, fertilizer use efficiency was not higher.

Major nutrients used on cereal crops such as barley/wheat in French and German plains increased appreciably during 1930–2000. Periodic revision of crop yield goals upwards, rapid depletion of nutrients and lack of appropriate residue recycling procedures necessitated supply of higher quantities of fertilizers. In France, nutrient supply increased from 6 kg N:19 kg P:10 kg K ha^{-1} in 1936 to 95 kg N:40 kg P and 56 kg K ha^{-1} in 2000 A.D. In Germany, during the same period, fertilizer inputs to cereals increased from 25 kg N, 32 kg P 50 kg K ha^{-1} to 110 kg N, 26 kg P and 40 kg K in 2000 (Isherwood, 2003). Field evaluation of fertilizer schedules, formulations, in relation to soil types, location, environment, and yield goals, now seems almost a continuous exercise in all subregions of European steppes. Fertilizer optimum keeps changing as soil-N depletions and removal via grains /forage increases season after season. After all, soil-N needs to be replenished appropriately, whether it is high-intensity cropping zone in France or subsistence fields in southern Russian steppes. According to Yara International ASA (2012), based on over 122 field trials, it was concluded that winter wheat grown in different soil types in the French plains need, on an average 124 kg N ha^{-1}. It could be preferably supplied as ammonium nitrate to reap best harvests. An additional 27 kg ha^{-1} of UAN is needed if wheat farms need to be economically highly profitable. In Germany, wheat fields need to be supplied with 210 kg N ha^{-1} to reach optimum grain productivity. The German loess soils needed an additional 15 kg UAN ha^{-1} to reach higher economic advantages. Appropriate timing of fertilizer-N is essential, if the intention is to avoid undue accumulation and loss in efficiency. Usually, fertilizer-N inputs are split in order to match the crop's requirement at various stages. Fertilizer-N inputs are timed at tillering, stem elongation, ear head emergence, and grain fill stages. A study of the ratios of major nutrient used in the plains shows that use of P and K has decreased in relation to N, otherwise ratios are 1: 1.5: 1; N:P:K.

Fertilizer N formulations used in the European plains are Ammonium sulfate, Calcium ammonium nitrate (34% N), Ammonium nitrate (27% N), Potassium nitrate and Urea. Urea is among the most common fertilizer-N sources. It contains about 46 percent N and is perhaps cheapest (Isherwood, 2003). Fertilizer-N sources can be grouped into; (a) Nitrogen derived from nitrates. Nitrates are easily soluble and available to crop roots. Applying nitrates ensures availability of N to crops; (b) Nitrogen is derived from NH_4 salts. This moiety is directly absorbed by crop roots although in small quantities. Some crops such as rice grown in wet lands and under submerged conditions prefer NH_4-N. NH_4+ ion is relatively less mobile in soil than NO_3. NH_4-N may get immobilized into microbial component of soil in good quantities, but could be released at a later stage; and (c) Nitrogen is derived mostly from Urea fertilizers. Crop roots do absorb sizeable quantity of N in ureide form. Urea upon hydrolysis releases NH_3 (See Yara International ASA, 2012).

According to Krauss (2002), farmers in the steppes of Eastern Europe and Russia have drastically revised fertilizer usage in their farms. Since the economic reforms of mid-1980s, fertilizer consumption by Eastern Europe and Russia has declined by 14–33 percent of original levels used during previous years. Ukraine, in particular has reduced the fertilizer consumption, in order to follow environmentally safe levels of nutrients in soil and aquifers. Further, it is said that lack of easy farm credit has subsided lavish applications of fertilizers in the farms. The decline in fertilizer use between the years 1980 and 2005 in the Western European plains was 140 kg ha^{-1}–118 kg ha^{-1}, in Central and Eastern Europe decrease was 48 kg ha^{-1}–8 kg ha^{-1}; in Russia decline was from 130 kg ha^{-1} to 119 kg ha^{-1} and in Ukraine it was from 65kg ha^{-1} to 8 kg ha^{-1} (Krauss, 2002). As a sequel to reduced fertilizer inputs, cereal crop productivity in Ukraine, Russia, and Eastern European Steppes either stagnated or declined. For example, wheat yield in Russia stagnated at 1.5–2.0 t ha^{-1}, and it declined from 3.2 t ha^{-1} to 2.3 t ha^{-1} in Ukraine. A compilation by Rosas (2011) suggests that during past three years, application of N, P, and K into crop fields in Russian Steppes and Ukraine has decreased marginally, owing to environmental concerns (see Table 4.1).

TABLE 4.1 Fertilizer supply to crops cultivated in the Steppes of Russia and Ukraine

Nation/Nutrient		Wheat	Corn	Barley	Oats	Rye	Rapeseed
		kg ha^{-1}					
Russia	N	20	66	16	17	18	85
	P	8	20	9	2	3	27
	K	3	13	6	1	2	35
Ukraine	N	27	73	17	19	18	105
	P	10	22	9	3	3	37
	K	4	14	7	1	2	23

Source: Rosas (2011)

Ukraine is the "Bread basket" of Eastern Europe. Its agriculture thrives on fertile Chernozems and congenial climate. However, since 1991, productivity of wheat expanses dipped several folds because of lack commensurate replenishment of fertilizers. In some locations, soil nutrient depletion was rampant. For example, during 1960–1985, annual fertilizer-based nutrient consumption increased from 1.4 million t year^{-1} to 5 million t year^{-1}. As a consequence, wheat productivity increased from 2.2 t ha^{-1} to 4.7 t grain ha^{-1}. However, this advantage was lost, during 1990s due to lack of fertilizer supply. Depreciated soil fertility was the major reason for low wheat production in Ukraine (FAO, 2005). Ukrainian farmers apply both organic manures and inorganic fertilizers. About 66 percent of Large Corporate Farms supply major nutrients, using a large doze of organic manure and inorganic fertilizers. Private farms and small farmers too use inorganic complex fertilizers to supply nutrients (FAO, 2005). Winter wheat responds to supply of major nutrients. If the grain yield is about 0–0.8 t ha^{-1} with no fertilizer supply, 150 kg NPK results in 2.5 t grain ha^{-1}. Reports by FAO (2005) suggest that winter wheat crop in Ukraine is supplied with 80 kg N, 70 kg P$_2$O$_5$, and 70 kg K$_2$O ha^{-1}; barley is provided with 60–90 kg N, 45–60 kg P$_2$O$_5$, 45–60 kg K$_2$O ha^{-1}; forage crop 30 kg N, 60 kg P$_2$O$_5$, 60 kg K$_2$O ha^{-1}; sugar beet 170 kg N, 140 kg P$_2$O$_5$, 190 kg K$_2$O ha^{-1} and maize 130 kg N, 100 kg P$_2$O$_5$, 100 kg K$_2$O ha^{-1}. Oilseed such as sunflower is supplied with 60 kg N, 60 kg P and 60 K ha^{-1}. Over all, whatever the basic soil fertility status, we have to replenish the extent of nutrients removed by the crops.

Nitrogen dynamics in the agrarian zones of Europe, that predominately includes plains/steppes and forests has been studied and reported in great detail by many agencies and national research institutes. One of the recent reports by Sutton and Reis (2012) provides details on crop response to fertilizer-N supply, nitrogen emissions and GHG effect, nitrogen budget in fields and nitrogen cycle.

Atmospheric deposits add N and other nutrients to crop fields. Such N deposits are influenced by dust settling on the plains and precipitation pattern. According to Machon (2011) soils may derive 11.2–137 kg N ha^{-1} from atmospheric deposits depending on season and precipitation pattern. During wetter years, Hungarian plains receive even high N deposits ranging from 14 to 142 kg N ha^{-1}. The N budget studies indicated that N input from atmosphere could be higher than loss due to emissions. Nitrogen loss due to emissions is less than 1 kg N ha^{-1} as NO or NH$_3$. Of course, there are other modes of soil-N loss traced in the Hungarian plains. They are soil-N percolation to subsurface and vadose regions, loss via seepage to channels and loss via removal of residues (Krishna, 2003; Sutton and Reis, 2012).

In the German Plains, intensification of agricultural cropping has led to use of large amounts of inorganic and organic manures, irrigation, and pesticides. This has caused a variety of soil maladies. It has induced soil erosion, emissions of NH$_3$, N$_2$O, and N$_2$O, nitrate leaching, eutrophication of water bodies and groundwater contamination. Farmers in German plains have often applied high amount of fertilizer-N, about 84–128 kg N ha^{-1} surplus. Hence, agricultural agencies have been concentrating on devising as many computer models and simulations to understand the N dynamics as accurately as possible. They aim at most appropriate fertilizer-N schedules and timing. Kusterman et al. (2009) describe one such model that integrates, organic farming aspects, conventional fertilizer supply aspects, livestock, and recycling of nutrients. It also considers N-derived from Biological Nitrogen fixation and rates of N inputs from atmosphere. Nitrogen surplus under organic farming systems was estimated at 38 kg N ha^{-1}and 44 kg N ha^{-1} under the conventional system. Organic-N accumulation ranged from 32 to 35 kg N ha^{-1}, but declined by 28 kg N ha^{-1} under conventional system. Fertilizer N supply aimed at long term upkeep of crop fields usually considers aspects such as yearly accumulation/depletion of N per year.

Again, soil N dynamics in the Eastern European Plains has been studied in great detail by agricultural agencies. For example, in Czeck Republic, models have been used to study a range of parameters related to N dynamics and their influence on productivity of crops such as wheat, barley, maize potato, winter rape, alfalfa, and pasture grass (Hlavinka et al., 2012). Crop growth model known as HERMES was used to devise crop rotations that offer high grain yield with better N efficiency.

4.2.1.3   SOIL NUTRIENT BALANCE

European plains and steppes support large-scale agricultural enterprises as well as small individual family farms. Inorganic fertilizers are the major sources of N and P in the European agrarian zones (Pau Vall and Vidall, 2003). Their excessive use can be deleterious to the soil and environment. Hence, it is useful to integrate mineral fertilizer use with environmental concerns while devising fertilizer supply schedules. Monitoring N and P surpluses in the soil periodically is almost essential. Accurate knowledge about nutrients that enter the system/field that lost via natural and man-made factors, that absorbed into crop phase of the ecosystem and that recycled in residues and stubbles is necessary. Nutrient inputs, if they match the outputs in terms of quantity and timing, then undue accumulation and pollution could be reduced in the plains. The crop phase of European plains should extract nutrients supplied as rapidly and efficiently as possible. Major avenues of N supply to cereals fields in the plains is via, inorganic fertilizer-N, FYM and crop residue. Atmospheric deposits may be relatively small, however it needs to be accounted under total N inputs. The nitrogen balance is also affected by losses due to volatilization as NH_3, N_2O, NO_2, N_2, NO_3 leaching, percolation, and erosion along with soil particles. Microbial and chemical transformations in the soil phase may affect the N balance in the field (OECD, 2008). Nitrate accumulation in the soil profile can indeed be problematic. On a long term, soil NO_3 levels could be beyond what is permissible. For example, in the Czeck plains, consistent supply of fertilizer-N enhanced soil NO_3 levels from 20 mg /L in 1976 to 56 mg NO_3-N during 2003 (Holas and Klir, 2003). Wide range of measures that included change in land use, cropping systems, reduction in fertilizer supply and efficient management of fertilizer-N is required. Nitrogen accumulation in the soil phase, that retained in the subsurface and that percolated to groundwater are all dependent on efficiency with which crop roots scavenge and deplete soil-N. Groundwater pollution gets minimized, if roots absorb nutrients added, as rapidly as possible without allowing it to move deep into profile. Nutrient balances can be judged for individual plots, entire field, large farms, areas or even a nation. Following is an example that depicts N balance in some European nations that are encompassed by the plains/steppes (Table 4.2).

TABLE 4.2 Nutrient balance in the European Plains, a few examples pertaining nitrogen and phosphorus

Country in the European Plains	Nitrogen Balance		Phosphorus Balance	
	000' t year^{-1}	kg N ha^{-1} year^{-1}	000' t year^{-1}	kg P ha^{-1} year^{-1}
France	1,589	54	114	4
Germany	1,926	113	68	4
Poland	797	48	45	3
Czeck Republic	300	70	7	2
Hungary	217	37	−1	−1

Source: OECD (2008)

The physicochemical transformations of soil-P and fertilizer-P added to soil during crop production are different from those known for soil-N. Yet, most of the general soil environmental and man-made factors that affect N balance in the agrarian zones of Europe are almost equally applicable to Phosphorus balance studies. A few aspects such as emission studied for soil-N loss does not apply, because P emissions are nonexistent. However, chemical fixation of a portion of fertilizer-P added and immobilization of P into organic phase of the soil profile occur and it needs to be considered, with regard to P dynamics and balances. According to a report based on

Eurostat (2013), the average surplus supply of phosphorus into fields in the Northern European plains could be 2 kg P ha^{-1}, but it is much less or even nil in most regions of Eastern Europe. It is generally agreed that such surplus in soil-P is indicative of environmental risks in the long run. Excessive concentrations of N and P in water bodies such as lakes, reservoirs, ponds, and slow flowing rivers can induce eutrophication. Soils supplied with such water sources may easily get deteriorated. Such soils may emit higher amounts of N to the environment and P loss from the soil profile too may reach unmanageable levels. Most importantly, excessive N and P may contaminate the groundwater sources and aquifers. In the Ukrainian Steppes, available-P levels of the Chernozems seem to dictate the wheat /maize grain yield to a certain extent. Reports by FAO (2005) suggest that at <50 mg available-P kg^{-1} soil, wheat grain yield ranges from 2.6 to 3.3 t ha^{-1} without fertilizer and 3.4 to 4.8 t ha^{-1} with fertilizer. Grain yield is highest if available-P pools are above 100 mg kg^{-1} soil.

Regarding soil potassium dynamics in the Steppes, Kraus (2002) believes there is a clear tendency to supply lower levels of K-bearing fertilizers to soil. Since, soil-K is absorbed by crops and its sources are depleted at higher proportions, K dearth may appear sooner or later. Soil K mining seems to occur in many locations within the steppes. It seems until early 1980s, fertilizer supply to crops in steppes were more than optimum. It resulted in positive-K balance. In parts of Czeck Republic, in the absence of repeated high dosages of fertilizer-K, organic manures became the major source of K to crops. A negative balance for soil-K may affect the K pools in soils that supply K to roots/rootlets. Reports by Losack et al. (2009), suggest that about 40 percent of agrarian regions in Central Europe needs fresh supply of K each year, of course based on crop and yield goals. The situation is similar with regions in the Russia steppes and they too need extra supply of fertilizer. Yet in many locations, K input is still at the levels adopted in mid 1950s, that is, 10 kg K ha^{-1} (Cermak and Torma, 2006; Cermak, 2010). According to Madaras et al. (2010), annual K refurbishment needed to stabilize soil K pools is at least 108–203 kg K ha^{-1} for locations afflicted with severe K deficiency. Kovacevic et al. (2012) have clearly shown that, if P and K are applied in optimum quantities, then grain yield of maize improves by 14 percent and that of soybean by 32 percent over fields not fertilized with K. In case of potatoes grown on Chernozems of Ukraine, it is said the crop removes about 8 kg K t^{-1} tubers. Hence, farmers are advised to supply soils with commensurate quantities of fertilizer-K based on yield goals. Ukrainian steppes are supplied with fertilizer-K produced within the region as well as that imported from regions.

4.2.1.4   SOIL ORGANIC MATTER AND CARBON SEQUESTRATION IN THE PLAINS

As stated earlier, Polish soils are not endowed richly with organic matter. Further, sandy soils with oxidative conditions easily loose inherent SOM. Therefore, even if soils are supplied with optimum levels of FYM, it is depleted rapidly due to CO_2 evolution. Long term trials at 33 different locations in Poland suggest that depletion of organic matter content in the plains is dependent on cropping systems practiced. Rotations with mainly wheat and barley lost SOM relatively rapidly. Despite consistent residue recycling and FYM inputs, SOM reduced from 0.78 to 0.58 percent from 1970 to 2008. However, in locations/fields with legume-grass mixtures included into crop rotations, SOM got stabilized at 0.75 percent all through the 3 decades (Rutkowska and Pikula, 2013). The quality of SOM is also important. Since it decides the rate at which mineralization/immobilization processes occur in the soil profile. There is no doubt that Polish farmers may grow crops that deplete different fractions of SOC preferentially. For example, long term studies using different crop rotations suggest that quality of SOM clearly depended on the manure applications schedules and crops selected for rotation. Humification processes were also dependent on crop production (Mercik et al., 2005). Rutkowska and Pikula (2013) state that crop rotations devised shrewdly are most important with regard to stabilization of SOM in the plains. High inputs of mineral fertilizers may not influence SOM stabilization processes advantageously.

Reports about Hungarian plains suggest that SOM content of Haplic Luvisols are influenced by precipitation, fertilizer supply trend and cropping systems. High fertilizer dosage, especially N, P, and K increases biomass, which upon incorporation adds to SOC. Over a 20 year period, SOC decrease was noticed. It decreased by 16 percent of original estimations. Low precipitation and low fertilizer supply was the main cause (Van and Van, 2009).

The soils of Russian steppes are moderately endowed with organic matter. They possess moderate levels of humic substances. Soil organic matter is predominately derived from natural flora in the steppes (Stolbolov, 1999). Chernozems of Russian Steppes have been incessantly cropped. Repeated tillage and arable cropping has led to oxidative conditions and loss of soil-C via CO_2 evolution. Soil degradation process and loss of SOM has occurred in the Chernozem belt of Russia for at least two centuries now (Rozanov, 1984). There is ample proof in various parts of steppes that indicate loss of SOC. For example, it seems humus content of Chernozems during mid-nineteenth century when soils were analyzed by Dokuchev, it ranged from 10 to 13 percent humus equivalent to 39,000 t km^{-2}. Now, in the same area, the upper 0–30 cm depth of soil profile shows up only 7 percent of humus, equivalent to 21,000–30,000 t km^{-2}. The total loss of humus as CO_2 or through soil erosion was estimated at 90 t km^{-2} $year^{-1}$. In a different region called Ulyanovsk, SOC loss was much high at 270 t km $year^{-1}$. In the Stavropol region, SOC loss was 70–80 t km^{-1} $year^{-1}$. Average loss of humus in the Chernozems of Central Russia ranged from 25 to 30 percent during last year. In the Volga region loss of humus was much greater at 50–60 percent of the levels known in nineteenth century (Chesnyak et al., 1983).

4.2.1.5   TILLAGE AND ITS RELEVANCE TO CROP PRODUCTION IN EUROPEAN PLAINS

Tillage is among the most conspicuous man-made effects on the European plains. Tillage begins with the clearing of forests, shrub vegetation and natural sod if any. Tillage intensity is related intricately to soil properties such as aeration, oxidative nature, nutrient release, emissions, and moisture retention in the profile. Conventional tillage was adopted as a matter of rule in most regions of European plains and Russian steppes. Tillage involved use of draught animals, tractors, and other mechanical devises. Mechanical tillage using disks to achieve deeper disturbance to soil has been in vogue for the past several decades in the entire plains/steppes. However, during past three decades, conservation tillage or no-tillage systems have gained in importance and popularity among the entire European agricultural belts. Let us consider a few examples about advantages and disadvantages of adopting no-tillage system, from each of the nations in European plains.

French plains support production of wheat, barley, and grapes in large expanses. Farmers in the cereal based prairies adopt intensive farming techniques. They supply high amounts of inorganic fertilizers and organic manures, irrigate, and use high-yielding genotypes. Cereal productivity is commensurately high. Tillage systems adopted by farmers has changed during recent past. For the past 2 decades, No-tillage systems are popular in the French plains. Recent evaluations have shown that No-tillage is easily accepted by farmers in the French plains because it reduces labor costs and allows better profits. Currently, about 17 percent of 3 million ha cereal belt in France is adopting conservation tillage. Since 1999, no-tillage system has expanded to over 1 million ha. Further, Khaledian et al. (2007) state that no-tillage system adopted in French plains consumes lower amounts of fuel, energy, and requires less labor hours to generate 1 t wheat grains (see Table 4.3). No-tillage essentially involves leaving stubbles and residues from previous crop. This procedure thwarts loss of surface soil and nutrients. It also helps in better seed germination. However, it has been argued that low temperatures created by mulches and residues has affected growth rate of seedlings and even reduced final grain productivity under no-tillage. Development of mechanical resistance to root growth is another reason for slightly lower growth and grain formation under no-tillage systems. Yet, we should realize that benefit from no-tillage out -weighs marginal decrease in wheat grain yield.

No-tillage or reduced-tillage systems are among the most preferred procedures in the Czeck plains. They are suited for cereals grown in rows. Novack et al. (2013) opine that among various combinations of tillage, mulches, and crop species examined, no-tillage with mulches are best with regard to thwarting loss of surface soil. No-tillage reduces soil loss by 6–7 folds compared to conventional tillage. Conventional tillage may produce a uniform profile, but it is prone to loss of soil and nutrients.

TABLE 4.3 A comparative evaluation of No-Tillage and Conventional Tillage on Durum wheat yield at CIRAD experimental station, in Montpellier, France

Tillage System	Fuel Consumption	Energy Consumption	Work Duration	Wheat Grain Yield
	Litres ha^{-1}	MJ ha^{-1}	h	t ha^{-1}
Conventional tillage	64	2,476	9.35	5.94
No-tillage	37	1,431	3.14	2.75

Source: Excerpted from Khaledian et al. (2007)

Note: Energy consumption to prepare land and harvest 1 t grain ha^{-1} is 233–261 MJ under conventional tillage, but under No-tillage energy requirement is only 100 MJ.

4.2.1.6 PRECISION FARMING IN EUROPEAN AND RUSSIAN STEPPES

Precision farming in the German plains is yet to make a major impact on cereal farming. Precision technique has been in use since past 5–8 years in some form. Surveys indicate that farmers adopting Precision farming in the plains increased rather slowly during 2001–2006. There were difficulties in obtaining all necessary gadgetry, data, maps, and fertilizer laden variable dispensers in one place as needed (Reichardt et al., 2010; Wagner, 2011; and Reichardt and Jurgens, 2012). Reichardt et al. (2009) have reported that majority of German farmers in the plains are selective and adopt portions of Precision techniques that are easily amenable. They are data collection techniques, soil sampling using GPS referencing, site-specific seeding, calculations of fertilizer supply based on appropriate computer-based crop response models, and so on. However, use of variable rate applicators mounted on to GPS guided tractors are yet to pick up in a big way. About 6.5–11 percent of farmers situated in different regions within German plains use Precision techniques. Young farmers were adopting precision techniques more easily. The German plains, as such, it is filled with many large corporate cereal farms. Such large farms are highly amenable to precision techniques since large areas can be quickly covered using GPS guided traction and implements. It is generally believed that Precision techniques offer greater economic advantages to large farms and when the product (grain) is valued higher. More interesting is the fact that fertilizer supply using soil maps and variable rate applicators reduce the quantity of fertilizer needed to reach the same yield goal. Farmers tend to gain by delaying accumulation of unused fertilizer-based nutrients. Reports by Yara (2012) clearly show that farmers in France, Germany, and several of the Eastern European nations such as Poland and Hungary have been choosing from different aspects of Precision Farming (Swinton and Lowenberg-Deboer, 2009). For example, many of the Farmers in Eastern Europe preferred to use N-sensor (chlorophyll meters), Hand-held N testers, obtain data, soil maps and fertility variations from satellite and internet based companies. For example, agencies such as "Astrium" or "RapidEye" (GIM International, 2009) provided periodic soil maps, crop growth maps and prescriptions about agronomic procedures, fertilizer inputs and harvest dates after careful evaluation of the fields (Astrium, 2005). Precision techniques have also been adopted to judge soil moisture and supply water suitably using computer guided variable spray nozzles and applicators.

Precision farming techniques that involve mechanized sampling of large number of soil samples from a field using GPS, rapid analysis using sensors, preparation of soil fertility maps and devising fertilizer supply based on yield goals are in vogue in the Russian Steppes. Hitherto, many of the large farms were using slow and tedious sampling procedures. Tsirulev (2010) reported about the influence of precision farming on wheat cultivation and productivity. Soil fertility variations in the fields within Stavropol district were studied in detail using GPS guided techniques. Soil nutrients such as P and K, organic matter content and pH were analyzed using precision techniques. Adoption of Precision techniques resulted in better yield at 2.5 t grains ha^{-1} compared to conventional techniques that gave 1.7 t grains ha^{-1}. Since fertilizer requirements were reduced under Precision techniques economic advantages were higher. There are many large farms in Ukraine and Russian steppes that are highly amenable for adoption of Precision techniques. Its adoption allows state agencies to regulate soil fertility and thwart degradation. In fact, it offers a good chance to avoid undue accumulation of nutrients in the soil profile that could be vulnerable to loss or that may induce contamination of groundwater in the Steppes. Precision techniques are also easy to adopt in small individual farms that are recent in inception. Precision techniques to assess variability of soil fertility and monitor crop growth are available. They are in easy reach of single farms. Adoption of precision techniques helps in removing drudgery, since much of the activity is computer guided and automatic. It may also bring uniformity to crop productivity within fields. Recent reports indicate that Precision techniques that create uniform soil fertility across individual fields or even larger regions are sought in Ukraine. Computer guided decision support systems and GPS guided variable-rate applicators are also sought. Satellite monitoring systems such as " *Mapexpert*" in Ukraine and "*Vega*" in Russia are useful in providing the field maps, soil fertility variations and water resources to farmers in the steppes. It is said that use of precision techniques easily brings about a 10 percent savings to farmers in the steppes of Ukraine and Russia. This is in addition its effect on fertility, quality, cropping systems and regulation crop production systems (IEASSA, 2013).

4.2.2   *WATER RESOURCES, IRRIGATION, AND CROP PRODUCTION TRENDS*

The upper Rhine aquifer is an underground system of North European plains. It is a large body of water held beneath the soil profile. It is a major source of water to cities in this zone. The aquifer also supplies water to industries and agricultural zones. It harbors on an average 450,000 km^3 of fresh water. It is among major water sources for human populations in Germany and France (Wikipedia, 2013a).

Hungary is situated in the Carpathian basin and is exposed to severe floods. Excess water that flows into rivers may often flood the plains. Management of water output from rivers is important, if not it leads to water logging of the crop fields. Low lying plains that constitute 10 percent of cropping zones are vulnerable to water logging (Varallyay, 2007; 2011; Birkas et al., 2012). The plains are congenial for rain fed (dry land) cropping systems. Hungarian plains also experience water deficit and droughts. Annual deficit of water occurs because precipitation is only 450–600 mm year^{-1}, but evapo-transpiration caused due to rapid crop growth and warm summer is higher at 680–720 mm. Water deficit is answered by Hungarian farmers using lift irrigation and riverine sources.

Ukraine is served by a natural precipitation pattern that is variable with regard to intensity and distribution within the plains. Annual precipitation in the southern steppe regions is relatively low at 300–500 mm. Sometimes it is scanty and may affect cereal production severely. In the North-western plains of Ukraine, where cereal production is rather intense, the precipitation ranges from 600 to 700 mm. Ukrainian Plains are endowed with large number of rivers and innumerable tributaries that irrigate the crop, whenever natural precipitation is insufficient. Major rivers that serve to irrigate crops are Dnieper and its tributaries such as Prypyat and Desna; Danube, Dniester, Pivdenny Bug, Zahidny Bug, and Siversky Bug. There are several dams across these rivers that accumulate water. They are used in the off season and summer months. Ukrainian steppes are also served by several lakes that help in irrigating crops (Ministry of Environmental Protection and Nuclear safety of Ukraine, 2000).

4.2.3 CROPS, CROPPING SYSTEMS AND ECOSYSTEM PRODUCTIVITY

European Plains and steppes support production of variety of field crops, pastures, and horticultural species that serve to feed a very large human population with varied tastes and habits, plus a large population of farm (livestock) (see Table 4.4).

TABLE 4.4 Crop species cultivated in different countries encompassed by European Plains and Steppes

Country/Region	Field Crop Species
France plains	Soft red wheat, barley, maize, lentils, rapeseed, pastures species, grape vines
German plains	Wheat, barley, maize, rapeseed, grape vines
Poland	Wheat, barley, maize, millets, lentil, oilseeds

TABLE 4.4   *(Continued)*

Country/Region	Field Crop Species
Czeck republic	Wheat, barley, maize, millets, legumes, oilseeds
Hungary	Wheat, barley, millets, rape seed, sugar beet, grape vines
Ukrainian steppes	Wheat, barley, maize, sugar beet, vegetable, fruits such as grape vines, apples, cherries
Russian steppes	Cereals such as wheat, barley, and maize; millets such as panicum, sorghum, setaria, legumes such as chickpea, peas, lentil, vetch, rapeseed, sunflower, sugar beet, potato, vegetables, and variety of pasture grass/legumes

Sources: Blagoveschchenskii et al. (2007)

French Plains generates a posse of crops such as soft wheat, durum wheat, barley, triticale, oats, sorghum, rye, and rice. French plains also generate large quantity of grapes. These crops create a large agricultural prairie zone extending into all across the nation. France is among the foremost countries producing wheat. Wheat is in fact the major cereal cultivated in the entire North European plains region. Winter wheat is common and it is sown in October/November and harvested next July/August. Picardie and Centre regions support intense cultivation of wheat. These two regions within France contribute about 26 percent of total wheat grain harvest. Common soft wheat is predominant type in the French plains. Wheat cultivars that dominate the plains have changed based on preferences of farmers of each region. In general, wheat genotypes that offer high quality grains have progressively occupied more area. During past decade, 77 percent of the wheat belt is filled with cultivars of high quality grain (e.g., Apache, Isengrain, Charger, Orvantis, etc.) (Schafer and Ferret, 2003). At present, French prairies support four different types of wheat. The ecosystem generates (a) Soft wheat; (b) Superior bread wheat; (c) Standard breadmaking wheat; and wheat for other uses such as pastries. Following is the wheat production trend during past three decades in the French plains:

	1980	1990	2000	2010
Total Production (m t year^{-1})	23.7	33.3	37.6	39.3
Harvested Area (m ha year^{-1})	4.6	5.1	5.3	5.8
Productivity (kg ha^{-1})	5.2	6.5	7.1	7.1

Source: FAO (2010)

Winter wheat is the main type of wheat cultivated in the plains of Germany. Its production is concentrated in regions of Bayeren and Niedersachen. Together, these regions contribute 36 percent of Germany's wheat produce derived from the plains. Winter wheat, as usual is sown in October/November and harvested in next April or May. The hard red winter wheat is most prominent wheat variety grown in the Russian plains. Wheat culture is concentrated in the region between Moscow and Black sea and entire Ukrainian plains. It is also a major crop in the Volga region. Wheat crop is sown in August to October. Panicles emerge in next May and the crop is harvested in July. Wheat production extends into entire steppe region of Ukraine. Wheat culture is feeble only in the deep north of the nation. Winter wheat dominates the landscape and constitutes about 95 percent of total whet area. About 80 percent of wheat produce is used for milling. Bakery industry, bread making and pastries account for 15 percent of wheat produced in the Ukrainian plains. Wheat belt extends into Poland, but its production is intense in the central region called Scheldlce. Scheldlce accounts for 6 percent of nation's wheat grain harvest. Wheat planting begins in September and crop reaches harvest by June (USDA Economic and Statistic Survey, 2012).

Barley is a staple cereal in European plains and Russian steppes. Russia, Ukraine, and Germany are major barley producing countries of the plains region. Barley was originally domesticated in the Fertile Crescent region of Middle-East. It was introduced into European plains and steppes perhaps during ancient period through human migration. Barley cultivation is restricted more to the plains of Northern European nations, although it is grown in other regions of European Union equally intensively. Barley production zones in the French plains fluctuate based on several factors related to agroclimate, soil fertility, yield goals and demand. During past 3 years, barley belt in the French plains has ranged from 0.9 m ha to 1.05 m ha (Rutenberg, 2012). Barley is an important forage cum grain crop in Ukraine. About 90 percent of barley is sown in spring season. At present, much of barley cultivated in Ukraine is spring barley. It is planted in April and harvested by August/September. Spring barley is relatively less tolerant to winter, hence it is confined more to southern Ukraine. Ukraine generated about 9.23 m t barley grains during 2010. Barley is cultivated in large expanse in the Central European nations such as Hungary, and Czeck Republic. Selection of cultivars could be based on the end product that is, either for grains or for malting or forage; based on crop duration and height, also based on productivity and grain quality. A few examples of barley varieties popular in Czeck Republic and Hungary are Amulet, Akcent, Kompakt, Krona, and Tolar. Spring barley production is relatively intense in the Steppes. Crop matures in about 110–125 days. It tillers profusely and produces about 2,000 tillers m². Soils that are loamy, shallow, with large quantity of water and nutrients are preferred. Acid soils are avoided for barely production, but soils with pH 6.2–7.2 are preferred. Barley prefers long day light period of 12–13 h day. Spring barley needs about 145–160 mm from heading stage until grain maturity. The heat units required ranges from 1,300 to 1,800 for a crop season.

Maize is predominately a fodder crop in the European plains. Maize is grown in Ukraine for both fodder and grain. About 25 percent of maize production is meant for grains, rest is for fodder/silage. Maize cultivation in Ukraine is concentrated more in the eastern and southern regions. Maize for fodder is sown in April and harvested by Mid-September. It withstands water scarcity to certain extent during the season. Despite, constraints to its production, maize cultivation zone in Ukraine is expanding steadily.

4.2.4   CROPPING SYSTEMS

Cropping systems currently in vogue in France and Germany involve wheat mono-crops, wheat-fallows, wheat-oilseeds or wheat-legumes. In the Polish plains, wheat and barley based cropping sequences have been in vogue since many decades. During past three decades, Polish farmers have shown priority to sequences such as potato-winter wheat-spring barley and maize for silage; potato-winter wheat-spring barley and clover/grass mixed pastures. Monocrops of wheat or barley are also common in the plains. Depending upon soil tests and yield goals, fertilizer inputs have ranged from 45 to 135 kg N , 54 kg P_2O_5, and 85 to 160 kg K_2O ha^{-1} for potato; 120 kg N ha^{-1} 54 kg P_2O_5 and 85 to 160 kg K_2O ha^{-1} for winter wheat and mixed pastures.

Arable cropping systems in the Hungarian plains have evolved gradually in response to soil type, its fertility status, irrigation facilities and agroclimate. Crop species and availability of suitably adapted genotypes and yield goal set by farming communities have also influenced cropping systems. According to Birkas and Schmidt (2004) and Birkas et al. (2012), historically, only low intensity cropping occurred on plains. Conventional integrated farming with deep tillage was in vogue between 1860 and 1950s. Since 1990s, Hungarian farmers have been adopting at least three different types of crop production tactics. They are: (a) modern intensive with high fertilizer input; (b) modern low intensity farming commonly known as subsistence or dry land farming with low yield goals; and (c) organic farming that utilizes residue recycling and FYM in greater proportion, to supply nutrients and add quality to soils. Cereal-fallow system is the most common rotation that was practiced during eighteenth–twentieth century. Depending on frost and temperature, cereal-fallow with cover crop or cereal-vegetable-fallow systems is adopted.

Primitive agricultural communities in Russian steppes, like several other prairie zones practiced crop production based solely on inherent soil fertility and water resources. Later, slash-burn system that cleared preexisting forests was common. During first 5–6 years after clearing forests, crop productivity was relatively higher. However, with repeated farming, fields lost their fertility levels, and nutrients got depleted. In the absence of cover crops, top soil loss due to erosion too became prominent. Crops such as barley, oats, millets, legumes, and vegetables were cultivated in mixtures.

Long fallow system was common in Ukrainian and Russian Steppes. Here, virgin soils were utilized to grow staple cereals, gourds, oilseeds, and flax. Again, continuous cropping led to depreciation of soil fertility and grain yield. Such fields were left for long fallow of 6–8 years. It allowed refurbishment of productivity. Farmers then moved to new locations with virgin soils after a stretch of cultivation of particular field. Fields left for long fallow for 15 years were again cleared and cereals were cultivated on it. Such shifting cultivation practices persisted in Ukraine and southern Russian Steppes during middle ages—that is, fifteenth and sixteenth century A.D. and until nineteenth century (Vorobev, 1979). Moon (2005) states that early settlers in the Russian steppes practiced extensive farming with long fallows. They often burned the natural vegetation in the virgin zones and started crop production. They followed crop-long fallow system. The fallow period extended 12–15 years. German settlers pioneered crop rotations that included cereal-short fallow and they pioneered four-field rotations. German settlers also practiced cereal-pasture rotations that kept soil loss at low levels (Tulaikov, 1910).

Cereal grain-fallow system was common for many decades during nineteenth and twentieth century. Fields with long cropping history were often allocated to mixed pastures. This system helped cattle ranching in the steppes. Waste lands were also converted and tilled to establish pastures with perennial grasses. Soon, many famers practiced pasture grass-cereal crop rotations. Traditional cropping system in Russia involved crop-fallow. It prevailed all through the feudal period until revolution. Multiple cropping with no fallows was advocated in Russia, quite early during nineteenth century. However, multiple cropping got confined to large farms with good infrastructure. During past decades, multiple cropping based on soil fertility status and fertilizer supply is common. In the drier steppes of Ukraine, row crops such as wheat, sugar beet, sunflower, and corn are grown in variety of sequences. Cropping systems are dependent on irrigation facilities. Cereal-vegetable rotations are possible in regions with irrigation and suitable frost-free period. Modern techniques such as shelter belts, contouring, agroforestry, strip cropping and mulching are all common in the steppes. According to the reports by FAO (2005), Ukrainian cropping systems has been affected by the change in land holdings and reduction in state owned large crop production enterprises. Majority of the private small holdings have preferred to grow wheat and vegetable. Major cash crops that are popular are sunflower, sugar beet and rape seed. These are highly profitable to farmers. Yet, total cropping area in the Ukraine reduced from 32 million ha in 1991 to 30.4 million ha in 2004. A few major crops grown in Ukrainian Steppes and their productivity during 2004 is as follows:

Wheat ($3.16 \, t \, ha^{-1}$), Maize ($3.83 \, t \, ha^{-1}$), Sugar beet ($23.6 \, t \, ha^{-1}$), Sunflower seeds ($0.9 \, t \, ha^{-1}$), potato ($13.3 \, t \, ha^{-1}$), vegetable ($14.8 \, t \, ha^{-1}$), grapes ($4.5 \, t \, ha^{-1}$), and fruits/berries ($5.84 \, t \, ha^{-1}$).

Farmers in Ukraine adopt several variations of crop rotations. Most common rotation followed since several decades is wheat-wheat-fallow. Fallow is meant mainly to refurbish soil moisture. In the regions with paucity of water, winter wheat-sunflower-spring barley and maize is a rotation that helps in economizing on water resources. There are other crop rotations that include a fallow, two seasons of wheat and four seasons of pasture grass/legume. A typical 7 year rotation includes winter wheat, winter barley, sugar beet, winter wheat, winter barley, sunflower, and maize. It is believed that a large section of small farmers in Ukrainian steppes still do not practice appropriate crop rotations. They do so despite a clear knowledge that non-observation of crop rotations may lead to degradation of soil fertility, buildup of disease and insect pests (FAO, 2005).

4.2.5 GM CROPS IN EUROPEAN PRAIRIES

The North European plains and Russian Steppes generate food crops that satisfy needs of a very large populace in the entire Europe and several other nations, wherever the grains are exported. It is a large biome of prairies filled with variety of cereals, legumes, oilseeds, vegetables, cash crops and fruit plants. Any change in the crop genetic make-up may have several types of consequences, some that are easily forecasted and many which were never anticipated. GM crop is different genetically and its reaction to agroclimate, soil, and biotic factors in the surroundings such as insects, pathogens, and abiotic stresses may differ. Its advantages and disadvantages in the plains that help it to successfully flourish or perish are different. At the bottom line, even though genetic modification might have been tailored in the previously well accepted and highly productive genotype, a GM genotype is still a different entity that fills the ecosystem, if we grow it all through the plains of Europe. Within the context of the European plains, France were one of the earliest regions where GM crops such as soybean, maize, cotton, and rapeseed was grown and tested for productivity. During 1996–1999, major effort was confined to test the BT-cotton and BT-maize in the southern France. However, soon GM genotypes had to be withdrawn because of skepticism and possible adverse effects related to excessive use of herbicide and possible effects on humans who consumed BT-maize (Bock et al., 2002).

In the agricultural cropping belts of Europe, GM crops and non-GM crops (i.e., conventional genotype) both coexist. Field that supports GM crops may be utilized in the next season to cultivate a non-GM conventional genotype. In the market yard, grains of GM and non-GM crops might get mixed at various stages of transport, grading, and sales. Such mixing may affect the dispersal and spread of GM/non-GM crops based on their ratio within such mixtures. Several scenarios of coexistence andintermixing of GM seeds and non-GM seeds were studied in great detail by a Joint Research Committee. They argue that intermixing is a matter of concern to natural settings in the European prairies and human health (Bock et al., 2002). Some of the major concerns were what would be the effect if share of GM crops increase beyond 10 percent of cropping area, in case of three crop species studied namely maize, potato or cotton. What is the extent of presence of GM seeds in the fields if seeds were mixed up in the market yard. What should be the threshold of GM crops in an area. What is the effect of planting a non-GM cultivar after a GM genotype of the same crop in the same field? Introduction of GM crops may also affect the emergence and survival of natural vegetation and common weeds in the European plains and Steppes. We need to understand the fate of higher number/quantity of herbicide sprays, seepage, and persistence of herbicides in the ecosystem.

Hungary does not seem to produce GMO crops, yet. Hungary does import several commodities such as soybean derived from GMO crops (Wikipedia 2013e). However, some reports suggest that Bt-Maize and herbicide-tolerant Maize were accepted in the plains of Czeck Republic and Hungary with certain clear advantage with regard to herbicide/pesticide use and economic gains (Demont et al., 2003). There are peer reviewed data and white papers prepared that argue several aspects of GMOs in Hungarian plains (EuropaBio, 2011). There are reports that in the Hungarian plains, standing GMO crops such as GM-Maize (MON810) was destroyed in the field, before flowering, pollination, and seed set. This was to ensure that pollen from GM crops would not pollinate conventional maize crops in the vicinity. In some areas, it is said a kind of propaganda is in place to reject GMO crop genotypes in the European plains.

During past decade, a few of the major commercial seed producers who supplied GM maize (MON810) had to withdraw from some of the regions of European plains, for example from French plains. However, in certain other regions of the same continent, GM crops were grown. There seems to be lingering controversies about the possible effects of cross-pollination of conventional and GM crops genotypes. Effect of over use of herbicide and pesticide upon acceptance of GM crops has to be ascertained. Hence, it may require further experimentation and confirmation before GM crops enter into the plains/steppes of Europe. Yet, it may be a matter of time when European steppes are filled with GM crops.

4.3 AGRICULTURAL CROPPING AND ENVIRONMENTAL CONCERNS

French plains have been utilized for crop production since, Neolithic period. It supported a wide range of crop species, each with its specific requirements of soil and water management. Yet, it is said, farming was a well amalgamated enterprise and did not bring drastic changes to soil, water, and other natural resources. However, intensification of crop production in the plains using high dosages of fertilizers, irrigation, and lack of commensurate soil conservation procedures has led to deterioration of agricultural zones (Wichireck and Bossier, 1995). In fact, intensive farming practice has led to soil degradation, erosion of top soil, and decline in soil quality. These aspects have immediate and long term ill effects on the soil and agroclimate. During past 50 years, extensive and deep tillage has induced soil erosion. Soil erosion in the French agrarian regions is attributable to both wind and water.

It is estimated that wind erosion causes high loss to exchequer, much above 50 billion French francs annually. Top soil loss and depreciation in crop productivity due to wind erosion is severe in most regions of the plains. Soil erosion due to water is also important. It affects more than 2.7 million ha in the French plains. Soil erosion rate due to water is estimated at 30 to 50 t ha^{-1} annually. Wichereck and Bossier (1995) suggest adoption of appropriate soil and water management measures. Soils with clay, organic matter and high percentage of stable aggregates usually resist erosion. Whereas, sandy soil exposed to splashing water and high-intensity wind is susceptible to erosion. Topography of the field and slope also plays a role in extent of erosion. Regarding water-induced erosion, it is said that high-intensity rainfall events with big droplets, that are impinged with high kinetic energy, cause more erosion. As usual, agronomic measures such as contour bunds, wind breaks with tall trees, mulching, residue recycling, intercropping, and no-tillage systems are useful in reducing the erosion in the plains.

Row-crops grown in Czeck plains are vulnerable to loss of top soil due to erosion (Novak et al., 2013). Much of the cereal belt (53%) occurs in undulating regions with an average gradient of 3° which is sufficient to induce soil erosion. Erosion occurs mainly due to water and wind. Soil erosion depends on physicochemical properties, mineralogy, and organic matter content. Cereal crops grown on widely spaced rows are prone more to erosion due to water. Soil mulching with organic residues and adoption of reduced tillage are among the foremost techniques to reduce soil erosion. Surface runoff is among the lowest if no-tillage system is adopted. Soil loss due to surface runoff is greatest if fields are treated under conventional deep tillage system (58 g m^{-2}) or if the fields are left to fallow without a cover crop (83 g m^{-2}) (Novak e al., 2013). Soil loss reduces to 10–18 g m^{-2} if reduced tillage or no-tillage system is adopted.

4.3.1 SOIL EROSION IN STEPPES

The rapid expansion of agricultural zones in the Russian Steppes during eighteenth and nineteenth century had necessitated removal of natural grass land vegetation, deep plowing and repeated fallows without cover crops. Trees and shrub vegetation was being increasingly removed to serve fuel needs of human settlements. Hence, steppes became increasingly vulnerable to soil erosion and gullying. Erosion through wind too became prominent and created conditions of dust bowl, similar to ones traced in other continents (Moon, 2012). This phenomenon had been recorded in 1880s and 1890s. It prompted the famous soil specialist Dokuchaev to suggest that gullying and period of dry spells were causing the groundwater levels to fall. Then, crop production became less than optimum. It was argued that human activity was responsible for inefficient crop production (Moon, 2005). Reports by Mott Macdonald (2011) suggest that steppes region from Ukraine, through Hungary up to Russian Urals had been supporting broadly diverse vegetation. However, major part of this natural biome was converted into agricultural steppes. Ploughing and grazing resulted in soil erosion. Soils were eroded and deteriorated irreversibly in many regions of the steppes, biodiversity was reduced and biomass productivity too got reduced progressively.

Chernozems of Ukraine, otherwise called Mollisols in US soil classification system are most important of the natural resources in the steppes (Dudekin, 2013). They are rich in SOM and minerals. They are highly productive. However, they are prone to deterioration because of natural and man-made factors. Erosion affects Chernozems at varying intensities and for different lengths of period. Erosion seems to enhance soil pH and carbonate content. It decreases humus, clay, and nutrient contents of Chernozems. Periodic supply of fertilizers along with soil conservation practices restored soil productivity (Kharytonov et al., 2004).

Chibiliyov (1984) states that the European steppes are highly susceptible to gully formation and top soil erosion. It definitely leads to loss of SOM and mineral nutrient that ultimately depreciates crop productivity. The physical geography seems to be a major cause, since parent material is mostly loess and chalk that are easily erodible, if an intense rainfall event occurs. Freezing and thawing that occurs as seasons progress also induces erosion. Anthropogenic removal of natural vegetation if any, cropping, fallows without a cover crop, repeated plowing may all accentuate surface runoff and top soil loss in the steppes (Stebelsky, 1974, 1983). While discussing the history of soil erosion in the Russian plains, Stebelsky (1974, 1983) observes that early agricultural efforts during Neolithic period and right up to medieval period, farmers did not even perceive sporadic erosion of soil as a problem. Incessant cropping that involves repeated deep digging, plowing, and removal of natural vegetation for several decades until mid-nineteenth century has made erosion to be considered a major problem in the steppes. Soil survey for erosion, gully formation and extent of top soil loss was conducted extensively all through the steppe region of Russia during 1930s and 40s by the erstwhile Federal Soviet Union. Soil maps that depict the erosion and gully formation in relation to crop production and/or establishment of natural vegetation are available. Chernozem region of the Steppes suffered maximum due to erosion and gullies. About 10 percent of land area in this region was left unaffected by gullies/erosion. Gullies of moderate length (0.2–0.4 km) covered about 30 percent of Chernozem region in steppes. Intense gullies of greater length (0.5–1.5 km) occurred in 35 percent of steppes (Guzhevaya, 1948; See Chibiliyov, 1984).

In the Hungarian plains, nitrate directive has been implemented in most areas. Nitrate excess, if any, in the groundwater and aquifers is associated with large-scale intensive farming that employs high fertilizer-N dosage. Animal husbandry, use of liquid fertilizers and sandy soils with high percolation rates are other causes of nitrate contamination in the plains. Birkas et al. (2012) suggest that there are clear zones prone to nitrate contamination. Hence, agricultural fields in these specific zones need to be suitably managed.

Soil phosphorus is not utilized efficiently by the crop roots. In the general course, only a small portion of 20–23 percent P is utilized and rest is allowed to accumulate in the profile. This fraction is vulnerable to loss via leaching, seepage, percolation in dissolved or particulate state into lower horizons and groundwater. It pollutes the water sources, if PO_4 content exceeds threshold levels. It also induces eutrophication of water bodies that collect such P polluted water (Berenkova, 2011). Major sources of P in the agrarian zones of Eastern Europe and Russian steppes are soil-P that is augmented with fertilizer-P sources. Soil erosion is greater from locations that are rich in P. Runoffs that contain high P and subsurface seepage of P in dissolved state is also common. Phosphorus that leaches moves into subsurface and collects at higher quantities. It may also contaminate groundwater. Soil erosion is said be the main mode of P transport in the agrarian steppes. High-intensity precipitation events, absence of soil conservation procedures and high P supply may all cause higher P loss and its transport to different locations. Berenkova et al. (2011) state that proper standards and threshold levels for loading P into channels and streams is required. More importantly, P in irrigation water and in water bodies in the surroundings of a farm needs to be regularly noted and appropriate measures should be adopted to reduce P loading. Inadequate precaution surely leads to eutrophication of water sources and contaminated groundwater beyond limits. There are several computer-based models that help us to study the P dynamics in the ecosystem and forecast P levels at different points in the farm. At the bottom line, in any farming region, Sharpley and Beegle (2001) opine that "Best Management Practices" that balance P inputs and recover by crop (output) should be adopted; sources that contaminate irrigation channels and water bodies with P should be curtailed and threshold levels of P in the subsurface, vadose layer, groundwater, and irrigation channels needs to be monitored periodically.

4.3.2   LOSS OF SOIL NUTRIENTS THROUGH GASEOUS EMISSION

Fertilizer-based N supply to soil is exposed to various natural and manmade factors. Fertilizer-N applied as NH_4 salts or inherent soil N is vulnerable to be emitted as NH_3. Incidentally NH_3 emissions are primarily caused by agricultural fields. Much of NH_3 emissions are derived from organic fraction of soil. NH_3 volatilization leads to reduction in fertilizer-N efficiency. About 8–20 percent of N inputs could be lost as NH_3 depending on geographic location and cropping systems. In addition, NH_3 volatilization also affects environmental quality. At present, there are stipulations to curb excess of NH_3 volatilization in agrarian zones. Fertilizer-N formulations such as Urea and UAN that are highly popular in the European plains are most vulnerable to loss via NH_3 volatilization (Yara International ASA, 2012). Incorporation of such fertilizers deeper into soil profile is essential. We should note that fertilizer-N loss as NH_3 is greater from grasslands and pastures than regularly cultivated cropland. They say a grass matt has higher urease activity that leads to higher levels of NH_3 evolution. Following is an example:

Fertilizer Formulation	Loss of N via NH$_3$ Volatilization	
	Arable Cropland	Grassland
Calcium Ammonium Nitrate	0.6%	1.6%
Urea AN	6%	12%
Urea	11.5%	23%

Source: Yara International ASA (2012)

Agricultural soils are the major source of N$_2$O in the European Plains. Based on results from several field trials in Europe, Sutton and Reis (2012) have opined that N$_2$O emissions could be regulated using alternative tillage systems, organic farming systems, appropriate fertilizer-N supply schedules, land use pattern and adoption of appropriate field drainage systems. The average N$_2$O emissions in different locations ranged from 0.4 to 16.85 kg N$_2$O-N ha^{-1} year^{-1}. In the arable belts, it ranged from 1.0 to 4.9 kg N$_2$O-N ha^{-1} year^{-1}. In the natural prairies and shrub lands of steppes, it is said, drought, wetting and drying cycles and hydrological changes affects the N dynamics, especially N$_2$O emissions.

In the European plains, atmospheric-N deposition is often offset by the N loss as NH$_3$, N$_2$O, NO, and N$_2$ (Sutton and Reis, 2012). Intensive study of N-flux suggests that, N loss from agricultural fields increases as fertilizer-N input is enhanced. Nitrogen loss through water and as gaseous emission increases with N inputs. Soil moisture status, environmental parameters such as temperature and relative humidity, cropping systems may all affect N emissions. As stated earlier, agricultural soils are among the most important N-emitters. Further, based on study of different simulations and scenarios proposed by IPCC, N$_2$O emissions are projected to increase by 4–8 percent on the present levels by 2030. It seems, just improving the nitrogen efficiency of crop production system using appropriate agronomic procedures and crop genotype may reduce N emissions by 10–30 percent of the present levels in the same area. One of the other estimates suggests that, if fertilizer-N supply is channeled exactly in quantity required by crop at different stages, then N$_2$O emissions could be drastically reduced. For example, in a particular location such well-timed fertilizer-N supply reduced N-loss from 8.8 kg N$_2$O-N ha^{-1} to 6.9 kg N$_2$O-N ha^{-1}. It amounts to reduction of 22 percent in N emissions (Sutton and Reis, 2012).

According to Birkas et al. (2012), high-intensity farming in the Hungarian plains aims at generating more biomass and grains per unit land. It requires use of high amounts of fertilizer-N, organic manures and residue recycling. As a consequence, excess of soil-N still unutilized becomes vulnerable to loss via emissions. Soil-N loss through NH$_3$ volatilization and NO emission is most conspicuous for N loss. Soil-N emissions are of course common in the Hungarian plains. Most common forms are NH$_3$, NO, and N$_2$O. Nitrogen emissions are affected by several factors such as weather pattern, especially soil and ambient temperature. Soil-N loss is not high. It is <1 percent in many parts of the Hungarian Plains (Machon, 2011).

4.3.3 CARBON DIOXIDE AND METHANE EMISSIONS

According to Smith et al. (2004), agricultural expanses that occupy about 50 percent of European land surface are among CO_2 emitters to atmosphere. They suggest that accurate forecasts about CO_2 emissions are required. Regarding croplands in the plains from the west coasts in Brittany to Urals in Russia, it is said that farming and crops induce a net loss of 300 MT C year^{-1} (Janssens, 2003; Sluetel et al., 2003). Estimates of GHG emissions from European nations that occur in the steppes show a marked increase in CO_2 emission due to agricultural crop production. According to Smith et al. (2004), some estimates of GHG that occurs due to change from natural grassland to cropland could be high. Net GHG emission, exclusively from cropland is estimated at −78 MT C year^{-1}. It occurs mainly as CO_2 and smaller fraction as CH_4. During recent years, GHG emissions have shown a decline due to change in agronomic practices, especially adoption of no-tillage systems, residue recycling and mulching. Greenhouse gas emissions in the Steppes are also generated through farm animal management. Such emissions are mainly confined to CH_4 and N_2O. Estimates in individual expanses vary based on number, intensity, and type of farm animal management procedures adopted. In general, cropland and animal farming that often exist together in the mixed-framing enterprises in most regions of Europe could be classified as high GHG emitters. Farming practices that affect GHG emissions from prairies are inorganic fertilizer supply trends, organic manure, biological nitrogen fixation and crop residue recycling. Farming systems and management, especially soil tillage, ridging, interculture, irrigation and water management, animal grazing, and residue burning also affect GHG emissions (Smith et al., 2004). Highest emissions of CH_4, CO_2, and N_2O are recorded in mixed farms with high number of ruminants. Among the factors, synthetic fertilizers, animal emissions and crop residues induce relatively higher emissions ranging from 56 to 228 Gg N_2O year^{-1}, if entire European agricultural prairies are considered (Smith et al., 2004).

Evaluation of over 80 farms located in German Plains has shown that emissions of CO_2 and CH_4 were much lower in organic farms compared to conventional farms that utilize inorganic fertilizer and FYM application based on soil test and yield goals. Kustermann and Hulsbergen (2008), have stated that organic farms emitted almost three times lower quantity of CO_2 (785 kg CO_2 eq ha^{-1} year^{-1}) compared to conventional farms (2,165 kg CO_2 eq ha^{-1} year^{-1}). Following is an example that depicts N_2O emission from agriculture expanses of two most important European nations located within the northern plains region, namely, France and Germany:

	N_2O Emissions from Agriculture (million kg N_2O year^{-1})		Methane Emissions from Agriculture (million kg CH_4 year^{-1})	
	1990	2000	1990	2000
France	17,721	17,228	1,667	1,591
Germany	15,709	13,287	1,605	1,205

Source: Smith et al. (2004)

4.3.4 CLIMATE CHANGE AND CROP PRODUCTION

Let us consider some forecasts about possible influence of climate change on crops in the European plains and Russian steppes during a few decades later. Olesen et al. (2011) have made plausible forecasts using computer based weather models, crop production data and crop models that assess various climatic parameters.

Impact of climate change may not be congenial for higher productivity of winter wheat. Higher temperatures may make farmers to switch to late maturing genotypes and this will increase crop duration. Soil conditions for seeding may become more congenial. Damage due to frost is expected to be less. Computer simulations suggest that climate variability within a season may be either neutral or negative on grain productivity. Forecasts suggest buildup of insects and pests owing to higher temperature and moist conditions. Heat stress may also appear in some parts of European plains that are used to produce winter wheat. Precipitation, if it becomes intense may induce soil leaching and NO_3 leaching.

Spring sown barley too may suffer growth and grain yield depression due to climate change. Again, higher ambient temperatures may prolong crop duration to maturity. Frosty conditions may get delayed and risk to barley harvest may be less. Changes in season may ask for versatility with regard to advancing or postponement of sowing barley crop. However, damage caused to growth pattern and seed-set due to higher temperature and extra heat units received may be prominent. Other forecasts are accentuation of weeds, disease, and pests in the fields due to extra number of warmer days. Soil erosion and nitrate leaching is expected to increase in the barley production zones of North European plains and Steppes.

Climate change may have several effects on maize production in Northern Europe and Steppes. Generally, C_4 cereal species are expected to perform better. Lengthier growth period is said to affect growth and productivity positively. The window for harvest is expected to increase by few days, since frost occurrence gets delayed in the plains. Maize may encounter or adapt itself better to drought and cold stress in the steppes. Intense rainfall events and hails may induce loss of soil and nutrients from maize fields. As usual, warmer climate may accentuate weed population, pest, and diseases. Over all, crop genotypes resistant to abiotic and biotic stresses are needed to overcome ill effects of climate change.

Regarding pastures, grasses, and natural grazing stretches in the plains and steppes, Oleson et al. (2011) state that climate change reduces biomass formation marginally. However, warmer climate may induce disease, pests, and weed growth. The magnitude of climate change effects in the Northern Plains could be felt more severely than those in other regions such as Pannonian plains/ Steppes.

Shorter winters and delayed frost onset will be congenial to grape orchards in the French plains and steppes. However, frosty days may increase in some areas leading to shortening of growing season and fruit development period. Factors such as drought, disease, and pest pressure may be enhanced marginally. Grape vine production may suffer due to hail and high-intensity rain fall events (Oleson et al., 2011).

4.5   LAND USE CHANGE

Land use change in the European continent, like most other agricultural zones, began with domestication and spread of crops or due to introduction of crop species from other centers. Natural vegetation that included native grass/legume species and others gave way to food crops. Kusters (2000) states that introduction of domesticated cereals into European loess region and Steppes totally changed the landscape. Forests were cleared rapidly by the migrating human population. It was almost an irreversible alteration of natural vegetation. It might have affected the prevailing environment, perhaps drastically that is still not understood well. Initially, during Neolithic period, the crop package derived from "Fertile Crescent" included only few species such as cereals, herbs, tubers, and spices. Cereals such as wheat, oat, barley, and setaria seem to have replaced natural prairies and forest belts of Northern Europe.

Boonman and Mikhalev (2008) state that, Russian steppes with natural vegetation have taken too many migrants into its fold since the Neolithic period. Many of the migratory onslaughts have left indelible mark by replacing natural vegetation and fauna with cropland. Most recent and last of such drastic influences on Russian Steppes occurred in 1950s due to huge agricultural expansion supported by Soviet Governmental Agencies. During this campaign, about 43 million ha of Steppes were converted to cropland. They say, it almost eradicated naturally evolved vegetation in Ukraine and Russia (Maslov, 1999). Further, Maslov (1999) states that Russian steppes stretched into 83.6 million ha, out of which 21.6 million ha was hay meadows and 62 million ha was pastures. About 17 million ha of natural grass land in the steppes were converted into cropping expanses through large-scale land use change campaigns since 1954.

It is opined that Russian Steppes are a formidable natural resource that has evolved through the ages. Yet, agricultural cropping and pastures have made considerable incursions into the vast expanse. Pastures and fodder crop production systems have literally replaced waste lands, sparsely vegetated patches and even fallows after a crop rather conspicuously. Permanent and temporary pastures that include tall or short statured grasses have become common in areas prone to cattle farming.

Russian steppes are among terrestrial habitats with rich plant biodiversity. It has however faced degradation during past few decades both due to natural and man-made factors, rather perceptibly (Verheut, 2012). The land use change from steppe grass lands to crop production is among the major factors that have affected the vast steppe ecosystem. It seems over 41 million ha of steppe zone was converted into cropping expanses in just 6 years between 1954 to 1960, by the erstwhile Soviet Union's Virgin land Program. It left only a small stretch of steppes still unused and in natural condition. At present, conservation agencies are trying to induce Russian farmers in the steppes to intensify production of cereals in the steppes, in some selected pockets. This will allow them to harvest grains at same levels but with less area of natural grass land being converted. It mainly allows a stretch of natural grass lands unused and in pristine form.

It seems large mechanized farms of the farmer Soviet Union, namely Kolkhozy and Sovkhozy type croplands and livestock farms are being preferred less. According to Boonman and Mikhalev (2008), Russians in the Steppes have preferred to retain large state-owned farms only for production of cereals such as wheat, barley, and maize plus legumes. These may form the core of their crop production strategy. Livestock and pasture management has been progressively made responsibilities of family farms. Natural grass lands have been used by farmers in steppes.

4.6 CROP PRODUCTION AND PRODUCTIVITY TRENDS IN THE PLAINS

The European plains generate large quantity of cereal grains, such as wheat, barley, millets, legumes, oilseed rape and forage. Agricultural prairies in France, Germany, and Poland contribute about 50 percent of wheat grains derived from entire 27 member nations of European Union; about 60 percent of oilseed rape and half of barley forage/grains. It is easily attributable to high-intensity crop production trends using high dosages of fertilizers, organic manures, irrigation, and pesticides. The monocrops of wheat or rapeseed allows governmental agencies and farmers to concentrate and improvise on production procedures. Reports suggest that average wheat yield in the Northern European plains is 5.27 t ha^{-1}. Higher productivity in parts of German and Polish prairies has resulted in better grain exports. Studies on wheat grain yield progress, in general, in the European Plains suggests a steady increase by 0.123 t ha^{-1} year^{-1} for the past 70 years (Brisson et al., 2010). However, mild fluctuation and stagnation of wheat productivity at about 7 t ha^{-1} has been noticed during past 7–8 years. Wheat grain yield increase in the plains has been attributable to progress in plant genetics, mainly harvest index and total biomass, irrigation, and pesticides. Wheat grain yield stagnation has been attributed to marginal reduction in fertilizer-N usage, no-tillage systems, and change in previous crop in the sequence from a legume to rapeseed.

Wheat is the main cereal crop of the French Plains. It covers about 50 percent of area of the plains in that nation. Wheat cropping zone in France has ranged from 4.72 to 5.07 m ha during past 3 years. Frost and cold front are the major factors that deter expansion of wheat cropping zones (Rutenberg, 2012). Soft wheat is common. The average productivity of soft wheat is 6.4 t ha^{-1}. France is the fourth largest producer of wheat and its commodities in the world. Total wheat produce during 2010–2012 averaged at 37.06 m t year^{-1}. France is also a major consumer of wheat grains. Wheat import by France during past 5 years averaged 1.09 m t year^{-1}. However, France exports soft wheat. The soft wheat exports averaged at 1.73 m t year^{-1}. Wheat buffers held in France ranks fifth in the world (USDA Economic and Statistic Survey, 2012). Major importers of soft wheat from French plains are Egypt, Algeria, Morocco, Yemen, Senegal, Ivory Coast, and so on.

Germany is again among major wheat producing nations from European Plains. It contributed 19.23 m t of grains during the year 2012. Germans consumed about 15.9 m t year⁻¹ of winter wheat grains. Average imports were 1.57 m t year⁻¹. However, annual wheat grain exports are high at 5.40 m t year⁻¹. Poland generated 8.59 m t year⁻¹ of wheat grains in the plains and consumed about 9.34 m t year⁻¹ of wheat grains during 2012. Poland imported 0. 85 m t year⁻¹ of wheat grains year⁻¹. The average productivity of wheat and even other crops is relatively low because individual farms are small and inefficient in terms of resource utilization.

Cereal belt in Czeck Republic dominates other crop species. It extends into 1.6 m ha, but it has stagnated during past 3 decades. The productivity of wheat belt is 5.6–6.2 t ha⁻¹. Ukraine is an important wheat bowl for Eastern European region. Its contribution was 15.33 m t year⁻¹ during 2012. Average annual consumption was 14.11 m t year⁻¹. Ukraine is said to be 10th largest consumer of wheat grains and products in the world. (USDA Economic and Statistics Survey, 2012). Russia is the fifth largest wheat producing region of the world. During 2012, it generated 34.65 m t and consumed about 38.01 m t. Russia also imports wheat from Western Europe and Argentina. Russia exports a small amount of wheat at 0.78 m t year⁻¹. Russian wheat buffer stock was estimated at 3.37 m t year⁻¹. It ranks eight in the world wheat grain stocks.

Durum wheat is produced by several different countries that are encompassed by the European plains and Steppe region. Durum wheat generated in the European plains is largely utilized to make semolina. The productivity of durum wheat is low at 2–3 t ha⁻¹ in southern Europe and Steppe regions. In other places, where it is provided with fertilizer and other inputs the grain yield reached 3.4 t ha⁻¹ (USDA Economic and Statistic Survey, 2012). Following is the area and total production of Durum wheat grains in the northern Europe and Russian plains:

Country	Area	Production
	(x1000 ha)	(x1000 t year⁻¹)
France	363	1,031
Germany	6	336
Russia	1,220	2,240

Source: Babini (2006); USDA Economic and Statistic Survey (2012)

Ukrainian and southern Russians steppes are important wheat producing regions in Europe. Red winter wheat or Bread wheat is the dominant crop. About 80 percent of 17.5 million t red wheat grain produced annually is consumed domestically in Ukraine. The productivity of wheat is about 3.62 t grain ha⁻¹. The area under wheat for fodder has reduced since 1990. Consequently, total fodder production from wheat crops has declined from 12 million t in 1990 to 5 million t in 2004 (FAO, 2005).

Barley is among staple cereals produced in Russia. Barley is also the major cereal feed generated in the Ukrainian steppes. Spring barley accounts for 90 percent of barley area. Winter Barley is restricted to extreme southern regions of Russia. Barley is grown in over 415,000 ha in the Central European plains. Total grain harvest is about 1.52 m t year^{-1} at an average productivity of 3.62 t grain ha^{-1}. About 0.52 million t grains were used for malting and rest consumed as grains. Central European farmers grow several different cultivars of barley. Russian federation generated about 17.5 m t barley grains during 2007. During the same year, Germany produced 11.78 m t, Spain 11.2 m t and Ukraine 6 m t, Hungary 14.7 m t (FAOSTAT, 2011).

Maize is grown mostly to serve as animal feed and it covers about 19 percent of the French cereal belt. Maize farming zones are concentrated in the Northeast and southwest France. The average grain yield is about 7.4 t ha^{-1} Barley cultivation is also wide spread in France. It is a crop sown in spring. Grain yield stated during 2008 is about 5.6 t ha^{-1} (Agraste, 2008). Maize production in the Ukrainian region reached 8.9 million t year^{-1}. It amounts to productivity of 3.83 t grain ha^{-1}. Maize is also an important fodder crop in the Steppes of Ukraine. The maize planting for forage has shown steady increase in Southern and Eastern regions of Ukraine (FAO, 2005).

Rape seed production in the European plains is intense and equally expansive, covering almost all nations of European Union. Productivity of rapeseed ranges from 3.0 to 3.8 t ha^{-1} depending on agroclimate and soil fertility status. France is the largest rapeseed producer among nations of European plains. The rapeseed expanse ranges from 1.57 to 1.69 m ha and it generates about 6–7 m t rape seed year^{-1}. Rapeseed cultivation extends into 1.5 m ha in the German prairies. It generated on an average 5.2–5.9 m t seeds year^{-1} during 2010–2012. Polish rape seed belt contributed 2.1 m t year^{-1} grown on 0.8–1.0 m ha area. In the Czeck region, rape seeds are cultivated in relatively larger scale compared to other oilseed crops such as soybean and sunflower. Rapeseed production zones extended into 373,000 ha and contributed 1.76 m t seeds during 2011. During past 2 years, rapeseed export from Czeck Republic has marginally decreased. It is attributed to shifts in biofuel sector (Mikulasova, 2011). Soybean and sunflower area is small. Together, they occupy 35,000 ha and contribute 90.8 thousand t seeds. They do not seem to compete with rape seed in the plains.

Rapeseed cropping belt is prominent in Southern Russian steppes, particularly Stavropol region. During 2012/13, rapeseed area in the Russian steppes is expected to decrease by almost 10–15 percent to 0.9 m t year^{-1}, owing to winter frost (USDA GAIN, 2012). Frost may also reduce productivity. Russian rapeseed planting and production trend is partly driven by demand from other European nations for cooking medium and for biofuels. Other oilseed species such as linseed, flax, and mustard is said to supplement for decrease in rapeseed area. Russian steppes exported about 250,000 t of linseed to USA during past year. Sunflower and soybean are other oilseeds of importance to Russian steppes. Soybean is produced on 1.2 m ha, but grain yield is low. Total soybean production during 2012 was 1.8 million t. Low grain yield has been attributed to allocation of soybean, nitrogen fixing legume to consistently low fertility zones within the farms (USDA GAIN, 2012; USDA Joint Agricultural Weather Facility, 2012).

Sunflower is an important oilseed crop in Eastern European plains and Russian steppes. Reports suggest that area expansion, along with a 5 percent increase in grain productivity, improved sunflower harvest by 11 percent in many regions of Europe such as Hungary and Ukraine. Plentiful summer rainfall and optimum soil moisture are said to be major causes for better sunflower yield during 2011 (USDA Economic and Statistic Survey, 2012). Reports by FAO (2005) suggest that sunflower is among major oilseed crops in the Ukraine. Ukraine contributes about 3.05 t year^{-1} at an average productivity of 0.9–1.3 t seeds ha^{-1}.

Permanent grasslands and pastures are well spread out in the uplands and undulating meadows of Czeck Republic. During past three decades, permanent meadows have occupied 6.24 m ha and pasture about 0.3 m ha. Fodder producing zones has declined from 10.08 to 7.2 m ha during past 3 decades. The productivity of meadows and pastures has ranged from 2.15 to 2.95 t ha^{-1}, but fodder crops have generated 5.66–7.56 t ha^{-1} (Kralovec, 2006).

KEYWORDS

- **Chernozems**
- **Climate change**
- **European plains**
- **Inorganic fertilizer**
- **Natural vegetation**
- **Organic manure**
- **Topography**

REFERENCES

1. Agraste; Agricultural Statistics of France. **2008**, 1–2, http://www.deeurope.net/en/french-cerals-production.asp
2. Alcamo, J.; et al. Europe. In: In Impacts, Adaptation and Vulnerability. Contribution of Working Group II. To the Fourth Assessment Report of the Intergovernmental Panel on Climate Change. Cambridge, United Kingdom: Cambridge University Press; **2007**, 541–580 p.
3. Astrium, Precision Agriculture. **2005**, 3–4, August 1, 2013, http://www.astrium-geo.com/en/1003-precision-agriculture.htm
4. Babini, V.; Soil Tillage and Crop Rotation Effects on *Triticum Durum* Yield and Mycotoxin in Its Grain. Faculti Bologna, University of Bologna; **2006**, 127 p.
5. Berenkova, T.; Managing non-point source Phosphorus—A literature review. *J. Landscape Stud.* **2011**, 4, 45–57.

6. Beyer, L.; Soil geography and sustainability of cultivation. In: Soil Fertility and Crop Production. Krishna, K. R.; ed. Enfiled, New Hampshire, USA: Science Publishers Inc.; **2002**, 33–64 p.

7. Birkas, M.; Jolankai, M.; Mesic, M.; and Bottlik, L.; Soil Quality and Land Use in Hungary. **2012**, July 5, 2013, http>//wwwbib.irb.hr/datoteka/572537.1. 9-23.pdf

8. Birkas, M.; and Schmidt, R.; Land use systems approach to soil and environment conservation: Regions-country side-environment. Proceeding of 11th International Science Conference. Belajova, A.; ed. **2004**, 1–7 p.

9. Blagoveschchenskii, V. G.; Popovtsev, V.; Shectsova, N.; Romanenkov, V.; and Komarov, K.; Country pasture/forage Resource profiles: Russian Federation. **2007**, 1–43, July 1, 2013, http://www.fao.org/as/AGP/AGPC/doc/Counpro/Russia/russia.htm

10. Bock, A.; Lheureux, K.; Libeau-Dollos, M.; Nilsagard, H.; and Rodriguez-Cerezo, E.; Scenarios for co-existence of genetically modified, conventional and organic crops in European Agriculture. Institute of Prospective Technologies. European Commision: Joint Reaserch Centre. Report No EUR 20394EN **2002**, 1–123.

11. Bogucki, P.; The Neolithic Mosaic on the Northern European Plains. **1992**, 1–6, http://www.princeton.edu/-bogucki/mosaic.html

12. Bogucki, P.; How agriculture came to North Central Europe. In: Europe's First Farmers. Price, D.; ed. England: Cambridge University Press; **2000**, 197–218 p.

13. Boonman, J. G.; and Mikhalev, S. S.; The Russian Steppe. Food and Agricultural Organization of the United Nations, Rome, Italy. **2008**, 1–27, July 4, 2013, http://www.fao.org/docrep/008/y6344e/y6344e0h.htm

14. Brady, N. C.; Nature and Propeties of Soil. New Delhi: Prentice hall of India; **1975**, 548 p.

15. Brisson, N.; Gate, P.; Gouche, D.; Charmet, G.; Oury, F. X.; and Huard, F.; Why are wheat yields stagnating in Europe? A comprehensive data analysis for France. *Field Cops Res.* **2010**, *119*, 201–21.

16. Cappers, R. T. J.; and Raemaekers, D. C. M.; Cereal cultivation at Swifterbant? Neolithic Wetland farming on the North European plain. *Current Anthropol.* **2008**, *49*, 385–388.

17. Cermak, P.; Report on Division of Field Experiments. Drnovska, Czeck Republic: Crop Research Institute; **2010**, 18, http://www.vurv.cz/index.php?p=odbor_polnich_pokusu&site=vyzkum_en

18. Cermak, P.; and Torma, S.; Importance of Balanced Fertilization for Sustainable Crop Production in the Czeck and Slovak Republics. Switzerland: International Potash Institute; Annual Report, **2006**, 1–28.

19. Chesnyak, G. V.; Gavilyuk, F. Y.; Krupenevik, A.; Laktionov, N. I.; and Shilikhina, I. I.; Condition of Humus in Chernozems. In: Chernozemy Rossii. Kovda, V. A.; and Somoliova, I. M.; ed. Moscow; **1983**, 453.

20. Chibiliyov, A.; Steppe and forest steppe. In: The Physical Geography of Northern Eurasia Shahagedanova, M.; ed. New York: Oxford University Press; **1984**, 248–266 p.

21. Christensen, J. H.; and Christensen, O. B.; A summary of prudence model projections of changes in European climate by the end of this century. *Climate Change.* **2007**, *81*, 7–30.

22. Demont, M.; et al. Impact of GM crops in Hungary and the Czeck Repbulic. Katolieke Universitat Leuven, European Union Welfare effects of Agricultural Biotechnology. **2003**, 1–28, July 21, 2013, http://www.agr.kuleuven.ac.be/aee/clo/euwab.htm

23. Devaney, E.; Soil types in Europe. **2013**, 1–2, July 1, 2013, http://www.ehow.com/list_7220717_soil-types-europe.html

24. Dezes, P.; Schmid, S. M.; and Ziegler, P. A.; Evolution of the European Cenozoic Rift System: Interaction of the Alpine and Pyrenean orogens with their foreland lithosphere. *Tectonophysics.* **2004**, *389*, 1–33.

25. Dickenson, R. E.; Germany: A Regional and Economic Geography. London; **1964**, 1–84.

26. Dobrovolsky, G. V.; Russia's Soil Resources During the Last 150 Years. **2009,** 1–3, July 1, 2013, http://www.rus-stat.ru/eng/ index.php?vid =2002&page=2.htm

27. Dubrovsky, M.; Trnka, M.; Zalud, Z.; and Semeradova, D.; Estimating the crop yield potential of the Czech Republic in present and changed climates. **2004,** 1–12, July 4, 2013, http://www. ufa.cas.cz/dub/crop/ecac2004.ppf

28. Dudekin, O.; Bringing Ukraine's steppe back to life-interview. **2013,** 1–3, June 30, 2013, http://euukrainecoop.com/2013/03/01/steppe-biodiversity/

29. Encyclopaedia Britannica. European Plain. 2010, 1–2, July 10, 2013, www.britannica.com/ EBchecked/topic/196332/European-Plain

30. Eurostat; Agri-environmental indictor-risk of pollution by Phosphorus. **2013,** 1–11, July 20, 2013, Epp.eurostat.ec.europe.eu/statistics_explained /index.php/Agri-environmental_indicator_-_risk_of_pollution_by_phosphorus. htm

31. Evers, J.; and West, K.; Europe: Physical Geography. National Geographic Association. **2013,** 1–5 July 1, 2013, http://www.education. nationalgeographic.com/education/encyclopedia/ europe-physical-geography?ar_a=1

32. EuropaBio. Plain facts about GMOS-Hungary White paper. **2011,** 1–3, July 19, 2013, http:// www.eurpabio.org/agricultural/positions/plain-facts-about-gmos-hungary-white-paper

33. FAO. Fertilizer use by crop in Ukraine. Food and Agricultural Organization of the United Nations. Rome, Italy; **2005,** 1–56.

34. FAO. European Wheat Producing Countries: France-basic Statistics. 2010, 1–2, June 7, 2013, http://www.fao.org/ag/AGP/AGPC/doc/field/ Wheat/europe/france.htm#agrozones

35. FAOSTAT. Barley production statistics. Food and Agricultural Organization of the United Nations. Rome, Italy; **2011,** 1–3.

36. Ghuzhayeva, A. F.; Gullies in Central Russian Uplands. Proceedings of Institute of Geography of the USSR Academy of Science. **1948,** *42,* 37–74.

37. GIM International. Improving French Precision Farming. **2009,** 1–2, August 1, 2013, http:// www.gim-international.com/news/id3998-improving_ French_Precision_Farming.htm

38. Giorgi, F.; Bi, X.; and Pal, J.; Mean, interannual and trends in a regional climate change experiment over Europe. 2. Climate Change Scenarios (2071–2100). Climate Change Dynamics. **2004,** *23,* 839–858.

39. Haase, D.; Mapping the most fertile soils. Climate change monitoring. Science Daily. **2007,** http://www.sciencedaily.com/ releases/2007/11/071115113328.htm

40. Haase, D.; Fink, J.; Haase, G.; Ruske, R.; Pecsi, M.; Ritcher, H.; Altermann, M.; and Jager, K. E.; Loess in Europe-its spatial distribution based on a European Loess map. *Q. Sci. Rev.* **2007,** 26, 1301–1312.

41. Helmoltz Association of German Research Centre. New European Loess map. Liepzig, Germany: Helmoltz Centre for Environmental Research; **2007,** 1–5, December 24, 2013, http:// www.ufz.de/index.php?en=15536

42. Hlavinka, P.; et al. Modelling of yields and soil nitrogen dynamics for crop production by HERMES under different climate and soil conditions in the Czeck Republic. The Journal of Agricultural Science. Cambridge, UK: Cambridge University Press; **2012,** 1–2, July 4, 2013, http://dx.doi.org/10.1017/ S0021859612001001

43. Holas, J.; and Klir, J.; Diffuse pollution in the Czeck republic at the example of selected watersheds case study. Proceedings of Pollution Conference. Dublin, Ireland; **2003,** 3, 1–5.

44. IEASSA. Precisiona Agriculture and the Ukrainian Reality. **2013,** 1–3, August 1, 2013, Ieassa. org/en/precision-agriculture-and-the-ukraine-reality/

45. Ingold, T.; "Time, social relationships, and the exploitation of animals: anthropological reflections: anthropological reflections on prehistory." In: Animals and Archaeology: 3. Early Herders and their Flocks. Clutton-Brock, J.; and Grigson C.; eds. Oxford: BAR International Series 202; **1985,** 3–12.

46. Isherwood, K. F.; Fertilizer Use in Western Europe: Types and Amounts. Encyclopedia of Life Support Systems (EOLSS). **2003**, 1–11, July 21, 2013), http://www.eolss.net/Eolss-sample-AllChapter.aspx

47. Janssens, I. A.; et al. Europe's biosphere absorbs 7–12% of anthropogenic carbon emissions. *Science.* **2003**, *300,* 1538–1542.

48. Jones, P. D.; and Moberg, A.; Hemispheric and large-scalesurface air temperature variations: An extensive revision and an update to 2001. *J. Clim.* **2003**, *16,* 206–223.

49. Khaledian, M. R.; Maihol, J. C.; Ruelle, P.; Forest, F.; Rollin, D.; Rosique, P.; and Delage, L.; The impacts of no-tillage on grain yield of durum wheat and energy requirement in the mediterranean climate. *Check J.* **2007.**

50. Kharytonov, M.; Bagorka, M.; and Gibson, P. T.; Erosion effects in the Central Steppe Chernozem soils of Ukraine! Soil Properties. *J. Agric.* **2004**, *3,* 12–18.

51. Kjellström, E.; Bärring, L.; Jacob, D.; Jones, R.; Lenderink, G.; and Schär, C.; Variability in daily maximum and minimum temperatures: recent and future changes over Europe. *Climatic Change.* **2007**, *81,* 249–65.

52. Kovda, S. M.; Biosphere, Soils and Their Use. Moscow: Grasslands of Northern Hemisphere; **1974,** 383 p.

53. Kosivart, Ukraine: Climate. **2013a,** 1–2, July 1, 2013, http://www.kosivart.com/eng/index-cfm/do/ukraine.climate

54. Kosivart, Ukriane: Land. **2013b,** 1–2, July 1, 2013, http://kosivart.com/eng/indexcfm/do/ukraine.land htm

55. Kosivart, Ukraine: Natural Resources. **2013c,** 1–2, July 1, 2013, http://www.kosivart.com/eng/index.cfm/do/ukraine-natural-resources/.htm

56. Kovacevic, V.; Rastija, M.; Josipevic, M.; and Loncaric, Z.; Impacts of liming and fertilization with phosphorus and potassium on soil status. *Soil Plant Food Interact.* **2012**, *13,* 190–197.

57. Kralovec, J.; Czech Republic: Country Pasture/Forage Resource Profiles. Grassland Research Station, Zavisin, Czeck Republic; **2006,** 1–23 p.

58. Krauss, A.; Balanced Fertilization, Essential for Sustained Soil Fertility and Efficient Crop Production. Norcross, Georgia, USA: International Potash Institute; **2002,** 1–9, July 19, 2013, http://www.ipipotash.org/en/present/bfefsfs.php

59. Krishna, K. R.; Agrosphere: Nutrient Dynamics, Ecology and Productivity. Enfield, New Hampshire, USA: Science Publishers Inc.; **2003,** 346 p.

60. Krupennikov, I. A.; Landscapes of Chernozem zone In: Russia's Chernozem 100 Years after Dokuchev. Moscow: Nauka Publishers; **1983,** 150–163.

61. Kumhalova, I.; Matejkova, S.; Fifernoca, M.; Lipavsky, J.; and Kumhala, F.; Topography impact on nutrition content in soil and yield. *Plant Soil Environ.* **2008**, *54,* 255–261.

62. Kustermann, B.; Christen, O.; and Hulsbergen, K.; Modelling Nitrogen cycles of Farming systems as basis of site and Farm specific nitrogen management. *Agricu. Ecosyst. Environ.* **2009,** 1–7, July 20, 2009, Doi:10.1016/j.agee.2009.08.014

63. Kusterman, B.; and Hulsbergen, K. J.; Emisssion of climate—relevant gases in organic and conventional farming systems. 16th, IFOAM Organic World Congress. Modena, Italy; **2008,** 1–2, July 20, 2013, http://orgprints.org/veiw/projects/conference.html

64. Kusters, H.; Northern Europe Germany and Surrounding regions. In: The Cambridge World History of Food. Kiple, K. E.; and Ornelas, K. C.; eds. **2000,** *2,* 1226–1231, July 5, 2013, http://www.cambridgge.org/us/books/kiple/northerneurope.htm

65. Kusters, H.; Sammelfruchte des Neolithikums. Abhandlungen aus dem Westfalischen Museum fur Naturekunde. **1986,** *48,* 433–440.

66. Kusters, H.; Neue Planzen fur die Alte Welt. Kartifel und Mais als Karrner der Industriellen Revolution. Kultur und Technik. **1992,** *16,* 30–35.

67. Losack, T.; Hlusek, J.; and Popp, T.; Potash sulphate and Potassium chloride in the nutrition of Poppy in realtion to nitrogen supply. International Potash Institute, E-ifc Research Findings. **2009,** 19, 1–43.

68. Machon, A.; Measurement and simulation of Nitrogen exchange between a Grassland and the atmosphere in landscape scale. Szent Istvan University, Godollo, Hungary, Abstract of PhD thesis. **2011,** 1–23.

69. Madaras, M.; Koubova, M.; and Lipavski, J.; Stabilization of available potassium across soil and climatic conditions of the Czeck Republic. *Arch. Agaron. Soil Sci.* **2010,** *56,* 433–449.

70. Marachi, G.; Sirotenko, O.; and Bindi, M.; Impacts of present and future climate variability on agriculture and forestry in the temperate regions: Europe. *Climate Change.* **2005,** 70, 117–135.

71. Masalsky, V. I.; Gullies in the Black Soil Region of Russia: Their Extent, Development and Impacts. St Petersburg, Russia; **1897,** 146, (in Russian).

72. Maslov, B. S.; Otcherki po istorii melioratsii w Rossii. Moscow, Russia; "Meliowodinform"; **1999,** 238 p.

73. Mercik, S.; Stepien, M.; Stepien, W.; and Sosulki, T.; Dynamics of organic carbon content in soil depending on long term fertilization and crop rotation. Roczniki Gleboznawcze **2005,** 13, 4–54.

74. Mikulasova, J.; Oilseeds Update Czeck Republic—Second Best Harvest of Rapeseed. USDA Foreign Agricultural Research Services. GAIN Report No EZ1111. **2011,** 1–5 p.

75. Ministry of Environmental Protection and Nuclear Safety of Ukraine. Natural conditions and a history of the use of Nature in Ukraine. **2000,** 1–3, July 19, 2013, http://enrin.grida.no/biodiv/ biodiv/national/ukraine/1_ind/ index.htm

76. Moon, D.; The environmental history of Russian Steppes: Vasilli Dokuchev and the Harvest failure of 1891. Transactions of the Royal Historical Society 6th Series. **2005,** *15,* 149–174.

77. Moon, D.; Agriculture and Land Use on the Russian Steppes Since ca 1800. The Environmental Histories of Europe and Japan. Ngoya, Japan; **2008,** 77–90 p.

78. Moon, D.; The Russian Steppes: An Environmental History, 1700-1914. CRCEES Working Papers, WP2008/07, **2012,** 1–11, 1–15, http://www.gla.ac.uk/media/media_209913_en.pdf

79. Moon, D.; the Plough that Broke the Steppes: Agriculture on Russia Grasslands: Agriculture and Environment on Russia's grasslands, 1700–194. Oxford, England: Oxford University Press; **2013,** 312.

80. Moore, J.; "Forager/farmer interactions: information, social organization, and the frontier." In: The Archaeology of Frontiers and Boundaries. Green, S.; and Perlman, S.; eds. Orlando: Academic Press Inc. **1985,** 93–112 p.

81. Mott Macdonald. Eurasian Steppe- Encouraging sustainable Use,Moldova,Ukraine, Russia. **2011,** 1–2, July 20, 2013, http://www. environment. mottmac.com/ projects/?mode=type&id=249615

82. Novack, P.; Masek, J.; Hula, J.; and Petrasek, S.; Aspects of growing wide row crops on slopes. Proceedings of Conference on Engineering for Rural Development. **2013,** 203–207, July 14, 2013, http://tf.llu.lv/conference/proceedings2013/ Papers/037_Novak_P.pdf pp. 203-207

83. OECD. Environmental Performance of Agriculture in OECD Countries Since 1990. Organization for Economic Cooperation and Development Paris, France. **2008,** 1–62, July 20, 2013, http://www.oecd.org/tad/env/indicaors

84. Olesen, J. E.; and Bindi, M.; Consequences of climate change for European agricultural productivity, land use and policy. *Euro. J. Agron.* **2002,** 16, 239–262.

85. Olesen, J. E.; et al. Impacts and adaptation of European crop production systems to climate change. *Euro. J. Agron.* **2011,** *34,* 96–112.

86. Pashkevich, G. A.; Agriculture in the Steppe and Forest-Steppe zones of Eastern Europe in the Neolithic and Bronze age (Paleo-ethnic and Botanic evidence). *Stratum Plus.* **2000,** *2,* 404–418.

87. Pau Vall, M.; and Vidal, C.; Nitrogen in Agriculture. The European Commission: Agriculture and Environment. **2003,** 1–7, July 20, 2013, Eu.europa.eu/agriculture/enviro/report/en/nitro_en/report.htm

88. Reichardt, M.; and Jurgens, C.; Adoption and future perspective of precision farming in Germany: Results of several surveys among different agricultural target groups. Precision Agriculture. **2012,** 1–2, August 1, 2013, doi.10.1007/s11119-008-9101-1.

89. Reichardt, M.; Jurgens, C.; Kloble, J.; Huter, J.; and Moser, K.; Dissemination of Precision Farming in Germany Acceptance and Training Activities. **2010,** 1–2, August 1, 2013, http://www.agriexchange.org/results/literature/dissemination-precision-farming-germany-acceptance-adotpion-obstacles-knowledge-transfer-and-training-activities.

90. Reichardt, M.; Jurgens, C.; Huter, J.; and Kloble, U.; Precision Farming Education in Germany-Obstacles and solutions. **2009,** 1–8, August 1, 2013, http://www.preagro.de/Veroeff/PF-Education-in-Germany.doc

91. Rode, A. A.; Explanatory Dictionary of Soil Science. Moscow, Russia: Nauka Publishers; **1975,** 264 p.

92. Romermann, C.; Dutoit, T.; Poschold, P.; and Buisson, E.; Influence of former cultivation on the unique mediterranean steppe of France and consequence for conservation management. *Biol. Conserv.* **2008,** *121,* 21–33.

93. Rosas, F.; World Fertilizer Model-The World N,P,K modle. College of Agriculture and Rural Development. Ames, Iowa: Iowa State University; Working Paper No. 11-wp 520 **2011,** 1–23, July 28, 2013, http://www.ageconsearch.umn.edu/bitream/103223/2/11-wp_520-NEW.pdf

94. Rozanov, I. G.; Principles and Doctrines on the Environment. Moscow: Moscow University Press; **1984,** 115 p. (in Russian)

95. Rutenberg, R.; France's wheat area to slump to nine year low on winter kill. **2012,** 1–3, July 6, 2013, http://www.bloomberg.com/new/2012-04-06/france-s-wheat-area-to-slump-to-nine-year-low-on-winter-kill-1.htm

96. Rutkowska, A.; and Pikula, D.; Effect of Crop Production and Nitrogen Fertilization on the Quality and Quantity of Soil Organic Matter. **2013,** 249–267, July 4, 2013, http://dx.doi.org/01.5772/53229

97. Schafer, W.; and Ferret, M.; French Wheat classes. **2003,** 1-7, July 6, 2013, http://www.muehlenchemie.de/downloads-future-of-flour-/FoF-cap_11.pdf

98. Shahgedanova, M. I.; The Physical Geography of Northern Eurasia. New York: Oxford University Press Inc.; **2008,** 553.

99. Sharpley, A. N.; and Beegle, D. B.; Managing Phosphorus for Agriculture and the Environment. Pennsylvania State University Extension Service. **2001,** *16,* 1–22.

100. Sluetel, S.; De Neve, S.; and Hofman, G.; Estimates of carbon stock changes in Belgian cropland. *Soil Use Manag.* **2003,** *19,* 166–171.

101. Smith, P.; Ambus, P.; Amezquita, C.; Andren, O.; and Arrouays, D.; Greenhouse Gas Emissions from European Croplands. Carbo Europe-GHG, France; **2004,** 1–64.

102. Stebelsky, I. V.; Environmental deterioration in the Central Russian Black soil region: The case of soil Erosion. *Canad. Geogr.* **1974,** *18,* 232–249.

103. Stebelsky, I. V.; Agriculture and soil erosion in the European forested steppes. In: Studies in Russian Historical Geography. Bater, J. H.; and French, B. A.; eds. London; Academic Press Inc.; **1983,** 45–63.

104. Stolbovol, V.; Carbon Density and Pools. **1999,** 1–3, July 1, 2013, https://nsidc.org/data/docs/fgdc/ggd601_russia_soil_maps/russian_soil-desc.html

105. Stolbovol, V.; Description of Russian soils. **2000,** 1–3, July 1, 2013, https://nsidc.org/data/docs/fgdc/ggd601_russia_soil_maps/russian_soil-desc.html

106. Stolbolov, V.; Savin, I.; and Sheremet, B.; Soil Reference Profiles. **2000,** 1–3, July 1, 2013, https://nsidc.org/data/docs/fgdc/ ggd601_russia_ soil_maps/russian_soil-desc.html

107. Sutton, M.; and Reis, S.; The Nitrogen cycle and its influence on the European greenhouse gas balance. Entre for Ecology and Hydrology, European Science Foundation. **2012,** 1–43, June 24, 2013, http://www.nitroerope.edu

108. Swinton, S. M.; and Lowenberg-Deboer, J.; Global Adoption of Precision Agriculture Technologies: Who, When Why. **2009,** 557–562, http://www.edu/user/swinton/078_swintonECPA01.pdf

109. Tarasov, K. G.; Central European Plains. The Great Soviet Encyclopedia. **1979,** 1–2, July 1, 2013, http://www.encyclopdeia2thefreedictionary. com/Central+European+Plain.htm

110. Tsirulev, A.; Spatial variability of Soil fertility parameters and efficiency of variable rate fertilizer application in the Trance-Volga Samara region. *Better Crops.* **2010,** *94,* 26–28.

111. Tulaikov, N.; "Sukhoe Zemledelie (sistema Kembellya)." Polnaya entsiklopediya russkogo sel'skogo khozyaistava I saprikasayushchicksya s nimi nauk. **1910,** *12,* 1262–1267.

112. USDA Economic and Statistic Survey. Wheat world supply and demand summary. **2012,** 1–4, July 1, 2013, http://www. spectrumcommodities. com/education/commodity/statistics/wheat.html

113. USDA GAIN. Russian Oilseeds and Products Annual 2012. **2012,** 1–14, http://www.the cropsite.com/reports/?category=39&id=345

114. USDA Joint Agricultural Weather Facility. Global Production Review 2011. **2012,** 1–10, July 1, 2013, http://www.usda.gov/oce/weather/pubs/ Annual/GlobalCropProductionReview2011.pdf

115. Van, E. B. W.; and Van, K. O.; Organic carbon changes over 40 years in a Haplic Luvisol type Farm land in Hungary. *Global Change Biol.* **2009,** *15,* 2981–3000.

116. Varallyay, G.; Risk of extreme soil moisture regime and possibilities of its control in the Carpathian Basin. In: Environment Management. Trends and Results. Koprivanac, N.; and Kusic, H.; eds. Zagreb; **2007,** 153–168.

117. Varallyay, G.; Water-dependent land use management in the Carpathian basin. Noventermeles. **2011,** *60,* 297–300.

118. Verheut, W.; Steppe change. Mott Macdonald. **2012,** 1–2, June 30, 2013, http://www.mott-mac.com/steppechange.htm

119. Vorobev, S. A.; Crop cultivation system. The Great Soviet Encyclopedia. London: Gale Group; **1979,** 689.

120. Wagner, P.; The Future of Precision Farming-The Development of a Precision Farming Information System and Economic Aspects. **2011,** 1–11, August 1, 2013, http://lb.landw.uni-halle. de/publickationen/pf/pf_efita99.htm

121. Wichereck, S.; and Bossier, M. O.; Impact of Exploitive farming practices on soil, Water and Crop Quality: A need for Remedial measures. **1995,** 1–8, July 18, 2013, www.infrc.or.jp/english/KNF_Data_Base_Web/.../C4-4-118.pdf

122. Wikipedia. Upper Rhine Valley. **2013a,** 1, https://en.wikipedia.org/wiki/Upper_Rhine_Plain

123. Wikipedia. Steppe. **2013b,** 1–3, July 4, 2013, http://en.wikipedia.org/wiki/Steppe

124. Wikipedia. Poland. **2013c,** 1–41, July 4, 2013, http://dictionary.sensagent.com/Poland/en-en.htm

125. Wikipedia. North German Plain. **2013d,** 1–4, July 1, 2013), en.wikipedia.org/wiki/North_German_Plain

126. Wikipedia. Geography of Hungary. **2013e,** 108, July 4, 2013, En.wikipedia.org/wiki/Geography_of_Hungary. Htm

127. Wikipedia Geography of France. **2013f,** 1–6, July 1, 2013, En.wikipedia.org/wiki/Geography_of_France

128. Wyss, R.; Technik, Wirtschaft and handel. In: Archaologiie de Schwiez. Schweizerische Gessellscharft fur Archaologie. **1971,** *3,* 123–144.

129. Yara International ASA. Nitrate Fertilizer: Optimizing yield and Preserving Environment. **2012,** 1–13, July 19, 2013, Htttp://www.yara.com/doc/ 35521_Nitrate_pure_nutrient.pdf

130. Yara; Precision Farming. **2012,** 1–2, August 1, 2013, http://www.yara.com/sustainability/how_we-engage/green_growth/precsion_farming_tools/ index.aspx

CHAPTER 5

SAVANNAHS OF WEST AFRICA: NATURAL RESOURCES, ENVIRONMENT, AND CROP PRODUCTION

CONTENTS

5.1   THE SAVANNAHS OF WEST AFRICA: AN INTRODUCTION

5.1.1   PHYSIOGRAPHY AND AGROCLIMATE OF WEST AFRICAN SAVANNAH

The West African region considered within the context of this chapter encompasses area between 15°W to 15°E latitude and 5°N to 25°N longitude. Atlantic Ocean forms the western and southern edge; northern border is roughly the beginning of the Sahara desert coinciding with the upper reaches of river Niger. The eastern edge is generally recognized as the Lake Chad, river Benue, or even eastern border of Nigeria but including small regions in Cameroon. The sixteen countries that occur within the West African zone are Benin, Burkina Faso, Cote de Ivoire, Cape Verde, Gambia, Ghana, Guinea, Guinea-Bissau, Liberia, Mali, Mauritania, Niger, Nigeria, Senegal, Sierra Leone, and Togo. West Africa occupies about 6,140,000 km², which is equivalent to 20 percent of total area of African continent. The terrain is undulating with small sandy hills but mostly sandy plains. The highest altitude recorded does not exceed 300 m.a.s.l. However, there are some highland locations such as Fouta Djallon, Guinea Highlands, Sierra Leone Mountains, Nimba Mountains, Jos Plateau, and Plateau of Djado (FAO, 2001). Sahel is a transitional dry zone. It occurs between Sahara in the north and Sudanian-Savannah region in south. Natural and agricultural Savannahs are 160–240 km in breadth and extend from west coast of Senegal to Cameroon in the east.

Savannahs of West Africa are served by several rivers. They can be classified into five river systems. They are (a) first set is rivers that emanate from Fouta Djillon/Nimba mountains; (b) second set includes rivers such as Bandama, Oueme, and others that flow into gulf of Guinea; (c) third set comprises rivers Senegal and Volta; (d) fourth set includes rivers that drain into Lake Chad; and (e) fifth set is represented by river Niger, which is the longest. It emanates from Guinean Highlands, flows toward northeast to Mail, then takes a southward turn, and drains into Atlantic in southern Nigeria (FAO, 2001).

Geographically, Sahelian zone in the African continent spreads into 5,400 km from Atlantic coast in the west to Red sea in the east. Its width is <1,000 km and it covers an area of 11,78,850 km². It occurs in the transitional area between Sahara and Guinea Savannahs (Figure 5.1). Sahel is predominantly a semiarid grassland compared to steppes of Russia or Pampas of Argentina, but it is drier and warmer than other grasslands. Sahelian zone in West Africa is mostly flat terrain and altitudes range from 200 to 400 m.a.s.l. Hilly regions that experience colder climate do exist in the Northeast of Niger around Agadez. Senegal, Gambia, and Benue are major rivers in addition to several tributaries and rivulets that drain the Sahelian region. There are several lakes that support vegetation, crops, and dependent population of human settlements and domestic animals.

West African region supports natural grasslands and cropping expanses of diverse composition in terms of plant species. Agricultural prairies include dry land crops such as pearl millet, sorghum, cowpea, and forage species. Wetlands that support production of rice are conspicuous near the west coast and parts of tropical low land region in Burkina Faso, Ghana, and Nigeria. Following is the total land area and percentage cropping zone for each of the nations in West Africa:

Benin 113,000 km², 16%; Gambia 11,000 km², 27%; Ghana 239,000 km², 12%; Guinea 246,000 km², 6%; Guinea Bissau 36,000 km², 10%; Cote de Ivore 323,000 km², 12%; Liberia 111,000 km², 4%; Mali 1,240 km², 17%; Mauritania 1,030,000 km², 2%; Niger 1,267,000 km², 3%; Nigeria 924,000 km², 33%; Senegal 196,000 km², 27%; Sierra Leone 72,000 km², 25%; Togo 57,000 km², 26%; and Burkina Faso 274,000 km², 9% (FAO, 2001).

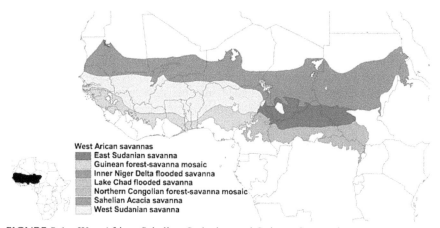

West Arican savannas
East Sudanian savanna
Guinean forest-savanna mosaic
Inner Niger Delta flooded savanna
Lake Chad flooded savanna
Northern Congolian forest-savanna mosaic
Sahelian Acacia savanna
West Sudanian savanna

FIGURE 5.1   West African Sahelian, Sudanian, and Guinean Savannah zone.
Note: West African Savannah zone supports large areas of hardy, drought-tolerant agricultural crops in the semiarid Sahel, and luxuriant cereal/legume/vegetable mixtures in the humid tropical regions. Natural Savannah vegetation shows wide diversity in species and growth pattern based on agroclimate and soil fertility.
Source: Natural Resource Ecology Laboratory, Colorado State University, Fort Collins, Colorado, USA
Note: Within the context of this chapter, area spanning between Atlantic Coast and Lake Chad is only considered, although the above picture depicts extended entire Savannahs stretching until western borders of Tanzania and Kenya.

5.1.2   AGROCLIMATE

In general, the agroclimate of large tracts of West Africa is not congenial to high crop productivity. Atmospheric parameters near the fringes of Sahara are harsh enough to curtail any worthwhile agricultural inputs. Currently, at the best, it allows only nomadic cropping and some pastoralism. Sahelian region is endowed with low and erratic precipitation pattern along with sandy soils of poor fertility. Subsistence farming with *in situ* recycling of crop residues is the most common option for farmers in this region. However, efforts to device most appropriate cropping systems and to improve soil fertility and water resources are in place. Next, the tropical agroclimate of the Sudanian and Gambian region is congenial for lush growth of natural vegetation and to produce crops of moderate to high productivity (see Figure 5.2).

Regarding atmospheric phenomena that regulate or shape the general agroclimate of West Africa, it is said that interaction of two migrating air masses creates the wet and dry periods. The hot tropical air mass from northern high pressure system give rise to dry, dusty storms that blow from Sahara desert zone and spread across most agricultural areas of West Africa. High-intensity dust storms that lash the Sahelian and Sudano-Sahelian zones are called "*Harmattan*." Harmattan storms occur from November to February. The second air mass system is derived from maritime location in the South West or Equatorial regions. It produces moisture laden tropical winds that occur between May/June and October (FAO, 2001). West Africa is characterized by high sunshine and high temperature regime. The average temperature is 26°C with fluctuation of 1.7–2.8°C. Mean temperature in the Saharo-Sahelian zone is high at 30°C. Summer temperature is very high and intolerable both for flora and fauna. Natural vegetation is either stunted, may perish or hibernate. Ambient temperature during summer months of May till July may reach 53°C during day time and reach as low as 4°C in the cool nights. In the Sahel, warmer regions experience a mean temperature of 10–35°C depending on the season. The day-length period does not fluctuate much in the Sahelian and Guinean zones. It ranges from 11.5 to 13 h. The annual biomass productivity ranges from 4,000 g m^{-2} in the wet Savannah/forest zones of southern Nigeria and reduces to as low as 500 g m^{-2} in the dry Savannah (FAO, 2001).

Senegal is a major agricultural country in the western region of African continent. It experiences a variable climate and soil conditions that determine the cropping pattern and productivity. The cropping season is restricted to rainy period that lasts from May/June to October. A typical sequence of crop production activities includes *Nor* that involves manuring and clearing the fields during February to April. It is a dry period. *Thioron* involves sowing crops and lasts from May to June end. Next, *Navet* extends from July to October. It involves parts of sowing, gap filling or repeat sowing, weeding and interculture, and harvest. Lastly, *Lolly* that spans November till January includes stacking, threshing, winnowing, and transport.

The average annual precipitation received in the West African nations such as Niger, Burkina Faso, Togo, Mali, and Senegal ranges from 400 to 600 mm. The rainfall period lasts for 60–100 days in the Sahelian zone. In the Sudanian region, natural grassland/crops receive precipitation that ranges from 600 to 900 mm, and the rainfall period stretches from 90 to 120 days. In the Guinean-Savannah region, precipitation ranges from 900 to 1,200 mm and the rainy period lasts for 210–240 days in a year (Bertlessen and Traoure, 2002; Buerkert et al. 2002). In the Eastern region of Sahel including Burkina Faso, Niger, and parts of northern Cameroon, precipitation ranges from 430 to 578 mm. The rainy period begins during June 2nd or 3rd week and lasts till October 1st week. The crop growth period ranges from 80 to 110 days (Buerkert et al. 2002; Figure 5.2).

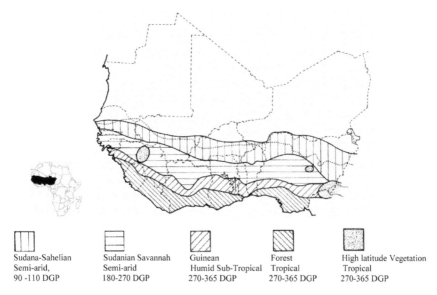

Sudana-Sahelian	Sudanian Savannah	Guinean	Forest	High latitude Vegetation
Semi-arid,	Semi-arid	Humid Sub-Tropical	Tropical	Tropical
90 -110 DGP	180-270 DGP	270-365 DGP	270-365 DGP	270-365 DGP

FIGURE 5.2   Agroclimatic zone of West Africa.
Source: FAO (2001) (Redrawn based on Ferguson, 1985).
Note: Days of Crop Growth Period (DGP).

As stated earlier, Senegal is an important agricultural country in West Africa. Precipitation pattern decides crop species and cropping intensity. Following is the precipitation pattern encountered in different regions of Senegal and adjoining nations:

Northern Senegal: It encompasses regions such as Theis, Diuorbal, and Louga. Northern Senegal is dry belt that receives 400–600 mm precipitation. Precipitation season is short and ranges from 45 to 60 days during June to September/October. Millet is the major crop since it withstands soil moisture deficits better than other crops.

Central Senegal: It encompasses Theis, Dourbal, Kaolack, and Mibour. The precipitation ranges from 600 to 800 annually. However, most precipitation occurs within 45–60 days from June to September. Major crops that suit the precipitation pattern are millet, cowpea, and cassava.

Southern Senegal: It includes Sine and Saloum regions. Precipitation ranges from 800 to 1,000 mm annually that lasts for 55–70 days. Major crops grown are pearl millet, sorghum, maize, cowpea, cotton, and cassava.

Senegal River valley: It includes the Alluvial delta zones. The precipitation pattern is erratic and ranges from 600 to 800 mm. Crops thrive more on canal irrigation. Major crops grown are rice, vegetables, and cassava.

Eastern Region: The precipitation pattern is fairly congenial for good rainy season crop. It ranges from 600 to 1,500 mm. Major crops are rice, maize, cassava, and cotton.

Cassamance: Precipitation in this region ranges from 700 to 1,600 mm. Crops grown on the Ferralitic Alfisols are rice, maize, and cotton.

(*Source*: Coura Badiane 2001; Tappan et al. 2004)

Sahelian climate is often described as semiarid, but its aridity is more than that stipulated according to Koppen's classification of 390–850 mm year^{-1}. Sahel receives 100–600 mm precipitation annually. Precipitation pattern is erratic but concentrated more during the months June to September. Since mean annual temperature is again higher at 30°C, it results in accentuation of aridity. Based on precipitation pattern, Sahel in West Africa can be further classified into three subregions. They are:

Saharan-Sahelian climate is dry with a mean annual precipitation of 100 mm. Here, vegetation is feeble and cropping locations occur sporadically. Nomadic agriculture and tribes are common. Locations in Northeast Niger and northern Mali support such nomadic farming tendencies;

Sahel *in strict sense* receives 200–400 mm annual precipitation. In locations near upper reaches of river Niger (e.g., Tillaberi), the precipitation is scanty. Yet, it supports pearl millet production, although a low yielding crop and few farm animals and cattle. Agriculture is totally subsistent and dependent entirely on inherent soil fertility and rainfall pattern.

Sahelian-Sudanese zone receives slightly higher rainfall annually at 400–600 mm. It supports intercrops of pearl millet and cowpea during rainy season and fields are fallowed during winter and summer. Summer crops are confined to riverine area or those with irrigation. Locations such as southern Mali, Niger, Burkina Faso, Gambia, and Southern Senegal experience Sahelian-Sudanese climate.

5.1.2.1 CLIMATE CHANGE IN WEST AFRICAN SAHEL AND SUDANIAN ZONES

Reports by International Food Policy Research Institute at Washington, DC, clearly suggest that climate change has far fetching effects on agricultural productivity and livelihoods of large section of population residing in West African nations. The detrimental effects are more pronounced in areas harboring famers practicing subsistence methods, marginal and poor in terms of economic status. Weather-related crop failures can be devastating on the economies of individual small farms in the Sahel and Sudanian regions of West Africa (Jalloh et al. 2013a). Climate change may also affect the livestock, its population and market value. In some instances when intensity of drought or high-intensity rainfall devastates the farms, it can even lead to difficulties for survival of farmers and their livestock. Crops may get wiped out entirely if climate is uncongenial. Therefore, policy makers for West Africa consistently advise the researchers and farmers to adopt as many agronomic and other integrated procedures that help them in avoiding climate change effects. Some of them are, timing the crop, seeding appropriately using weather forecasts, changing crop genotype, adopting soil conservation procedures, harvesting, and preserving the grain in buffers.

Wittig et al. (2007) state that climate change is an important aspect to consider among various factors affecting the West African agriculture, natural flora and fauna. During past few decades, there has been notable spread of Sahara desert into Sahelian zones. Further, there has been southern shift of Sahel-like conditions to the productive regions of Sudanian and Guinean Savannahs. It seems anthropogenic activities have caused climate change and induced Sahelization of areas considered as Sudanian region. Using satellite mediated observations of terrain, vegetation, and cropping pattern, Wittig et al. (2007) traced that land cover changes similar to one noticed during 1950 and 60s due to humid conditions did occur even in 1970s, but to a lower extent. Arable field and fallows increased from 580 to 2,870 km^2 during the past 30 years. There were also redistribution and migration of fauna between Sahel and Sudanian region. Over grazing by cattle was one of the main factors affecting the soil quality. Precipitation showed some remarkable changes in isohyets toward south and it did translate to lower rainfall in parts of the Sahelian region.

Reports by global climate networks suggest that since 1880s till 2000 A.D., temperatures have fluctuated between -1.5 and $+2°C$ from the established mean. Land use change has, it seems, induced fluctuation in precipitation pattern to a small extent. Otherwise, precipitation pattern has fluctuated marginally since 1900 till 2000 A.D. (Appinsys, 2010). Regarding temperature and precipitation changes in West Africa, it has been observed that Sahelian West Africa has registered a temperature fluctuation of $+2°C$ since 1970s. Nations such as Liberia, Niger, and Nigeria have reported a change of $3.5°C$ in temperature during past decade. Some of the weather models and forecasts suggest temperature change of $1–4°C$ during next few decades. It seems, well directed measures can save about 10–15 percent of animal diversity and restrict crop loss, if any, due to temperature changes (Jalloh, 2013a, b). Fluctuation in the precipitation pattern and total amount has induced changes in cropping pattern. Countries such as Nigeria and Togo have experienced a net reduction of rainfall by 50–100 mm in cropping zones. There has also been undue increase in rainfall in some locations within Sahel that has increased total precipitation by 100–200 mm. A few high-intensity rainfall events can be devastating, leading to loss of topsoil, depletion of soil fertility, delayed or sometimes total loss of crop stand, higher incidence of disease and insects, and loss of grain harvest, etc. High-intensity rainfall and general increase in total precipitation have certain effect on natural flora, the grasslands, and fauna dependent on the region. According to Jalloh et al. (2013a, b), there is no doubt that climate change affects the net harvest of staple cereals such as pearl millet, sorghum, and maize in West Africa. It also affects growth and harvest of other crops such as groundnuts, cowpea, cocoa, yam, and cassava.

Previous experiences regarding climate change effects on Sahel suggest that, during 1969, high-intensity drought caused almost a total loss of cereal/legume crop and 30 percent reduction in livestock. In Burkina Faso and Togo, droughts in 1874 and 1875 resulted in drastic reduction of forage and livestock production. Hence, most of the agricultural agencies situated in Sahel suggest formation of highly diversified farms that includes several staple cereals and livestock along with other avenues of deriving exchequer for farm families (World Bank, 2009). Augmenting irrigation in the Sahel using rivers will bestow great advantages to farmers when rain fails. However, there are also reports that climate change may affect the water flows in West African rivers. For example, water flow in Volta river may depreciate perceptibly by 30–45 percent, if the climate change effects are felt drastically. It may deprive water for agricultural crops in Burkina Faso and adjoining area (Cooke, 2013). The influence of climate change on agricultural productivity and exchequer and loss of labor opportunity may also bring about social and political instability in the Sahel and Sudanian regions. Descroix et al. (2009) believe that land use and climate changes have affected the hydrology of West African Savannahs. Human pressure, increase in cropping area and intensity, and repeated droughts in several locations within Sahel have led to soil crusting and desertification in West African Savannahs.

5.1.3 NATURAL VEGETATION OF WEST AFRICAN SAVANNAHS

Savannahs, in the widest sense include landscapes supporting pure grassland to dense woodland with the presence of an understory of thick grass and herbaceous layer. Now, let us consider some historical aspects of the development and perpetuation of savannahs in West Africa. Archaeological studies at several sites in the Northeast of Nigeria suggest that during Holocene, swampy forests and Guinean vegetation thrived in low lying areas. Around 11500 B.C., Savannahs supported open grasslands and Sahelian zone had vegetation tolerant to aridity (Goethe University 2009b; Salzman, 1998, 2000; Salzman and Waller, 1998). Vegetation nearer swamps and depressions were filled with *Syzgium* and *Uapaca*. However, by 6–7th millennium B.C., Guinean vegetation showed subsidence in the dune and depressions of Sudano-Sahelian zones of West Africa. Archaeological evidences derived from rock shelters and areas around sand stone massif in Eastern Burkina Faso and Mali suggest that around 7th millennium B.C., dry forests, savannahs, and few domesticated crop species were intermingling along with domesticated animals and pastoralism (Goethe University, 2009e). Anhuf (1997) has reconstructed the vegetation changes in the West African Savannah zones during prehistory spanning from

8500 to 2500 B.C. It has been suggested that around 2500 B.C. human influence on tree and grassland composition within the parklands were clearly noticed. Most common trees species encountered in the archeological sites were *Acacia albida, Acacia laeta, Balanites aegyptiaca, Sclerocaryea, birrea, Adansonia digitata, Parkia biglobosa, Danilla oliveri,* and *Elaeis guineensis.*

By 3500 B.C., aridity got intensified and it eventually led to Sahelian vegetation and crops that withstand scanty soil moisture and low precipitation pattern (Goethe University, 2009b). This situation is comparable to one that exists currently in the West African Savannahs. Almost similar trends were noticed in the Sudanian zones and transitional areas of Nigeria. By 8000 B.C., aridity intensified leading to vegetation changes. Dominance of species tolerant to low soil moisture and high temperature was noticed. Savannahs were dominated by Sudanian taxa, replacing swampy species. By 4200 B.C., water levels depreciated in the lakes and groundwater, leading to dominance of vegetation that tolerates dry climate. Some of the effects noticed could be anthropogenic, while rest are attributed to natural succession in vegetation due to climatic changes.

Salzman's (1998, 2000) and other research reports from Institute of Archaeology at Goethe University suggest that West African savannahs, especially as we notice it today is derived out of anthropogenic activities more than slow natural events. Anthropogenic activities during past 5 millennia had replaced forests and swamps with vegetation adapted to semiarid climate. Woodcutting, bushfires, and cropping have all contributed to formation of Sudano-Sahelian savannahs.

Forest patches are integral to West African savannahs. They develop near the water bodies and moist zones. Dry forests too persist in small areas. However, grassy savannahs and tall cereals dominate the landscape. Plant communities traced in the forest patches encompass trees, shrubs and herbaceous species that cover the floor. Moist zones near to water bodies support species such as *Mimusops kummel* and *Melacantha ainfolia.* In the drier regions and well drained soils, *Fagaria zanthoxyloides* and *Pterocarpus eninaceous*species are common. Forest species are often traced on soils not very suitable for crops. They are shallow. It is said dense shrub layer below the trees avoids bushfires from engulfing the whole vegetation (Goethe University, 2009d).

Regarding vegetation of West African Sudano-Sahelian zone, Wezel et al. (2000) opine that changes that have occurred are partly due to climatic variations, population expansion, agricultural cropping systems and woodcutting. Precipitation pattern has been a key factor deciding the vegetation in most regions within Sudano-Sahelian zones of West Africa. For example, shrub Savannahs are common on sandy flat terrain, valleys and fixed dunes that receive 400–600 mm. Grassy Savannah and pastoral regions are confined to areas with 200–400 mm precipitation annually (Wezel et al. 2000). Vegetation diversity has been stable in regions with consistent aridity (i.e., Sahara-Sahel borders and even Sudano-Sahel borders). In the Northern Niger, Benin, Mali, and Burkina Faso prevailing vegetation is Savannah interspersed with trees. Tree species belonging to Mimosaceae, Combretaceae, and shrubs are common. Perennial grass layers are filled with species such as *Andropogan gayanus, Loudetia spp,* and *Hyperrhania spp.* In the transitional zones of Sudano-Sahel, Fontes, and Guinko (1995) recognize vegetational communities such as *Chrozophora brocchiana, Aristida sieberiana, Tribulus terrestris,* and *Limeum pterocarpum.* Since past few years, vegetational changes in the West African Savannahs have been studied in greater detail using remote sensing. Satellite images derived from MODIS and GPS have helped us in ascertaining fluctuations in boundaries of vegetation, in response to various natural and man-made factors (Rian et al. 2009). According to Rian et al. (2009) and Opoku-Duah et al. (2013) Savannah flora and cropping zones have been affected more severely by precipitation, droughts and receding water table than any other factor. Shallow roots of grass and crop species are not well adapted to receding water table. Trees with deep roots do adapt well to water scarcity in the Sahel and Sudano-Sahelian zones. Following is the representative vegetation zonation of West Africa based on ambient climate and precipitation:

Saharo-Sahelian zone (200–500 mm year⁻¹ precipitation): This region is confined mostly to north and is characterized by hardly drought-tolerant species such as perennial tussock grass (e.g., *Aristida pungens and Panicum turgisum*). Shrubland species common to dry Sahel are *Acacia ehrenbergiana, A. senegal, A. tortilis, Balanites aegyptiaca, Pilostigma reticulatum,* and *Combretum micranthum.* The northern most part of this zone is the beginning of Sahara desert where vegetation is much lower in density and crops are almost nonexistent. Annual rainfall in this area does not exceed 100 mm year⁻¹. Sandy soils allow only stunted growth of natural vegetation, if any.

Sahelo-Sudanian zone (500–700 mm year⁻¹ precipitation): This region includes riverine regions of Niger and Senegal. It supports a mosaic of vegetation that includes trees, dense shrubland as well as agro forestry parks and parkland Savannah. For example, Parkland in Senegal supports well distributed tree population along with Savannah grass and agricultural crops, such as pearl millet or groundnut. Gallery forests too occur in this vegetation zone, but it is confined to locations nearer pond, lakes, and rivers. Tree species traced are *Acacia digitata, A leiocarpa, Combretum spp, Pterocarpus spp,* and *Terminalia sp.*

Sudanian zone (700–1,000 mm year⁻¹ precipitation): The vegetation is composed of large number of herbaceous grassy species, thick shrubs, and even a number of woody species. This region supports natural wooded Savannahs and parkland vegetation. Tree species that thrive well are *Adansonia digitata, Ficus species, Parkia biglobosa,* and *V paeadoxa.* Many of the tree species are deciduous. Grasses like *Andropogan psuedapricus* dominate the landscape. Vegetation thrives mostly on sandy soils.

Sudano-Guinean zone (≤1,200 mm year⁻¹ precipitation): Sudano-Guinean region in West Africa is characterized by Savannah forests and woodland. This region is commonly classified as Doka woodland and Guinea woodlands. They support a mosaic of vegetation that receives perhaps, highest annual precipitation in the West African Savannah zones. Tree species common to this region are *Anogeissus leiocarpus, Danialla oliveri, Isoberlinia doka, Monotes kerstingii, Khaya senegalensis, Ptreocarpus sp,* and *Terminalis macroptera.*

(*Source*: Rian et al. 2009)

Fallows are common occurrence in the Sahel and Sudanian regions. The cropping season last only during the rainy period and fields without irrigation support are left as fallows until next year. The residual moisture, scanty precipitation if any, and nutrients still available allow development of natural flora in the fallows. Most commonly, traced flora within the fallows in the Sudano-Sahelian regions of Niger, Burkina Faso, Mali, and Senegal are as follows (see Roose et al. 1996):

Herbaceous species: *Acanthospermum hispidum, Andropogon guyenensis, Boerhavia diffusa, Brachiaria ramose, Cassia obtusifolia, Chorchorus olitorius, C. tridens, Cyperus esculentus, Dactylochtenium aegypticum, Erogrostis tremula, Euphorbia hirta, Ipomea eriocarpa, Marischus cylindrachyus, Molluga nudicaulis, Panicum laettum, Setaria pumila,* and *Zornia glocidiata.*

Shrubs and trees: *Acacia albida, Acacia nilotica, Adansonia digitata, Balanites aegypticum, Combretum aculeatum, C. micrantheum, Piliostigma reticulatum,* and *Ziziphus mauritiana.*

5.1.4 DEVELOPMENT OF AGRICULTURAL PRAIRIES IN WEST AFRICA

The West African terrain, human dwellers and animals have traversed through evolutionary stages common to many other regions of the world. The hunter-gatherer communities in the Savannahs evolved into sedentary farming and nomadic pastoralism. Crop domestications and invention of implements that help in digging and tillage aided the development of farming (See Breuning and Neumann, 2002; Hohn and Neumann, 2011).

It is believed that agricultural practices were invented perhaps *independently* in the upper reaches of the river Niger during early Neolithic period. Pearl millet and other small-seeded cereals such as sorghum were, perhaps the earliest of the crops grown by native West African populace in the Savannahs. Neolithic settlements (3200 B.C.) in the upper reaches of river Niger and on the banks of Senegal river have been credited with domestication of several important food crops, such as pearl millet, sorghum, African rice (*Oryza glaberima*), and cowpea. In fact, much of the agricultural expanses in West African dry and wet Savannahs are currently composed of genotypes of these crops (see Kahlheber and Neumann, 2007). There are, of course several reports about natural history, domestication and genetic diversity of major crops that were supposedly domesticated in West Africa. Yet, Manning (2010) believes that there are still too many details we need to know, especially regarding early domestication events, locations, and spread of domesticated land races within West African Sahelo-Sudanian regions. Domestication of cereals might have occurred up in the northern reaches of Niger and Telemsi valley, and it may predate domestication of crops to 4000 B.C. Crops such as pearl millet, sorghum, and cowpea were grown throughout the West African Savannah regions by 1000 B.C.

It seems, initially, pearl millet was cultivated as mono-crop in the Sahel, right from 3000 B.C. to 1000 B.C. Crop mixtures began to appear in the landscape much later.

It seems about thousands of years ago, the present-day dry Sahel and Savannah regions of West Africa and large areas of entire Sahara supported lush green Savannah grass and woodland. Fruits trees, vegetables, cereals, and animals flourished in the area. Sedentary cultivation practices too had existed. The climate change in Sahara led to loss of vegetation. Vegetation that thrived was scanty and drought-tolerant type. Dwindling arable regions and dry conditions, it seems induced human migration with consequent reduction in agricultural crops (Spivey, 2003).

During Stone Age, around 2nd millennium B.C., the human inhabitants in the regions within present-day Burkina Faso, Mali and Niger practiced small scale sedentary farming using pearl millet. Hunting, gathering, and nomadic pastoralism were also in vogue. It is believed that rapid change over to farming must have affected composition of vegetation, soils, and environment to a certain extent (Goethe University, 2009c; Watson, 2010). Manning (2010) states that agropastoral techniques became prominent during 2nd millennium B.C. and it brought diversity to crops grown in the Sudano-Sahelian zones of West Africa.

Reports from Goethe University (2009c) suggest that during Iron Age, Sudano-Sahelian farmers adopted sedentary crop production systems centered around dwellings. Mixed farming that involved drought-tolerant cereals such as pearl millet, sorghum, legumes such as cowpea and Roselle (*Hibiscus sabdariffe*) was practiced in the vicinities of their settlements. Production of crops using known folklore procedures, cattle, and nomadic pastoralism were in vogue in the entire stretch of West African Savannah during Iron Age (100 B.C. to 100 A.D.). Migrations and introduction of crop packages within the Sudano-Sahelian regions occurred. It led to adoption of agricultural systems in tune with the dry Sahelian or marginally wet Savannah environment.

Archaeological studies in the Dogon country of Mali suggest that by 1000 A.D., rigorous woodcutting was made using iron implements. Areas under cropping were extended by clearing woody vegetation (Goethe University, 2009a). During fourth century A.D., extensive contacts with Middle East allowed traders and travelers along the Sahara to disseminate several crop species that were native to West African Savannah and also to obtain and to introduce exotic species into the sandy Savannahs (Goethe University, 2009c).

During medieval period, crops such as groundnut, cocoa, and vegetables were introduced into West Africa by the Spanish and Portuguese arriving from locations in South America. During the colonial period, extending up to mid-twentieth century, agricultural development and cropping pattern were influenced by the need to export grains, groundnuts, and palm oil. During 1880s, Senegal was cultivating groundnuts in large scale and Nigeria exported palm oil. The cash crop production was largely tuned to support France and other European powers. The cropping and yield targets were largely decided by the needs to satisfy export requirements. Large scale groundnut production in the Sudanian region, in fact reduced rice area and productivity. This led to a situation, wherein natives of Senegal, Nigeria, and other nations had to import rice grains at higher cost . In Senegal and Gambia, cultivation of groundnut led to deficit of rice during 1911–1915.

At present, major crops that fill the West African Savannahs are pearl millet, sorghum, maize, beans, cassava, yams, bananas, cocoa, and oil palm. Rice is confined to wet regions. Yams, cassava, and maize are among staple diet for Yaruba population. Pearl millet, sorghum, and cowpea-based food items sustain most of the population in the Sahelian belt. Migrants from non-Yaruba locations are utilized in the fields and plantations, since the native population has become settled and own the farms. Farm laborers from different parts of the Savannahs are periodically recruited. Pearl millet-cowpea, pearl millet-groundnut, maize-yam, and maize cassava are some of the most popular crop combinations preferred in the West African agrarian zones.

5.1.5 HUMAN MIGRATION, SETTLEMENTS AND AGRICULTURAL DEVELOPMENT IN WEST AFRICA

Archaeological evidences from human settlements in the West African region dates back to 12,000 B.C. They adopted hunting and gathering as well as pastoral way of life in the Savannahs. Evidences about Guinean tribesman suggest that development of sedentary agricultural villages in the savannahs took place around 5th millennium B.C. Crop production, mainly pearl millet and cattle rearing occurred surrounding the early Neolithic settlements (3000 B.C.). It seems there was a large migration within the West African region, mostly out of the present-day Sahara due to rapid and final desiccation of Sahara around 3000 B.C. (Donnelly, 1993; Wikipedia, 2013).

Migration of agricultural population from Sahelian zone to Guinea and further into forests in search of appropriate occupation is common. Equally so, return migration from tropical Savannahs, coasts and plantation zones to Sahel and Sudano-Sahel is frequent. Migration away from Sahel is predominantly induced by crop productivity. It is said a good harvest of cereal crop such as pearl millet or sorghum can keep Sahelian population *in situ* without migration. Adoption of improved agronomic procedures and economic prospects has often induced intense human flows within the West African zone. Return migration to location in Sahel has been most conspicuous during the years of better grain harvests (Konseiga, 2004).

The region comprising the present-day nations such as Senegal, Guinea, Mauritania, Mali, Burkina Faso, Ghana, Niger, Nigeria, and Cameroon was known to medieval Byzantines and Islamic regions of Middle East as "Sudan." Regular trade routes supported human migration in both directions between Sahelo-Sudanean West Africa and Arab region. In the process, it allowed dissemination of new crop species, genotypes, and trade in livestock. Movement of Islamic religious groups was marked during sixteenth century. It allowed excellent interaction and trade of agricultural products between those in Sahelian West Africa, especially Housa land and Byzantines. There were also large movement of Arabs into West Africa during eighteenth century, specifically into various settlements and their surroundings such as Agades, Kano, Katsina, Zaria, and right up to Futa Djalion in Guinea and into Senegal (Archnet, 2013). Such migrations allowed introduction of various geno-types of crops into West African Savannah region. There were also migrations from other continents into West Africa that resulted in new introductions such as Cocoa from South America. The plantations of rubber and cocoa that developed in due course had immense effects on human settlement pattern. It also induced periodic migration of peasants from Sahel into Guinea regions of Ghana and Nigeria.

Migration of West African peasants to North American Plantations and the plains region occurred during 1600 to mid-1800s. About 69 percent of migrants, who moved out of Gold coast, Ivory Coast, and Slave coasts, during 1700s, were originally inhabitants of Western and Central Africa. During 1800s, it seems about 32 percent of African migrants were constituted by native West African tribes. This phenomenon must have influenced the spread and intensification of agricultural Sa-vannahs in the native zones of West Africa (North Carolina Digital History, 1972). Migration was also caused by trade in salt from desert, copper from Sahelian zones, and gold from Guinea and forest zones of West Africa. Yet, this did not undermine or relegate trade and migration related to agricultural goods such as pearl millet, sorghum, and kola nuts.

The development, sustenance and perpetuation of cropping pattern in the "Groundnut Basin" of Senegal, was largely due to human migration from North-ern Senegal during nineteenth century (Coura Badiane,2001). It seems migration of settlers into "groundnut basin" in groups, driven by poverty and search for better lively hood was prominent during 1930s. Most of the early settlers were *Wolofs* from northern fringes of Sahel. Although early settlers opposed continued migra-tion of farming communities into "Groundnut basin," the spree continued. Later, severe droughts, lack of food items, and economic insecurities in other parts of Sa-hel, brought in more of migrants. For example, in 1970s, Fulanis from Mali and Ni-ger traveled into "groundnut belt" for settling. There were also settlers from ethnic groups such as *Serere, Toucoulcur, Diola , Mandingue,* and other small herdsman. Currently, 44 percent of groundnut farmers are *Wolofs*, 17.5 percent are *Fulani*, 9 percent *Toucoulcur*, 9 percent *Diola* and few small groups (Coura Badiane, 2001). This is a clear example where specific crop has induced human migratory behavior and led them to depend on crop production for their survival.

Tiffen (2004) states that West African human population has increased, rather markedly in the Sahelian and Savannah regions during past 4 decades. It has created a large section of peasants and farm laborers that engage themselves in the crop production. Development of industries and higher remuneration in the towns/cit-ies has of course induced migrations away from rural districts and farms. Although percentage of West Africans engaged in farm development and crop production has decreased, total number still dependent on farming has increased steadily and is expected to show markedly higher increase during next decade (see Table 5.1; Figure 5.3).

TABLE 5.1 Population of West Africa in Rural farms and Urban settlements 1960–2020 (in millions).

Year	Total	Rural	Percent	Urban	Percent
1960	87	75	86	12	14
1990	194	117	60	78	40
2020	430	160	37	270	63

Source: Cour (2001),Aregheore (2009)
Note: Values for 2020 are forecasts

We should note that slave labor movement and migrations out of West Africa is almost negligible since second half of twentieth century. Demand for labor *in situ* within West African regions for cereals, legumes, oilseeds, and livestock production has increased, in tune with population density in the cities and rural farming belts. Human population density has shown an increase both in rural and urban centers of Savannahs. Following is the trend noticed for human density in three different regions of West Africa:

Nigeria		Niger		Senegal	
Year	Rural Human Density (number km^{-2})	Year	Rural Human (number km^{-2})	Year	Rural Human Density (number km^{-2})
1952	77	1960	13	1960	42
1991	169	1977	22	1976	70
		1988	33	1988	94

Source: Tiffen (2001), Barry et al. (2000)

According to reports by CILSS, population of Sahelian and Sudanian zones of West Africa has increased at fairly rapid rate. About 40 percent of the populace is engaged in core agricultural cropping sector. A smaller percentage is involved in processing, transshipment, and marketing sectors. There are at present over 53 million humans to feed just in the Sahelian zone and it is expected to cross 100 million by 2020 and may reach 200 million by 2050. These forecasts clearly necessitate rapid development of agriculture, in terms of expansion, productivity and food grain distribution. The West African agricultural prairies may need mechanization to a greater extent. Irrigation too has to be augmented to reap better harvest. The genetic diversity available within crops is high. Crop improvement programs of various agricultural research institutions such as International Institute for Tropical Agriculture (IITA), International Crops Research Institute for the Semi-arid Tropics (ICRISAT), and national programs of different nations have created genotypes with higher grain/forage yield. The yield gaps need to be constricted further using better agronomic procedures to arrive at better total grain output.

Farmers tend to increase cropped zones, but grazing land may decrease. Fallows too decrease, leading to loss of soil fertility. Soil maladies such as erosion, loss of topsoil, and nutrient depletion occur. Crop residue recycling *in situ* decreases because greater fraction of crop residue gets allocated to livestock (Tiffen, 2004). Further, Tiffen (2004) argues that as need for crops and livestock increases, it induces proportionate changes in farm labor allocated to each section. Greater demand for livestock moves a certain section of farmers to change professions and migrate from subsistent regions. One other consequence noticed is the intention of farmers in Savannahs to increase productivity of crops and forage species. It is believed that closer integration of crops and livestock is a necessity in West Africa in order to regulate the demand versus supply equations in rural and urban centers, for grains, forages, and livestock products. According to Tiffen (2004), enhanced human population in West African Savannahs means expansion of crop production zones and changes in farming techniques. It also means higher number of farm animals being supported in farming zones, nearer to dwellings compared to pastoral zones. Domestic livestock get confined to areas closer to human settlements. In fact, settlement pattern itself may change as pastoral methods reduce and sedentary human dwellings and *insitu*crop/livestock production gets accentuated. Nomadic tribes that practice pastoral methods or shifting cultivation may decrease. In fact, there are instances when nomadic Fulani tribes have taken up settled farming along with Hausas and developed an integrated crop-livestock system in the locations near Maradi, Niger (CARE, 1997). In Senegal, settlements of Wolofs were actually good examples of integrated systems of crop and livestock production. The Serere and Wolof village cultivated crops such as pearl millet, cowpea, groundnuts, vegetable along with livestocks (Faye and Fall, 2000).Following is an example from an agricultural location of Northern Nigeria, in Sahelo-Sudanian region that depicts the fraction of human population that are nomadic and those thriving as settled farmers:

1925			1948		1966	
	Number	Percent	Number	Percent	Number	Percent
Nomadic farmer	29,060	33.8	14,367	13.4	14,162	12.7
Settled farmer	56,021	65.0	92,177	86.6	91,552	81.5

Source: Tiffen (2004)

Human population, its distribution, density, and major preoccupation, has direct relevance to the West African Savannahs. People engaged in agricultural cropping have a major say in the type of vegetation and productivity of the Savannahs. The

West African Savannahs are not uniformly populated. Instead, area in the fringes of Sahara and dry Sahel harbors only 1–5 people km^{-2}. Most regions in Sahel support 5–20 people km^{-2}, while Sudanian and Guinean regions have 20–100 people km^{-2}. There are patches in Nigeria, Gambia, and Senegal in the Sudanian region that support 100–400 people km^{-2} (Jalloh, et al. 2013a, b; Figure 5.3). The population in West Africa increased by 2.9–5.1 percent over previous during 1988 to 2008. Nations such as Gambia, Niger, and Nigeria recorded 5 percent increase in population, but others such as Mali and Senegal registered 3.5–3.7 percent increase in population. Population density in states such as Katsina, Kano, and Jigawa in Nigeria, Southern Coastal states of Nigeria, parts of Gambia and Senegal has also increased. Currently, West Africa is demarcated into 16 nations that totally occupy 5 million km^2 and supports 290 million people. About 60 percent of the population is actively engaged in agriculture or related occupations. Agriculture generates a gross domestic product of 35 percent in the West African region (Jalloh, et al. 2013a). A sizeable fraction of Sahelian farmers still follow subsistence farming tactics and are often exposed to poverty, degraded soils, droughts, and lack of resources.

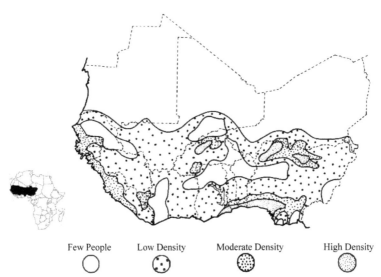

FIGURE 5.3   Population density of West African Sahelian, Savannah, Guinean and Tropical forest zones.
Note: Human population is not entirely related to cropping pattern of the Agricultural Prairies. Higher density of population is encountered in the West coast, comprising regions within Liberia, Sierra Leone, Senegal, and Gambia. Southern Nigeria with tropical Savannahs too supports higher population density. Much of the Sahelian and Guinea Savannah zones that support production of crops such as pearl millet, sorghum and cowpea have moderate to low density human population. There are several patches within Sahel and Sudano-Sahel that has very few human inhabitants.
Source: FAO (2001) (Redrawn based on Ferguson, 1985).

5.2 NATURAL RESOURCES OF WEST AFRICAN SAVANNAHS

5.2.1 SOILS

One of the earliest and comprehensive soil survey and descriptions of the West African Savannah region was done some 50 years ago by Charreau (1974). Agencies such as IRAT and ORSTOM have published soil descriptions, especially of those encountered in the Sahel and Sudanian regions. The West African soil map has 12 classes divided into sub classes, families, series, and types (Fall and Faye, 1999). However, based on Charreau's (1974) descriptions of soil types, West Africa is classified into five large soil zones. They are:

(a) Sand and Sand dunes of Senegal, Mali, Niger, and Chad;

(b) Alluvial deposits of pure sand and clay in the arc region of river Niger and Chad Basin;

(c) Ferralitic sandy Oxisols of Niger, Burkina Faso along the river Niger, and Volta, Senegal, and banks of lake Chad,

(d) Vast areas of Ordovician sandstones in southern part of river Niger in Nigeria, regions in southern Mali, and parts of Burkina Faso, and

(e) Soils derived from plutonic rocks, metamorphic rocks, and volcanic rocks in the southern Mali, Burkina Faso and Chad.

Regions in Sahel, Sudan and even Guinea possess soil types classified as ferruginous Alfisols and Oxisols that have red or brownish yellow tinge. The ferruginous layer is a mixture of three minerals Kaolins, amorphous hydrated oxides, and Fe, and Al sesquioxides. They are characterized by water holding capacity of 15–20 percent (w/v). The wilting point for crops is reached at 7–9 percent of soil moisture (v/v) (Ducreax, 1984). The ferruginous soils are deep and show the accumulation in subsurface, but surface layers show coarse sandy texture or at best loamy.

Soils of West Africa traced in the highlands are derived from crystalline rocks resistant to weathering and erosion. Sedimentary rocks such as sand stones are traced in Fouta Djillon in Gambia, Banfora and Hombori in Mali and Manpong in Ghana. Younger sedimentary rocks are the source of soils traced in the vast expanses of Nigeria, Togo, Benin, Burkina Faso, Senegal, and Senegambia (FAO, 2001; See Ref. Aregheore, 2009). Soils found in the West African agrarian zones are highly weathered and low in fertility. Major soil types encountered in West African Savannahs are Alfisols, Ultisols, and Oxisols. Small areas with Entisols, Vertsols and Inceptisols also occur. The Sahara in the north is predominantly endowed with Aridisols.

Soil fertility status and its suitability for arable cropping is of course an important criteria. Evaluations in the Southern Guinea, Savannah of Nigeria indicate that most of these soils possess an Argillic layer of silt and clay. They are termed loamy or sandy loams. Ploughing, harrowing and ridging are needed to achieve good tilth, and drainage. Most soils from this region show low K content which has to be replenished (Adeboye et al. 2009). A combination of crop residue and inorganic fertilizers keeps soil fertility at optimum level.

According to Onyekwere and Ezenwa (2009), soils found in Southern Guinea Savannah of Nigeria are low in SOC, total N, available-P, exchangeable-K, Mg, Ca and Na. They are classified as Typic Haplustolls, Dystric Ustochrepts, Aquic Ustochrepts, and Typic Tropoquepts. These soils are amenable for both arable cropping systems and wet land rice production.

Physicochemical properties of different soil types encountered in Sahel, Sudanian, and Guinean regions of West Africa—a few examples:

Peanut Basin in Senegal:

Ferrugenous Paleudalf:- Clay- 31 g kg^{-1}, C-0.36 g kg^{-1} 0.36, Total N- 180 mg kg^{-1}, Total P -68 mg kg^{-1}, Ca- cmol kg^{-1}, CEC- 3.3 cmol kg^{-1}, Base saturation- 83%, pH -4.9.

Compound fields with Millet/sorghum:-Carbon- 6.2 g kg^{-1}, N - 0.5 g kg^{-1}, P- 0.50 µg kg^{-1}, K- 0.23 meq 100g^{-1}, Ca- 2.3 meq 100 g^{-1}, Mg- 0.34 me q 100 g^{-1}, CEC-3.62meq 100 g^{-1}, pH 6.5.

Bush fields with peanuts:- Carbon- 4.07 g kg^{-1}, N- 0.33 g kg^{-1}, P- 2.15 µg kg^{-1}, K- 0.04 meq 100g^{-1}, Ca- 1.2 meq 100 g^{-1}, Mg- 0.85 me q 100 g^{-1}, CEC- 2.4 meq 100 g^{-1}, pH- 5.6.

Sahelian Southern Mali:

Clay- 30 g kg^{-1}, C-0.48 g kg^{-1} 0.36, Total N- 142 mg kg^{-1}, Total P-87 mg kg^{-1}, Ca -0.2 cmol kg^{-1}, CEC- 1.3 cmol kg^{-1}, Base saturation- 83%, pH- 4.9.

Northern Burkina Faso:

Shallow Red Aflisol:-Sand 376 g kg^{-1}, Loam 384 g kg^{-1}, Clay 243 g kg^{-1}, Total Carbon- 11 g kg^{-1}, Total N 0.7 g kg^{-1}, P 0.32 mg kg^{-1}, Ca 4.3 cmol kg^{-1}, Mg 0.7 cmol kg^{-1}, K 0.11 cmol kg^{-1}, CEC 5.7 cmol kg^{-1}, Saturation rate 89%, pH 5.5

Deep Brown Eutropept:- Sand 258 g kg^{-1}, Loam 370 g kg^{-1}, Clay 374 g kg^{-1}, Total Carbon- 7 g kg^{-1}, Total N 0.6 g kg^{-1}, P 0.31 mg kg^{-1}, Ca 14.1 cmol kg^{-1}, Mg 4.6 cmol kg^{-1}, K 0.09 cmol kg^{-1}, CEC 20.3 cmol kg^{-1}, Saturation rate 93%, pH 5.5.

Sahelian zone, Sadore, Niger:

Clay- 30 g kg^{-1}, C 1.57g kg^{-1}, Total N 74 mg kg^{-1}, Total P- 72 mg kg^{-1}, Ca- 0.4 cmol kg^{-1}, CEC-1.3 cmol kg^{-1}, Base saturation- 86%, pH -4.5.

Sudanian zone, Gaye-Bengou, Niger:

Clay- 130 g kg^{-1}, C- 3.3 g kg^{-1}, Total N-89 mg kg^{-1}, Total P-43 mg kg^{-1}, Ca-0.5 cmol kg^{-1}, CEC- 1.9 cmol kg^{-1}, Base saturation-66%, pH4.2

Sudanian zone, Kaboli, Togo:

Clay- 160 g kg^{-1}, C- 6.5 g kg^{-1}, Total N-128 mg kg^{-1}, Total P- 52 mg kg^{-1}, Ca-1.3 cmol kg^{-1}, CEC- 3.3 cmol kg^{-1}, Base saturation- 71%, pH 4.7

Sudano-Sahelian Cameroon:

Clay- 34 g kg^{-1}, Organic Carbon- 2.4%, Total N- 0.14%, Availale-P-12 ppm, K- 0.01 cmol$^+$ kg^{-1}, Ca- 4.09 cmol$^+$ kg^{-1}, Mg 0.5 cmol$^+$ kg^{-1}, CEC- 0.72 cmol$^+$ kg^{-1}, pH- 5

Sources: Buerkert et al. (2002), Roose et al. (1996), Yost et al.(2002), Manlay (2002b), Tabi et al. (2013).

Soil variability with regard to several characteristics and their consequences on natural vegetation and crops is a phenomenon noticed conspicuously in the Sahel and some regions of -Sahel. Soil under the canopy of a leguminous tree is relatively fertile and supports good crop growth and productivity. This phenomenon is attributed to the exudates of tree roots; stem, and leaf flow and leaf/twig fall which add to organic matter and minerals. It improves soil quality as well as productivity. Topographically, dunes and troughs show sometimes marked variation in the seedling establishment and cereal growth. In fact, most of those who visit Sahel, accept that soil variability is a trait well associated with subsistence farming practices. Bouma (1997) opines that soil variability is caused by factors such as;

(a) Soil crusting that affects water infiltration and its redistribution;
(b) Soil having macro- and microflora that affects local nutrient content and its transformations;
(c) Local distribution of natural vegetation especially herbage, shrubs and trees and their leaf drop, root exudates, and below-ground biomass degradation;
(d) Soil fertility gradients created by anthropogenic events, location of farm dwellings, and soil fertility-related practices adopted by farmers;
(e) Uneven depletion of soil nutrients by crops that are not uniform; and
(f) Uneven water availability that results in variable nutrient mobility in the soil profile.

During recent years, precision farming techniques have been in vogue in different continents. Precision techniques utilizing GPS guided vehicles and use of fertilizers at high rates may not be feasible, both logistically and economically in the vast expanses of sandy regions of Sahel. However, there are variations of precision techniques, such as coarse analysis of soil nutrients with fewer grid points and hand-held chlorophyll meters that help in rapid detection of soil-N fertility. Satellite pictures of crop stand could be obtained and repeated seeding done manually with greater ease and accuracy. These steps will add uniformity to soil fertility and crop growth. Consistent adoption for several years will automatically enhance uniformity with regard to soil fertility and crop production in Sahel.

5.2.2   AGRONOMIC PROCEDURES IN THE WEST AFRICAN PRAIRIES

Tillage is an important anthropogenic factor that influences nutrient dynamics, crop growth, and productivity of West African agrarian regions. Soils are sandy or loamy and they require loosening before planting. This situation is unlike the one encountered in Sahel, where sandy soils are utilized directly by dibbling seeds in small pits (hills). Hills are placed wide apart with 20–30 seedlings at each hill. There are several types of tillage practiced in the Savannahs. They are zero-tillage, manual clearing, plowing, plowing plus harrowing. It seems zero tillage causes least detriment to soil fertility (Agbede, et al. 2009). Zero tillage conserves SOM and moisture, better than plowed fields. Soil fertility related parameters such as pH, SOM, N, P, K, Ca, and Mg declined over time as tillage intensity increased. Productivity of crops such as maize, sorghum or cowpea was higher, if zero tillage or low intensity conservation tillage systems were practiced. In the sorghum belt of West African Savannahs, tillage plus mulching seems necessary to preserve and/or enhance soil fertility. Agronomic procedures like mulching, ridging, and heaping all preserved SOM and soil moisture better. It induced better rooting, nutrient acquisition and grain productivity of sorghum crop (Ojeniyi et al. 2009; Bankole and Ojeniyi, 2005).

Land preparation is an important aspect of Sahelian and Sudanian farming zones. Soils are often highly sandy and fragile. Ploughing and disturbing soil prior to seedbed preparation is said be a major agronomic procedure that leaves soils exposed to environmental vagaries. It causes loss of nutrients and organic matter. It severely destroys the soil structure and aggregation which is already poor in Sahelian zone. Soils in West African dry Savannahs are prone to loss of surface layers due to wind and water mediated erosion. Yet, farmers have adopted several variations of tillage/plowing and land preparation. In the Sahelian region, farmers who practice subsistence techniques do not resort to plowing or disturbing the soil profile. They routinely adopt no-tillage system on the fields/large areas that were left for fallow after previous year's harvest of crops and Stover. They usually dibble pearl millet or sorghum seed with the onset of first rain. Next, cowpea/groundnut is dibbled in the interrow space. The hills are placed widely at 1×1 m. To thwart soils erosion, stubbles and stover is placed on the surface or mixed lightly at the top layer of soil. This acts as mulch and reduces topsoil loss. Mulching also protects soil from reaching high temperature of 40–42°C in the afternoons. Direct seeding is widely practiced from groundnut basin of Senegal to pearl millet cropping zones of northwest Niger/Chad.

According to Fall and Faye (1999), plowing light soils found in West Africa has its share of difficulties. Lack of soil stability and sandy nature makes furrows and ridges fragile. They are easily destroyed due to wind and water. First rains that occur in June usually distort the furrows and any marking made using tillage implements. Tillage equipment like tines cut only furrows but does not turn the soil. In areas that receive higher levels of precipitation of 600–800 mm year^{-1}, if soils are relatively less sandy, deep plowing is a possibility. However, light plowing reaching 6–10 cm is the common practice. Deep plowing opens the surface and makes soil vulnerable to erosion by both wind and water. First rains with large droplets have enough strength to break the soil aggregates and remove large amount of sand grains plus nutrients. Since sandy soils are very poor in organic matter, soil aggregation is very low. If soil is not protected by application of Stover on the surface or by mixing it into soil, soil erosion could be severe. Estimates show that runoff and topsoil loss is rampant and on an average 27 t soil ha^{-1} is lost annually (Ducreax, 1984; Khatibu, 1984).

Scarification of soil surface is a procedure done using different shapes of harrows. It aims at subsurface tillage and cutting stubbles and removing weeds. The main advantage lies in mixing of stubbles removed from the land into upper layers of soil profile. Sahelian farmers adopt shallow subsoiling. It involves plowing with chisels placed at 60°. It requires less draught energy and is supposed to preserve soil moisture and structure better than other tillage systems. It reduces surface runoff. It is a common tillage practice in the groundnut basin of Senegal (see Le Thiec, 1996; Le Thiec and Havard, 1996).

Soils of West African Savannahs are not among the congenial types for crop production. They have several maladies inherently and a few that develop as we utilize them incessantly. Development of hard pans is one of them and is related to tillage systems adopted by farmers. Asiamah and Dwomo (2013) state that formation of plinthite and iron pan (petro plinthite) affects crop establishment and nutrient acquisition. Plinthite iron pan is an iron rich, humus deficient clayey material that accumulates and hardens into a pan. Actually, wetting and drying causes the hard pan. Such hard pans are common in agrarian zones of West Africa. Farmers have adopted different agronomic measures to overcome such hard pans. They prepare stone bunds, vetiver bunds, anti-erosion measures such as contouring, mulching, cover crops, and agroforestry trees. Zai method is also practiced in some area to overcome hard pans. Following is an example of influence of formation of stone bunds, contour planting, and Zai plus manure application on sorghum grain yield in the Sudanian West Africa:

	Sorghum Grain Yield (kg ha^{-1})
Control	331
Stone bunds	397
Contour planting plus manure	789
Zai system plus manure	805

Source: Bationo et al. (2005)
Note: Soil erosion and loss of fertility is effectively curtailed by stone bund, planting system, and Zai system. The higher productivity under Zai system has been attributed to better water harvesting and manure that aids seedling establishment and its rapid growth.

Zai system is a traditional tillage cum soil conservation system standardized by the native West African intelligentsia, perhaps since ages. The word "Zaiegere" in the Moore language means to wake up early morning and go to fields to prepare seedbed and plant. The Zai system currently adopted in the Sudano-Sahel region is a much improvised version of what was known to natives some two centuries ago. Zai system is popular in Mali, Burkina Faso, and Niger (Roose, et al. 1996; Fatondji, et al. 2012). In its bear minimum, Zai system involves dibbling seeds of sorghum or pearl millet, a handful in one spot, by creating a pit/trough in the sandy soil and placing a small quantity of organic matter along with seeds. The organic matter helps better rooting and keeps soil structure intact at least in the immediate surroundings of roots. Zai system aids better seedling establishment and early growth. However, there are several variations of Zai systems currently in vogue. One of the more complex Zai systems described by Roose, et al. (1996) involves soil restoration through placement of organic matter along with seeds, runoff capture in microwatersheds, and allowing termite networks to aid water infiltration and seep to adjoining areas. Zai system has been adopted in the dry belts (400–700 mm) of low precipitation levels with great advantage. For example, in Yatenga province, it resulted in 11 percent enhancement of sorghum growth and biomass accumulation (Marchal, 1983). Here, Zai system was utilized by subsistence farmers. Adoption of Zai system in Mossi plateau and several other locations of Burkina Faso have been beneficial to sorghum farmers. On degraded Alfisols and deep brown Inceptisols, it resulted in marked increase of sorghum grain yield. They say, grain production ranged from 500 to 1,000 kg ha^{-1} and forage 2 to 4 t ha^{-1}. The concentration of water into small pits and application of organic matter improved biomass production by sorghum. The fallows under Zai system allows development of a wide range of natural flora that adds to biomass and recycling of SOM. Roose, et al. (1996), for example, recorded over 22 species of herbaceous weeds and 12 forage species in the fallows that could add to SOM. A more recent study by Fatondji et al. (2012) using crop models such as CERES-Millet has shown that Zai system is good enough for farmers who reside in low fertility regions and those practicing subsistence tactics. Zai system seems to support better crop establishment and biomass accumulation despite, scanty precipitation trend, and highly variable soil fertility. It is highly advantageous when rainfall pattern is erratic in the initial stages after seeding and when soil crusting is a major problem for seeds to sprout. Over all, we should appreciate that indigenous knowledge and ingenious methods have their utility in sustaining the crops and natural prairies of West Africa. It seems, during recent years, because of high population density, higher yield goals require better production efficiency. It has therefore meant relegation of Zai system. Currently, it is not a very common or mandatory system in the Sahel and Sudanian regions.

5.2.3 SOIL FERTILITY AND NUTRIENT DYNAMICS IN THE PRAIRIES

There are several reports by experts dealing with productivity of West African prairies that suggest that soil fertility and its management is among most important factors. For example, Bationo, et al. (2005) believe that amending sandy soil deficient in major nutrients, low in organic matter, and possessing low water holding is a challenge to soil fertility specialists. The main constraint, it seems, is devising optimum combinations of inorganic and organic manures. High cost of chemical fertilizers and unavailability of FYM or organic manures is the major reason that drives West African farmers to practice subsistence farming. Organic residues are not effectively recycled to replenish soil organic carbon, because much of it is allocated as forage to feed farm animals. After all, supply of fertilizers and FYM commensurate with yield goal and intensity of crop production are essential. Bationo et al. (2005) have discussed several options and variations of soil fertility management that are amenable to farmers in Sahelian and Sudanian regions of West Africa (see Refs. Bationo, et al. 1987, 1993, 2003, 2004a, 2005).

The West African savannahs receive nutrients into natural vegetation and cropping expanses from various sources. Atmospheric deposits, precipitation, thunder storms, dust storms, and natural erosion may bring in certain quantity of nutrients into an area, i.e., crop field (see Ref. Krishna, 2008). Geographic location and its surroundings from where the atmospheric dust or storms arise may hold the key regarding nutrients received. Nutrients derived from atmosphere hold relatively greater significance in West African Sahel and even Sudanian regions where farmers are adopting low input subsistence farming methods. Atmospheric inputs could be small or at best moderate, but they are highly valuable sources for crop growth and its sustenance. In West Africa, farming systems involve both short and long fallows during which the open fields may receive nutrients from atmospheric sources. "*Harmattan* dust" from Sahara adds 3.0 kg N, 1.0 kg P, and 0.7 kg ha^{-1} year^{-1} (Romheld, et al. 2000). Applying mulches made of crop residues to entrap and collect *harmattan* dust in farmer's fields is a traditional practice in Sahel (Keitchi and Toshiyuki, 2002). In fact, it is generally known that *harmattan* dust is an enriched source of nutrients to crop fields in Sahel. Further, cropping zones nearer to the industrial region and sea coast (e.g., fields near Atlantic coast in Senegal or Liberia) may receive slightly higher quantities of nutrients through enriched atmospheric dust/ deposits. Bird deposits in the Sudano-Sahelian region is said to add 40 kg dry matter ha^{-1} year^{-1}. Elemental analysis suggests that bird deposits add 2.2 kg N, 0.4 kg P, and 0.7 kg K ha^{-1} year^{-1}. The Guano sources are more pronounced in the Atlantic coast. During fallows, farmers allow leguminous crop species or natural vegetation to proliferate until the advent of next rainy season. Such vegetation upon recycling adds 5–0 kg N ha^{-1} in addition to biomass. As stated above, nutrients added to each field may appear small, but if extrapolated to large expanses such as Sahel and Sudanian ecoystem, it is a very large supply/recycling of nutrients. The vast expanses of groundnut in Senegal's "Groundnut basin," or Pearl millet zones in Mali, Niger, and Burkina Faso, in fact derive advantages from atmospheric deposits that may after all go unnoticed. Similarly, in intensive cropping belts of Guinea region, atmospheric sources of nutrients may not be accounted. Farmers neglect atmospheric deposits while devising fertilizer schedules such as maize, cassava, or plantain.

West African soils are not inherently rich in nitrogen content. The sandy Oxisols, if cultivated incessantly, become deficient for N in a matter of two or three seasons. Soils are replenished with N through several sources, such as inorganic fertilizers, organic manures, crop residue recycling, and modifying cropping systems to include a legume that adds to soil-N. Herbaceous legumes such as cowpea, groundnut, Stylosanthes, or woody legumes such as Sesbania, Leucana, and Macuna add to soil-N upon incorporation. The extent of N benefit derived ranges from 40 to 70 kg N ha^{-1}(Sanginga, 2003). Despite knowledge about the usefulness of including legumes in the crop rotation or as intercrop, their cultivation frequency is not high. Legume crops of short duration can be excellent green manure crops. They add soil-N and improve soil quality. During recent years, soybean has been touted as a leguminous crop that adds to soil-N in addition to protein rich grains, especially in the Guinean regions of Nigeria, Togo, and Benin (Sanginga, 2003). Soybeans may add 38–126 kg N ha^{-1} through symbiotic association with *Bradyrhizobium*. Legume crops also affect the crop grown in succession through the residual-N in soil profile. Farmers in Sudanian region are in fact advised to grow soybean–maize in rotation to extract full advantage from the legume. The carry over-N derived from soybean ranges from 10 to 24 kg N ha^{-1}. It satisfies 14–30 percent of maize crop's require-ment of N, given the low yield goals possible in rain fed conditions. Nitrogen de-rived from atmospheric N fixation is highly valuable to farmers in the dry Sahelian belt. The cereals grown under subsistence conditions on inherently poor soils may not be supplied with optimum fertilizer-N. In fact, use of inorganic fertilizer-N is feeble in Sahel. Therefore, cultivation of legume and deriving maximum N-benefit from them are essential.

Soil fertility is an important constraint to higher grain/forage yield, perhaps all through the Sahel and many subregions within the Sudanian and Guinean regions of West Africa. It is generally agreed that crops respond to application of fertil-izer, provided soil-P, which is the most limiting fertility factor available at optimum level. Series of experiments by international agricultural agencies such as ICRISAT and International Fertilizer Development Center (IFDC) has shown that sufficiency level of P in the acid sandy soil is 7.9 μg P g^{-1} soil (Bray-1 P) for pearl millet. The productivity of pearl millet achieved is about 1,520 kg ha^{-1}. West African region is endowed with several rock phosphate mining zones. Farmers often try to use powdered rock phosphates or partially acidulated rock phosphates as substitute for fertilizer-P such as SSP or TSP, which is cost prohibitive to subsistence farmers. Since soil-P is considered as a major constraint to productivity of natural grass lands and crops alike, it is worthwhile to spread use of rock phosphates. It seems, if rock phosphates are used along with crop residues, cereal grain and forage yield are sig-nificantly high. In some regions of Niger, this combination has provided spectacular gains in crop productivity (Bationo and Mokwunye, 1991; Akponipke, 2008). Inte-grated nutrient management procedures that involve application of FYM, recycling at least a portion of crop residue and use of rock phosphate may help subsistence farmer in Sahel to achieve better productivity.

Nutrient loss from agrarian zones of Sahel and Sudanian region is perceptibly high. Soil nutrient loss occurs through several avenues. They are: erosion of sur-face soil due to runoff or intense winds, leaching to lower horizons of sandy soils, seepage to channels, emissions, and loss to weeds, etc. Estimates of soil nutrient loss from different nations of West Africa indicate that they differ based on several natural and man-made factors. Such nutrient loss affects both natural vegetation and crop fields. In the crop belt, aspects such as soil type, nutrient supply trends, cropping systems and agronomic procedures that ensure higher nutrient absorption influence nutrient loss.

Bationo et al. (2005) reported that West African nations suffer loss of large quantity of soil nutrients. For example, extrapolations suggest that during 1990s, about 2813,000 ha of Nigerian crop belt lost 1.1 mt N, 0.14 m t P, and 0.8 m t K year[-1]. In the Sahelian belt, 8015,000 ha in Mali lost 61,707 t N, 7,817 t P, and 55,382 t K year[-1]. Estimates by Stoorvogel and Smaling (1990) suggest that at a rate of 49 kg N,P and K ha[-1] loss, about 9.3 million tons of major nutrients were lost by the sub-Saharan West Africa. During the year 2000, it seems that at a rate of 60 kg N,P and K ha[-1] loss, Burkina Faso and adjoining regions might have lost about 6.6 million tons of major nutrients. Nutrient loss from the cropping belts is high. Agronomic measures that reduce even small quantities of nutrient loss, through soil management, contour bunding, mulching, and placement of fertilizers near seedlings can be very effective, if the entire Sahel or Sudanian region is considered. On a crop field basis, Manlay, et al. (2002a, b) opine that traditional farming procedures cause loss of certain quantity of nutrients. Estimates suggest that 20–25 kg N and 2.5 kg P ha[-1] are lost from groundnut fields in the Sahelian zone within Senegal. Loss of SOM is also high at 2–4.7 percent of original levels at the beginning of cropping season.

Nutrient budget that indicates the difference between nutrient supply to the agrarian region through several sources and that lost without being used by crops or preserved *in situ* in the profile is an important indicator of the ecosystem status. It is interesting to note that for almost all nations or subregions within West Africa, nutrient budgets have been negative since several years (Bationo et al. 2005). Clearly, farmers and native population are recovering soil nutrients in excess. They need to curtail such nutrient loss plus improve nutrient use efficiency.

Pearl millet sole crop is most common in the Sahelian region. Pearl millet adapts better than other crops to low precipitation zones (<500 mm). Pearl millet is sown in hills with a group of 20–25 seeds. It is aimed at ensuring seedling survival and establishment. Hills are spaced widely at 1×1 m apart. Pearl millet roots profusely, forms as many as 30–35 tillers hill[-1] and bears long panicles. These features allow the mono-crop to efficiently explore soil, harness photosynthetic radiation, and form forage/grain. Pearl millet sole crop responds to fertilizer supply with greater biomass and grain yield. Together with FYM or crop residue application, pearl millet shows marked increase in foliage, stem growth, and grain formation. Following are few examples of pearl millet sole crop's response to fertilizers and combination of fertilizer plus crop residue:

Inorganic Fertilizers	Grain Yield kg ha[-1]	Total Dry matter kg ha[-1]
0N, 0P, and 0K	217	1,565
30N, 30P and 30K	930	3,012
120 N, 30P and 30 K	1,000	3,300
0N, 30 P and 30 K	710	2,300

Source: Bationo et al. (2005)

Most soil fertility researchers dealing with the Sahelian and Sudano-Sahelian soils opine that phosphorus is indeed a major constraint to higher crop production. Several sources of phosphorus such as Triple super phosphate, Single super phosphate, and Rock phosphates have been used by farmers. Incidentally, West Africa abounds with phosphate mines and hence powdered and partially acidulated phosphate rocks are highly recommended for use on sandy soils poor in available-P (Olsen's P < 3.0 ppm). Fertilizer-P added is generally not utilized efficiently. Usually, less than 20–22 percent P applied as soluble source is recovered by the crop, rest is held as residue, or a fraction is lost via soil erosion, runoff, and leaching. Phosphate rocks that are partially soluble in acid soils are preferred. They may actually release P into available pool slowly. Combinations of soluble-P fertilizers and phosphate rocks have also been in vogue. Such a combination provides rapid seedling growth and allows time for P from rock phosphates to get solubilized. Soluble-P source provides the priming effect. Farmers often use fertilizer-P or phosphate rock sources along with crop residues/FYM. Such a combination provides better crop response. Following are few examples from ICRISAT Sahelian Centre, Sadore, Niger:

	Pearl Millet		Cowpea	
	Grain yield (kg ha^{-1})	P use efficiency (kg grain kg^{-1} fertilizer-P)	Grain yield (kg ha^{-1})	P use efficiency (kg grain kg^{-1} fertilizer-P)
Control (Zero P)	634		1,688	
Single Super Phosphate	887	19	2,375	134
Tilemsi Phosphate Rock	1,039	24	3,375	161
Kodjari Phosphate Rock	745	9	2,469	141

Source: Bationo et al. (2005)

Influence of Crop residue, Fertilizer-N and Phosphate supply on Pearl millet grain yield and Phosphorus-use efficiency:

	Grain Yield (kg ha^{-1})	P Use Efficiency (kg Grain kg^{-1} Fertilizer-P)
Control	**889**	33
13 kg P ha-1 + 60 kg N ha-1 + 5 t Crop Residue	**1,471**	98

Source: Bationo et al. (2005)

Soil fertility is prone to change periodically based on the cropping pattern, nutrients removed from the fields via grains/residue, nutrients lost due to natural factors, and regular additions of fertilizers and organic manure. At the same time, farmers' yield goals are dependent on economic advantages, agricultural agency stipulations, and human preferences. The fertilizer prescriptions prepared usually take note of such fluctuations. Yet, we have to realize that most of the Sahelian and Sudanian zone still supports subsistence farming. Soil fertility replenishment is achieved primarily through *in situ* residue recycling, FYM supply, and perhaps small amounts of inorganic fertilizers. Whatever be the popular manuring practice, we must recognize that a combination of inorganic fertilizer containing major nutrients N, P, and K and organic manures seems most congenial and useful recommendation.

West African Sahel is characterized by large expanses of millet/cowpea intercrops that thrive on undulating terrain, with dunes and troughs. The crop stand noticed is often uneven in such regions. Such undulations are known to induce variations. Such variations are attributable to key soil characteristics such as profile characteristics, depth of top layer, texture, pH, CEC, soil organic matter, and mineral content in the summits of dunes compared with low troughs. Hence, crop stand and even productivity varies based on the location of a pearl millet hill. Soil fertility variation is also rampant in flat terrain. Obviously, uneven extraction of water and nutrients leads to erratic crop growth and grain formation.

During past 5–8 years, one of the latest and most advanced agronomic procedures adopted and that relates to precise management of soil fertility status is known as "Precision Farming." It involves the use of soil sampling and analysis, mapping of soil fertility variations accurately, use of computer-based crop growth models, and supply of nutrients to remove the variation using GPS guided farm vehicles. In the Sahel, resource poor farmers may not adopt the precision techniques in entirety, using high sophistication and costly tractors fitted with variable rate applicators. The uneven germination of seeded hills and crop stand is most common in Sahel. A satellite picture of farm, made available at subsidized rates on the internet helps farmers to locate regions that need reseeding accurately. Large farms with uneven germination and seedling growth can be effectively covered rapidly using satellite pictures of individual farms. Hand-held low cost leaf N monitors can help farmer to rapidly and accurately decide on amount of fertilizer-N/FYM needed at each spot. Fertilizer-N dispensation can then be accomplished manually. Precision methods could also be used to get a rough estimate of biomass and grain yield from Sahelian farms using GPS and satellite pictures. Fallows and their vegetation intensity, soil erosion pattern if any, and loss of topsoil in different regions within large farms can be assessed easily using satellite pictures. Precision farming in its low cost version and as per the requirements of farmers is perhaps the best proposition in West African Savannahs (see Ref. Krishna, 2013). Precision farming can potentially alter the Sahelian landscape, especially uneven growth of natural grasslands and cropping expanses. It may avoid undue application of nutrients right at the initial stages of intensification of a farm in the Sahel. It clearly reduces loss of fertilizer-based nutrients via erosion, seepage, and emissions.

5.2.4 SOIL ORGANIC MATTER, RESIDUE RECYCLING, AND CARBON SEQUESTRATION

Soil organic matter content in the West African savannah soils is relatively low at 0.65 percent (Jones, 2006). Major factors that influence the SOM content are the clay percentage and moisture. In addition, anthropogenic activity has also influenced SOM content. Multiple linear regression analysis suggests that SOM content in West African soils is influenced by clay portion to the extent of 45 percent and rainfall pattern of 57 percent. Actually, a combination of low SOM, scanty and erratic precipitation, and sandy nature of soils in West Africa causes low crop yield.

Soil organic carbon content reported for areas supporting forest is 24.5 g organic carbon (OC) kg^{-1} soil; that for Guinea Savannah is 11.7 g OC kg^{-1} soil and for Sudanian Savannah it is 3.3 g OC kg^{-1} soil. The Sahelian dry areas possess much less organic matter which is <1 g OC kg^{-1} soil. Soil organic matter content also depends on type of usage and location of the field. For example, home gardens in West Africa possessed 11–22 g OC kg^{-1} soils, but a village field used more commonly for cereal production had 5–10 g OC kg^{-1} soil. Bush fields that are fallows and with native vegetation had only 2–5 g OC kg^{-1} soil. Location of the fields from farm households and livestock seems to affect organic matter inputs (Bationo et al. 2006).

Soil organic carbon content plays an important role in crop production in West African savannahs. It buffers soil nutrient, supplies mineral nutrients upon mineralization, enhances cation exchange capacity, improves soil aggregation, water retention, and supports microbial activity. Bationo et al. (2006) report that compound fields with higher SOC levels equivalent to 2.5 to 3.2percent have consistently provided higher maize grain yield (3.5–4.5 t grain ha^{-1}) compared to other locations with SOC of <1.0 percent that gave 1.5–2.3 t grain ha^{-1}. Further, Bationo et al. (2006) have reported that pearl millet grown in Sahel gave higher yield in fields that had higher SOM content. Mulching with crop residues from previous crop seems to improve SOC and consequently induces higher grain yield of pearl millet. Following is an example from a location in Sahelian zone, namely Sadore, near Niamey in Niger:

	Productivity (kg ha^{-1})	
	Grain	**Stover**
Control	160	1,100
Crop residue (No fertilizer)	770	2,950
Fertilizer (no residue)	1,030	3,540
Crop residue plus fertilizer	1,940	6,650

Source: Bationo et al. (2006)

Senegal's "Groundnut basin" is an important agricultural belt to that nation in terms of vegetation, groundnut production, and export. Soil fertility is an important factor that decides productivity. Paucity of soil organic matter is the prime reason for low productivity of soil. McClintock and Diop (2005) point out that 70–80 percent of Senegalese soil classified as Dior (or *Joor*) are sandy soils with low water retention and only 0.3–1.0 percent SOM. It is necessary to apply SOM through FYM, compost, or crop residue to enhance soil fertility. Composts (*Sentaare*) derived from kitchen wastes, crop residues, and foliage can enhance SOM, microbial activity, and nutrient availability of the sandy soils. Application of FYM/composts to soils in the Groundnut basin is said to improve, stability, quality, and productivity of sandy soil. The quantity of compost applied by farmers in the Groundnut basin to sandy soils ranges from 160 to 440 kg ha^{-1} depending on the crop. However, agricultural extension agencies recommend at least 2–4 t compost ha^{-1} to refurbish the sandy soils (TRI, 2002; Ouedraogo, et al. 2001). According to McClintock and Diop (2005), application of *sentaare* has consistently improved several soil fertility traits, especially organic-C, N, P, K, and micronutrients. Following is an example from different locations in the "Groundnut basin" of Senegal:

Treatment	Bulk Density	Total-C	P	K	Ca	Mg	Na	Al	CEC
Sentaare	1.35	4,283	36	120	439	92	81	183	4.31
Control	1.34	2,809	26	105	342	67	68	182	3.69

Source: McClintock and Diop (2005)
Note: Nutrient content of soil supplied with *Sentaare* is higher than those left unattended

Application of *Sentaare* or compost has generally provided higher pod yield in many locations within the "groundnut belt." It has in fact restored soil fertility and sustained it for longer stretch of time compared to fields without organic matter supply.

5.2.4.1   CROP RESIDUE RECYCLING C-SEQUESTRATION

Crop residue recycling is perhaps one of the important phenomena in the natural grassland of West Africa. It is the mainstay for regeneration and fresh growth of vegetation in dry regions of Sahel, since it offers organic-C and mineral nutrients upon decomposition. Crop residue recycling sequesters a sizeable portion of C, whenever, above-ground biomass is incorporated into soil profile. Decaying roots from previous season's flush also adds to organic-C and nutrients into soil. In the dry sandy soils of Sahel and Sudanian region, soils are generally very low in organic matter content. Therefore, crop production for 2–3 seasons depreciates SOC to levels below threshold. Crop residue recycling is perhaps the best suggestion to maintain SOC at appropriate levels of 3–5 percent. There are several reports, which prove usefulness of residue recycling to crop productivity. The sandy soils of Sahel are actually low in both organic-C and available-P. Field trials at Sadore, in Niger has shown that application of crop residue along with 13 kg P ha^{-1} as TSP results in excellent nutrient recovery and biomass accumulation by pearl millet. Crop yield on an average was improved by 25 percent over control plots (Bationo et al. 2003, 2004b). It is believed that application of crop residue firstly enhances organic-C content and supply of P results in prolific root growth and its spread. A well spread out root absorbs more nutrients. Of course, extra root biomass gets recycled *in situ*. A series of location trials across Sahel has clearly demonstrated the advantages of application of crop residue plus mineral nutrients. Sahelian crops such as pearl millet, cowpea, and groundnut could derive immense benefits from the synergistic effect of crop residue plus fertilizers. Field trials in Nigeria have shown that combination of crop residues derived from peanut at 6–12 t ha^{-1} and a combination of fertilizers (N, P, and K) enhanced maize grain yield markedly (Nottfige et al. 2005). In case of groundnut grown in the "Groundnut basin" of Senegal, Buerkert et al. (2002) argue that incessant cropping without much residue recycling has resulted in depreciation of soil fertility. Farmers have generally allocated greater share of fodder from pearl millet/groundnut intercrops to feed domestic animals. This apart, it has also induced soil deterioration in many locations within Senegal. Reports suggest that farmers in Senegal recycle at best about 2 t haulms ha^{-1} back to soil.

Application of crop residue is known to function as mulch that restricts loss of topsoil and nutrients. Crop residue improves organic-C content that eventually enhances soil moisture and nutrient buffering capacity. According to Buerkert et al. (2000), a combination of factors such as improved soil moisture status, low resistance to root growth, greater nutrient availability, and stimulation of soil microbial process at attributable to mulch mediated beneficial effects on crop productivity in Sahel.

Kato et al. (2011) have summarized the effects of Integrated Soil Fertility Management (ISFM) on soil fertility, crop production, and its importance to West African farmers. They have compared the influence of applying only chemical fertilizers and integrated measures that included, FYM, inorganic fertilizers, microbial inoculants, rock phosphates, mulching, and intercrops. Kato et al. (2011) found that ISFM is beneficial in many ways. It affects soil fertility aspects positively. Crop production on sandy soils is increased significantly. It reduces soil erosion and nutrient loss, and preserves soil productivity. It has been pointed out that ISFM is useful in mixed farms that include livestock farming, cereal production, and agroforestry. Study recommends adoption of ISFM all through the West African savannahs from Senegal to Nigeria. Let us consider an example from Sudanian region of Togo. Wopereis et al. (2006) selected fields situated close to farmer's dwellings ("infield") and those with a history of consistent supply of organic manures/FYM and compared it with those away from village and very low in SOM content ("outfield"). Maize grain/forage productivity was consistently higher in "infields" due to interaction of SOM and mineral N inputs (Table 5.2). Further, soil moisture and precipitation too had its influence on maize productivity. During seasons with erratic or low rainfall "infields" performed much better than "outfields." We should note that higher SOM helps in conserving soil moisture, buffers nutrients, and aids soil nutrient transformations better.

TABLE 5.2   Maize production in Sudanian region of Togo is positively influenced by Integrated Nutrient Management Methods involving Inorganic chemical fertilizers and Organic manures levels.

	SOC content g kg^{-1}	Nutrient Supply kg ha^{-1}	Nitrogen Recovery kg kg^{-1}	Maize Yield Increase (t ha^{-1})	
				Normal	Low/Erratic
In Field	13.4	100 N; 50 P	0.41		
Out Field	6.3	100 N; 50 P	0.33	1.0–1.5	0.8–2.0

Source: Wopereis et al. (2006)
Note: "In Fields" are those located nearer farm dwellings and are richer in SOM due to repeated FYM inputs. "Outfields" are situated away from villages and are low in SOC content. Normal and Erratic in the last column refer to precipitation pattern. Maize grain yield increase refers to enhanced grain yield in the Infields compared to out fields.

5.2.5   WATER RESOURCES AND IRRIGATION IN WEST AFRICA

Water is in fact the most crucial factor that affects crop sustenance and productivity in the entire Sahelain and Sudanian region. These regions are endowed with low levels of precipitation. Further, an erratic precipitation pattern and intermittent droughts really depress crop yield to very low levels. Farmers are driven to adopt subsistence farming due to paucity of water resources. Scanty precipitation ranging from 200 to 400 mm year^{-1} allows production of only pearl millet and cowpea in the Sahel. In the Sudanian region, slightly higher precipitation level ranging from 500 to 800 mm secures good grain/forge yield of crops such as sorghum, maize, cowpea, and vegetables. However, in the Gambian region higher rainfall allows for more than one crop and enhanced productivity. Grain/forage yield goals are generally higher in the Guinean region. Clearly, natural vegetation and crops that constitute the savannahs are highly regulated by the water availability either through precipitation and/or supplemental irrigation (see Table 5.3).

TABLE 5.3   Water resources of major West African Nations

Country	Senegal	Guinea	Liberia	Mali	Burkina Faso	Togo	Ghana	Niger	Nigeria
Precipitation (mm year^{-1})	686	1,651	2,391	282	748	1,168	1,187	151	1,150
Surface water (SW) (km^3 year^{-1})	23.8	226	200	50	8.0	10.8	29	12	14

TABLE 5.3   *(Continued)*

Country	Senegal	Guinea	Liberia	Mali	Burkina Faso	Togo	Ghana	Niger	Nigeria
Groundwater (GW) (km³ year⁻¹)	3.5	38	45	20	9.5	5.7	26.3	2.5	87
SW and GW overlap	1.5	28	45	10	5	5	25	0	80
Total area of the country	19,672	24,586	11,137	1,24,019	27,442	5,679	23,854	126,700	92,377

Source: Aquastat (2013), http://www.fao.org/nr/aquastat

Major rivers that support agricultural activity along their banks and those irrigated by them are Senegal, Gambia, Niger, Volta, and Benue. Several rivulets and tributaries of major rivers also supply water to cropping zones. Reports suggest that there are over 150 small dams constructed across rivers that supplement crop production in West Africa. There are several more in the process of completion during next decade (Skinner et al. 2009). Development of such water resources will obviously induce both expansion and intensification of cropping systems in the West African Savannahs. Dams may also induce perceptible changes in soil nutrient dynamics, grain, and forage productivity in the region of its influence. Diarra (2011) states that large dam would induce ecological changes in the region. For example, several dams have been coordinated in the river Niger, and its influence spreads into large areas within the Sudano-Sahelian regions of Mali, Burkina Faso, Nigeria, Benin, and Cote d'Ivore. We should also note that climate change effects on dams and water resources could be detrimental. For example, Diarra (2011) suggests that water storage in the dams may fall by several cm height, if precipitation pattern is erratic in a particular year. This affects availability of both irrigation water for crops and potable source for human dwellings.

The potential irrigable area in West African savannah is above 2,660,000 ha. So far, mere 25–27 percent of potential irrigable area is utilized, depending on the subregion and nation. The intensity of cropping achieved using supplemental irrigation ranges from 52 to 75 percent in different countries of West Africa. Farmers in West Africa utilize both traditional small scale irrigation and modern well designed channels from reservoirs to irrigate their crops. During past two decades, the area under modern irrigation systems has increased to 229,000 ha and that under traditional irrigation is about 1059,000 ha. No doubt, traditional irrigation facilities still dominate the agricultural zones.

Groundwater sources are widely used in the West African savannahs. Reports by international agricultural agencies suggest that farmers extract groundwater from aquifers located in different geological zones. Following is an example (UNO, 1988; Purkey and Vermillion, 1995):

Geological Aspects	Sand stone	Lime stone	Schist	Gniess	Granite	Continen- tal Shield	Alluvium
Country in West Africa	Cam- eroon/ Senegal	Mauri- tania/ Senegal	Mali/ Mauri- tania	Burkina Faso/Mau- ritania	Mali/ Chad	Mali/ Mau- ritania	Cameroon/ Mauritania
Yield of Water (m³ year⁻¹)	10–20	0.1–4.0	0.5–20	1–4	1–4	50–100	10–80

According to Descroix et al. (2009), factors such as land use change, natural vegetation, and cropping intensity affect the catchment areas, surface, and subsurface water flow in the Sahelian/Sudanian region. The plutonic bedrock seems to reduce chances for enlargement of groundwater extraction. Whereas, sedimentary basins are good sources of groundwater. Surface runoff seems to be an important source for replenishment of aquifers in Sahel. We may note that careful regulation of extraction of groundwater and maintenance of aquifers are important. There is also "Sahelian paradox" that pertains to large runoffs and droughts. Runoffs are known to cause a rise in groundwater.

Several types of water resources that are utilized by farmers in West Africa have induced wide diversity in the crops grown in the agricultural land scape. Demographic pressure too has resulted in exploration of different water resources and matching crops. Survey in the Sudano-Sahelian region has shown that farmers in regions with erratic precipitation pattern and high risk crops prefer to ensure against failure and sustain productivity by opting for supplemental irrigation through pumps (Takeshima and Futoshi, 2012).

Microirrigation systems have also been tried in the savannahs. Basically, dry agroclimate, paucity of storage of rainwater, and depreciated groundwater have all made microirrigation less plausible. It seems, even substantial investments have not produced good results with regard to microirrigation (Saa and Akuriba, 2010). However, during recent years, small holder farms that prefer to cultivate high value cash crops have adopted low cost, drip irrigation systems.

Among crops, rice, sorghum, and maize receive irrigation supplements in preference to others such as vegetables, pulses, and fruit crops (Purkey and Vermillion, 1995). The extent of area under a particular crop that gets irrigated depends on several factors related to availability of water, crop species, its productivity, and economic advantages. For example, percentage area under irrigation for rice ranges from 30 in Niger, 82 in Mali, to 97 in Gambia. About 16 to 78 percent of area under other cereals may get irrigated. Water resources and their efficient use along with soil fertility improvements may hold the key with regard to crop productivity.

5.3 AGRICULTURAL CROPPING AND ENVIRONMENTAL CONCERNS

5.3.1 SOIL EROSION AND NUTRIENT LOSS FROM AGRICULTURAL FIELDS

Soil erosion is defined as dislodgement of soil particles from a location, then its transport by wind and/or water. Soil erosion due to wind is usually gradual compared to that caused by high-intensity precipitation events and rapid surface surges. Soil erosion due to wind is rampant in Sahel and Sudanian regions of West Africa. Wind erosion, indeed engulfs a very large area of natural grasslands and agricultural cropping zones of West Africa. It is a conspicuous natural event of larger proportion occurring year after year. Sterk et al. (1998) point out that soil erosion is related to intensity of farming and length of fallows. Short fallows may not be effective in soil amelioration. Fields unattended after crop harvest are most vulnerable to loss of topsoil due to blowing wind. Wind erosion is severe when soils are dry, loose with low aggregation, and the wind velocity is high. Wind velocity less than 12–19 km hr^{-1} at 1.0 m height above the field surface does not create soil erosion. We may note that wind mediated erosion is not influenced much to be gravity. Reports suggest that wind mediated soil erosion is wide spread in the Sahelian region. Sahelian agricultural prairies and natural grass lands are severely affected by the "*Harmattan*" winds generated by Sahara. The effect of *Harmattan* on topsoil and crop stand is felt severely during May to July. It sometimes coincides with the thunderstorms and moves westwards from Niger till Senegal coast and perhaps into Atlantic sea. It occurs mainly during November to April in Mali and Senegal (Michels et al. 1995). Farming belts in Niger experience wind erosion during the crop season and in between two crop seasons. It is more intense between October and April (Sterk et al. 1996, 1998; Shivakumar, 1989) when fields are without crops. The intensity and period for which wind mediated storms occur are important, since it decides the amount of sand lifted from a place, its movement in the atmosphere, and deposition at a different location. More crucial is the nutrient carrying capacity of the sand/dust particles that drift in the air (Table 5.4.).

TABLE 5.4   Sand storms of Southwest Niger in Sahelian West Africa, their characteristics and Nutrient enrichment ratios of sand/dust particles

Storm event Date	Duration seconds	Storm Height m	Wind Speed m S^{-1}	Mass Transport kg m^{-1}	Nutrient C	Enrichment Ratio			
						C	N	P	K
Mid June	1,481	0.05	10.3	102.7		1.18	1.33	0.83	0.98
		0.50				2.06	3.02	2.08	1.14
Early July	3,004	0.05	9.2	149.8		1.27	1.47	0.83	0.91
						4.15	4.74	2.25	1.98

Source: Sterk et al. (1996), (1998), Shivakumar, (1989)

Note: Nutrient enrichment ratio is defined as ratio of nutrient element held/trapped in sediment to total nutrientelement in the topsoil/soil particle

There is no doubt that crop productivity is affected due to wind mediated loss of topsoil. According to Bielders et al. (2002a, b), crop loss due to wind erosion is comparable to drought. Investigation in several locations within Sahel, wherein farmers practice traditional cropping systems, has shown that topsoil loss averages at 27 t ha⁻¹ year⁻¹. The dust/sand particle deposited in fallow fields averaged 24 t ha⁻¹ year⁻¹. Storm deposits increase the height of soil layer by 1.5–2.0 mm year⁻¹. The extent of soil erosion and wind created deposits received may vary based on season, crops grown, and agronomic procedures adopted (Beilders 2000c). Ridging and mulching were the common measures adopted by local Sahelian farmers. Ridging does not effectively contain the topsoil loss and sand drift. It creates a sort of resistance to movement of sand particles. Mulching with crop residues helps to curtail erosion and confines the sand particles to the location. Mulching intensity is important. For example, at 2 t ha⁻¹ crop residue, the enrichment of soil particles reduced to 0.93–1.8 for organic matter, 0.9–2.3 for N, 1.5–1.8 for P and 1.5–2.9 for K. However, at 500 kg ha⁻¹ mulch, soil erosion and enrichment ratios were unaffected. Clearly, farmers in the Sahel need to be informed about the threshold levels of mulch required to thwart loss of soil fertility and avoid degradation of crop fields.

Soil erosion due to water is a common occurrence in most, if not all regions of West Africa. Soil erosion is more pronounced in Sahel and Sudanian savannahs that are left bare without a cover crop or natural vegetation during fallows. Soil erosion is marked immediately after a high-intensity rainfall event. The extent of topsoil loss and nutrients carried with it depends on the cover crops and nutrient carrying capacity of sand/silt particles. There are several detailed studies on this aspect that emanate from different nations of West Africa. For example, evaluation by Anikwe (2007) in Eastern Nigeria suggests that species of cover crop grown affects the extent of protection against soil erosion. No doubt, topsoil loss is greater in fields/regions left without cover vegetation. Following is the effect of cover crop on topsoil loss in one of the representative locations of the Savannah zones of Nigeria

Crop Cover	Groundnut	Soybean	Bambara Groundnut	Bare Field
Topsoil loss height (cm)	0.87	1.83	1.95	2.4
Quantity (t ha⁻¹)	134	283	302	372

Source: Anikwe et al. (2007)

Soil organic carbon loss is rampant in the Sahel. The SOC loss occurs due to erosion of topsoil and organic fraction. It is mediated by rain water and runoff. High-intensity precipitation events that cause severe runoff also remove large amounts of organic matter both in dissolved and particulate form. Therefore, nutrients held in the upper layer of soil decreases as erosion/runoff increases. The quantity of soil sediments and particulate organic matter eroded and their nutrient carrying capacity is to be noted. Following is an example that depicts SOC loss due to erosion, runoff, and leaching in the Sudano-Sahelian zone of West Africa.

Soil Organic Carbon Loss (kg ha^{-1} year^{-1} from top 30 cm soil)				
Korogho, Cote D'Ivore			**Saria, Burkina Faso**	
Factor	Sudanian Savannah (Natural Vegetation)	Cereal Field	Sudanian Savannah (Natural Vegetation)	Cereal Field
Erosion	6	65	9	150
Runoff	2	18	1	5
Leaching	13	3	2	1
Total	21	86	12	156

Source: Excerpted from Roose and Barthes(2001),Bationo et al. (2006)
Note: Natural vegetation that acts as good soil cover throughout the year protects surface soil against erosion and runoff. The soil in cereal field is disturbed due to plowing and other agronomic procedures. Weeding removes natural cover for soils and exposes the interrow space to severe erosion and runoff.

The cropping system that dominates the landscape and the clay/silt content of the region seems to affect erosion and runoff. Characteristics of upper layer of soil especially clay/silt content too has its influence on SOC loss. For example, fields with sorghum mono-culture suffered 1.5–2.6 percent SOC loss; those with cotton+cereal lost 6.3 percent, SOC; millet+groundnut fields lost 4.3–7.0 percent, SOC; cereal+legume intercropped fields lost 3.8–4.7 percent, SOC (Bationo et al. 2006). Factors such as tillage and its intensity also affect loss of SOC. Coura Badiane (2001) states that soil erosion is a major environmental constraint to crop production in the Senegal's Groundnut Basin. It seems, in the Central Groundnut basin, removal of vegetation has accentuated soil erosion. Central and Sine Saloum regions have experienced rampant loss of permanent vegetation leading to erosion. In Southern Soloum, about 75 percent of natural tree growth has disappeared due to loss of topsoil. Further, it is said that persistent and scanty rainfall has again induced soil erosion. Soil erosion in the "Groundnut basin" has also depreciated SOM and quality. In some locations, soil maladies such as alkalinization and acidification have appeared.

Mulching with crop residue is a very important agronomic procedure in all regions of Sahel. Long-term trails at ICRISAT Sahelian Centre at Sadore have clearly shown that intensity of mulching affects SOC loss. For instance, at 2 t ha^{-1} crop residue mulch, SOC at 0.1 m depth was 1.7 g SOC kg^{-1} soils and at 4.0 t ha^{-1} mulch it was 3.3 g SOC kg^{-1} soil (Bationo and Buerkert, 2001). Mulches help in preserving SOC by avoiding high soil temperature that may otherwise induce higher microbial activity, soil respiration, and CO_2 loss. Mulches reduce loss of SOC through erosion and runoff.

5.3.1.1   EMISSIONS

Savannahs of West Africa are basically a combination of vast stretches of grasses/legumes and other short statured annual species interspersed with trees. Like other biomes, West African Savannahs too are prone to loss of mineral nutrients and organic-C through several natural and man-made factors. West African Savannah is a large tract in the continent that can easily influence the agroclimate. Actually, West African Savannah vegetation is in constant interaction with other aspects of the ecosystem. This biome exchanges gaseous and dust material with atmosphere. Emissions of CO_2, CH_4, N_2O, NO_2, N_2 and water vapor are most common forms of exchange (Grote et al. 2009). Basically, natural vegetation and agricultural crops that fill the biome has a major influence on gaseous emissions from soil phase to atmosphere. There is a large body of knowledge accrued regarding nutrient dynamics in the soil phase and its impact on emissions. Yet, we strive to attain greater details about the gaseous interactions and nutrient loss from the vegetation to atmosphere. Measurements in the Southwest Burkina Faso indicate that NO_2 emissions reach 20 µg N_2O-N m^{-2} h^{-1} after rainfall event, and on a warm day N_2O-N emission may even reach 150 µg N_2O-N m^{-2} h^{-2}. Soil respiration leads to emission of 10-350 mg CO_2 m^{-1} h^{-1}. Emissions do fluctuate within a day and on day to day basis. There are several computer-based models that try to rationalize and predict gaseous emissions from the biome to atmosphere. They are essentially biogeochemical models that consider several parameters related to nutrient dynamics within the biomes and its impact on emissions. Evaluation of simulation models and actual manifestations at a location in Burkina Faso have shown that usual biogeochemical models that analyze the nutrient dynamics, especially N, C, and water fluxes, are amenable. However, it is argued that detailed knowledge about soil moisture in the profile and fluctuations in the ambient temperature needs due consideration, while judging the emissions from Savannah. Energy balances too should be considered while arriving at forecasts (Grote et al. 2009). It is believed that coupling of such parameters with biogeochemical cycles that proceed within the Savannahs will provide better insights into mechanisms of emissions.

5.3.1.2   FIRE IN THE SAVANNAHS

Based on series of micro and macro analysis of archeological sample from Dogon in Mali and adjoining areas, anthropologists from Goethe University (2009a) have reported that Sahelian and Sudanean regions were not exposed to periodic bushfires during Holocene. Fires did not affect diversity of natural vegetation that existed then. However, during the period after 8000 B.C., dense Sudano-Sahelian savannahs and open bamboo forests were periodically subject to fires. The aridity of the general climate and fires together affected the Sahelian agricultural prairies as well as natural vegetation during second millennium B.C. Fires must have been the cause for large scale emission of CO_2 and loss of organic-C from the terrain. Further, phytolith analysis of charred grains of Pearl millet at Verves Ouest has indicated that fires have indeed affected crops. Several other reports suggest that fires must have been altering the natural vegetation and domesticated crops during Neolithic period in the Sudano-Sahelian zone. Anthropogenic activity such as shifting agriculture, woodcutting, bushfires, and grazing seems to have affected the natural vegetation and crops alike during past 4–5 millennia (Salzman, 1998; Goethe University, 2009b).

Fire that is man-made or that erupting out of natural factors both have played an important role in shaping the natural vegetation in many biomes, including the West African Savannah. Historically, fires have occurred in the vegetation even during geologic timescale from the pre-Quaternary period. It continues to occur till date. Perhaps, it may occur perpetually but sporadically (Ballouche, 2004; Bird and Cali, 1998; Diouf et al. 2012). Archaeological examination of charred or burnt preservations of grains, grass blades, pollen, and other parts of vegetation from Holocene sites has proved that fire occurred periodically through the ages. It has shaped the dynamics of vegetation in West Africa. In the samples from Dogon valley in Mali, Biu Plateau in Nigeria, sites on the banks of river Yame have all proved that fires occurred during prehistoric period. Archaeological samples from 8th millennium to 4th millennium have shown that fires have affected the savannahs of West Africa (Ballouche, 2004). It is believed that since Neolithic period (3rd millennium B.C.) fires have affected in intervals. Alternating layers of burnt soil and organic debris and decomposed organic matter without charring indicates that fires have shaped the vegetation and climate in intervals. Fire in the savannahs must have affected the C emissions and loss of C from the biome to atmosphere. According to Ballouche (2004), fires have caused aridification of West African agricultural prairies and natural savannahs during the past 2 millennia. It is proved by the presence of large number of charcoal remains of artifacts in the human dwellings and fields. Anthropogenic fire has been common occurrence in this period and has been among the main cause of C-emissions, loss of organic matter, and climate change.

Fire in natural savannahs, fields with natural vegetation, or those with crop residues/stubbles adds to soil nutrients through the ash it generates. Minerals added are of course proportionate to quantity of residue or biomass burnt and the mineral content. Let us consider an example from northern Nigeria that ordinarily supports dry savannah vegetation. Fallow vegetation consisted of *Chromolena odorata, Tridax procumbens,* and *Imperata sp.* Crop residue and stubbles burnt were derived from a previously grown maize crop. The fertilizer application trends, inherent soil fertility, and amount of crop residue burnt decided the net increment in mineral content of soils. Soil-N, P, and K levels were augmented whenever crop residues were burnt. Azeez et al. (2007) believe that application of fertilizers, ash derived through burning of crop residue, and incorporation of FYM, all together had positive influence on crops and natural vegetation that develops during fallows.

5.3.1.3   DROUGHTS, DUST BOWLS AND CROP PRODUCTION IN WEST AFRICA

Droughts and dry spells are natural phenomenon in the Sahel. It is understood, imbibed and experienced as a matter routine by the Sahelian farming community. Droughts and dust storms are among the most serious environmental concerns that affect savannah vegetation, its growth, and perpetuation. Droughts basically reduce growth of natural grasslands and agricultural prairies. Their occurrence is conspicuously felt in the warmer rainy season. Droughts that strike the West African savannahs vary with regard to timing, length of period, and intensity. Droughts that afflict the cropping zones during critical stages obviously are most devastating. Crops are stunted, do not complete normal life cycle, formation of foliage, photosynthetic rate, nutrient translocation, panicle initiation, seed set, and grain filling is affected detrimentally. Droughts, in addition to affecting crop growth and grain yield, also affect water resources of a region and hence affect farm animals and human population. It seems a mega drought hit the Sahelian region spanning from 1450 to 1700 A.D. During past five decades, Sahel has consistently experienced droughts of varying length of period and intensity. Droughts during 1914–1916 caused severe famine and induced human migration out of Sahel into more congenial locations. A long drought occurred from 1968 to 1974. Droughts reduced crop production during early 1980s. Droughts that struck Niger and Mali during 2010 caused water scarcity, high ambient temperature of 43–46°C, and resulted in famine for over 350,000 people (Henry, 2010). Reports about above normal and high temperatures are common in many locations within Sahel and Guinea Savannah regions of West Africa. Natural vegetation and crop stretches in Sahel were almost decimated. Droughts, no doubt cause crop loss, reduction in biomass, food grain scarcity. Droughts also cause loss of topsoil, loss of fertility, soil nutrients, and loss in soil structure. Drought accompanied with wind causes large dust storms. Dust storms erode crop fields, shift topsoil, and destroy soil aggregation. Water holding capacity is also reduced.

Sahara desert to the north of West African savannahs is world's largest repository of dust. Here dust that emanates all through the year drifts into different directions. A large share of sand particles and dust from Sahara traverses into cropping zones and natural grasslands of West Africa. The Sahel receives dust flux periodically during November till March. Dust fluxes are also created within Sahel. They deplete a sizeable quantity of nutrient enriched sand from surface layers and deposit them at a different location farther away. Dust fluxes whether derived from Sahara or locally created within Sahelian belt also adds to soil nutrients. Therefore, a dust bowl situation in West African savannah is not entirely deleterious to soil fertility. Mulitza et al. (2010) state that accurate measurements of dust fluxes, their intensity and distances that they traverse before settling have been possible since 1970s. Such dust bowls and droughts in Sahel and Sudan regions are related to global climatic variations. The African dust bowl is actually a phenomenon occurring since several millennia. Mulitza et al. (2010) have reconstructed the dust characteristics, frequencies, and intensities using historical data and simulations of several weather related parameters. They have inferred that tropical rainfall pattern in West African savannahs are among important factors that induced dust storms. Most importantly, they suggest that at the beginning of nineteenth century, Sahel/Sahara experienced intense and widely distributed dust flux. This period coincides with removal of shrub land and expansion of agricultural cropping. The onset of commercial farming in the Sudanian region too has been listed as factors that generated dust fluxes. During the past 200 years, dust flux generated through anthropogenic activity in Sahel and other natural factors related to this region has loaded the region with thick layers of dust mass.

Dust storms created during crop season that last from May/June till September and intense sand drifts received from Sahara during October/November till March/April are important meteorological phenomena that affects West African Savannahs and cropping belts. *Harmattan* that generates in Sahara brings in large amounts of sand/dust. The sand storms within Sahel too remove topsoil and nutrients. However, what Sahelian/Sudanian farmers experience is the net difference between topsoil lost and amount of deposits received from *harmattan* and sand storms during crop season. The actual difficulty that farmers face is the rampant removal of topsoil, leading to loss of fertility, lack of proper seedling emergence, formations of crusts, and poor crop stand. It is said that a good quality mulch of optimum intensity say >5 t ha^{-1} can effectively stop loss of topsoil and simultaneously act as a trap to collect the Saharan nutrient rich dust in the webs created by crop residue. Such ingenious measures can effectively overcome soil degradation and other environmental constraints.

5.3.1.4   LAND USE CHANGE IN WEST AFRICA

West African savannahs and cropping zones are among the agroecosystems vulnerable to climate change and other factors that can potentially alter the expanse and productivity. There are several reports that confirm the harmful trends in natural vegetation and food crops due to climate change (Tieszen et al. 2004; Liu et al. 2004). For example, series of crop failures and shrinkage in the area of groundnut basin in Senegal has been attributed to higher temperatures experienced. It seems that the crop genotypes were not well adapted to such rapid changes in climatic parameters. Next, fluctuation in carbon cycle has been attributed to large scale changes in land use from forests and shrub land to agriculture (Smith, 2008). Investigations of West African savannahs for land cover and usage using satellite imagery from LANDSAT reveals some important trends and fluctuations of agroecosystems (Tieszen et al. 2011; Table 5.5)

TABLE 5.5   Major land cover and usage changes in the West African countries during past 3 decades—a few examples.

Land Use Class	Total Area km²	Land Use Change (percent)
Forest	40,280	−17.5
Savannahs	841,744	−10.9
Agriculture (Total)	320,080	48.0
Irrigated agriculture	3,584	50.1
Sandy zones	66,884	38.1
Bare soil	20,576	34.8
Human settlements	6,024	45.3

Source: Tieszen et al. (2011), Aregheore (2009)
Note: Total area and land use change depicted refer to West African nations such as Senegal, Guinea, Mauritania, Burkina Faso, Ghana, Benin, Togo, and Niger. Most prominent changes that occurred were reduction in forest and natural grass lands, with a simultaneous increase in agricultural prairies. Minus (−) sign indicates reduction in area

West African savannahs and agricultural prairies together constitute an important biome that affects global carbon cycle, especially CO_2 emissions and C sequestration within the vegetation. Large scale C emissions are often associated with land use change pattern, mainly from forest/savannahs to crop fields. Land management following conversion to crop production also affects the C emissions. It is said that in Sahel, land continues to loose organic fraction even after conversion to crop fields. This is attributed to loss of organic matter via erosion, and emissions mediated by oxidative status of soil and microbial activity. The cropland may then experience reduction in fertility and crop productivity. As time lapses, in the absence of proper land management, the area turns into a degraded zone fit for abandonment. Tieszen et al. (2011) state that bare soil region, which is in fact bad land unfit for cropping, has steadily increased in West Africa. Therefore, programs to resurrect them and convert them through reforestation are in place. In some areas, the sandy unusable zones have been utilized to raise settlements (e.g., in Mali). There are currently Environment Management and Information Systems that monitor and record changes in C sequestration pattern, emissions, and cycling. There are also computer models that consider future trends in land use change, organic matter, and nutrient cycling in the savannahs. They are computer-based models that utilize historical data and future trends in land use, and then simulate biogeochemical cycles in different agricultural belts and natural vegetation. For example, General Ensemble Biogeochemical Modeling System (GEMS) has been used to study the forests, savannah, and cropland changes and C emissions in several West African locations (Liuet al. 2004; Tan et al., 2009; Tieszen et al. 2011).

5.4   CROPS, CROPPING SYSTEMS AND SAVANNAH PRODUCTIVITY

Pearl millet (*Pennisetum glaucum*) was domesticated in the upper reaches of river Niger, sometime during early Neolithic period in 3rd millennium B.C. Evidences derived from archeological sites and Kintampo culture locations in West Africa suggest that pearl millet cultivation and consumption were in vogue by 3500–2500 B.C. Pearl millet domestication might have taken place in the lower Tilemsi valley in Eastern Mali (Manning et al. 2011). Pearl millet samples derived from northern Ghana indicate that pearl millet cultivation occurred in Sahel during 1700 B.C. (D'Andrea et al. 2007; Hirst, 2013). Since then, native West African tribes have cultivated this hardy cereal to derive their carbohydrate and energy requirements. The spread of pearl millet cultivation into other regions of West African Sahelian and Sudano-Sahelian region was mediated mainly by the human migration within the region. Pearl millet is an excellent cereal for subsistence farming communities of West Africa. It survives and offers grains/forage in the harsh environs where most other species perish. Pearl millet is a drought-tolerant crop. It withstands intermittent drought more common to Sahel and survives on scanty erratic precipitation pattern of 400–500 mm. Pearl millet cultivation spreads into entire West African Sahelian region from Atlantic coast of Senegal to Sudano-Sahelian regions of Cameroon. The West African landraces are tall at 3–4 meters. They put forth tillers and foliage profusely. Panicles are long at 1.5 to 3 feet. Roots are deep and profusely branched. This aids roots to explore sandy soil efficiently. Pearl millet is sown with the onset of rainy season during May/June and harvested by September end /October. It is a fairly early maturing cereal well adapted to complete its life cycle well ahead of dry season. Several landraces have been grown by the native West African tribes for the past 3–4 millennia. They differ with regard to tolerance to abiotic and biotic stress factors within the regions where they dominate the landscape. Pearl millet is often intercropped with cowpeas in the Savannahs. The productivity of pearl millet is low, since it is grown mostly under rain fed dry land conditions. Fertilizer inputs are feeble, if any, and organic matter is predominantly derived from crop residues or FYM. A pearl millet land race grown in Sahel may offer 0.5–0.6 t grain ha^{-1} and 3–4 t forage. During recent decades, several improved cultivars, composites, and hybrids have also been released into the Sahelian pearl millet belt. Some of the newly introduced genotypes are also resistant to diseases and pests. The composition of pearl millet agroecosystem, the genotypes that dominate the land scape, soil fertility, and water management, together determine the productivity of pearl millet productivity. During 2011, international agencies such as ICRISAT have released several cultivars/hybrids of pearl millet that literally determine several aspects of ecosystem, in addition to grain harvest. Reports by IFPRI-ASTI (2013) suggest that pearl millet-based savannahs in West Africa was introduced with 12 genotypes in Senegal, 35 in Mali, 26 in Burkina Faso, 32 in Niger and 11 in Nigeria (Plate 5.1).

PLATE 5.1: A village in the midst of Pearl millet based Agricultural Savannah.
Note: Above location is near Berni-N-Konni, Southern Niger, Western Africa.
Source: ICRISAT Sahelian Centre, Niamey, West Africa and Dr. Krishna, K.R. 1986

Sorghum (*Sorghum bicolor*) was independently domesticated in the West African savannahs along the river Niger during early Neolithic period around 3200 B.C. (Harlan and DeWet, 1972). However, some experts prefer to denote West African savannahs as region of secondary center of genetic diversity and consider Ethiopian highlands as the primary center. Sorghum landraces of West Africa are taller and form panicles of different shapes and sizes. Sorghum land races grown under subsistence conditions in the Sudanian region offer very low productivity of 0.8–1.2 t grain and 3–5 t forage ha⁻¹. Since 1960s, landraces have given way to sorghum hybrids. During 1980s, several sorghum hybrids were introduced into Niger, Mali, Nigeria, and Burkina Faso. They were derived mostly from ICRISAT'S Hybrid Adaptation Trials that cover all the 16 nations of West Africa (Atokple, 2012). Currently, Sorghum research centers in West Africa are aiming at spreading genotypes with greater grain/forage yield stability into the savannahs. In addition, sorghum genotypes tolerant to drought and low soil fertility conditions may find their way into the savannahs. Sorghum cultivars well suited to soil fertility stress and variable agroclimate is a necessity in West African savannahs. Therefore, genotypes that show better growth and grain formation despite experiencing fluctuation in precipitation pattern, intermittent droughts, high ambient temperature, incidence of pest and diseases are most sought. Genotypes with greater plasticity for grain yield, despite expression of climate change factors are useful in Sudanian savannahs (Hausmann et al. 2012). Clearly, sorghum cultivars that dominate the landscape and affect grain productivity need greater monitoring. Long-term effects of a sorghum genotype on the savannah landscape and environment could also be simulated using computer programs.

Maize is an important cereal crop in the Guinea savannah of Southern Nigeria, Ghana, Togo, Benin, and Ivory Coast. It is also grown in other regions of West Africa where Guinean or Sudanian climate prevails. Reports by IITA, Ibadan, Nigeria suggest that maize cultivation zone has spread rather rapidly during past 2 decades. It has also replaced traditional cereals such as sorghum in wetter savannahs. Maize is preferred in moist savannahs with higher photosynthetic radiation levels, low night temperature, and reduced incidence of pest and diseases. Maize grain yield potential is higher in Nigerian Guinean region compared to traditional belts. Maize cultivation is spreading into Sudanian region because of availability of short season cultivars that mature within 90–100 days. According to Kamara (2013), at present, maize production in Guinean region experiences several constraints that restrict biomass/grain yield. Major constraints listed are poor soil fertility, drought, weeds such as *Striga hemonthica*, and pests. Attempts to intensify maize production zones are hampered by soil degradation and depletion of nutrients from the profile. Fertilizer-N replenishments are needed for almost every season, since it limits the productivity. Wherever drought occurs along with soil fertility constraints, reduction in grain yield is indeed severe (Banziger et al. 1999).

African Rice was domesticated in the West African savannahs around river Senegal, and Atlantic coastline encompassing Liberia, Senegal, and Sierra Leone. The African rice is starchy and cooks soft into pastes. It is known as *Oryza glaberrima*. The Sudano-Sahelian zone harbors wide genetic diversity of rice. It encompasses germplasm lines of species such as *O. glaberrima, O. barthi, O. sativa,* and *O. longistaminata* (Africa Rice Centre, 2013; Sow et al. 2013). However, *O. glaberrima* dominates the wet prairies of West Africa. Rice is not the major cereal in West Africa. Yet, agencies such as African Rice Centre could potentially alter the landscape of the wetter regions by releasing as many new rice genotypes/hybrids. The new germplasm and cultivars that they release may dominate the wet prairies of West Africa periodically and then fade away as preferences decline. Rice is cultivated in four different ecosystems within West Africa. They are drylands (30%), rainfed wet lands (20%), deepwater mangrove regions in Nigerian and Senegal (9%) and irrigated wet lands (20%) (Balasubramanian, 2007). Recent reports suggest that with greater demand for rice in many of the West African regions, potential for its expansion and intensification is quite high (Takeshima et al. 2012; Nin-Pratt et al. 2009).

Rice is grown in West African agrarian regions wherever precipitation is congenial, say 1,000 mm and above, and/or water resources permit. As stated earlier, the rice growing regions are concentrated in Senegal, Gambia, Guinea, and Sierra Leone. The flood plains of river Senegal and its tributaries are important rice production zones in West African savannahs. Rice is also grown in the tropical zones of Ghana, Togo, Benin, and Nigeria. The flood plains of river Benue and Lake Chad also support rice crop production (Tabi et al. 2013). During past three decades, consumer preference to Asian rice has shown marked increase. Hence, West African farmers have introduced and grown many Asian rice varieties. Soil fertility and water resources are major constraints to rice production in Sudano-Sahelian regions (Abe et al. 2010; Tabi et al. 2013). Balanced fertilization with major and micronutrients is almost essential. In acid soils, with higher levels of Al and Mn, soil amendment such as phosphor-gypsum is required. Research updates from IITA at Ibadan in Nigeria suggests that Wet prairies that support rice production may undergo certain improvements with regard to productivity. A change over from traditional West African rice cultivation systems common in the valleys of river Senegal and Niger seems preferable. A new system, more popular in Southeast Asia, if adopted, results in 3.5 t grain ha^{-1} compared to current 1.5 t grain ha^{-1} under traditional systems (Crop Biotech Update, 2008; Bado et al. 2010).

Cowpea is a legume native to West Africa. It was domesticated during early Neolithic period in the upper reaches of river Niger and Sudanian savannahs. There are clear archeological evidences that cowpea cultivation and consumption occurred in the southern coasts of Ghana and right up to northern fringes of Sahelian zone during 3600–3200 B.C. Domesticated cowpea was produced in the semideciduous forest/savannahs that occur in Ghana and Nigeria during 1700 B.C. (D'Andrea et al. 2007). Currently, cowpea cultivation has spread into all across Guinean, Sudanian and even Sahelian zone. Its cultivation as mono-crop or intercrop is common in Sudanian and Sahelian region. There are several different genotypes with wide array of genetic traits that occupy West African savannahs. Cowpea productivity is moderate to low in most of the regions, since it is commonly allocated to fields with low fertility. Productivity ranges from 325–800 kg ha^{-1}. Cowpea is a legumecrop that nodulates with promiscuous *Bradyrhizobium* strains. It fixes atmospheric nitrogen through symbiosis with *Bradyrhizobium*. Cowpea crop may add 20–130 kg N ha^{-1}. Cowpea is also preferred because it provides forage rich in protein and minerals.

Groundnut was introduced into the West African savannahs by Spanish and Portuguese travelers and slaves returning from American continent. Groundnut cultivation got foothold in the Sudanian regions of Senegal in the medieval times. Later, groundnut found its way into many regions within Sahel and Sudanian savannahs. Groundnuts are often intercropped with cereals such as pearl millet or sorghum. It is confined to low fertility sandy soils. Hence, productivity is low. Groundnut belt in northern Nigeria that was well entrenched had to be phased out because of virus diseases that lashed the belt in 1970s. Currently, groundnut cultivation is pronounced in the Groundnut basin of Senegal.

Cassava was introduced into West Africa during seventeenth century. During nineteenth and twentieth century, its cultivation spread into the Guinean regions extending from Gambia to Nigeria. Human migration played its role in the spread of cassava within the Savannahs. For example, Igbo migration into eastern states of Nigeria during seventeenth century, it seems resulted in introduction of Cassava into parts of Guinean region, encompassing areas in Nigeria, Chad, and Cameroon. During eighteenth century, Abeokuta, Lagos, and Ijeba were centers of cassava production in the Guinean region of Nigeria. It spreads northwards into Sudanian and

Sudano-Sahelian region and was restricted because of weather, especially precipitation pattern. Cassava is a perennial shrub that has adapted to tropical savannahs of Nigeria, Ghana, Togo, Ivory Coast, and other regions in the Atlantic coast. It grows rapidly, reaches a height of 1–1.5 m, and puts forth carbohydrate rich bulky storage roots. Leaves are rich in protein and could be used as vegetable. Cassava cultivation is spreading into several locations within Guinea savannahs of West Africa.

Agroforestry trees are an integral part of West African savannahs. Some of the most frequently encountered agroforestry trees species in the Sahelian/Sudanian region of West Africa are *Faidherbia albida, Prosopis spp, Adansonia digitata, Balanites spp, Accacia laeta, Luecana leucocephala, Glyricida spp, Accacia tortilis, Azadirachta indica,* and few others. Many of the agroforestry tree species that occur in the Sahel and Sudanian zones are not just rapid biomass accumulators and forage species. Those belonging to leguminosae form symbiotic association with *Rhizobium sp* and actually add to soil nitrogen through biological nitrogen fixation (BNF). Quantification of rhizobial species within the soil profile suggest that tree species vary with regard to extent of nodulation and nitrogen fixation. Hence, advantage derived by farmers in terms of soil-N due to BNF depends on tree species. For example, *Gliricidium sepium* had 24.6 cpf g^{-1} soils, *Accacia mangium* has 19 cfu g^{-1} but *Leucana leucocephala* had only 13 cfu g^{-1} in the root zone (Oyun, 2007). The spread of nodulating bacteria in soil, soil type, and its physicochemical traits may all affect the N benefits derived from tree species.

5.4.1   GM CROPS IN WEST AFRICAN SAVANNAH

The West African Savannahs currently support natural vegetation that has diverse plant species. Regarding agrarian zones, West Africa supports large number of land races, selections, cultivars, composites, and hybrids of various crops such as pearl millet, sorghum, maize, cowpea, groundnut, cassava, yam, and bananas. Several international agricultural agencies and crop improvement programs of each nation within West Africa have periodically released and established new genotypes of different crops. The crop genotype that dominates the landscape keeps changing with time. During past year itself, the West African Agrarian zone was introduced with several new genotypes of crops. For example, IFPRI-ASTI (2013) estimates that 101 genotypes of pearl millet, 151 genotypes of sorghum, 162 genotypes of maize, 190 genotypes of rice, 112 genotypes of cowpea, 96 genotypes of groundnuts, 188 genotypes of Cassava, and 51 genotypes of yams were released into West African cropping zones.

During recent years, there has been a trend to introduce "Genetically Modified" crops into different agrarian zones. West African cropping zones too have been exposed to new crop genotypes that are genetically modified. Crop genotypes with herbicide tolerance may not find place in the subsistence farming zones of Sahel, because farmers are still poor and investment on chemical herbicides are low. Seeds packed with genetic traits for tolerance to abiotic stress such as high temperature, drought, and biotic factors such as major diseases that afflict the crops are preferable. Large farms with mechanization, high input technology, and high yield goals may be able to utilize the currently available herbicide-tolerant maize or soybean. Reports from West African region indicate that a few nations have refrained from GM crops, yet others are still testing and buying time to ascertain the net effects of GM crops in their ecosystems. A few other nations within Savannahs feel that right now West African agricultural prairies could be made more productive by mending soil; improving its fertility; and adopting appropriate water management methods. The genetic potential of genotypes now grown by Sahelian or Sudanian farmers is sufficiently higher than what they harvest. Genetic diversity with regard to other traits such as diseases or pest tolerance too is available (UNEP,2011; Hinneh, 2012; Hirsh, 2013; Jinukun, 2006; Kuyek, 2004; Makinde, 2010). Perhaps, GM crop genotypes that suits their economy and offers excellent advantages may find its way into Sahel and savannahs of West Africa some time in future. Whether it is now or later, we ought to understand that GM crops introduce new genes into the ecosystem and it may have some influence on the naturally evolved crop diversity, because several crops are freely cross pollinated in natural settings.

5.4.2 CROPPING SYSTEMS

Farmers in West Africa have adopted a series of farming systems commensurate with natural resources, their own knowledge about crops, and their productivity. Earliest was pastoralism that involved movement of farmers and herds from one location to other in search of greener pastures. Sedentary and nomadic pastoralism was in vogue for a long period during prehistory. Shifting cultivation was practiced in forested regions, shrub lands, and even in scantily vegetated Sahelian locations. Shifting cultivation involved removal of native vegetation through cutting and/or burning. Farmers shifted after few years of cropping the same land to another location and left the previous field for fallow. Long fallows were common. Crops were also rotated with bush fallows. Such rotations are common in Senegal and Burkina Faso. Rotations involving crop-grass lands are common in Savannahs and dry Sahelian regions. Permanent cultivation using crop rotations involving cereal, legume, vegetables, and cash crops are feasible with farmers possessing permanent dwellings and land holdings. They practice regular cropping by allocating fields to different crops. In areas prone to floods and stagnating water, rice-based rotations are common. Mixed farming is perhaps most common in West African agrarian region. It involves cultivation of different species of crops simultaneously within the farm. Agroforestry tree species for forage and fuel, cash crops such as groundnut, tobacco, or cotton, plus perennial trees are cultivated simultaneously in a farm (FAO, 2004; See Ref. Aregheore, 2009).

West African farmers adopt several combinations of cropping systems. Many of them have been evolved through trial and error for, perhaps, several centuries. The cropping systems adopted obviously try to adapt best to the prevailing soil fertility conditions, water resources, weather pattern, and economic advantages. During past centuries environmental concerns received less consideration while deciding on crop rotations/mixtures. The use of chemical fertilizers, herbicides, and pesticides were feeble and continue to be the same. Cropping systems, therefore, were predominantly decided by the agroclimate, soil type, and vegetation/crop productivity possible naturally in the region. The cropping system in the dry Sahelian region with low precipitation levels of 400–500 mm year[-1], relatively high temperature and intermittent drought is primarily pearl millet-based. Pearl millet is hardy cereal that grows and produces grains in adverse soils and weather. The companion legume is often cowpea or groundnut. In the Sudanian region where in precipitation is relatively higher at 600–800 mm year[-1] and grasslands put forth good biomass, the cropping systems is predominantly sorghum-based. Sorghum is the staple cereal in Sudanian region. Sorghum mono-crops, sorghum/legume intercrops, and sorghum sequences that last 3–4 years are followed in the Sudanian region (Bationo et al. 2005).

Maize is an important cereal in Guinean region. It is grown as sole crop and in combination with legumes and vegetable. Sole crop of maize has generally performed less efficiently in terms of land use, total grain harvest, and C sequestration. Studies in the Nigerian Guinean region suggest that maize+cowpea, maize +groundnut or maize +soybean always performed better than sole crop of maize. Sole crops of maize depleted soil-C, N, and P but intercrops did not. It seems interplanted legumes had positive influence on soil nutrient and productivity of maize (Osundare, 2007).

Traditional West African methods involve crop-fallow systems. Farmers grow cereal/legume intercrops for 4–5 years and then fallow the land for 10–15 years to refurbish the depleted nutrients and organic matter. Fallow periods are generally long ranging from 10 to 15 years after which staple crops are cultivated. The length of the fallow is an important aspect that depends on farmer's household economic conditions, number of people in the household/village, immediate need for cereal grains, fallow conditions, and vegetation if any during fallows (Manlay et al. 2002a, b). During past three decades, length of the fallows has decreased progressively from 10 or 15 years to 3–4 years. It is attributed to higher population density, need for higher quantity of cereal grains, and famers intentions to achieve better land use efficiency and economic advantages. Fields cannot be under fallow without crops for long. During fallow period, farmers in the Sahelian belt of Senegal usually allow herbaceous species to grow, but semipermanent shrubs and short trees may grow rapidly and establish themselves if the fallow period is more than 2–3 years. The vegetation that thrives during fallow has its influence on soil physicochemical properties. According to Manlay et al. (2002 a, b) soil bulk density, porosity, aggregation, pH, organic-C, N, P, and K are some of the prime indicators of beneficial effects of vegetation during fallows. Fields kept under continuous cropping with either millet or groundnut accumulated low amounts of C, N, and P compared to short fallows or long fallows with tree vegetation. Constant removal of crop residues and stubbles out of the crop fields causes reduction in organic-C component in the fields. In addition, natural factors such as mineralization, emissions, erosion, and seepage may induce steady loss of soil fertility in fields with annual crops (see Table 5.6).

TABLE 5.6: Influence of crop-fallows and length of fallows on Organic Carbon, Nitrogenand Phosphorus in the surface layer of a sandy soil found in the Sahelian zone of Senegal

Cropping system	Organic- C t ha^{-1}	Nitrogen kg ha^{-1}	Phosphorus kg ha^{-1}
Groundnut/Millet fields	12.2	1.04	6.47
Young fallows (1–5 years)	15.8	1.31	8.47
Long fallows (15 years)	14.9	1.28	6.19

Source: Manlay et al. (2002a, b)

According to Manlay et al. (2002a), rapid increase of organic-C, N, and available-P in the fallows is attributable to woody vegetation that develops and sheds leaves, twigs, and root exudates into upper layers of soil profile. It definitely adds to soil fertility and quality. During fallows, the below-ground organic-C increased by 12.6 tha^{-1} annually in young fallows and about 3.6 t ha^{-1} in long fallows. Tropical conditions with high temperature induce rapid decomposition of even the large pieces of woody material that accumulates during fallows. It seems in the Sahelian conditions, Manlay et al. (2002a, b) observed decomposition of entire succulent and hardy material from fallow vegetation within 6 months. Rapid recycling causes refurbishment of organic-C and nutrients in the upper horizon of soil. Evaluation of fallows suggests that N and P replenishments are higher in the young fallows during first 1–2 years compared to 10 years and beyond. Soil nutrient recycling would have stabilized in long fallows and hence fluctuation in soil fertility is marginal. It is interesting and useful to note that burning the woody vegetation of fallows developed in Sahel or using it later as firewood or allowing it to decompose on surface may result in emission of 27 t C ha^{-1} annually. Nearly 12.3 t ha^{-1} of organic matter could be reincorporated *in situ* in long fallows.

PEARL MILLET-BASED CROPPING SYSTEMS IN WEST AFRICA

Pearl millet cropping zones are almost confined to dry Sahelian belt. Here, pearl millet outcompetes other cereals because of its excellent adaptation to scanty precipitation, short rainy season, and intermittent droughts. It is also a crop tolerant/resistant to disease and pest pressure. Sahelian farmers grow pearl millet during rainy season and fallow the fields during off-season until next year. Pearl millet-fallow system is the most common cropping system adopted in Sahel, especially in Niger, Mali, Burkina Faso, and Senegal. The productivity of such pearl millet-fallow systems is low because subsistence farming systems are practiced. Fertilizer-based nutrient supply is low or nil. Soil fertility replenishment is mainly through *in situ* crop residue recycling. Even this share of nutrient cycling is getting smaller because larger portion of crop residue is allocated to live stock. Productivity of pearl millet-fallow rotation is low at 0.8 to 1.2 t grain ha^{-1} and 3–5 t ha^{-1} (Bationo et al. 2003, 2004b). Incessant cropping has led to depletion of soil fertility in many regions within Sahel. Further, in areas with high population density, fallows are getting shorter because the fields are allocated to vegetable or legume production. Therefore, farmers are suggested to supply nutrients through chemical fertilizers and crop residues in certain proportions. Retaining crop residues in the field after pearl millet harvest and growing N-fixing legumes is the most common advice to farmers in Sahel. This way, agricultural prairies of Sahel could improve its productivity. The biomass and grain generation could be improved. In addition, soil-N status could be enhanced because of legume grown, instead of leaving the fallow without crop. Further, reports from Cinzana Research Station near Segou in Mali suggest that crop residues retained add P, K, and small amount of micronutrients. Soil quality is generally better, if crop residue from legumes such as cowpea or groundnut is recycled (Caulibaly et al. 2012). The average productivity of pearl millet-fallow or pearl millet/cowpea-fallow systems is 700–800 kg ha^{-1} pearl millet grains, 4–5 t fodder ha^{-1}; 250–300 kg cowpea grains ha^{-1} plus 2,800 to 1,000 kg ha^{-1} fodder if pearl millet is intercropped with cowpea. Over all, we should note that pearl millet-fallow or pearl millet/cowpea-fallow system is the most preferred cropping system. It still sustains the Sahelian population to great extent.

GROUNDNUT-FALLOW SYSTEM IN SENEGAL'S "GROUNDNUT BASIN"

Groundnut is a major cash crop in the Sahelian/Sudanian regions of Senegal. It is grown as sole crop or in combination with pearl millet during rainy season and the field is left fallow during off season. Such a practice continues for 3–5 years and then the fields are left for a long fallow. Long fallows are said to recuperate and refurbish lost soil fertility. The same fields are brought back to cultivation after 10–15 years. During groundnut fallow systems, major nutrients are recycled efficiently through reincorporation of residues. Soil-N is added through Biological N fixation (BNF). Manlay et al. (2002a, b) have attempted to understand the nutrient dynamics and productivity pattern of groundnut-fallow systems in the "Groundnut Basin" of Senegal. For convenience, they grouped the zone into (a) Natural savannah, which is least integrated with other aspects of farming; (b) "Bush Ring" containing groundnut fields and a mosaic of herbaceous plants on the fallows and

(c) "Compound Ring" devoted to production of cereals such as pearl millet and sorghum. Cereal fields are under cultivation continuously and supplied with fertilizers and organic manures. On a long-term basis, say 25 years, total biomass accumulated in the groundnut-fallow system was 49 tha^{-1}. It equates to 18 t C, 250 kg N and 21 kg P ha^{-1}. The carbon sequestration was rapid in young fallows. The below-ground biomass of groundnut fields, especially coarse and fine roots increased 5–6 folds from initial levels. In a location with groundnut crop, initial levels of 5.5 t C, 106 kg N, and 5.9 kg P increased to 17.7 t C, 231 kg N, and 19.6 kg P in 9 years (Manlay, 2002a, b). It seems that clearing vegetation from fallows is not a good practice, since it induces rapid loss of topsoil and nutrients. Soil erosion could account for 50 percent of soil-N loss in a fallow field without vegetation. In the Bush ring, presence of groundnut crop helps in replenishing soil-N to certain extent via BNF. Root biomass too adds to soil nutrients upon decomposition. However, removal of nutrients through pod harvests and forage could be significant. In a 10 year period, recycling groundnut crop residue caused significant improvements in soil-C, N, and P. In case of "Compound Ring," recycling pearl millet or sorghum crop residue amounts to accumulation of 63 kg N ha^{-1} and 5.2 kg P ha^{-1} each season. It almost counterbalances the nutrients removed out of the field through grain harvest. We may note that groundnut-fallows systems in the Bush ring and that of cereals in the Compound ring is now a well stabilized system with regard to grain/forage harvest, nutrient recycling, and water usage. Yet, it is not the best situation with regard to level of intensification and yield goals. A great deal of improvements with regard to crop genotypes, soil management, cropping systems, precipitation/irrigation use efficiency needs to be achieved.

Bush fallows are common in the entire West African agrarian regions. Bush fallows are practiced for different lengths of period. Length of the fallow is dependent on farmer's choice of fields and refurbishment of soil nutrients that bush fallow bestows. In Nigeria, for example, bush fallows practiced by 30 percent of farmers are continued for only 4 years. About 26 percent of farmers allow bush fallows only for one year. Whereas about 13.5 percent allow Bush fallows for 6–7 years, after which they initiate cultivation of maize or sorghum (Ogban et al. 2005). Bush fallows affect soil physicochemical properties in several different ways. Bush fallows affect soil bulk density, porosity, water stable aggregates, and hydraulic conductivity positively. Soil characteristics such as acidity, organic carbon, CEC, and mineral nutrients were all improved due to Bush fallows.

Mixed farming is a common occurrence within the agrarian regions of West Africa. Mixed farming is adopted more frequently in areas with low population density. Farmers adopt subsistence techniques involving low intensity cropping, low or no chemical fertilizers, and the crops rely on *in situ* residue recycling and FYM inputs. Mixed cropping or intercrops offer certain definite advantages to farmers. A few of them are;

(a) Mixed crops offer higher stable grain/forage yield. Its advantage is realized better in low fertility subsistence farming zones.
(b) It allows better and efficient use of land resource per unit time.
(c) Intercropping or mixed cropping involves wide range of crop combinations that are well adapted to West African agroclimate.
(d) Mixed farming minimizes risk of failure in the event of natural calamities such as formation of soil crust that restricts seedling emergence and establishment, drought, high-intensity rain that erodes soil and affects soil fertility etc.
(e) Mixed crops offer better advantages in case pests or diseases afflict the region. Total wipe out of productivity is avoided. Soil fertility inherent in a field and that supplied through fertilizers/manure are efficiently utilized by an intercrop. Let us discuss a few intercrops or mixed cropping systems common to West African agrarian region.

MAIZE-GROUNDNUT ROTATIONS IN SUDANIAN AND GUINEA REGIONS OF WEST AFRICA

Maize is a preferred cereal in the Guinean region and it offers better biomass/grain advantages compared to other cereals. In the Guinean region, tropical thick vegetation is interspersed with maize sole crop or as mixed crop with legumes/vegetable, plantains, or cassava. Maize is commonly rotated with groundnuts in the Guinean zone (Buerkert et al. 2001). Groundnut–Maize rotation is also common in the Sudanian regions of Senegal (Sene and Niane-Badiane, 2001b). Groundnuts are preferred more often in fields with low N. Maize that follows utilizes the residual-N efficiently. Groundnut could also effectively reduce the fertilize-N supply needed to achieve set yield goals of maize. Maize grain yield is generally higher, if the previous crop is groundnut or a groundnut/maize intercrop. For example, continuous maize produced 2,100 kg grain ha^{-1}; but a maize in rotation with groundnut gave 2,700 kg grain ha^{-1} (Frank et al. 2004; Naab and Alhaasan, 2001; Buerkert et al. 2001). Estimates of soil-N gain due to BNF through groundnut-*Bradyrhizobium* are as high as 60 kg N ha^{-1} (Dakora et al. 1987; Kaleem, 1989). Long-term effects of maize–groundnut rotations have also been investigated. It is said that periodic replenishment of soil-P through soluble P fertilizers or phosphate rock source is essential. Application of phosphor-gypsum is recommended since it adds to soil-P, Ca, and enhances the pH by 0.2–0.5 from the acidic range of 4.5. Maize-peanut rotations that substitute for fallow are advantageous with regard to land use efficiency and exchequer. It also curtails soil erosion/emissions that are otherwise rampant in the Guinean region. On an average, maize–groundnut rotation in the Guinean Nigeria offers 1,201 kg maize grain ha^{-1} and 1,325 kg groundnut pod ha^{-1} (See Ref. Krishna, 2008).

PEARL MILLET-GROUNDNUT INTERCROPS AND ROTATIONS IN THE SAHEL AND SUDANIAN REGIONS OF WEST AFRICA

Sahelian West Africa is a region with relatively low rainfall and dry sandy soils of low soil fertility regime. Here, pearl millet-based cropping systems are dominant. Pearl millet mono-crops are rotated with legumes such as groundnut or cowpea. Agricultural prairies are sown with pearl millet at the beginning of the rainy season and a relay intercrop of groundnut is seeded with a 15–30 days staggering. This supposedly helps in many ways. It improves precipitation use efficiency. Pearl millet draws soil moisture from first rains, while groundnut extracts water from precipitation events that occur later in the season. Soil nutrients are very efficiently scavenged first by the cereal then by groundnut. Cereal puts forth foliage rapidly flowers and forms mature grains early by end of September. Groundnut continues for a month and utilizes residual soil moisture and nutrients for its growth/maturity. Pearl millet–groundnut rotations are best suited to low input subsistence farming in Sahel (Bationo et al. 2002a, b, c). Inorganic fertilizer supply is low in Sahel, but in Sudanian region, about 40–80 kg N is supplied to millet to achieve the set yield goals. Traditional low input systems fetch 350 kg groundnut pod ha^{-1} and 850 kg pearl millet grain ha^{-1}. Pearl millet–groundnut rotations are also practiced with moderately higher inputs of fertilizer and irrigation, depending on yield goals. Soil test values are considered prior to supply of fertilizers (Diagana et al. 2001). The yield decline experienced under continuous pearl millet–fallow system could be effectively avoided if groundnut is either intercropped or rotated in the same field (Bagyoko et al 2000a, b). Long-term studies indicate that groundnut/pearl millet intercrops improved soil-physicochemical properties. It increased soil pH from 4.2 to 4.5, organic matter increased from 2.8 to 4.1 percent, total N increased from 168 mg kg^{-1} to 193 mg kg^{-1}, CEC increased from 1.2 to 1.4 cmol kg^{-1} soil, base saturation also improved, and land equivalent ratio improved from 1.29 to 1.48 (Bationo et al. 2004b). The legume effect on succeeding sorghum or pearl millet is often useful. The cereal grain yield increases 8–10 percent over that derived under continuous cereal systems. Multilocation trials in the Sahel suggest that annual increments in soil-N was 9 to 39 kg N ha^{-1} due to atmospheric N fixation during legume phase of the rotation.

SORGHUM-BASED CROPPING SYSTEMS

Sorghum-based cropping sequences are most common in the Sudanian region. Farmers prefer to cultivate sorghum in the rainy season. Farmers in Sudanian region grow several different genotypes of sorghum based on local preferences, its adaptability, soil fertility, and yield goals possible. Sorghum-fallow is common in regions with poor soil fertility and paucity of water resources. Sorghum is rotated with legumes, groundnut, vegetables, and cash crops. Sorghum intercrops with cowpea is perhaps dominant in the Sudanian region of Nigeria, Ghana, and Togo. Sorghum grain productivity depends on inherent soil fertility and fertilizer addition. Sorghum grain yield is often 1–1.5 t ha^{-1} and 5–6 t forage. Cowpea produces 300–350 kg grain ha^{-1} plus 1 t fodder ha^{-1}.

The Guinean region of West Africa is a relatively high rainfall region. It supports several different cropping systems that offer better grain/forage productivity. Crops grown too are diverse. Maize and sorghum are predominant cereals. They are intercropped with cowpea, French beans, vegetables, yam, etc. Several combinations of intercrops and rotations are in vogue in the Guinean regions. For example, Yam-based cropping systems are in vogue in Wet savannah and Guinean regions of southern Nigeria, Ghana, Togo, Sierra Leone, Liberia, and Gambia. The tropical climate and relatively higher precipitation received allows cultivation of yams and intercrops grown with it. Soybeans could be relay cropped with yam/maize intercrop. This way, it intensifies crop production and enhances land use efficiency. Soybean productivity is said to increase from 0.45 t ha^{-1} to 0.6 t ha^{-1} without affecting the yam/maize yield (Ano, 2005). Soybean variety utilized seems to influence grain and haulm production. Haulm production increased by 12 t ha^{-1} based on cultivar used. We should appreciate that recycling of soybean haulm adds to soil-N and C, in addition to enhancing general soil quality. Haulms also recycle minerals such as K, S, Ca, and Mg to a certain extent.

Faideherbia is an important agroforestry species common to many regions of African continent. Its spread is accentuated in West African zones. It is traced interspersed in the Guinean-Savannah and lower fringes of Sahelian belt. Its distribution is restricted to sandy soils receiving 500–700 mm annually. It is also common to Parkland regions of Senegal and adjoining regions that receive very low precipitation of 300–450 mm. *Faidherbia* withstands low relative humidity and dry conditions for longer stretch. It periodically sheds leaves, especially at the beginning of rainy season but continues to produce green leaves even in dry periods. Hence, it is valuable as forage suppliers in the drought afflicted regions of West Africa (Wood, 1992).

5.4.3 CROP-LIVESTOCK INTEGRATION IN WEST AFRICAN SAVANNAHS

The agricultural prairies of West Africa seem to have undergone a considerable intensification in production. This is in addition to whatever expansion that has occurred with regard to food grain crops. It has been attributed to growth of human population and livestock. Major constraints to intensification of fodder production in West Africa are poor soil fertility, pests, diseases, parasitic weeds, drought, and competition between grain crops and forage crops. Soil deterioration and depletion of fertility has restricted farming zones nearer to urban human dwellings. Hence, it is worthwhile to integrate livestock and crop production. Integration of crops and livestock is most desirable nearer to urban dwellings and livestock production centers (Kamara et al. 2005; Tarawali, 2004; Tiffen, 2004). In fact, Kamara et al.(2005) infer that intensification of fodder production is attributable to human population pressure and enhanced preference for livestock farming. Forecasts indicate that during 2005 to 2020, livestock number may increase rapidly by 200–250 percent (Delgado et al. 2001). Therefore, it is a major challenge to grow fodder for enhanced livestock intensity. The crop species whose production has been intensified are sorghum, cowpea, and groundnut. In the Guinea region, maize and soybean have been preferred for intensification. These crops have been integrated with livestock. Crop residues provide the forage. Animal manure, in turn, is used effectively to improve soil fertility and quality. Traditional fallows have become shorter to accommodate short season forage crop.

In the Guinean savannahs, cassava has become an important feed for livestock. During recent years, its production has increased from 3 to 10 percent in different parts of West Africa. Production of cassava received a boost due to ban on import of maize forage from other regions. Currently, it is used as animal feed in Nigeria, Ghana, and Cote D'Ivore. In Ghana, use of cassava-based feed increased by 2 percent and that in Cote D'Ivore increased by 5 percent, and in Liberia by 1percent (FAO, 2002, 2005).

5.5   AGRICULTURAL PRODUCTION TRENDS IN WEST AFRICAN AGRICULTURAL PRAIRIES

West African savannahs shelter a large population of farmers in villages who are scattered and endowed with insufficient natural resources and economic capital. Hence, agricultural productivity is relatively very low at <1.0 ton cereal grains ha^{-1}. The cereal production zones in the Sahel stagnated during first half of twentieth century. It was about 8 m ha during 1960 to 1972. However, cereal cropping zones increased steadily each year by 0.25 m ha $year^{-1}$ since 1970. Currently, cereal production zones in Sahel occupy 18 m ha^{-1} (Appinsys, 2010). Based on FAO reports, Bationo et al. (2005) have remarked that agricultural infrastructure and grain harvests increased steadily in Sahel between 1961 and 1996, but the per capita food availability has decreased since 1996. It has been attributed to rapid increase in population density that has actually doubled during past 3 decades. Many researchers believe that present soil and crop management systems are still insufficient to enhance yield goals that meet the requirements of enlarged population. Further, it should be noted that grain harvests in West Africa increased during 1980s till 2005, mainly due to expansion of the crop area and much less due to enhanced productivity. In some of the nations, cereal yield in the Sahelian belt actually decreased marginally from 1.1 t ha^{-1} to 0.89 t ha^{-1}. Farmers try to overcome cereal grain yield depressions by adopting mixed cropping that includes legumes such as groundnut or cowpea. The productivity of legumes has increased by 50–300 kg ha^{-1} in West Africa during 1970s till 2005. Groundnut productivity in Senegal and other regions of Sahel/Sudan has crossed 850–1,000 kg ha^{-1} mark during past three decades.

5.5.1 MAJOR CROPS OF WEST AFRICA

Pearl millet is indeed the most important cereal food consumed by the West Africans, specifically those residing in Sahel and Sudanian regions. During 2007, pearl millet cropping extended into 16.02 million ha in the West African Sahel and parts of Sudanian region. The productivity of pearl millet has increased perceptibly since 1950s from 0.5 t ha^{-1} to 1.2–1.5 t grain ha^{-1} depending on soils and rainfall pattern. Forecasts by IFPRI, Washington D.C. suggests that pearl millet yield may increase from 2.2 to 2.3 t ha^{-1} by 2050 (Jalloh et al. 2013a, b).

Sorghum is a dominant crop in Sudanian belt. During 2007, it occupied 14.2 million ha in West Africa. The productivity is again low or moderate depending on soil fertility and inputs. The average productivity of sorghum in West Africa ranged from 0.63 t ha^{-1} to 1.64 t ha^{-1} (Gambia, Sierra Leone). Forecasts indicate that productivity of sorghum in West Africa may reach from 1.95 to 3.25 t grain ha^{-1} depending on the region (Jalloh et al. 2013a,b).

Average productivity of maize in the Sahel, not entire West Africa, remained at 0.6 to 0.7 t grain ha^{-1} from 1960 till 1972, but registered steady increase in grain productivity during the next 40 years. The maize grain productivity has increased from 0.7 t ha^{-1} in 1972 to 1.2 t ha^{-1} in 2010. According to reports by IFPRI at Washington D.C., maize cropping extended into 7.7 m ha during 2007. The productivity of maize, which was 0.78 t ha^{-1}–1.98 t ha^{-1} during 2010 is expected to increase gradually, depending on the location from 1.9 to 3.10 t grain ha^{-1} by 2050 (Jalloh et al. 2013a,b). Kamara (2013) has pointed out that maize cultivation has spread rapidly in the savannahs, but average grain yield is still low at 1–2 t ha^{-1}. It leaves a large yield gap of 3–5 t grain ha^{-1} compared to productivity achieved in experimental farms, which is 5–7 t grain ha^{-1}.

West African rice-based wet prairies extend into several countries in the region. The wetlands are concentrated in Liberia, Sierra Leone, Gambia, Senegal, Burkina Faso, Nigeria, Ghana, and other nations. It is mostly confined to locations with riverine irrigation, lakes, or groundwater resources. West African rice is mainly composed of *Oryza glaberrima*. During 2007, rice production zones in West Africa occupied 5.7 million ha. Nigeria, Sierra Leone, Guinea, and Mali possess relatively larger rice cropping areas. However, during recent years, rice from Southeast Asian nations (*Oryza sativa*) has made a small dent in the zone. Forecast for rice production in some of the West African nations are as follows. Most of them have registered an increase in rice production compared to previous year:

Country	Senegal	Burkina Faso	Mali	Cote de Ivore	Liberia	Sierra Leonne
Rice Production t year^{-1}	443,000	180,000	1300,000	456,000	140,000	128,000
Percentage increase	8	15	13	10	7	13

Source: USDA GAIN (2013)
Note: Percentage increase refers to increase in 2012 compared to 2011.

Rice production in Burkina Faso occurs in West and Central regions, where it is not affected much by the erratic precipitation pattern, because rice culture occurs under irrigation. The rice production zones are expected to increase further because of fresh support from governmental agencies. The rainfed belt may increase from the 1,500 ha to 7,500 ha during next couple of years (USDA GAIN, 2013). In addition to enhanced fertilizer supply (150 kg NPK ha^{-1}), infrastructure to process rice harvests have been developed. Cote de Ivoire has also revised rice production goals both in terms of area and productivity. It aims at reaching 2.0 million t year^{-1} although difficult, since it involves a fourfold increase in rice production. Rice production in Mali is deficient compared to targeted 2.7 million t year^{-1}. The Senegalese rice belt too is expected to increase in area with better irrigation facilities. Senegal aims to harvest over 1.0 million t rice grain year^{-1} during 2015 and later (USDA GAIN, 2013). At the bottom line, it is clear that West African savannah region is expected to support higher productivity and area with regard to wetlands. This involves augmenting, irrigation, commensurate fertilizer supply, and designing appropriate cropping systems. The per capita rice grain consumption by West African population has increased. In some countries of this region, it has replaced pearl millet in the diet to a certain extent. Agronomic procedures that are more efficient and enhance total grain yield have been experimented in West African wet prairies. Sawah system of rice production upon introduction in Senegal and Nigeria may enhance productivity of rice by 1.5 to 2 tha^{-1}. It is believed that expansion of rice cultivation zones and adoption of improved techniques will lead to higher grain production in West Africa. Estimates suggest that Nigeria alone may harvest about 10 million tons of rice by 2020.

Cowpea is perhaps the most important vegetable protein source for West Africans. It is a predominant legume in all agroclimatic regions of West Africa. It is mostly intercropped with millets in the Sahelian zone. During 2007, cowpea crop extended into 10.29 million ha in West Africa. Nigeria (4.3 million ha), Niger (4.7 million ha), Burkina Faso (728,000 ha), and Mali (245,000 ha) are major cowpea producing nations (FAO, 2010).

Groundnuts were introduced into West African region by the European settlers and travelers during medieval period. Spaniards who traveled to and fro between South American Coasts and West Africa are responsible for its introduction into savannahs. Today, its cultivation zone spreads into most regions within Senegal. It is a major cash crop for Senegal. It is grown extensively in the "Groundnut basin". Other regions of West Africa such as Burkina Faso, Niger, Northern Nigeria, and Ghana also support groundnut production. Groundnuts occupied 5.2 million ha of the West African Sudanian zones. Groundnut cropping zones are conspicuous in Senegal, Nigeria, Ghana, Mali, and Burkina Faso. Productivity is low at 0.8 to 1.2 t pods ha^{-1}. It is expected to increase to 2.4 t ha^{-1} based on soil fertility, fertilizer amendments, and supplemental irrigation.

Soybean is also grown in West Africa, but it is confined to Guinean zones of southern Nigeria. Soybean extends into 661,000 ha in Nigeria and 19,000 ha in Benin. The productivity is 2.2 t grain ha^{-1} (Jalloh et al. 2013a, b).

Cassava is an important crop in wet savannahs and Guinean regions of Nigeria (3.8 million ha), Ghana (797,000 ha), and Cote d'Ivore (339,000 ha). Cassava grown under traditional systems using local genotypes provides 6–7 t tubers ha^{-1}. However, improved cultivars released by International Institute for Topical Agriculture grow rapidly and accumulate starch at higher rates. Improved varieties yield 30–40 t roots ha^{-1}. Yams are cultivated on 4.3 million ha. It is predominant in Nigeria (3.0 million ha). Cote d'Ivore and Ghana are other nations that have cassava production zones. Cocoa cropping zones occur in Nigeria (1.1 million ha), Ghana (1.67 million ha), and Cote d'Ivore (2.15 million ha). Cotton, Coffee beans, and plantains are other major crops that occur in the Wet savannahs and Guinean regions of West Africa.

KEYWORDS

- **Anthropogenic activities**
- **Drought-tolerant**
- **Genotype**
- **Pastoral method**
- **Savannah**
- **Soil types**

REFERENCES

1   Abe, S.S.; Buri, M.M.; Issak, R.N.; Kiepe, P.; and Wakatsu, T; Soil fertility potential for rice production in West African Low lands. *JARQ.* **2010,** *44,* 343–355.

2. Adeboye, M.; Osunde, A.; Ezenwa, M; Odofin, A.; and Bala,A;Evaluation of the status and suitability of some soils for arable cropping in the Southern Guinea Savannah of Nigeria. *Nigerian J. Soil. Sci.* **2009,** *19,* 144–153.

3. Africa Rice Centre, Rice research for development in Africa. Coutonou, Benin, **2013,** 1–96, August 22, http://www.africarice.org/

4. Agbede, T.; Ojeniyi, S.; and Adekayode, F; Effect of Tillage on soil properties and Yield of Sorghum (*Sorghum bicolor*) in Southwest Nigeria. **2009,** *19,* 85–92.

5. Akponipke, P. B. I.; Millet response to water and soil fertility management in the Sahelian Niger: Experiments and Modeling. University of Catholique de Louvain, Doctoral thesis, **2008**, 1–187.

6. Anhuf, D.; Paleo-vegetation in West Africa from 18000 B.P. and 8500 B.P. Eiszeitalter u gagenwart 47, **1997**, 112–119, http://quaternary-science.publiss.net/system/articles/pdfas/683/original_vol47_no1_a07.pdf?1284108093

7. Anikwe, M. A.; Ngwu, O. E.; Mbah, C. N.; Onoh, C. E; and Ude, E. E.; Effect of groundnut cover by different crops on soil loss and Physico-chemical properties of an Ultisol in Southeastern Nigeria. *Nigerian J. Soil. Sci.*, **2007**, *17*, 94–97.

8. Ano, A. O.; Effect of soybean relayed into yam miniset/maize intercrop on the yields of component crops and soil fertility of yam-based system. *Nigerian J. Soil. Sci.* **2005**, *15*, 20–25.

9. Appinsys; Sahel, Africa. Global Warming Science. **2010**, 1–13,August 5, 2013, http:/www.appinsys.com/global warming/RS_Sahel.htm.

10. Aquastat; Computation of Longetern annual renewal water resources by country: Senegal. Food and Agricultural Organization of the United Nations. **2013**, 1–3, December 22, 2013, http://www.fao.org/nr/aquastat/

11. Archnet; Dictionary of Islamic Architecture. Digital library, **2013**, 1–8, August 3, 2013. http://www.archnet.org/library/dictionary/ entry.jsp?entry_id=DIA1010&mode=full.htm

12. Aregheore, E. M.; Nigeria: Country pasture/Forage resource prfiles. Food and Agricultural organization of he united nations, **2009**, Rome, Italy, 1–42.

13. Asiamah, R. D.; and Dwono, O; Ethno-management of Plinthic and Iron pan soils in the savannah regions of West Africa. *Ghana. J. Agric. Sci.* **2013**, 23–29. http://www.ajol.info/index.php/gjas/article/veiw/60641

14. Atokple, I. D. K.; Sorghum and Millet Breeding in West Africa in Practice. CSIR-Savannah Agricultural Research Institute, **2012**, Tamale, Ghana. Internal Report, 1–12.

15. Azeez, J. O.; Adetunji, M. T.; and Adebusuyi, M.; Effect of residue burning and fertilizer application on soil nutrient dynamics and dry grain yield of maize (*Zea mays*) in an Alfisol. *Nigerian. J. Soil. Sci.* **2007**, 17, 71–80.

16. Bado, B. V.; Aw, A.; and Ndiaye, M.; Long term effect of Continuous cropping of irrigated rice on soil and field trends in the Sahel of West Africa. Second Africa Rice Congress, **2010**, Bamako, Mali, 2.10.1–2.10.9.

17. Bagayoko; M.; Alvey, S.; Buerkert, A; and Neumann, G; Root–induced increases in pH and nutrient availability to field grown cereals and legumes on acid sandy soils of West Africa. *Plant. Soil.* **2000a**, *225*, 117–127.

18. Bagayoko, M.; Buerkert, A.; Lung, G.; Bationo, A.; and Romheld, V; Cereal/legume rotation effects on cereal growth in Sudana-Sahelian West Africa: Soil mineral nitrogen, Mycorrhiza and Nematodes. *Plant. Soil.* **2000b**, *218*, 103–116.

19. Balasubramanian, V.; Increasing rice production in sub-Saharan Africa: Challenges and Opportunities. *Adv. Agron.* **2007**, *94*, 55–133.

20. Ballouche, A; Fire and burning in West Africa Holocene Savannah Paleo-environment: Anthropogenic and Natural processes in environmental changes. In: Proceedings of a conference on Environmental catastrophe in Mauritania, the desert and Coast, **2004**. August 3, 2013, 1–4. http://www.at.yorku.ca/c/a/m/u/05.htm

21. Bankole, E. A.; and Ojeniyi, S. O.; Comparative effect of ridge-furrow and zero tillage on cowpea at Abeokuta and Ahure, Nigeria. *Nigerian J. Soil. Sci.* **2005**, *15*, 54–59.

22. Banziger, M.; and Edmeades, G. O.; and Lafitte, H. R; Selection for drought tolerance increases maize yield across a range of nitrogen levels. *Crop Sci.* **1999**, *39*, 1035–1140.

23. Barry, A.; Ndiaye, S.; Ndiyae, F; and Tiffen, M; Region de Diuorbel: les aspects demographiques. Dry lands research working paper 13 Dry lands Research Centre, Crewkene, United Kingdom, **2000**, p. 49

24. Bationo, A.; and Buerkert, A.; Soil organic carbon management for sustainable land use in Sudano-Sahelian West Africa. *Nut. Cycl. Agroecosyst.* **2001**, *61*, 131–142.

25. Bationo, A.; Christensen, B. C.; and Klaij, M. C; The effect of crop residue and fertilizer use on Pearl Millet yields in Niger. *Fertil. Res.* **1993**, 34, 251–258.

26. Bationo, A.; Christensen, C. B.; and Mokwunye, A. U.; Soil fertility management of the millet-producing sandy soils of Sahelian West Africa: The Niger experience. Paper presented at the workshop on Soil and Crop management systems for rain fed agriculture in the Sudano-Sahelian zone. International Crops Institute for the Semi-arid Tropics, Niamey, Niger, **1987**, 234–239.

27. Bationo, A.; Kihara, J.; Vanlauwe, B.; Waswa, B.; and Kimetu, J.; Soil organic carbon dynamics, functions and management in West African agro-ecosystems. *Agric. Syst.* 2007. **2006,** *94*, 13–25

28. Bationo, A.; Kihara, J.; Waswa, B.; Ouatara, B.; and Vanlauwe B; Technologies for sustainable management of sandy Sahelian soils. Food and Agricultural Organization of the United Nations, **2005**, 1–19.www.fao.org/docrep/010/ag125e /AG125E42.htm

29. Bationo, A.; Koala, S.; and Bado, V; Sources and Management of N fertilizer and the effect of different cropping systems on land productivity, **2002a**, 1–3, August, 22, 2007. http//www.icrisat.org.CCER.htm

30. Bationo, A.; Koala, and Bado, V; Organic amendment of Soil in the Sudano-Sahelian zone, **2002b**, 1–4, August 22, 2007. http://www.icrisat.org CCER.htm

31. Bationo, A.; Mokwunye, W; Role of Manures and Crop residue in alleviating soil fertility constraints to crop production: With special reference to the Sahelian and Sudanian zones of West Africa. *Fertl. Res.* **1991**, *29*, 117–125.

32. Bationo, A.; Ndjeunga, J.; and Koala, V; On-farm evaluation of different soil fertility management options in different agro-ecological zones of Niger, 1–4,August 22, 2007. http://www.icrisat.orgCCER.htm

33. Bationo, A.; Mokwunye, U.; Vleck, P. L. G.; Koala, S.; Shapiro, B. I. and; Yamaoh, C; Soil fertility management for sustainable land use in West African Sudano-Sahelian zone. In: Soil Fertility Management in Africa-A regional perspective. Eds. Gichuru, M. P., Bationo, A.; Bekunda, M. A., Goma, P. C., Mafangoya, P. L., Mugendi, D. N., Murwira, H. M., Nandwa, S. M., Nyathi, P.; and Swift, M. J.; Tropical Soil Biology and Fertility/Centro Internacional Agricultura Tropicale, **2003**, 438.

34. Bationo, A.; Nandwa, S. M.; Kimetu, J. M.; Kinyagi, J. M.; Bado, B. V.; Lompo, F.; Kimani, S.; Kihana, F.; and Koala, S.; Sustainable intensification of crop-livestock systems through manure management in Western Africa. In: Sustainable Crop-livestock Production for improved livelihoods and natural resource management in West Africa. Eds. Williams, T. O.; Tarawali, S. A.; Kieneaux, P.; and Fernandez-Rivera. Proceedings of an International Conference. IITA, Ibadan, Nigeria, **2004a**, 19–22.

35. Bationo, A.; Traore, Z.; Kimetu, J.; Bagayoko, M.; Kihara, J.; Bado, V.; Lompo, M.; Tabo, R.; and Koala, S.; Cropping systems in the Sudanain-sahelian zone: Implications on Soil Fertility Management. Syngenta Workshop, Bamako, Mali, **2004b**, p. 209.

36. Bertlessen, M.; and Traoure, G; The NRM InterCRSP for West Africa: Context, Challenges and Approach, **2002**, 13–15, http:// www.oired.vt.ed/projects/current/projectsynthesis.pdf

37. Bielders, C. L.; Alvey, S.; and Cronyn, N; Wind Erosion: The Perspective of Grass Roots communities in the Sahel, pp. 5, ICRISAT@org CCER.htm

38. Bielders, C.; Michels, K.; and Rajor, J; On-farm evaluation of wind erosion control technologies. **2002b,** ICRISAT@org CCER.htm p. 5

39. Bielders, C.; Rajot, J.; Amadou, M.; and Skidmore, E.;On-farm quantification of Wind Erosion under TraditionalManagement Practices, p. 6, **2002c**. ICRISAT@org CCER.htm

40. Bird, M. I.; and Cali, J. A.; A million record of fire in Sub-Saharan Africa. *Nature.* **1998**, *394*, 767–776.

41. Bouma, J; Precision Agriculture: Introduction to the spatial and temporal variability of environmental quality. In: CIBA Foundation Symposium. New York: John Wiley and Sons; **1997**, 5–13.

42. Breuning, P.; and Neumann, K; From Hunters and Gatherers to food producers: new Archaeological and Archaeobotanical evidences from the West African Sahel. In: Ecological change and food security in Africa's later pre-history. Ed. Hassan, F; New York: Kluwer Academic Press Inc., USA. **2002**, 123–155.

43. Buerkert, A.; Bagayoko, M.; Bationo, A and Romheld, V; Soil fertility and Crop production in Semi-arid and Sub humid West Africa. In: Farming in West Africa: Issues, Potentials and Perspectives. Eds. Graef, F.; Lawrence, P and Von Oppen M.; Stuttgart, Germany: Verlag Ulrich; **2000**, 27–57

44. Buerkert, A.; Bagayoko, M.; and Romheld, V; Efficient phosphorus application strategies for increased crop production in sub-Saharan Africa. *Field. Crop. Res.* **2001**, *72*, 1–15

45. Buerkert, A.; Piepho, H. P.; and Bationo, A; Multi-site time trend analysis of soil fertility management effects on crop production in Sub-Saharan West Africa. *Exp. Agric.* **2002**, *38*, 163–183

46. CARE; Evaluation de la securite des conditions de vie dans le Department de Maradi. CARE-international au Niger, Niamey, Niger; **1997**, p. 24.

47. Caulibaly, A.; Bagayoko, M, Traore, S.; and Mason, S. C.; Effect of crop residue management and cropping system in Pearl Millet and cowpea, **2012**, p 13. http://www.ajol.info/indexphp/acsj/article/veiw/27681.

48. Charreau, C.; Soils of tropical dry and dry-wet climatic areas of West Africa and their use and Management. A series of lectures. Agronomy Mimeo 74–127. Department of Agronomy, Cornell University, Ithaca, New York, **1974**, 1–63.

49. Cooke, K; Warming climate could West Africa river levels. Climate New Network. **2013**, p 1–3. http://www.climatenewsnetwork. net/2013/07/warmer-climate-will-hit-volta-river-levels/

50. Cour, J. M.; The Sahel in West Africa.: Countries in transition to a full market economy. *Glob. Environ. Change.* **2001**, *11*, 31–47.

51. Coura Badiane; Senegal's Trade in Groundnut: Economic, Social and Environmental Implications.TED case studies number 646. **2001**, 1–32, August 3, 2013.http://www.Senegalcasestudy.htm

52. Crop Biotech Update; New production system to boost rice yield in West Africa. International service for the Acquisition of Agri-Biotech applications, **2008**, pp 1–2, August 5, 2013. http://iita.org/cms/details/new_details.aspxarticleid=1615&zoneid=81

53. Dakora, F. D.; Aboyinga, R. A.; and Mahama, Y.; Assessment of N fixation in groundnut (*Arachis hypogaea*) and cowpea and their relative N contribution to a succeeding maize crop in Northern Ghana. *MIRCEN. J.* **1987**, *3*, 389–399.

54. D'Andrea, A. C.; Kahleber, S.; Logan, A. L.; and Watson, D. J.; Early domestication of cowpea (*Vigna unguiculata*) from Central Ghana. *Antiquity.* **2007**, *81*, 686–698.

55. Delgado, C. L.; Rosegrant, M. V.; and Meijer, S; Livestock to 2020: The revolution continues. Paper presented at the Annual meeting of the International Agricultural Trade Research Consortium (IATRC). Auckland, New Zealand, **2001**, p. 39.

56. Descroix, L.; Mahe, G.; Lebel, T.; Favreau, G.; Gautier, E.; Olivry, J. C.; Albergel, J.; Amogu, O.; Cappelaere, B.; Dessouassi, R.; Diedhiou, A.; Le Beton, E.; Mamadou, I.; and Sighmnou, D.; Spatio-temporal variability of hydrology regimes around the boundaries between Sahelian and Sudanian areas of West Africa: Asynthesis. *J. Hydrol.* **2009**, *375*, 90–102.

57. Diagana, B. N.; Kelly, V. A.; and Crawford, E. W.; Dynamic analysis of Soil fertility improvement: A bio-economic model- Policy synthesis. USAID Office of Sustainable Development. No 58; **2001**, 1–8. http://www.aec.msu.edu/ageocon/fs2/ psynindx.htm. paper

58. Diarra, S. T.; West Africa: Niger river under pressure from dams, **2011**, p 1–3, http://www.globalissues.org

59. Diouf, A.; Barbier, N.; Lykke, A. M., Couteron, P.; Deblauwe, V.; Mahamane, A.; Saadou, M.; and Bogaert, J.; Relationships between fire history, edaphic factors and woody vegetation structure and composition in a semi-arid savannah landscape (Niger, West Africa). *Appl. Vegetat. Sci.* **2012**, *15*, 488–500.

60. Donnelly, F. G.; A History of West Africa: An Introductory Survey. Aldershot, United Kingdom, **1993**, p. 442.

61. Ducreax, A.; Characterisation mecanique des dols sableaux et sablo-agrileux zone Tropicale seche de l'Africe de L'Ouest. Academie de Montpellier, Univerisite des Sciences et Techniques du Landuedoc. Montpellier, France, **1984**, p. 38.

62. Fall, A.; and Faye, A; Minimum tillage for soil and water management with animal traction in the West African region. In: Conservation tillage and animal traction. Eds. Kaumbutho, P. G.; and Simelenga, T. E.; ATNESA, Harare, Zimbabwe, **1999**, 140–147.

63. FAO; The Environmental setting. Food and Agricultural Organization of the United Nations. Rome, Italy, **2001**, 1–10, August 3, 2013, http://www.fao.org/docrep/004/x6543e/x6543e01.htm

64. FAO; Strategic interventions for cassava in livestock production in Western, Eastern, Central and Southern Africa. **2002**, 1–8, http;//www.fao.org/docrep/007/j1255e/j1255e09.htm.

65. FAO, The progression from Arable cropping to Integrated crop and livestock production. Food and Agricultural Organization of the United Nations, Rome, Italy, **2004**, 1–21, http://www.fao.org/docrep/004/X6543E/X6543E03.htm

66. FAO; Use of Cassava in Livestock feeding in West, Eastern, Central and Southern Africa. Food Agricultural Organization of the United Nations. Rome, Italy, **2005**, 1–5.

67. FAO; Cowpea Statistics. Food and Agricultural organization of the United nations. Rome, Italy, **2010**, December 28, 2013, FAOSTAT.org

68. Fatondji, D; Bationo, A.; Tabo, R.; and Jones, J. W.; Water use and yield of millet under the Zai system: Understanding the processes using simulation. In: Improving Soil Fertility Recommendations in Africa using the Decision Support system for Agrotechnology Transfer (DSSAT). Netherlands: Springer, **2012**, 77–100.

69. Faye, A.; and Fall, A.; Makeuni district profile: Livestock management. 1990–1998. Dry lands Research Working paper no 22 Dry Land Research Centre, Crewkerne, United Kingdom, **2000**, p 33.

70. Ferguson, A. W.; Integrating Crops and Livestock in West Africa. Food and Agricultural Organization of the United Nations, Rome, Italy **1985**, p. 328.

71. Fontes, J.; and Guinko, S; Carte de la vegettaon et de l,occupation du sol du Burkina Faso. Institut de la Carte International de la Vegetation, university Toulouse, France; Institut du Development Rural, Universite Ouagadougou, Burkina Faso, **1995**, p. 66.

72. Frank, A. C.; Schulz, S., Oyewole, D.; and Bako, S.; Incorporating short-season legumes and green manure crops into maize-based systems in the moist Guinea Savannah of Wes Africa. *Exp. Agric.* **2004**, *40*: 463–479.

73. Goethe University; Paleoecology in the Dogon country (Mali), **2009a**August 4th, 2009. http://www.araf.studiumdigitale.uni-frankfurt.de/ index.php/en/paleoecology-in-the-dogon-country-mali.htm p. 1–3.

74. Goethe University; Holocene vegetation history of Northeast Nigeria-Sahel: Manga Grasslands, **2009b**, August 4th, 2013. pp. 1–3. http://www.araf. studiumdigitale.uni-frankfurt.de/index.php/en/research/results/88.

75. Goethe University; Late Holocene settlement and vegetation history of the Sahel in Burkina Faso, **2009c**. httP;//www.araf.studiumdigitale.uni-frankfurt.de/index.php/en/research/results/105

76. Goethe University; Dry forests in SW Burkina Faso, **2009d**, 1–2, August 3rd, 2013. http://www.araf.studiumdigitale.uni-frankfurt.de/index/ php/en/research/results/107.

77. Goethe University; Holocene settlement and vegetation history of the Chaine de Gobnangou, SE Burkina Faso, **2009e**, 1–2, August, 2013. http://www.araf.studiumdigitale.uni-frankfurt. de/index.php/en/-chaine-de-gonangou-eng.htm

78. Grote, R., Lehman, E., Brummer, C., Nicolas, B., Szarzynski, J.; and Kuntsmann, H.; Modelling and observation of biosphere-atmosphere in natural savannah in Burkina Faso, West Africa. *Phys. Chemist. Earth. 34.* **2009**, 251–260.

79. Harlan, J. T; and DeWet, J. M. J; A simplified classification of cultivated Sorghum. *Crop. Sci.* **1972**, *12*, 172–176.

80. Haussman, B. I. G.; Rattunde, F. H., Weltzein-Rattunde, E., Traore, P. S. C.; Vom Brocke, K.; and Parzies, H. K.; Breeding strategies for Adaptation of Pearl Millet and Sorghum to climate variability and change in West Africa. *J. Agron. Crop. Sci. 198*, **2012**, 327–339.

81. Henry, F; Millions face starvation in West Africa, **2010**, pp 1, August 10, 2013. http://www. gaurdian.co.uk/world/2010jun/21/millions-face-starvation-westafrica.htm

82. Hinneh, S. West African scientists reach consensus on GM cassava. Modern Ghana News. 2012; pp 1–3, August 21st, 2013, http://www. modernghana. com/print/24086/1/west-african-scientists-reach-consenses-on-gm-cass.html.

83. Hirsh, A; G. M.; Crops: Campaigners in Ghana accuse US of pushing modified food. The Guardian, **2013**, 1–4, August 21st, 2013http://www.the guardian.com/global-development/2013/jul/24/gm-crops-ghana-us-genetically-modified –food.htm.

84. Hirst, K. K.; Pearl Millet (*Pennesitum glaucum*). Domestication and History. **2013**, 1–3, August 5, 2013. http://www.archaeology.about.com/od/ domestications/qt/Pearl millet.htm

85. Hohn, A.; and Neumann, K.; Tilling the Dunes-shifting cultivation and development of a cultural landscape in the Sahel of Burkina Faso, West Africa. *Quat. Int. J.* **2011**, 251, 152–155.

86. IFPRI-ASTI; A consolidated database of crop varietal releases, adoption and research in Africa South of Sahara, 1–8, 2013, August 21, 2013. http://www.asti.cgiar.org/diiva#3

87. Jalloh, A.; Faye, M. D.; Roy-Macaulay, H.; Sereme, P.; Zougmore, R.; Thomas, T. S.; and Nelson, G. C.; Overview. In: West African Agriculture and Climate Change. International Food Policy Research Institute, Washington, D.C. **2013a**, 1–30.

88. Jalloh, A.; Faye, M. D.; Roy-Macaulay, H.; Sereme, P.; Zougmore, R.; Thomas, T. S.; and Nelson, G. C.; Summary and Conclusions. In: West African Agriculture and Climate change. International Food Policy Research Institute, Washington, D.C., **2013b**, 383–391.

89. Jinukun, Opening Remarks of Press Conference on Regional Project for Biosafety in West Africa. Coutnocou, **2006**, Benin, 1–2.

90. Jones, M. J.; The Organic matter content of the savannah Soils of West Africa. *J. Soil. Sci.* **2006**, *24*, 42–53

91. Kahlheber, S.; and Neumann, K; The development of plant cultivation in the semi-arid West Africa. In: Rethinking agriculture-Archaeological and Ethno-archaeological perspectives. Eds. Denham, T. P.; Iniarte, J. S.; and Vrydaghs, L. One World Archaeology, Left Coast Press, Walnut Creek, **2007**, 320–346.

92. Kaleem, F. Z.; Assessment of Nitrogen fixation by legumes and their relative n contribution to a succeeding maize crop. Annual Report 1988/89. Nayankapal Agricultural Experimental Station, Tamale, Ghana. Annual Report 78, **1989**, 23–28.

93. Kamara, A. Y.; Best practices for Maize production in the West African savannahs. International Institute for tropical Agriculture, Ibadan, Nigeria, R4D Review, **2013**, 1–5.

94. Kamara, A. Y.; Ajeigbe, H. A.; Omoigui, L. O.; and Chickoye, D.; Intensive Cereal-legume-livestock systems in West African Dry Savannahs. **2005**, 63–79. http://ciat.cgiar.org/wp-content/uploads/2012/12/chapter_5_eco_efficiency.pdf

95. Kato, E.; Nkonya, E.; and Place, F. M.; Heterogeneous treatment effects of Integrated Soil Fertility Management on Crop Productivity. International Food Policy Research Institute paper No 01089: **2011**, p. 48.

96. Keitchi, H.; and Toshiyuki, W; Sustainable soil fertility management by indigenous and scientific knowledge in the Sahel zone of Niger. Proceedings of 17th World Congress of Soil Science, Symposium No.15, paper 1251, **2002**, 611–612.

97. Khatibu, A.; Effects of tillage methods and mulching on erosion and physical and physical properties of a sandy clay loam in an equatorial warm humid region. *Field. Crop. Res.* **1984,** *8*: 239–254.

98. Konseiga, A.; Migration, Rural Development, and Food Security in West Africa. International Food Policy Research Institute, Washington, D.C. USA, **2004**, pp. 97–103, http://conferences. ifpri.org/2020AfricaConference/events/malich03.pdf

99. Krishna, K. R.; Peanut Agroecosystem: Nutrient dynamics and Productivity. Oxford, England: Alpha Science International Inc:, **2008**, pp. 75–129.

100. Krishna, K. R.; Precision Farming: Soil Fertility and Productivity Aspects. New Jersey: Apple Academic Press Inc., USA, **2013**, p 189.

101. Kuyek, D. G. M; Cotton to invade West Africa. Genetic Resources Action International (GRAIN), **2004,** 1–2, August 21st, 2013, http://www. cropwatch.org/article. php?id=9872&printsafe=1.htm

102. Le Thiec G; Agriculture africaine et traction animale. CIRAD, Collection Techniques, Montpellier, France, **1996**, p 362.

103. Le Thiec G.; and Havard M.; Les enjeux du marché des matériels agricoles pour la traction animale en Afrique de l'Ouest. *Agriculture et Développement*, **1996,** *11*, 39–52.

104. Liu, S.; Kaire, M.; Wood, E.; Dialllo, O.; Tieszen, L. L.; Impacts of land use and climate change on carbon dynamics in South-Central Senegal. *J. Arid. Environ.* 2004, *50*, 583–604.

105. Makinde, D.; New GM crops in the pipelane: These are staple crops that Africans love to eat several times a day, **2010**, 1–3, August 19, 2013. http://www.gmo-safety.eu/basic-info/1249. africa-genetically-modified-plants.html

106. Manlay, R. J.; Chotte, J.; Masse, D.; Laurent, J.; and Feller, C.; Carbon, Nitrogen and Phosphorus allocation in Agroecosystems of a West African Savannah. 3. Plant and Soil components under Continuous cultivation. *Agric. Ecosyst. Environ.* **2002a**, *88*, 249–264.

107. Manlay, R. J.; Masse, D.; Chotte, J. C.; Feller, C.; Kaire, M.; Fardoux, J; and Pontanier, R.; Carbon, Nitrogen and Phosphorus allocation in Agro-ecosystems of a West African Savannah II The soil component under semi-permanent cultivation. *Agric. Ecosyst. Environ.* **2002b**, *88*, 215–232.

108. Manning, K.; A developmental history of West African Agriculture. In: West African Archaeology: New Developments, new perspectives. Ed. Allsworth-Jones, P; **2010**, 42–53, August 12, 2013, http://discovery.ucl.ac.uk/1329100/

109. Manning, K., Pelling, R., Higham, T., Schweninger, J.; and Fuller, D. Q.; 4500-year old domesticated pearl millet (*Pennesitum glaucum*) from the Tilemsi Valley, Mali: New insights into an alternative cereal domestication pathway. *J. Archaeol. Sci.* **2011**, *38*: 312–332.

110. Marchal, J. Y.; Yatenga, dynamique d'un espace rural Soudanao-Sahelien (Haute Volta). Traveaux et a Documents 167 ORSTOM, Paris **1983**, p. 38.

111. Marchner, H., Rebafka, F. P., Hafner, H.; and Buekert, A.; Crop Residue management for increasing production of pearl millet on acid sandy soils in Niger, West Africa. In: Plant Soil Interactions at low pH. Ed. Date, R. A. Dordrecht, Netherlands: Kluwer Academic Publishers, **1995**, 767–770.

112. McClintock, N. C.; and Diop, A. M.; Soil Fertility management and Compost use in Senegal's peanut Basin. *Int. J. Agric. Sustain.* 2005, *3*, 79–91.

113. McCord, R. J.; Ancient Empires of Africa: Kingdoms of the Western Sudan. Xavier University of Louisiana, **2012**, 1–6, August 3, 2013, http://www.roebuckclasses.com/102/resources/ africa/westafricaptrade.htm

114. Michels, K.; Shivakumar, M. V. K.; and Allison, B. E.; Wind erosion control using crop residue. 1. Soil flux and soil particles. *Field. Crop. Res.* **1995**, *40*, 101–110.

115. Mulitza, S.; Heslop, D.; Pittauerova, D.; Fischer, H. W.; Meyer, I.; Stuut, J. B.; Zable, M.; Mollenhauer, G.; Collins, J. A.; Kuhnert, H.; and Schultz. M.; Increase in African dust flux at the onset of commercial agriculture in the Sahel region. *Nature.* **2010**, 466, 226–228.

116. Naab, J. B.; and Alhasaan, A. Y.; Effects of crop rotation, fertilizer nitrogen and manure on soil properties and maize yield in Northern Ghana. In: Impact, Challenges and Prospects of Maize Research and Development in Africa. Eds. Badu-Apraku, B.; Fakorede, M. A. B.; Oudraog, O.; and Carsky, R. J.; WECAM/IITA, Nigeria, Report No 23, **2001**, 269–280.

117. Nin-Pratt, A.; Johnson, M.; Magalheas, X.; diao, L.; You, L.; and Chamberlin, J. Priorities for realizing the potential and growth in Western and Central Africa. International Food Policy Research Institute, Washington, D.C. Discussion paper no 00876, **2009**; 1–12.

118. North Carolina Digital History; Africans before Captivity.**1972**, pp. 1–7, August 5, 2013, http://www.learnnc.org/lp/editions/nchrist-colonial/1972.htm.

119. Nottfige, D. O.; Ojeniyi, S. O.; and Asawalam, D. O.; Comparative effect of plant residues and NPK fertilizer on nutrient status and yield of maize(*Zea mays.L.*) in a humid Ultisol. *Nigerian. J. Soil. Sci.* **2005**, *15*, 1–8.

120. Ogban, P. I.; Ukpong, U. K.; and Essien, I. G.; Influence of Bush fallow on the physical and chemical quality of acid sand in the South eastern Nigeria. *Nigerian. J. Soil. Sci.* **2005**, *15*, 96–101.

121. Opoku-Duah, S.; Donoghue, D. M. M.; and Burt, T. P.; Vegetation and drought mapping in West Africa using remote sensing: A case study. On Line. *J. Soc. Sci. Res.* **2013**, *2*, 142–150.

122. Ojeniyi, S.; Odedina, S.; Odedina, J.; and Akinola, M. Effect of tillage and mulch combination on soil physical properties and sorghum performance on Alfisol of Southwest Nigeria. *Nigerian J. Soil. Sci.* **2009**, *19*, 72–77.

123. Onyekwere, I.; and Ezenwa, M. S.; Characterization and classification of Barkin sale, Nigerian Guinea Savannah soils for sustainable rice production. *Nigerian. J. Soil. Sci.* **2009**, *19*, 122–128.

124. Osundare, B.; Effects of interplanted legumes with major soil nutrients and performance of maize. *Nigerian J. Soil. Sci. 17*, 47–51.

125. Ouedraogo, E; Mando, A; and Zombre, N. P.; Use of compost to improve soil properties and crop productivity under low input agricultural systems in West Africa. *Agric. Ecosyst. Environ.* **2001**, *84*, 259–266.

126. Oyun, M. B.; Nodulation and Rhizobium population in root nodules of selected Tree legumes. *Nigerian J. Soil. Sci.* **2007**, *17*, 65–70.

127. Purkey, D. R.; and Vermillion, D.; Lift Irrigation in West Africa: Challenges for Sustainable Local Management. International irrigation Management Institute, Srilanka, Working paper No 33, **1995**, 1–75.

128. Rian, S., Xue, Y., Macdonald, G. M., Toure, M. B., Yu, Y., De Sales, F., Levine, P. A., Doumbia, S.; and Taylor, C. E.; Analysis of climate and vegetation characteristics along the savannah-desert ecotone in Mali using MODIS data. *GISci. Remote. Sens.* **2009**, *46*: 424–450

129. Romheld, D.; Marschner, H.; Becker, K.; Schecht, E.; Hulsebusch, C.; and Buerkert, A.; Effects of site factors on the efficiency of soil amendments and on nutrient fluxes in subsistence-oriented cropping of the West African Sahel, **2000**, http://www.troz.uni-hohenheim.de/reseearch/sfb308/Ebcont/bcheck.

130. Roose, E.; and Barthes, B.; Organic matter management for soil conservation and productivity restoration in Africa: A contribution from Francophone research. *Nutr. Cycl. Agroecosyst.* **2001**, *61*, 159–171.

131. Roose, E.; Kabore, V.; and Guenat, C.; Zai Practice: A West African Traditional rehabilitation system for Semi-arid degraded lands, a case study in Burkina Faso. *Arid. Soil. Res. Rehabilit.* **1996**, *13*, 343–355.

132. Saa, D.; and Akuriba, M. A.; Sustainable Micro-irrigation systems for poverty alleviation in the Sahel: a case for'Micro public-private partnership. In: Proceedings of Third Annual Conference of African Association of Agricultural Economists. Cape town, South Africa, **2010**, 1–22.

133. Salzman, U.; Zur holozanen Vegetations – und Klimaentwicklung der westafrikenschen savanen. Paleaookologische Untersuchengen in der Sudan zone NO-Nigeria's. Doktorbeit. Julius Maxmillans-Universitat, Wurzburg, Berische des Sonderforschungsberiechs., Germany, **1998**, pp. 268.

134. Salzman, U.; Are savannahs degraded forests?. A holocene pollen record from the Sudanian zone of NE Nigeria. *Veg. Hist. Palynol.* **2000**, *100*: 39–72.

135. Salzman, U.; and Waller, M.; The Holocene Vegetational History of the Nigerian Sahel based onmultiple pollen profiles. *Rev. Paleobot. Palynol.* **1998**, *100*, 39–72.

136. Sanginga, N.; Role of Biological nitrogen Fixation in Legume-based cropping systems: A case study of West Africa farming systems. *Plant. Soil.* **2003**, *252*, 25–39.

137. Sene, M.; and Niane-Badiane, A.; Effect of manure, P and Ca source on optimization of Soil, Water and Nutrient Use in the Corn/peanut rotation system in the peanut basin. Institute Senegalaise de Researchis Agricoles. Ministere de L'Agriculture, republic du Senegal, Senegal, **2001**, 17–30.

138. Shivakumar, M. V. K.; Agroclimatic aspects of rain fed agriculture in the Sudano-Sahleian zone. In: Soil, Crop and Water Management in the Sudano-Sahelian zone. International Crops Research Institute for the Semi-arid tropics. Patancheru, India. **1989**, 17–38.

139. Skinner, J.; Niassa, M.; and Haos, L.; Sharing the benefits of large dams in Wes Africa. International Institute for Environmental Development, London, United Kingdom Natural Resources issue No 19: **2009**, 1–146.

140. Smith, P.; Soil organic Carbon dynamics. In: Eds. Braimoh, A. K.; and Vleck P. L. G. Land Use and Resources. Heidelberg, Germany: Springer Science, **2008**, 234–248.

141. Sow, M.; Sido, A.; Laing, M.; and Ndjiondjop, M.; Agromoprhological variability of Rice species collected in Niger. *Plant. Genetic. Resour.* **2013**, *23*, 1–13. http://www.mendeley.com/c/5987876164/g/982971/sow-2013-agro-morphological-variability-of-rice-species-collected-from-niger/.htm August 22, 2013.

142. Spivey, D. West Africa, **2003**, 1–11, August 5th, 2003, http://www.enotes.com/west-africa-reference/west-africa

143. Sterk, G.; Herman, L.; and Bationo, A.; Wind blown nutrient transport and soil productivity changes in southwest Niger. *Land. Degrad. Dev.* **1996**, 7, 325–335.

144. Sterk, G.; Stroosnijder, I.; and Raats, P. A. C; Wind erosion process and Control techniques in Sahelian zone of Niger. **1998**, 1–14, http://www.wu.ksu.edu/symposium/proceedings/sterk.pdf

145. Stoorvogel, J. J.; and Smaling, E. M. A.; Assessment of soil nutrient depletion in sub-Saharan Africa: 1983–2000. Volume 1. Main report. The Winand Staring Centre, Wageningen. The Netherlands. **1990**, pp. 342.

146. Tabi, F. O.; et al. Soil Fertility classification for rice production in Cameroon lowlands. *Afr. J. Agric. Res.* **2013**, *8*, 1650–1660.

147. Takeshima, H.; and Futoshi, Y.; Irrigation pumps and milling machines as insurance against rainfall and price risks in Nigeria. International Food Policy Research Institute, Washington D.C. Paper series no 33, **2012**, 1–46.

148. Takeshima, H. Jimah, K.; Kolvalli, S.; and Diao, X.; Dynamics of Transformation: Insights from Rice Farming in K pongirration system (KIS). International Food Policy Research Institute, Transforming Agriculture Conference. Paper No 8: **2012**, 1–8.

149. Tan, Z.; Liu, S.; Tieszen, L. L.; and Tachie-Obeng, E.; Dynamics of Carbon stocks driven by changes in Land use, management and Climate in Tropical moist ecosystems. *Agric. Ecosyst. Environ.* **2009**, *130*, 171–176.

150. Tappan, G. G.; Sall, M.; Wood, E. C.; and Cushings, M.; Ecoregions and Land cover in Senegal. United States Geographical Service. EROS Data Centre. Centre De Sulvi Ecologicue, B. P.; 155532, Dakar, Senegal. 2004, 1–22.

151. Tarawali, G. A syntheses of the Crop-livestock production systems of the dry savannahs of West and Central Africa. **2004**, *old.iita.org/cms/details/crop-livestock/arti11.pdf.*

152. Tieszen, L. L.; Tappan, G. G.; Tan, Z.; and Tachie-Obeng, E; Land cover change, biogeochemical modelling of carbon stocks and climate change in West Africa. In: Proceedings of Conference on "Africa and Carbon Cycle- The CarboAfrica project" Accra, Ghana. **2011**, 25–27, 75–83.

153. Tieszen L. L.; Tappan G. G.; and Toure A.; Sequestration of carbon in soil organic matter in Senegal–an Overview. *J. Arid. Environ.* **2004**, *59*, 409–425.

154. Tiffen, M.; Profile of demographic change in the Kano-Maradi region, 1960–2000. Drylands Research Working paper 24. Dry lands Research, Crewkerne, United Kingdom, **2001**, p. 36.

155. Tiffen, M.; Population pressure, migration and urbanization: Impacts on crop-livestock systems development in West Africa. International Livestock Research Centre, Nairobi, Kenya, **2004**, 1–27.

156. TRI; The Projet Vandebilt. The Rodale Institute Rapport Annuel 2001–2002. This, Senegal, **2002**, 1–43.

157. UNEP, Genetically modified crops in Africa; **2011**, August 19th, 2013, 1–12. http://www.eoearth.org/veiw/article/152943

158. UNO; Groundwater in North and West Africa. Natural Resources/Water series No 18. Department of Technical Cooperation for Development. New York, **1988**, 1–51.

159. USDA GAIN; The CropSite-USDA GAIN- Oilseeds, cotton, Sugar, Grain and Feed. West Africa Annual 2013, **2013**, pp. 1–7, August 5th 2013, http://www.thecropsite.com/reports/?id=1942.

160. Watson, D. J.; Within Savannah and Forest: A review of late Stone Age Kintampo Tradition, Ghana. *Azania. Archaeol. Res. Afr.* **2010**, *45,* 141–175.doi:10.1080/0067270x2010.491361, August 3rd, 2013.

161. Wezel, A.; Bohlinger, B.; and Bocker, R.; **2000**; Vegetation zones in Niger and Benin- Present and Past zonation. (March 8th, 2013)

162. Wikepedia; History of West Africa. en.wikepedia.org/wiki/Hstory_of _West Africa. pp. 1–5. August 5th, 2013, http://www.uni-hohenheim.de/atlas308/a_overveiw/a3_1/html/english/a31ntext.htm pp 1-9

163. Witttig, R.; Konig, K.; Schmidt, M.; and Szarzynski, J.; A study of climate change and anthroporgenic impacts in West Africa., **2007**, pp 1–2, December 22, 2013, http:/www.ncbi.nlm.nih.gov/pubmed/17561777.

164. Wood, P. J.; The Botany and distribution of *Faidherbia albida*. In: *Faidherbia albida* in the West African semi-arid tropics. Ed. Vanderbelt, R. J.; International Crops Research Institute for the Semi-arid Tropics. Patancheru 502324, Hyderabad, India. **1992**, 9–17.

165. Wopereis, M. C. S.; Tamelkpo, K.; Ezui, D.; Gnakpenue, B.; and Breman, F. H.; Mineral fertilizer management of Maize on farmer fields differing in organic inputs in the West African savannah. *Field. Crop. Res.* **2006**, *96*, 355–362.

166. World Bank; Africa's Development in a Changing Climate. International Bank for Reconstruction and Development/World Bank. Washington, D.C., **2009**, pp. 239.

167. Yost, R.; Doumbia, M.; and Berthe, A.; Systems Diagnosis for Technology Adaptation and Transfer. 26–34, August 19, 2013, http:// www.oired.vt.edu/projects/current/project synthesis.pdf.,.

CHAPTER 6

PRAIRIES OF SOUTH ASIA: NATURAL RESOURCES, ENVIRONMENT AND CROP PRODUCTION

CONTENTS

6.1 SOUTH ASIAN PRAIRIE REGION: AN INTRODUCTION

South Asia is geographically a diverse region that encompasses massive mountain ranges, large plains with prairie vegetation, shrubs, trees, thick forests, semiarid regions with low input agricultural crops, highly intensely cultivated agricultural prairies, low land wet prairies, several rivers and their tributaries, deltas, long stretch of sea coasts; arid regions and deserts with scanty vegetation. The South Asian region extends into 44.4–44.6 million km^2 and hosts over 1.6 billion human beings plus large population of domestic farm animals. The South Asian region considered here in the present context is restricted to definition by SAARC. In the entire chapter, discussions are concentrated on agrarian regions of Pakistan, India, Nepal, and Bangladesh. There are only passing references to other SAARC nations such as Maldives, Sri Lanka, Afghanistan and Bhutan. Human population density of the South Asian Agricultural Prairie region considered here is among the highest at 350.6–400.1 km^{-2}. Hence, demand for food grains and nourishment is high. Much of the produce is consumed *in situ* by the local population. Natural vegetation and agricultural prairies of South Asia have traversed a long way in terms of evolutionary diversity, intensity of cropping and productivity, through the ages. During past 5 decades, since 1970s, this region has experienced rampant expansion of agricultural prairies followed by intensification of crop production in order to feed an enlarged population.

6.1.1 GEOLOGICAL ASPECTS OF SOUTH ASIA

In the context of this chapter on prairies of South Asia, we should note that geologically, "Indian plate" comprises present day Indus valley, Gangetic plains, Southern Indian peninsula and Delta region in the east. The Indian plate, it seems was formed out of an ancient continent known as Gondwana. The Indian plate collided with Asia during Eocene of the Cenozoic era some 55 million years ago. The great arc of mountains of Himalayas, Hindu Kush, and Patkai ranges were formed by the tectonic collision of India and Eurasian plates.

Geologically, Pakistan is classified into three basins namely, Indus, Axial belt, and Baluchistan (Siddiqui, 2011). The Indus river system probably got formed with the uplift of Tibetan region caused by India-Asia collision. Modern Indus river cuts across Indus suture in Ladakh-Karakoram region. The river appears to have not changed its course much during past 18 million years. At best, it has shifted in some places by <100 km since Eocene. Sediment fluxes from Karakoram and Lhasa have moved into Gulf of Oman since ages. The sedimentation rates have changed periodically (Clift, 2002; Inam et al., 2007; Kazmi and Qasimjan, 2011).

The Gangetic plains in India are attributed to geological processes that operated during Quaternary period sometime 1.8 million years ago. The Quaternary deposits can be classified into Indo-Gangetic basin and Peninsular basins. The Gangetic basin encompasses 8,50,000 km². The Quaternary alluviation of Ganga plains started with deposition of boulder conglomerates from Siwalik range (Prasad and Kar, 2005).

India's geologic features could be classified based on era of formation. Pre-Cambrian formations occur in Vindhyas and Cudappah in Deccan region. Paleozoic formations are traced in the Western Himalayas and Assam. The Mesozoic Deccan traps are common in the Central Indian plateau and Andhra Pradesh. The Black soils derived from Deccan traps are among fertile soils found in Peninsular India. Alfisols are common in Karnataka, Tamil Nadu, Andhra, and Aravallis. Laterites are common in Western Ghats. The cretaceous system occurs in the Indo-Gangetic plains. Gondwana system is seen in Narmada valley, Vindhyas and Satpura region.

The Ganges delta in the Eastern Indo-Gangetic plains lies at the junction of tectonic plates of India, Eurasia, and Burma. Large amounts of sediments are supplied by the Himalayan collision and erosion phenomena. The paleoshelf runs across delta from Kolkata to edge of Shilong plateau. The Delta has thick continental crust in northwest and thin oceanic crust in southeast. Sediment accumulation in Southeast Delta is high and it may cross 16 km in thickness in many locations (Humayun, 2013). The Delta soils are derived from alluvium made of Himalayan silt.

Massive eruption of Deccan traps around 67 million years ago in the Cretaceous period caused volcanic flow and formation of the Deccan plateau region. Series of volcanic activities during different periods induced several layers of highlands and plains. The region encompasses basalts and granites. Granites are rich in feldspar and quartz. Basalts are mafic in composition with pyroxene and olivine rich in Fe and Mg containing minerals. The Deccan plateau is rich in primary minerals. Basalts and igneous rocks have induced formation of large tracts of Black soils in the plateau. Granites have weathered into large tracts of Alfisols rich in Fe and Al oxides.

6.1.2 PHYSIOGRAPHY AND AGROCLIMATE OF SOUTH ASIAN PRAIRIES

Pakistan currently extends into 76.61 million ha. It comprises of four provinces namely Baluchistan, Northwest Frontier, Punjab, and Sind. Pakistan occurs within longitudes 23° 30' and 36° 45' North and 61° 00' and 75° 31' East latitude. Pakistan has areas that are classifiable as mountains found in north and western part. The semiarid and arid plains occur in south and southeast. Deserts and arid regions with low density natural shrub land are found east of lower Indus plains. At the center, fertile Indus plains fed by the river water and underground aquifer support a very large expanse of Agricultural prairies and natural shrub land vegetation. Pakistan experiences a wide range of climatic conditions from arid to humid, cold temperate to dry tropical conditions (Chaudhry and Rasul, 2004).

Deserts found in Sindh are called Nara and Thar. The dry arid belt in Punjab region is known as Cholistan. The desert extends into Rajasthan in India. Sand dunes and shifting sands are common in desert zone. Vegetation is really sparse, if any and precipitation is scanty, erratic, or absent for long stretch of years. The semiarid belt that fringes the deserts is a fairly fertile zone that supports crop production. Drought-tolerant crops such as millet, guar, and forage grasses are common. This region supports both permanent agriculture and nomadic pastoralism (Hasan and Raza, 2009).

Pakistan can be demarcated into at least 10 different agroecological regions based on physiography, climate, land use, and water resources (see Figure 6.1). They are (a) Indus Delta that extends around Hyderabad (Sindh) upto Arabian Sea, including Badin and Thatta. Here, monthly average precipitation is 75 mm in summer; (b) Southern Irrigated Plain includes lower Sindh plain from Jacobabad to Dadu. Monthly rainfall is 18 mm; (c) Sandy desert includes vast areas of Thar and Cholistan. Total annual rainfall is <300 mm; (d) Northern Irrigated plains consists of areas between two major rivers of Punjab – Jhelum and Sutlej. Mean annual precipitation is 300–500 mm; (e) Barani Lands includes regions in Potawar plateau, Himalayan piedmont, Attock, Rawalpindi, and Karak. Mean monthly rainfall is 200 mm; (f) Wet Mountains includes high ranges in Upper Hazara and Swat. Mean monthly precipitation is 235 mm; (g) Northern dry Mountains includes areas in Gilgit, Baltistan, Chitral, and Dir valleys. Mean monthly precipitation is 25–75 mm; (h) Western dry Mountains includes Bannu in Northwest frontier region, Zhob, Pashin, Quetta, and Kalat in Baluchistan. Mean monthly precipitation is 95 mm; (i) Western plateau includes Chegai and Makran regions. Monthly rainfall is 37 mm; and (j) Suleiman piedmont includes region from D.I. Khan in Northwest Frontier region to Dhera Ghazi Khan in Punjab and Dera Bugti, Nasirabad, and Kacchi in Baluchistan (See Chaudhry and Rasul, 2004).

FIGURE 6.1   Agroclimatic zonation of Indus valley and other regions of Pakistan that supports Natural Grasslands and Agricultural prairies.
Source: IRRI (2013a)

Nepal is predominantly a mountainous zone, except for a narrow stretch of tropical plains known as "Terai". The outer Terai ends as the foothills of Himalayas that is, Siwalik hills begin. This region is fertile with riverine silt and alluvium. The alluvium is called Bhabhar. In Nepal, inner Terai is *"Bhithri Madesh"* (Wikipedia, 2011). It is agriculturally a productive region. It generates cereal grains such as rice, maize, finger millet, and legumes such as pigeonpea. Incidentally, entire Terai region supports diverse natural vegetation and agricultural prairies. Terai region is supplied with water by large number small rivers/ tributaries that flow into mainstream Ganges. Soils are mostly Mollisols, Inceptisols, and Alluvial soils rich in organic matter and minerals. The altitude of Terai plains ranges from 67 to 300 m.a.s.l. The grasslands and savannahs are interspersed with tropical deciduous trees all through the length of the Ganges river. The Terai region in Brahmaputra area is called "Dooars." Terai region experiences a subtropical climate, with warm summers. Ambient temperature during winter reaches 3–5°C and 34–38°C during summer months May–July. Terai is a high rainfall zone, since it occurs just under the foothills that allow massive collection of monsoon clouds. Annual rainfall ranges from 1,800 to 2,300 mm. Crop production proceeds during all three season.

There are several different classifications possible that deal with physiography of India. Within the present context, we are more concerned with areas that support natural prairie vegetation and large agricultural expanses. India can be divided into five physiographic regions such as the Mountains of North India, Indo-Gangetic Plains, Peninsular India that comprise the Deccan and South Indian plains, The Thar Desert and Coastal Plains (Wikipedia, 2013b). Major mountain ranges are Himalayas that extend for 2,500 km from Jammu and Kashmir to Sikkim and Arunachal Pradesh in the east. Patkai, is predominant in the Eastern India and Burma. It includes the Gao-Khasi, Jaintia, and Lushai hills. The Vindhyas in Central India are lush green with vegetation. Average elevation of Central Plains is 300–600 m.a.s.l. The Satpura range in the West covers regions in Maharashtra, Madhya Pradesh, and Gujarat. The Aravalli range in Southern Rajasthan is a drier range. The Western Ghats or Sahyadri mountains extend into large area parallel to west coast of India. The Western ghats extend from Tapti river in Gujarat, entire Deccan and up to Annamalai and Nilgiri hills in the South. The average elevation of Western ghats is 1,000 m.a.s.l. The Eastern ghats is a discontinuous stretch of mountains with relatively lower elevation compared to Western Ghats. It is dissected by major rivers such as Mahanadi, Godavari, Krishna, and Kaveri. Eastern ghats extend from West Bengal till Tamil Nadu, up to the southern tip of peninsula. The mountainous terrain is home to diverse natural vegetation comprising prairies and forest. Cropping is also in vogue in the terraces.

The Landscape of Gangetic Plains is gentle slope of agricultural prairies and natural vegetation. The historic aspects of vegetation in this region have been described and discussed (See Singh, 2012). This region is classified using different criteria. Based on landscapes, it is classified into *Bhabar, Terai, Bhangar, Khadar.* The Bhabar region encompasses foothills of Himalayas, with pebbles, boulders, river streams, and thick vegetation of forest and underground shrubs/grasses. Soils of Bhabar region are homogeneous and moderately fertile. The Terai region is composed of alluvium from rivers that flow into plains from Shivalik hills. The region is moist with high rainfall and intense forest vegetation. The under story vegetation is sometimes filled with shrubs and tall grasses of *Saccharum* species. Bhangar region comprises of area filled with alluvium and terraces within flood plains of Ganges and its tributaries (Singh, 2012). Soils traced in Bhangar region are fluvial with soluble salt accumulation. They are neutral or slightly acidic. Soil types traced in regions closer to Ganges are sandy or sandy loams. Silt content is relatively high. Bhangar soils are deficient in N, P, and K (Tripathi, 2008). The Khadar region is lowland areas that appear after Bhangar. It is usually fresh with new alluvium and silt deposition by rivers flowing down the plains. Khadar soils are rich in nutrients and fertile with pH 6–8.

The Indo-Gangetic plain in India is served by rivers such as Indus and Sutlej, Ganges, Yamuna, Gomathi, Ghaghra, Dhamodar, Brahmaputra, Meghna, and Padma. These rivers flow perennially since they receive monsoon precipitation during rainy season and snow melt during summer from Himalayas (Wikepedia, 2013b).

The Deccan plateau is large table land in Southern India. It extends into several states with wide array of biomes, habitats, and agroclimatic regions. It supports evergreen forests, large-scale agricultural prairies, natural grasslands, shrub regions, and rocky arid zones with sparse vegetation. The Deccan Plateau and the plains are served by several rivers and their tributaries. They are Godavari and its tributary Indrāvati; Krishna and tributaries such as Tungabhadra, Bhīma, Mallaprabha, and Ghataprabha; and Cauvery and its tributaries such as Arkavati and Simsha. Deccan plateau is a fertile region that supports diverse crop species. Large expanses of major cereals like rice, maize, sorghum, pulse monocropping stretches in North Karnataka and Andhra Pradesh and oilseed crops are prominent. There are several projects that provide irrigation to crops. Vertisols that are clayey and rich in minerals are common in Maharashtra, Karnataka, and Andhra Pradesh. Alfisol plains that occur in the south, support large-scale production of rice, legumes, and oilseeds. The Western Ghats are home to large stretches of plantation crops such as coffee and spices.

The Eastern coastal plain is relatively a wider stretch of fertile sandy plains that supports some of the most prolific cropping zones of India. It extends from West Bengal to southern tip of India. Coastal plains possess fertile delta zones created by the alluvium of three major rivers, Godavari, Krishna, and Cauvery. The coastal plains are humid, with temperature ranging from 32°C to 36°C. It is served by both Southeast and Northeast monsoons. Geographically, Eastern coast can be demarcated into Mahanadi delta, Andhra plains, Godavari-Krishna delta, Coramandel coast, and Kanyakumari coast.

The West coast is a narrow stretch of fertile wetlands and dry arable region that extends from Gujarat to southern tip of Kerala. It occurs between Western ghats and Arabian sea coast. It is largely a plain with lateritic or sandy soils. It supports thick forest vegetation, coconut plantations, extended areas of wet land rice and arable crops. It is served by rivers such as Narmada and Tapti in Gujarat, Mondovi and Zuari in Goa, Sharavati and Netravati in Karnataka, and back waters in Kerala (Wikipedia, 2013a).

The agroclimate of India shows wide diversity based on geographic region, its topography, soil type, weather pattern, and cropping systems. Generally, we can identify three major climatic regions namely (a) Tropical wet climate; (b) Dry climate; and (c) Subtropical humid climate (Wikipedia, 2013c). The Planning commission of India identifies 15 agroclimatic zones within Indian union (Krishisewa, 2007). However, detailed analysis of agroclimatic patterns in relation to natural vegetation and crop production trends suggest that we can identify 20 major agroclimatic regions and 60 sub-agroecoregions in Indian subcontinent (Gajbhiye and Mandal, 2007; Krishisewa, 2007; and Krishna, 2010a).

Large expanses of natural vegetation and agricultural cropping areas in India are supplied water through various sources such as natural precipitation, rivers, lakes, reservoirs, and groundwater. Regarding precipitation, Indian agricultural prairies are supplied water through monsoon. The normal duration of South Asian monsoon is about 100–120 days during which much of the annual precipitation occurs. The monsoon begins during May/June and lasts till September end. This is called Southwest monsoon. Next, wind direction changes and Northeast monsoon brings rains during November and December. The precipitation pattern in entire Indian subcontinent is guided by wind systems. Hot trade winds get initiated during May. These are called southwest trade winds that induce Southwest monsoon rains. The wind direction changes during September and Northeast trade winds create cloud patterns during later phase. The long-term average rainfall for entire Indo-Gangetic plains and peninsular India is 1,160 mm (Kumar et al., 2005). The total rainfall is sufficient to support crop production all through the year. However, about 80 percent of precipitation is received during short period of 100–120 days (June–September). Hence, it creates unevenness in water availability. Precipitation ranges from 150 mm year^{-1} in the Thar region to 11,690 mm year^{-1} in tropical Assam. However, on an average, 21 percent of Indian cropped area receives 750 mm and is termed semi-arid belt. About 15 percent of agricultural belt receives 1,150 mm per annum. In the Sub Himalayan region (Terai), which is hot humid about 2,500 mm precipitation is received annually. Southern Indian dry lands receive about 600–800 mm annually. There are arid regions in Rajasthan and Northern Gujarat that receive really very low amount of precipitation ranging from 400 to 500 mm per annum. The Deccan plateau, especially the Western Ghats with its tropical forests and plantations enjoy torrential rains. Annually, about 2,500–3,200 mm precipitation occurs during June to December.

Bangladesh occurs at the extreme of the Indo-Gangetic plain, between latitudes 20° 34' and 26° 38' North and between longitudes 88° 01' and 92° 41' East. The plains are typically flat and predominantly a Delta zone formed by the rivers, Ganges, Brahmaputra, and Meghna. The Delta zone is 1,05,000 km^2 in expanse. It is triangular in shape and extends for 350 km in breadth near the sea coast. The Delta is composed of maze of water ways, lakes, swamps, flood plains, and sediments collectively known as "*chars*". There are sporadic upland areas, otherwise the area is mostly a lowland region fit for wetland crops like rice. Delta zone experiences at least three major seasons, namely winter, summer, and monsoon. The ambient temperature during winter ranges from a minimum of 7°C–12°C to 24°C–31°C. Maximum summer temperature reaches 36°C, with high relative humidity. Monsoon begins during June and tapers by October. About 80 percent of precipitation occurs during monsoon period. The average rainfall in Bangladesh is 1,400–4,300 mm. Some regions of Bangladesh with in densely forested areas and coasts receive very high rainfall. Cyclonic storms and highly intense precipitation events inundate fields and often are detrimental to paddy crop. Tidal waves and high-speed winds also affect standing crops. They also induce soil erosion and loss of fertility. Evapotranspiration from the tropical delta is quite high at 51–183 mm month^{-1} (See ICID, 2012).

6.1.3   CLIMATE CHANGE AND SOUTH ASIAN PRAIRIES

Climate change is a recent awareness among researchers and policymakers in South Asian region. However, climate change phenomenon has periodically occurred several times in the geological history of South Asia. It has affected the general conditions in the region to varying intensities and length of period (Saini, 2010). There have been periodic cooling/warming phases, ice ages, and glacial changes in the Asian region. Such climate change effects are felt strongly on soil quality and water resources. Hence, it has been suggested that detailed historical data and knowledge about climate change effects on vegetation/crops in the South Asian prairie and forest zones will be useful. It will be useful in devising agricultural programs on ecosystem basis. Climate change effects on regions that support vast expanses of crops and natural vegetation in South Asia can be drastic. Some of the most immediate consequences listed by experts relate to changes in temperature, precipitation pattern, droughts, and floods. It may result in depreciation of forest growth, degradation of natural vegetation, and reduction in crop yield. Projected changes and forecasts by Bhattacharyya and Werz (2012) are as follows:

Himalayas and Indo-Gangetic Plains	
Climate change	Increased glacial melt; increased temperature
Ecological impacts	Reduction in forest wood, flash floods, landslides, soil erosion; droughts, floods, soil erosion, reduction in agricultural area and productivity
Northeast Gangetic/Brahmaputra Region	
Climate change	Increased temperature; winter precipitation; increased intensity of summer precipitation

Ecological Impacts	Total cereal production may increase, but rice productivity may decrease; soil erosion may affect natural vegetation and plantation leading to reductions in area and productivity; increased runoff may cause loss of soil fertility and landslides; crop yield during rabi season decreases drastically
Western Ghats and Southern Indian Plains	
Climate change	Increased temperature, higher intensity of rainfall, depreciation and erratic distribution of precipitation in the plains
Ecological impacts	Severe depreciation of crop and plantation productivity; floods and soil erosion may increase
Coastal Plains	
Climate change	Increase in land and sea surface temperature; increased intensity of rainfall; soil erosion due to storms and surges; reduction in cropping zones
Ecological impacts	Decrease in rice and coconut production; salinity and salt accumulation in the farming zones; submergence of cropping area, reduction crop yield

Source: Excerpted from Bhattacharyya and Werz (2012); IGES-GWP (2012)

Cookman (2012) has pointed out that, during past 5 decades, Pakistan has experienced severe consequences of climate change effects. Shortages in precipitation, irrigation, and higher ambient temperature, flash floods due to glacial activity and soil surface erosion have been felt conspicuously. For example, massive flash floods during 2010/2011 caused by high monsoon rains have induced loss of soil fertility. It has led to reduced crop productivity in the entire Indus plains. Reduction of crop yield has been drastic in some areas of Multan and Sind province. Deforestation has been rampant in the upper reaches of Indus valley. Malik et al. (2012) have attempted to categorize the various agroecological zones of Pakistan based on their vulnerability to climate change. They suggest that greater chance of flooding and reduced grain production is a clear possibility due to climate change. Pakistan, it seems, still does not have a good flood warning system to overcome climate change effects, whenever they occur. Forecasts by IPCC suggest that average surface temperature increased by 0.6°C since mid-nineteenth century and in future temperature may rise by 1.1°C–6.4°C by end of twenty first century. Such a rise in temperature is expected to affect glacier activity, evapo-transpiration, and water flow in rivers. According to reports by IGES-GWP South Asia (2012) the Indus valley excessively depends on water resources derived from glacier melt in Hindu Kush and Karakoram region. These regions are vulnerable to changes in temperature. Global warming and carbon soot deposits in Himalayas may result in increased water flow in Indus and cause floods more frequently. Mahmood (2013) predicts that rise in temperature may actually depreciate rice grain yield formation from 4.3 t ha^{-1} to 2.6 t ha^{-1} due to a rise in 5°C. Regarding effect of higher CO_2 concentration in atmosphere, it is said that a rise from 350 to 375 ppm could increase rice grain yield from 4.15 t ha^{-1} to 5.0 t grain ha^{-1}. Regarding wheat, enhanced temperature may depreciate grain harvest from 4.2 t ha^{-1} to 3.4 t ha^{-1} due to increase of 5°C. Iqbal and Khan (2010) have pointed out that climate change may also affect the establishment, regrowth, sustenance, and perpetuation of several plant species. The natural vegetation itself may undergo a certain change due to climatic variations. The Indus valley is supposedly among the most vulnerable regions to climate change, especially with regard to natural vegetation and crop biodiversity (Rana, 2013).

The climate change effects on wet prairies of Bangladesh that supports rice production has been drastic during past 2–3 decades. Efforts to gauge and predict climatic change have been in place for some time. Several weather models and simulations have been adopted to forecast climatic change and its possible impact on agricultural cropping and human dwelling in the riverine zones of Ganges and Padma. According to Battacharyya and Werz (2012), periodic assessments and use of computer aided weather models suggest a possible increase in temperature by 2.3°F by 2030. It may increase further by 4.7°F by 2059. Fluctuations in precipitation pattern and drought are other changes that may affect cropping pattern. A few other climate-change-related aspects that will affect crops in riverine zones of Bangladesh are floods, salt water intrusion, rise of sea level, and erosion of coastal borders. According to IGES-GWP South Asia (2012), river bank erosion is a major problem anticipated in Bangladesh, as a consequence of climate change. Higher temperature that induces enhanced glacier melt and water flow may cause erosion of river banks. In greater detail, following is the projections on ambient temperature and precipitation changes in Bangladesh:

Year 2030	Annual Temperature Change	+1.0°C	Winter	−1.1°C	Monsoon	−0.8°C
	Annual Rainfall Change	+3.8%	Winter	−1.2%	Monsoon	+4.7%
Year 2050	Annual Temperature Change	+1.4°C	Winter	−1.6°C	Monsoon	−1.1°C
	Annual Rainfall Change	+5.6%	Winter	−1.7%	Monsoon	+6.8%

Source: Ahmed (2006)

Floods are a common occurrence during monsoon period. It affects crops to different extents depending on intensity. General forecast is that flood frequencies may increase due to climate change effects. Saline intrusion affects irrigation and crop growth. Alluvial runoffs and salt water may affect the Gangetic tidal flood plains, severely in terms of crop productivity. Saline intrusion into groundwater has severe consequences to crops and humans dwelling in the area. Salt water is known to destroy crops and make it useless for consumption, even if a portion is salvaged. Salt resistant rice and other crops have been of great help to local population. Soil erosion, it seems, is a perpetual problem in the flood plains and Gangetic Delta of Bangladesh. It results in loss of topsoil, fertility, and crop yield. It also causes human migration out of the zones that are highly prone to soil erosion.

Weather models suggest that sea level in the Bangladesh coast may increase by 5.5 inches by 2030 and by 12.6 inches by 2050. There are various predictions about its consequences on crop growth and productivity in the flood plains and coast (Battacharyya and Werz, 2012).

The climate change effects on the vast rain-fed cropping zones of South India could be severe, if unattended timely. Ashalatha et al. (2012) state that several aspects of climate affect the rain-fed agriculture. We can arrive at several different forecasts based on extent of change in a particular climatic factor. For example, increase of temperature by 2–4°C, may not have perceptible effects on crop yield in Vertisol zones of South India. However, a further increase of 5°C could affect by enhancing greenhouse gas emissions. Droughts caused due to climate change may reduce crop yield. Estimates suggest that erratic precipitation in North Karnataka caused 43 percent loss of grain yield of sorghum compared to normal, 14 percent loss in case of maize, 28 percent in case of pigeonpea; 34 percent in case of groundnut and 59 percent in case of cotton (Ashalatha et al., 2012). Further, it is believed that climate change effects could be felt with greater impact, if the crop production is at subsidence level. Subsistence farming relies more on natural resources, hence vagaries in rain, temperature, and soil fertility affects crop growth proportionately. Surveys in South Indian dry land zones suggest that small farmers who practice low input systems suffer greatest due to climate change effects (Ashalatha et al., 2012; Manickam et al., 2012).

6.1.4 NATURAL VEGETATION OF SOUTH ASIAN PRAIRIES

The Indus valley is host to a large expanse of grassland with diverse plant species and fauna. It has evolved through several millennia in response to biotic and abiotic factors that have impinged the fertile plains, mountainous areas and delta. Natural vegetation of this region has been dynamic with regard to composition of species, their dominance, the nutrients recovered into tissues, that recycled and that lost to environment. Natural genetic crossing and other processes that generate new biotypes and species have operated since long. Hence, we have an assortment of plant genera and species that are well adapted to Indus valley and its environs. Indus valley is also home to several wild species, ancestors, and domesticated crop species. Following is a list cereals, shrubs, and trees that are commonly traced during recent period, in the Indus Valley:

Grasses	*Leptoticha fusca* (Kallar grass), *Sporobolus arabicus* (Sporobolus grass), *Cynodon dactylon* (Lemon grass), *Hordeum vulgare* (Barley), *Sorghum halopense* (Baru), *Sorghum bicolor* (Jowar), *Panicum antidotale* (Bansi grass), *Echinochloa crusgalli* (Swank), *Polypogonum monspeliensis* (Dumbi grass), *Avena sativa* (Oats), *Lolium multiflorum* (Lollium grass), *Desmotachya binnata* (Dhib), *Panicum maximum* (Panicum grass)
Legumes and Shrubs	*Cajanus cajan* (Arhar), *Cicer arietinum* (Chana), *Suaeda fruticosa* (Lana), *Kochia indica* (Kochia), *Atriplex nummularia, Atriplex lentiformis, Atriplex undulate, Atriplex crassifolia, Sesbania formosa* (Jantar), *Sesbania aculeate* (Jantar), *Sesbania rostrata* (Jantar), *Lotus corniculatus, Trifolium alexandrium* (clover), *Trifolium resupinatum* (Shaftal), *Medicago sativa* (Lucerne), *Macroptylum atropurpureum* (Dolichos)
Trees	*Acacia sclerosperma, Acacia ampiliceps, Acacia victoria, Acacia nilotica* (Kikar), *Acacia salicina, Acacia calcicola, Acacia coracea, Acacia saligna, Acacia bivenosa, Acacia cunnighamii, Cassia nemophila, Cassia sturtii, Leucana leucocephala, Prosopis cineraria* (Jand), *Prospis juliflora* (Mesquite), *Prosopis chilensis* (Kejri), *Casuarina obesa* (casuarina), *Eucalyptus microtheca* (Safeda), *Eucalyptus striaticalyx*
Vegetables	*Brassica napus* (Toria), *Trigonella foenum-graecum* (Methi), *Spincea oleracea* (Spinach), *Medicago falcate, Brassica carineta, Brassica juncea* (Raya), *Lactuca runcinata, Brassica campestris* (Sarson), *Solanum melangena* (Brinjal)

Source: NIAB (2013)

Historical records suggest that Gangetic region was endowed with rich diversity of forest trees, grassland species, and herbaceous vegetation. It seems the flood plains of major rivers such Ganges, Yamuna, Saraswati, Gagra and Gomathi supported development of diverse plant species. Forest dominated with trees such as Sal, Babool, Khair, Sisham, and Semal have been recorded (Singh, 2012). Major tree species traced in the plains at present are Khair (*Acacia catechu*), Bael (*Aegle marmelos*), Bakli (*Anogeissus latifolia*), Palas (*Butea monosperma*), Sisham (*Dalbergia sisso*), Bargad, (*Ficus bengalensis*), Jamun (*Syzygyium cuminii*), Sal (*Shoria robusta*), Peepal (*Ficus religiosa*), grass species traced in the plains are Kilikkanal (*Arundo donax*), Kans (*Saccharum spontaneum*, Kush (*Desmostachya bipinnata*), Bhalai (*Imperrata cylingrica*), Khus (*Vetiver zizanoides*), Narkul (*Phragmites karka*) and Sens (*Chrysopogon fulvus*) (See Singh, 2012).

Based on phyto-geographical association, Rhind (2010) has classified the natural vegetation, especially tree/forest species into Moist Deciduous forest regions; Tropical Moist Deciduous Riverine forests; *Sporobolus-Chloris* saline scrubland and *Phragmites-Saccharum-Imperata* Grassland.

Natural grasslands are wide spread in the Indo-Gangetic plains and Peninsular India. Again, vegetation has indeed evolved through the ages in response to environmental changes. Periodic changes in the diversity and intensity of natural vegetation have been recorded in many regions of Indian subcontinent. Grasslands in India have actually evolved in response to droughts, periodic fire, grazing tendencies, interest in developing pastures, and loss of forest cover due to deforestation tendencies (Planning Commission, Government of India, 2012). For example, the "Terai" grasslands in the Upper Ganges zones of India and Nepal have developed as a consequence of large-scale clear felling of trees, soil deterioration, and erosion. The natural succession of vegetation has induced formation of grasslands after loss of climax vegetation that had forest trees. The Indian Council of Agriculture (ICAR), New Delhi, classifies the natural grassland vegetation found within different regions into five categories. They are:

(a) *Sehima-Dichanthium Type:* These types of grasslands are common in the Central Indian plateau region covering areas of Chota-Nagpur, Indore, and Aravalli. They extend into 17,40,000 km². We can trace about 24 species of perennial grasses, 90 species of annual grasses, 129 species of dicots, including 59 species of legumes.

(b) *Dichanthium-Cenchrus-Lasiurus Type:* This type of grass cover occurs in parts of Punjab, Rajasthan, Gujarat, and Uttar Pradesh. They occupy about 4,36,000 km². About 11 perennial grass species, 43 annual grass species, and 45 dicots including 19 legumes have been traced in this region of Indo-Gangetic plains.

(c) *Phragmites-Saccharum-Imperata Type:* Such grasslands are common in Gangetic plains, Brahmaputra valley, plains, and Haryana. They spread into 2,800,000 km². About 10 perennial grass species, 26 annual grass species, and 56 herbaceous species including 16 legumes are traced in this region of India.

(d) *Themeda-Arundinella Type:* These grassland are common in Himachal Pradesh, Jammu and Kashmir, Uttar Pradesh, West Bengal, and Assam. They extend into 2,30,000 km². About 37 perennial grass species, 32 annual grass species, and 34 dicots including 9 legumes are traced.

(e) *Temperate and Alpine Type:* These grasslands occur in cool temperate regions of Jammu and Kashmir, Himachal Pradesh, Uttar Pradesh, West Bengal and Assam. There are 47 perennial grass species, five annual grass species and six legumes traced in the temperate grassland area.

In addition to above categories, the Shola grasslands are common in the Western ghats. These are unique and confined to higher altitudes of Western Ghats of Peninsular India (Thomas and Palmer, 2007; Brilliant et al., 2012). The Western Ghats are a mountain range that runs parallel to western coast line of Peninsular India. The Shola grasslands are often interspersed by trees. They are efficient in absorbing and retaining rain water. They are efficient biomass accumulators. Shola grasslands were wide spread, but currently they occupy approximately 400 km² (Thomas and

Palmer, 2007). Some aspects such as succession of prairie plant species, climax community, dissemination, water and nutrient usage and recycling have been studied. Agricultural communities here use "slash and burn" type of cultivation. Large-scale *Eucalyptus globulus* plantations, *Acacia mearnsii,* and *Camelia sinensis* have appeared, making the grassland ecosystem an understory vegetation. It seems there are over 30 grass species that have categorized been as threatened with extinction in the Shola grass lands of Western ghats (Thomas and Palmer, 2007; Brilliant et al., 2012).

Natural vegetation of Ganges Delta features tropical forests with deciduous and evergreen trees, although in most regions forest have been cleared to make agricultural prairies that support staple cereals and legumes. Tall grass prairies made of *Saccharum* is common in Northern Bangladesh. The ecoregion with faintly saline conditions known as *Sundarbans* delta is home to variety of botanical species, including mangrove vegetation that exist along with crops. Most dominant tree species in the delta is *Heritiera fomes*, which is known locally as "Sundri." Several bamboo species, garyan, mangrove palm (*Nypa fruticans*) and mangrove date palm (*Phoenix paludosa*) are commonly encountered. The Delta sand dune region and thickly vegetated mangroves host a variety of fauna including tigers, leopards, crocodiles, river dolphins, and Asiatic elephants.

6.1.5 DEVELOPMENT OF AGRICULTURAL PRAIRIES IN SOUTH ASIA

Historically, seeds for an enlarged agricultural prairie in the fertile Indus plains were sown with domestication of cereals such as wheat and barley, sometime around 5,500 B.C. Archaeological evidences from Mehergarh (8,000–6,000 BCE), and Mohanjo-e-Dharo in Pakistan dating 2,600 B.C. suggest that natives of Indus valley cultivated domesticated species of Einkorn wheat, Emmer wheat, barley, jujube, and dates. They practiced sedentary and nomadic agriculture, as well as pastoral systems. Evidences from Indus valley that date to 4,500 B.C. suggest that crops were also irrigated using intricate canals that emanated from small reservoirs (Constantini, 1983; Gregory, 1996; and Asouti, 2006). Further evidences suggest that cotton cultivation began in Indo-Gangetic plains some time during 4th millennium B.C. (Baber, 1996). Rice was domesticated in Indo-Gangetic plains and its regular cultivation had begun during 2,500 B.C. (Vishnu-Mittre, 1961, 1976; Sharma et al., 1980). Sugarcane is native to Gangetic plains and its cultivation was in vogue during 2nd millennium B.C. (Sharpe, 1998). Examination of excavation sites in North Karnataka suggest that rice cultivation occurred during 1,500 B.C. Fuller et al. (2004, 2005a, b) believe that rice may have been domesticated independently in South Indian plains and swamps. Rice was also introduced into southern plains from Brahmaputra valley (Randhawa, 1983) during 2nd millennium B.C. Natives of Gangetic plains grew a package of crops that included several cereals such as rice, barely, setaria, wheat and millets; legumes such as chickpea, peas, lentils, cowpeas; sugarcane, cotton, jute, and vegetables. Agricultural practices were well organized. Fields were classified into Urvara (fertile field), Kshetra (subplot), and Kedhara (wetland). Cropping seasons were known as Graishmika (summer) and Haimana (winter). Trained farmers known as krishitantri carried out farming surrounding their villages.

Archaeological evidences suggest that agri-pastoralism was in vogue in the Shola grassland regions of Western Ghats by 3rd millennium B.C. Migrants who settled in this region could be from Northwest plains of Indian subcontinent (Thomas and Palmer, 2007). However, ethnic studies indicate that they are called Badagas and adopted "slash-burn" type of farming, moving from one location to other within the Western Ghats. Agricultural cropping and development of settlements in the Western Ghats took root sometime during early Neolithic era. This is also the region where several minor legumes of South Indian origin were domesticated and later spread to the plains, for example, *Cajanus* and *Vigna* species (Van Der Maesen, 1990; see Krishna and Morrison, 2010).

According to Fuller (2006), several staple crops were independently domesticated in the Western ghats and plains of Peninsular India. Crops such as rice, wheat, barley, setaria, sorghum, millets, pulses, oilseeds, gourds, and cucumber formed the basic food crop package of South Indian Neolithic agrarian regions. Many of these crop species diffused into Indo-Gangetic plains. On the other hand, Southern Indian plains received several crops from Fertile Crescent, European plains, Indus region, and China. Fuller (2005a, b; 2006) states that diffusion and counterdiffusion of Neolithic crop packages were caused by human migration. It aided development of diversified crop pattern in South Asia.

Initially, during Neolithic period, crop production must have confined to the vicinity of dwellings. Nomadic agriculture, shifting cultivation, and pastoralism were perhaps equally important to development of agricultural prairies, along with lush native grass land vegetation. It seems agrarian intensification in South Asia occurred during Iron age (first century B.C. to fifth century A.D.), when larger patches of farming belts were steadily developed in and around the citadels and human settlements. Agricultural intensification, it seems was well aided by the invention of plow, several types of iron implements, systematically sown line cropping and various postharvest procedures (See Morrison, 2008; Krishna and Morrison, 2010). Major crops grown during Iron Age included cereals, legumes, sesamum, sugarcane, jute, and vegetables. Medieval period in South Asia witnessed introduction of several crops from South America, via Europe and Africa. Crop production occurred in two main seasons – Kharif (monsoon crop) and Rabi (post rainy crop). During Modern period, farming was well entrenched in Indo-Gangetic plains and Peninsular India. Crop productivity was low commensurate with fertility of soil, cultivation practices, and scanty *in situ* recycling of residues. Major crops grown in Indo-Gangetic plains were cereals such as wheat, rice, pearl millet, setaria, panicum, chickpea, cowpea, lentils, sesamum, mustard, sugarcane, cotton, jute, and vegetables. Southern Indian plains supported crops such as rice, sugarcane, ragi, navanay, jola (sorghum), pani-cum, niger, sesamum, groundnuts, cotton, and vegetable. Cotton was a predominant crop in the Vertisol plains of North Karnataka (Buchanan, 1807; Watt, 1889).

At the time of formation of Pakistan as a nation in 1947, the region's economy was predominantly agrarian. About 53 percent of GDP was derived from agricul-ture and allied activity of the population. During the past 5 decades, agricultural crop production technology has improved enormously. Most of the agricultural op-erations have been modernized, mechanized, and several of them are electronically regulated. For example, application of irrigation and fertilizer supply is highly so-phisticated. Yet, the GDP derived from agricultural crop production has declined from 53 to 23 percent of nation's GDP. Mechanization has reduced the number of agricultural workers by over 20 percent during past 5 decades. We may note that despite reduction in investments, about 75 percent of nation's foreign exchange is derived out of agricultural exports (Amjad, 2004). Pakistan's agriculture depends mainly on riverine and canals network for irrigation and improvement in produc-tivity. Currently, out of 23 million ha of cultivated land 18.09 million ha of land is irrigated and the balance of 4 million ha is rain fed. Hence, potential to enhance crop productivity in the region is high. As such, improvement in agricultural infra-structure and subsidies, since mid-1900s, has helped farmers in the Indus valley to enhance crop production rather markedly. Farm mechanization has been highly suc-cessful in the plains. The use of tractors, cultivators, disk plows, drills, and threshers has increased by 4–8 folds between 1975 and 2000 (Amjad, 2004).

During 1800–1950, agricultural productivity stagnated at subsistence levels (Guha, 1992). Factors such as droughts, famines, floods, wars and laxity of admin-istrative set up all had their share of effects on crop production. Average productiv-ity of crops such as rice (930 kg ha^{-1}), sorghum (495 kg ha^{-1}), pearl millet (401 kg ha^{-1}), gram (319 kg ha^{-1}) and groundnut (690 kg ha^{-1}) depreciated by 20–25 percent between 1900 and 1947, just prior to formation of Independent India (Guha, 1992; Naidu, 1941). Since 1950s, agriculture in India has been prioritized and encouraged. During 1940s, coordinated projects known as "Grow More food" were popularized to induce farmers to intensify crop production. "Integrated Production Program" initiated during early 1950s, aimed at providing rural electrification, irrigation, im-proved seeds and machinery. During this period, crop production was primarily via organic farming. Soil nutrient refurbishment was entirely through residue recycling and organic manure supply. Grain yield levels were not high. During 1960s, con-certed use of inorganic chemical fertilizers, high-yielding varieties and irrigation, together improved crop production in the Northern plains and Peninsular India. High Yielding Variety Programs for various crops and All India Coordinated Project on Farming systems stabilized grain productivity. Further, during 1970s, crop geno-types with better grain yield potential, resistance to disease and pests and ability to tolerate drought stress were released for cultivation in the plains.

The agricultural prairies were intensified during 1970s using inorganic chemi-cal fertilizers based on soil fertility tests, adoption of high-yielding varieties and irrigation. Productivity of cereals such as wheat, rice, and sorghum was improved by 20–30 percent over previous levels. Rice-wheat cropping system has gradually replaced the erstwhile legume-wheat, wheat-fallow, wheat-oilseed rotations of In-do-Gangetic plains. Rice-based cropping system dominates Gangetic Delta, Coasts, and Southern plains. Vertisol plains in Central and South India are dominated by sunflower, cotton, and sorghum. Currently, agricultural prairies in India extend into over 195 m ha and grain production is over 212 m t year^{-1}. About 125 m ha (68%) is dry land prairies and 70 m ha (37%) is wet or irrigated cropland (The World Bank, 2012). Major constraints to productivity are soil fertility, droughts, floods, erosion, and salinity. During the period from 1950 to 2000, grain productivity of wheat and rice showed marked increase. However, prairie zones with coarse grains and legumes did not show impressive increases. Productivity of oilseeds, although conspicuous, still was insufficient to satisfy domestic needs (Library of Congress, 2013). Kerr (1996) states that gain in productivity has been pronounced in irrigated zones. However, a large share of Indian agricultural expanse is rain fed and it is classified as dry lands. Farming in this belt is generally performed with low levels of fertilizer and water supply. Hence, this dry land zone needs greater attention in future, so that crop productivity is enhanced.

6.1.6   HUMAN MIGRATION, SETTLEMENTS, AND AGRICULTURAL PRODUCTION TRENDS IN SOUTH ASIA

Human resource is a crucial component to development, sustenance and productivity of any of the several agrarian regions found in South Asia. The Indus Plains is host to 77 percent of Pakistan's human population. The per capita income, food consumption, and literacy rates were higher in this region compared to other zones. This situation acts as incentive to people in other areas situated far off to migrate to Indus plains and establish settlements. Historically, migration of peasants into Punjab and Sind plains were initiated during 1872 till 1929. As a consequence of human settlement about 4.5 million ha of desert and waste land in the plains were converted to produce crops, mainly cereals. Most of peasants were derived from Eastern deserts namely *Nara* and *Thar*. The settlers from Eastern Deserts soon established productive agricultural farms and relegated the natives called as *Janglis*. There was increase in population in Punjab plains by 18 percent over previous levels during 1901–1911. It was attributed to migration of peasants from east. During, next 2 decades, peasant settlements were consistent and their population increased by 9 percent. Agricultural produce and infrastructure development occurred at a commensurate pace. Agricultural markets (*Mandis*) were developed in many locations within the Indus plains. In due course, by early-1900s, older settlements of natives that stayed in mounds were phased out and new villages were developed. Hasan and Raza (2009) state efforts by British to colonize Indus plains with migrants from eastern deserts completely changed the agricultural cropping pattern. Original pastoral and nomadic tendencies of farmers got replaced by permanent agricultural settlements. During the next phase, British controlled bureaucracy helped in the development of land, irrigation dams, water collection points and market yards at several locations all through the Indus valley. Agricultural villages called *Chaks* were developed closer to cropping zones and at a reasonable distance from towns. Canal network, water use grids and *mandis* helped agricultural communities in Punjab and Sind to prosper. Railways induced them to produce higher quantities of grains that could be exported to other regions in Northwest Indo-Gangetic plains (Hasan and Raza, 2009).

In Nepal, major migratory trend is caused by the agriculturally rich Terai and Gangetic plains. Paharis from mid hills often move out into Terai plains during cropping seasons in search of jobs. There are several regions within Terai, where in high caste farmers have moved permanently into the plains from hills and are engaged in permanent agriculture. They grow rice, wheat, maize, and cash crops such as tobacco, groundnut, and fruits. Fertile soils, irrigation and profitable crops lure hill country population into Terai plains (Wikipedia, 2011).

The British Indian regions were divided into Pakistan and India. Partition induced large migration of rural farm and urban dwellers from both regions. The Indus valley witnessed vacation of over 4.5 million people from the plains and in their place 6.2 million traveled from Punjab and Gangetic plains into Indus region. It seems within a short period, the population of Pakistan had increased by 1.8 million. Most of them were agrarian workers and farmers. Farm settlements were developed fresh in Sindh and Baluchistan. Therefore, a spurt in agricultural cropping got initiated in the Indus plains with infusion of fresh immigrants from the Gangetic plains. Crops grown by refugees were not completely different. Major crops included wheat, pearl millet, brassicas, and legumes.

The basic concept in the agricultural belts of South Asia, like anywhere else, is that soil fertility and climate affects crop produce and it in turn has consequences on human settlements and migration trends. Cookman (2012) states that, most recently, climate change effects such as severe drought in the Indus plains and flash floods created by intense monsoon precipitation did affect the cropping pattern in the otherwise fertile plains of Pakistan. It also caused human migration from areas affected with drought or flash floods. For example, during 2010/2011, wide spread deforestation and loss of cropping zones affected lively hoods of 9 million people and induced a large section to migrate in search of agricultural land and earnings. Internal displacement of farm workers was felt in the Indus plains, Baluchistan, and even in Mountainous regions of Pakistan. Crop production trends were severely disarrayed due to floods. It seems over a million ha of cropped land was destroyed in the plains.

Population increase has been marked in all of the South Asian regions during past 5 decades. Total number of people engaged in agricultural cropping either as cultivators or farm workers too has grown steadily, but it has encountered internal alteration (RGI, 2012: Table 6.1). At present, majority of farmers (54.4%) are classified as medium in terms of socio-economic status, 28 percent as low and 17 percent as high (Kumbhar et al., 2012).

TABLE 6.1 Total human population and agricultural workers in India during 1951–2011 (in million)

Year	Total Population	Rural Population	Cultivators	Farm Labourers	Total Agricultural Workers
1951	361.1	298.6	69.9	27.3	97.2
1961	439.2	360.3	99.6	31.5	131.1
1971	548.2	439.1	78.2	47.5	125.7
1981	683.3	523.9	92.5	55.5	148.2
1991	846.4	628.9	110.7	74.6	185.3
2001	1028.1	742.6	127.3	106.8	234.1
2011	1210.2	833.1	183.3	142.1	325.4

Source: Registrar General of India (Demography), Planning Commission, New Delhi, India; http://www.ecostatassam.nic.in/ agri_population.html

Migration of agricultural farming community is in vogue since several centuries. It seems initiation of monsoons and seeding has induced migration of farm workers. Medieval literature in India, for example *Baburnama*, mentions that migration of entire villages to fertile zones occurred. Migration and shifting agriculture that induced entire communities to move from one location to other has occurred since ages. Deshingkar and Akter (2009) states that during recent decades, migration of farm labor to "high productivity cropping belts" such as Gangetic plains is common. Wages are better in high grain productivity zones. Human migration, especially that of farm labor is maximum during peak season when agronomic procedures such as fertilizer application, weeding, interculture, and harvesting is to be completed in time. During a crop season, about 8,19,000 migrant farm workers are hired in Uttar Pradesh and Bihar. Interestingly, a few of the agronomic procedures, it seems are exclusive domains of migrant labor from neighboring low productivity zones. Now, let us consider an example from Peninsular India. In Maharashtra, about 1.6 million farmers rely largely on migrant farmers to complete several agronomic procedures (Deshingkar and Akter, 2009).

Bangladesh is host to about 140 million human inhabitants. The delta zone is particularly densely populated with 200 inhabitants km^{-2}. In Bangladesh, nearly 66 percent per cent of population thrives on agriculture and related occupation to derive their lively hood. Large section of agricultural workers is poor and their population is increasing. It increased by 1.6 percent during past 2 years. Floods, droughts, and crop failures have induced human migration within the nation and to adjoining nations. Currently, poverty, lack of infrastructure, and agricultural employment seem to be important factors that induce human migration (Battacharyya and Werz, 2012).

Soil erosion is an endemic problem in the flood plains and coasts of Bangladesh. It affects about 160 thousand ha of cropland that occurs close to river banks and sea coast. As a consequence, it is known to displace large number of agricultural house-holds and affect their lively hood. Human migratory trends are also attributable to maladies such as salinization of soil, submergence due to rise in sea water level, and coastal erosion.

Reports by Asian Development Bank (ADB) suggest that climate change has induced alterations in demography of intensely populated regions in Gangetic plains (ADB, 2010). Enhanced atmospheric temperature, altered soil fertility, cropping systems, and water management are said to induce migration. Actually, climate change and its effect on agricultural productivity may induce further migration in future, rather significantly in India and Bangladesh. It is said that large section of rural population in South Asia is exposed to low wages from farm work and is vul-nerable for migration. Rapid urbanization is also an important factor for migration from rural farms (Mitra and Murayama, 2008). Report by UNFCC suggests that by 2010 about 50 million farm workers in Indo-Gangetic plains were categorized as displaced from their original place of residence (ADB, 2010). The interactive effect of climate change, agricultural cropping pattern, and human migration has been studied by several researchers. Yet, we need to accrue greater details on this aspect. It is believed that during next few decades, migration will still be the best strategy to adapt to wage differentials in farms across India and Bangladesh. The cereal belts of Indo-Gangetic plains and South India will decide the rate of migration (Vishwana-than and Kavikumar, 2012). Of course, economics of cereal production and profits will largely regulate farm labor and its income and hence migratory trends.

According to Deshingkar and Farington (2006), in addition to economic status of farm workers, rural work and earning opportunities affect migration pattern in India, Pakistan, and Bangladesh. Factors such as size of farmer's house hold, his skills, educational level, infrastructure, dwellings, transport, and communication facilities also induce shifts from one location to another. For example, in South India, rural to urban migration is common and is induced by infrastructure and labor market in nonagricultural aspects (Uma et al., 2013). Lack of dwelling and infra-structure induces a large number of skilled and unskilled farm workers from North Karnataka to to migrate to Southern regions of the State. In Bangladesh, landless, poor farm workers are most prone to shift from one region to another. For example, during 1984–2010, rural farm worker population decreased from 73 to 66 percent, but nonfarm workers in urban regions increased from 27 to 34 percent. This shows a clear trend to migrate from predominately agricultural villages to urban nonfarm settings (Deshingkar and Farrington, 2006). Farm wage is an important factor that decides the farming community's need to migrate or stay back in a location. For example, real wages for farm work increased by 2 percent in many areas of Indo-Gangetic plains and South India. Despite it, there are several other reasons related to labor organizations, economic disposition and education that induce migration. In the Deccan plateau, for example in Telangana, farmers are poor and are not endowed with sufficient infrastructure. Hence, they are prone to frequent migration from one location to another. Richer farmers in other parts of Andhra Pradesh are prone less to migration.

Overall, human resource is available in relatively higher number in South Asian region. However, its efficient deployment, periodic training, and efficient use on farm land are important, if agricultural prairies have to sustain and perpetuate. Following is the current status of human population, its density, rural population, contribution of agriculture to GDP, per cent area covered by agricultural prairies and share of irrigated area in South Asian nations:

	Bangladesh	India	Nepal	Pakistan	Sri Lanka
Total Area (km²)	1,44,000	3,287,260	1,47,180	7,96,100	65,610
Population (in millions)	146.2	1182.2	28.3	166.5	20.7
Population Density (no. km⁻²)	1015	360	192	209	316
Rural Population (%)	72	70	81	64	85
Agriculture's contribution to GDP (%)	18.6	19.0	32.8	21.2	12.8
Agricultural Area (% of total)	65	55	30	33	40
Irrigated Area (% of total)	35.1	18.9	8.0	25.1	8.9

Source: IGES-GWP South Asia (2012); FAOSTAT (2011); ADB (2011); World Bank (2011)

6.2 NATURAL RESOURCES OF SOUTH ASIAN PRAIRIES

6.2.1 SOILS OF SOUTH ASIA

Soil survey and cartography of Indo-Gangetic plains began during mid-1800s. Geomorphic studies indicate at least two layers younger than those belonging to 13,500 B.C. The plains, it seems experienced fluctuations in weather pattern after the glaciation. From a cold and arid plain it changed to semiarid and humid belt during Holocene, that is, 7,900 B.C. During this period soil profile was exposed to weathering and several other modifications. Tectonic changes too affected the soil profile. It seems weathering of clay material especially illuviation and de-calcification were important weather processes during soil formation. Soil analysis also depicts accumulation of smectite and kaolinite clays. Pal et al. (2006) state that weather patterns played important role in development of older soil types. Soils traced in drier regions of plains are generally impoverished of organic matter. During recent years, consistent adoption of rice-wheat cropping system has generated several effects on soils of Indo-gangetic plains. Currently, soils might have reached equilibrium with regard to organic matter and nutrients, because of 40–50 years of consistent cereal production (Pal et al., 2006).

Agricultural soils of Pakistan, in general, are fertile with alluvial deposits of the Indus and its tributaries. Major soil types utilized for crop production are moderately rich in organic matter and mineral nutrients. Still, productivity is low, because incessant cropping needs periodic replenishment. Major soil orders traced in the cropping zones of Pakistan are Alfisols, Aridisols, Entisols, Inceptisols, Molisols, and Vertisols (Brady, 1975; Khan, 2012; Figure 6.2). There are 21 sub orders of soils found in the agricultural prairies of Pakistan, 38 great groups, 112 subgroups, 353 families, and 894 soil series. Mollisols, Inceptisols, and Alfisols are among the most commonly used soils by the farmers in the rice-wheat belt.

The land use capability within Pakistan depends on various soil traits that make it useful for agricultural cropping or not so congenial. Pakistan has 86,909 thousand ha land that has been classified through series of surveys. The land capability has been classified into different classes such as class I—very good agricultural land; class II—good agricultural land; class III—moderate agricultural land; class IV—poor/marginal agricultural land; class V—good forest or range land; class VI—moderately good forest/range land; class VII—poor forest/range; class VIII—Barren land; Agriculturally unproductive land or nonagricultural land (Mian and Mirza, 1993).

Soils of agricultural prairies of Pakistan are affected by erosion induced by water and wind. It seems over 580 thousand ha of cropland is vulnerable to erosion through water and 4,760 thousand ha is afflicted by various degrees of wind erosion (Khan, 2012). Salinity/sodicity is among major soil-related maladies that affects Punjab and Sind agricultural areas. Pakistan has about 972 thousand ha of Gypsiferous irrigated and 713 thousand gypsiferous un-irrigated soil affected by Salinity. Porus salinity and Dense salinity are other types of salinity traced in agricultural soils of Pakistan. Together, salinity/sodicity is said have spread into 5,327 thousand ha of cropland in Pakistan. Draining fields with fresh water and growing relatively salt-tolerant crop species (e.g., rice) seem best alternatives to sustain the prairies (Khan, 2012).

FIGURE 6.2   Soil types of Indus Plains of Pakistan.
Note: Be = Eutric Cambisols; I = Lithosols; Jc Calcaric Fluvisols; Lo = Orthic Luvisols; Qc = Cambic Arenosols; Rc = Calcaric Regosols; Xh = Haplic Xerosols; Xk = Calcic Xerosols; Yh = Haplic Yarmosols; Yk = Calcic Yarmosols; Zg = Gleyic Solonchecks; Zt = Takyric Solonchaks; Wa = Water bodies; Gl = Glaciers.
Source: IRRI (2013a); FAO-UNESCO Soil maps

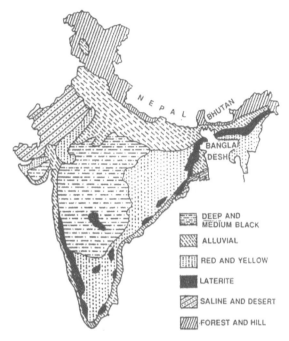

FIGURE 6.3   Soil Types of India.
Source: UNESCO-FAO maps

South Indian agricultural prairies and diverse natural vegetation thrive on several types of soils. These soils differ with regard to morphogenesis, physicochemical
traits, and fertility. Productivity of soil is highly variable. Southern Indian plains
and Coasts support intensive production of rice, sugarcane and arable crops such
as sorghum, maize, pigeonpea, groundnut, sunflower, and vegetables. Several species of cash crops and horticultural plantations also thrive with high productivity. Major soil types encountered in South Indian peninsula are Red Sandy loams
(Acrisols, Alfisols, Inceptisols and Entisols); Lateritic soils (Nitisols, Ferralsols);
Alluvial soils (Fluvisols); Black Clayey soils (Vertisols, Vertic Inceptisols); Forest
soils (Cambisols, Luvisols); Peats and marshy soil (Histosols); Saline and Sodic
soils (Solanchecks and Solonetz) (Krishna, 2010a; Figure 6.3). The Red Alfisols are
common in the plains and undulated regions of Southern Andhra Pradesh, Karnataka, and Tamil Nadu. Red Alfisols constitute 65 percent of cropping zones of South
India. Red loamy soils support wide range of crops such as wet land rice, arable
crops like sorghum, maize, pigeonpea, groundnut, cowpea, sunflower, and so on.
Black clayey soils or Vertisols are found in Central Indian Plateau region, North
Karnataka, and Andhra Pradesh. They are rich in Montomorillonite clays and possess high water and nutrient buffering capacity. They are sticky to plow immediately

after rain. They show shrink-swell reaction based on soil moisture and temperature conditions. In South India, Vertisols are popularly known as "Black Cotton soil" because it supports rich crop of cotton. Crops such as pigeon pea, sunflower, and maize are also grown on Vertisols. Lateritic soil is common in the Western Ghats. It is highly weathered and rich in Al and Fe minerals. They support rice, several dry land crops, and plantations. Forest soils or Histosols occur in the hill ranges and evergreen tropical forest region of South India. They are peaty and rich in organic matter content. Soloncheks and Solonetz are found in the Vertisol plains of North Karnataka covering regions in districts of Bellary, Kalburgi, Raichur, and Bidar. Salinity is attributed to high Na content of Black soils. Periodic draining using fresh water and cultivation of rice that tolerates salinity better is proposed.

Soils types found in Bangladesh that support wetland rice and arable crops are highly fertile due to incorporation of alluvium from the three major rivers. Based on geological aspects and fertility, soil types found in the delta and other regions of Bangladesh are classified into seven types. They are:

(a) Red Soil Tract or Madhupur Tract	Soils are clayey and ferruginous. They are deficient in N, P, organic matter, and lime. They are rich in Fe and Al. Phosphate fixation capacity is also high. Addition of FYM and organic manures is highly recommended prior to crop season. Soil pH ranges from 5.6 to 6.5. Such Red soils occur in Dhaka, Mymensingh, Comilla and Naokali.
(b) Barind Tract	Soils are alluvial and argillaceous. They are pale yellow or reddish brown in color. Kankar and ferruginous concretions are common. Soil is deficient in lime, N, and P. Soil fertility is generally poor and needs replenishments to achieve moderate grain yield. Soil pH ranges from 6 to 6.5. This soil type is traced in Rajshahi, Dinajpur, and Bogra region.
(c) Gangetic Alluvium Region	Soil types found are fertile with high Ca and P content. Soils are clay loams or sandy loams. Soils are generally fertile but crops do respond to fertilizer supply.
(d) Teesta Silt Area	Soils are fertile and supplied well with P and K. They are mostly silts and silt loams. Paddy, tobacco, and sugarcane are major crops found in this region. Soil pH is 6.0–6.5. Teesta silt is common in Dinajpur, Rangpur, Bogra, and Pabna districts.
(e) Brahmaputra Alluvium	Soils are sandy loams, fertile, and rich with deposits of river silt. Soil pH ranges from 5.5 to 6.8. Crops such as rice and jute are common. Winter paddy is provided with manure, N, P, and K fertilizers. This soil type is traced in Mymensingh, Dhaka, Chittagong, and Sylhet districts.
(g) Soils of Chittagong Hill Tracts	These soils are brown clay loams. They are acidic. They are sometimes shallow in depth and show up shale/sand stone bedrocks. Soils are utilized for "shifting agriculture."

Source: FAO-UNESCO Soil Maps; ICID (2012)

6.2.2 SOIL TILLAGE AND CROP PRODUCTIVITY

Soil tillage has been in vogue in the South Asian agrarian zones since the time when plow and plowshare was invented. Large tracts of Mollisols and Inceptisols of Indo-Gangetic region, Vertisol plains of Central India/Deccan; Alfisols of Southern Indian plains, sandy coastal soils and Aridisols of Rajasthan have all been subjected to tillage, since ancient period. Ploughing allows soils to be loosened and turned. It induces oxidative reactions and aerates the soil profile to a certain depth. The microbial activity gets enhanced releasing SOM as CO_2 into atmosphere. Ploughing reduces SOC content and reduces the extent of C-sequestration possible. Yet, plowing is routine exercise performed season after season in South Asia. Clearance of natural vegetation and plowing is perhaps the first agronomic activity in any field within the agricultural prairies. Aspects such as plowing implements, intensity of plowing, type of plowing and source of energy to perform the task has undergone, at times gradual and at other periods spurts of modernization. Initially, plowing was performed through human labor, it was followed by use of draught animal force (like Oxen, Buffaloes; Horses, and Camel). Since early 1900s, with invention of tractors and farm vehicles, plowing has been accomplished using mechanical power. Conventional deep tillage has been the main stay in most parts of South Asian Agricultural zones for many centuries, right until 1970s. Since past 3–4 decades, tillage intensity has been reduced. Conservation tillage or reduced tillage is practiced. Conservation tillage avoids excessive opening of soil and makes it less vulnerable to erosion and loss of SOC as CO_2. Conservation tillage and no-tillage practices affect several of soil physicochemical properties less detrimentally than conventional tillage.

Tillage was practiced since ages because of several advantages. Let us quote few of them. Tillage helped to loosen and soften soil, so that seed beds are easily formed. Tillage helped farmers to reduce weed intensity. It exposed weed seeds or induced them to germinate rapidly, so that weed seedlings could be uprooted easily. Tillage helped in releasing nutrients by hastening mineralization. Tillage aided during deeper placement of crop residues and organic manures. Tillage made soil profile more arable and loose without compaction. However, at a later stage, researchers and farmers alike, have observed that deep tillage induces loss of SOC, reduces soil quality and induces soil erosion (Hobbs et al., 2008). Hence, the tendency has been to reduce tillage intensity.

No-tillage is a popular agronomic suggestion in the Indo-Gangetic plains and other regions of South Asia. Zero tillage is supposed to preserve soil fertility better than other types of soil management procedures. It helps in conserving soil moisture in the profile. Fields under no-tillage are less prone to loss SOC via CO_2 emission and hence they sequester C more than other tillage systems. Zero tillage is advantageous because it avoids expenditure on equipment and fuel. Economic analysis suggests that farmers save about 36 lts of diesel ha^{-1} if they adopt no-tillage. However, fields under no-tillage are prone to infestation with weeds. Hence, timely weeding right at the early stages of the crop is a necessity. According to Erenstein (2009a), rice-wheat cropping belt has been experiencing stagnation in grain yield since mid-1990s. It has been attributed to soil deterioration, due to erosion, loss of soil nutrients, organic matter depletion, and reduced precipitation-use efficiency. Elaborate soil management procedures and tillage is difficult to perform because farmers are often allowed very short period after harvest of paddy and before planting wheat during early winter. Soil fertility and moisture conservation procedures are therefore sought everywhere in the plains from Pakistan to Bangladesh. Since, Indo-Gangetic belt is a major food generating biome for 40 percent of South Asian population, its upkeep with regard to soil fertility and crop productivity is essential. During past decade, no-tillage has been prescribed in many locations within Indo-Gangetic belt as a method that reduces loss of soil quality and stabilizes grain yield at higher levels. Currently, it seems over 2.2 million ha of rice-wheat belt is under no-tillage systems in India and Pakistan. Erenstein (2009a) opines that success of zero tillage in Brazilian Cerrados, Pampas of Argentina and Great Plains of North America has been a good example and impetus to farmers in South Asian plains. No-tillage requires a modified seed drill that dibbles seeds directly without recourse to soil tillage and land preparation. Currently, zero tillage equipment is being produced in large numbers to help farmers in the agricultural prairies.

Regarding economic advantages, in India, farmers adopting zero tillage derive higher profits equivalent to 97 UD\$ ha^{-1}. Large-scale surveys in Punjab, Haryana, and Uttar Pradesh suggest that farmer households have gained about 180–340 US \$ year^{-1} due to adoption of no-tillage. It is said cost saving on tillage equipment, fuel, and labor is the main contributor to profits. There are farmers who adopt double no-tillage system, which means that both rice and wheat are sown without prior tillage of fields. Erenstein (2009a, b) suggests that no-tillage system should be considered as part of a larger concept called "Conservation Agriculture." Conservation agriculture helps in maintaining soil fertility, its quality, delays or avoids deterioration through erosion and thwarts emissions. Consequently, it helps farmer in this biome to stabilize grain harvest at higher levels.

Conservation tillage or no-tillage systems are gaining in acceptance among farmers in the Indo-Gangetic plains, because of several reasons and clear benefits to soil fertility and productivity. According to Singh and Kaur (2012), rice-wheat cropping system is the main stay in most parts of Indo-Gangetic plains. It covers about 13.5–14.0 m ha encompassing large areas in Pakistan, India, Nepal, and Bangladesh (Gupta et al., 2003; Gupta and Seth, 2007; Timsina and Connor, 2001). This cropping system contributes a large share of cereals to South Asian nations. However, during past decade, productivity of the Indo-Gangetic plains has rather stagnated or even declined in some patches (Kumar et al., 2002; Ladha et al., 2003; Prasad 2005). The grain/forage productivity has to increase, if higher population is to be supplied with cereal grains, because expansion of area is not a possibility. There is a strong belief that excessive tillage, inadequate mulching, and rampant loss of soil fertility are some causes of grain yield stagnation and loss of fertility in Indo-

Gangetic plains. Conventional tillage systems led to yield stagnation. Some of the major ill effects of continuous rice-wheat culture stated by Singh and Kaur (2012) are: (a) it fatigues natural resources, (b) reduces soil organic matter, (c) salinity and sodicity reduces soil fertility, (d) groundwater levels get depreciated, (e) groundwater contamination due to seepage of NO_3 becomes a problem of serious proportions in many location in the plains, and (f) greenhouse gas emissions have decreased fertilizer-use efficiency. Most importantly, incessant tillage and alternate wetland/arable crop production have led to deterioration of soil texture, severe alterations in physicochemical and microbiological properties of soil and reduced soil productivity. Hence, most immediate step is to halt use of conventional tillage and in its place adopt shallow conservation tillage or no-tillage.

Major advantages due to adoption of conservation/zero tillage in the Indo-Gangetic plain are listed as follows: (a) Reduced soil erosion, (b) improved moisture content of soil, (c) nutrient rich soil, (d) balanced nutrients, (e) reduced consumption of fuel and usage of tractors/vehicles, (f) less runoff and sedimentation in the area, (g) reduction in wind erosion and reduced loss of soil-C to atmosphere (Singh and Kaur, 2012). According to Erenstein (2009b), adoption of no-tillage planting of wheat after rice harvest has been successful in North West Indian region, and to a lesser extent in the Indus Plains of Pakistan. There are advantages in terms of soil quality and productivity due to no-tillage (Table 6.2). No-tillage has indeed alleviated soil deterioration in many regions. No-tillage allows farmers to dibble seeds without the usual delays caused in land tillage and preparation. Hence, it permits rapid growth of wheat and development of wheat canopy prior to appearance of *Phylaris minor*, the major weed species of Indo-Gangetic plains. In fact, Gianessi (2013) states that, if zero tillage gains in popularity further and is adopted in almost all regions of rice-wheat belt of Indo Gangetic plains, then use of larger dosages of herbicide is inevitable. Weeds left out after rice harvest is to be cleared, either using herbicides or by hand.

TABLE 6.2 A comparison of no-tillage and conventional tillage with regard to soil properties and grain/straw yield

Tillage	Bulk Density	Infiltration Rate	Soil Temperature	Grain Yield	Straw Yield
	g cm^{-3}	cm min^{-1}	C	q ha^{-1}	q ha^{-1}
Zero tillage	1.44	8.0	31.4	37.2	56.6
Conventional tillage	1.46	12.0	31.8	36.9	58.1

Source: Erenstein (2009b)

Reports by FAO (2001) clearly suggest that, in the Indo-Gangetic Plains, rice harvest gets delayed beyond November, 20th, which is actually a date stipulated for seeding wheat. Wheat seeding therefore gets delayed further, leading to loss in grain harvests. However, with farmers adopting no-tillage, immediately after harvest of rice, they can clean up the stubbles and go for seeding wheat without recourse to elaborate plowing and land preparation. This system helps in conserving soil organic matter and reduces erosion of surface soil. Weed control needs to be pressed immediately since they could pose a problem.

As stated earlier, no-tillage systems are in vogue in large patches of rice-wheat belt of Indus Plains in Pakistan. It is generally accepted that zero tillage, plus application of crop residue mulch and fertilizer-N is highly effective in improving cereal yield in the plains (Mohammed et al., 2012). The water-use and fertilizer-N efficiency is enhanced significantly due to zero tillage. Application of crop residue mulch improved grain yield by 520 kg ha^{-1}. Crop rotations such as wheat-legume-wheat, wheat-fallow-wheat, and rice-wheat-fallow benefited most due to adoption of zero tillage.

Zero tillage practiced in the Terai region of Nepal was effective in reducing loss of SOC from the top layers of soil horizon. Zero tillage practiced along with crop residue application helped in enhancing SOC and quality (Ghimire, 2013). Fields with zero tillage had 76.1 t C ha^{-1} compared to 73.1 t C ha^{-1} in fields kept under conventional tillage.

Overall, agronomic procedures that need greater attention during rice-wheat production in the Indo-Gangetic plains are summer plowing, shallow plowing, puddling prior to transplantation of paddy seedlings, leveling, harrowing, conservation tillage prior to wheat seeding, preparation of ridges and furrows, and contour bunding. These soil tillage and land preparation methods do influence several soil physicochemical processes, crop growth pattern, and grain formation (Ikisan, 2013).

6.2.3   SOIL FERTILITY AND NUTRIENT DYNAMICS IN SOUTH ASIAN PRAIRIES

The inherent fertility of soils traced in the agrarian zones of South Asia varies enormously. Soil types and their nutrient status differ based on geographic location and hence farmers try to identify crops and pasture species that suit best. Major cereals such as wheat, rice, maize, and cash crops find their way into fertile plains. The agrarian belts in Pakistan, India, Nepal, and Bangladesh have been cropped incessantly for several 1,000 years now. The nutrients recycled through crop residues and fallows refurbished the fertility. Nutrients were replenished mainly using organic manures. This system allowed subsistence production systems with low level of grain/forage yield. Nutrients supplied via atmospheric depositions, irrigation channels, alluvium, dust storms, and precipitation is mostly small and often negligible. Yet on a large scale, Indus plains or South Indian Plains may receive large quantity of nutrients from the above sources.

Soil fertility deterioration, reduction in nutrient availability and crop productivity are common occurrences in regions that have been cultivated incessantly without appropriate nutrient replenishment schedules. Large areas of South Asian prairies are naturally deficient for various nutrients. Therefore, growth of natural vegetation and crops will be less than optimum, unless nutrient deficiency is corrected. Crop production trends and uneven removal of major and micronutrients from the profile is known to affect most agrarian regions in South Asia. This situation could be alleviated only by soil testing and devising nutrient inputs in proper ratios. Overall, it is clear that soil fertility depletion has to be corrected using fertilizers with accurate quantities, maintaining optimum ratios. Fertilizers should be dispensed at most appropriate time during crop growth.

The Indus valley is a major crop production zone in Pakistan. It supplies food grains to over 150 million people indigenous to the region. Soil types found in the Indus plains, mainly Mollisols and Inceptisols are fertile and moderately rich in organic matter. Crop productivity under organic farming systems may reach 1.2–1.5 t grains ha plus forage. However, incessant cropping depletes nutrients. Indus plains are currently-, moderately-, intensely-cultivated and grain yield expectations are higher at 4–5 t grains ha^{-1}. Hence, periodic soil tests and fertilizer supply schedules based on yield goal is mandatory all through the Plains of Punjab and Sind. Fertilizer supply is increasing commensurate with development of new crop genotypes, irrigation facilities, and higher yield goals. Fertilizer industry has also grown rapidly, yet nutrient produced is insufficient. Hence, Pakistan imports a certain small quantity of fertilizer-based nutrients from other regions.

Research on use of fertilizers, mainly N and P, to improve crop yield was initiated in the Punjab plains in 1909. During 1934–1952, fertilizer formulations were tested at Lyallpur and other locations for their efficacy in improving crop productivity. Standardized fertilized schedules prepared by FAO and other agencies were regularly adopted during 1960s. These steps helped in sustaining and improving the agricultural prairies of Indus region (FAO, 2007a). Fertilizer consumption in Pakistan, like other regions, is dependent on geographic region, weather pattern, season, availability of water/irrigation supplements, and support price. Punjab plain supports largest crop production in Pakistan. Crops are grown intensely with high fertilizer supply. Hence, maximum quantity of fertilizer is consumed in this region. Sind is also an important agrarian region that consumes large quantities of fertilizer-based nutrients. Following is the cropped area and fertilizer supplies in different regions of Pakistan during past decade:

Province	Cropped Area (m ha)	Fertilizer Supply (000 t)
Punjab	16.1	2,063
Sindh	3.2	674
NWFP	2.0	204
Baluchistan	0.9	77
Total (Pakistan)	22.12	3,013

Source: FAO (2007b)

Note: About 68 percent of total fertilizer-based nutrients generated in the industry gets spread in the agrarian regions of Punjab, followed by 22 percent into crop belts of Sindh. Small portions of fertilizers are used by other regions. About 7 percent of fertilizer is used by Northwest Frontier Province and 3 percent by Baluchistan. NWFP = North-West Frontier Province

Fertilizer supply varies depending on the season. During a year, fertilizer supply to crop belts gets accentuated just prior to seeding season. In the Indus valley, wheat sowing begins during late October/November. Hence, maximum quantity of fertilizer input into fields occurs during early October to late November (FAO, 2007b). Soil-N is depleted in higher quantities compared to other nutrients. Hence, fertilizers containing N are supplied in relatively higher quantities. For example, among the various fertilizer formulations, urea is supplied more. About 66 percent of all fertilizer used in Pakistan is urea. About 18 percent is Di-ammonium phosphate, 5.5 percent is Nitro-phosphates, 5.7 percent is Calcium ammonium Phosphate and 2.9 percent is Single super Phosphate.

Fertilizer allocation in the Indus plain is dependent on crop species and the area it occupies. Pakistan has 22.3 m ha of cropped land and consumes about 3 million t fertilizer annually. Crops that dominate the Indus region garner more of nutrients. Following data shows that wheat that dominates the prairies also garners largest fraction of fertilizer-based nutrients.

	Wheat	Rice	Cotton	Maize	Pulse	Sugarcane	Others
Area Occupied (%)	36.3	9.5	14.1	4.2	6.2	4.5	20.8
Fertilize Consumed (%)	45.4	5.4	23.2	2.3	1.1	10.6	7.4

Source: FAO (2007c)

Crop wise, wheat is supplied higher quantities of fertilizer-based nutrients because it is cultivated intensively with an aim to harvest more grains. Wheat receives about 90–170 kg N, 60–110 kg P, and 50–60 K ha^{-1} depending on the location. Wheat grown with irrigation support receives still higher quantities of fertilizers. Paddy is provided with 120–150 kg N, 67–100 kg P, and 50–60 kg k ha^{-1}. Cotton is an important cash crop. It receives 120–270 kg N, 60 kg P, and kg K ha^{-1}. Sugarcane is supplied with 90–300 kg N, 60–125 kg P, and 60–120 k ha^{-1}, depending soil type, location and genotype (FAO, 2007d). Fertilizer supply to sugarcane crop in Sindh and Punjab is about 70–110 kg N, 50 kg P_2O_5, and 50–60 kg K_2O (Sindh Agriculture Department, 2011). Sugarcane stays in the field for a longer duration. Pulses such as pigeonpea and chickpea derive a certain amount of their N requirement by establishing symbiosis with Rhizobium, that is, via Biological Nitrogen fixation. Hence, fertilizer-N supply is slightly lower. Pulses are supplied with 20–40 kg N and 60–90 kg P ha^{-1} (FAO, 2007d). During recent years, cereal crops are mostly supplied with fertilizer-N in split dosages, in order to improve fertilizer efficiency. Splitting fertilizer-N supply also avoids undue accumulation of nutrients in the soil profile. Split application of fertilizers reduces loss through leaching, percolation, and emissions.

Farmers in the prairies of Indus valley do refurbish soil nutrients depleted each season using organic manures. In fact, organic manures were the main stay for the crop productivity. Crop residues and leaf litter were other major sources that nourished natural vegetation and crops. Farmers in Pakistan utilize several different organic sources. Materials such as leaf litter, crop residue, animal manure, industrial waste, and city waste are used to prop up the prairies (FAO, 2007e). Green manure supply to rice fields prior to transplanting is a common practice in the wet prairies of Indus valley. Green manure application supposedly increases efficiency of inorganic fertilizer, if used in conjunction. Grain yield of rice improved by 150–200 kg ha^{-1}, if green manure and fertilizers are used together (Mian et al., 1988). Biofertilizers, such as *Bradyrhizobium, Azotobacter*, Blue green algae, and Phosphate Solubilizing microbes are also applied to crop fields in the Indus Valley. Legume crop and mixed pastures gain a certain amount of N from BNF. It may range from 20 to 40 kg N ha^{-1} season^{-1}. Soybean grain yield improved by 50–100 kg ha^{-1} if inoculated with rhizobium. Maize grain yield improved by 6–15 percent over control that produces 5.37 t grain ha^{-1} if soil is inoculated with *Azotobacter* (Hussain et al., 2010; see FAO, 2007e).

Micronutrient deficiencies restrict crop response to nutrients. Their deficiencies have appeared in Indus plains, since 1980s. This is attributable to depletion of micronutrients, especially Zn and B. Inadequate recycling of crop residues, reduced application of organic manures and soil erosion has caused micronutrient deficiency and imbalance in soil nutrient availability (Ahmad et al., 2013). Farmers in the prairies replenish micronutrients whenever they add FYM and other organic manures. Many others apply $ZnSO_4$ and Borax once in 3 years to correct the nutrient ratios (NIAB, 2013). Farmers in the Indus plain also practice integrated nutrient management. It involves supply of nutrients through several sources each contributing a certain amount of requirements of the crop. Inorganic fertilizers that are easily available and induce crop growth rather quickly are applied in larger proportion. Several other sources such as FYM, crop residues, organic manures, industrial wastes, and microbial inoculants are also supplied to crop fields (NIAB, 2013). Let us consider an example. Soils are eroded and low in fertility status in some parts of North West Frontier Province. Hence, integrated approaches are practiced that involve application of inorganic fertilizers (120 kg N, 90 kg P, and 40 kg K ha^{-1}), organic manures such as FYM (10 t ha^{-1}), plus mung bean residues as mulch. Fallows with legume crops are also in vogue in this region. Such a rotation helps in attaining wheat grain yield of 5.5 t ha^{-1}. It also thwarts soil degradation and loss of quality (Naeem, 2009).

Indo-Gangetic plain is a large fluvial plain that is cultivated intensively. It offers food grains (Rice/wheat) that is sufficient to feed 40 percent of India's population. During past 4 decades, this agrarian belt has been impinged with consistently higher levels of fertilizers and FYM, in order to achieve commensurately enhanced yield goals. The productivity of Rice-wheat system has reached 19.5–21.5 t grains ha^{-1} year^{-1} (Aggarwal 2000; Abrol et al., 2002). However, soil deterioration and reduction of fertilizer efficiency have affected the productivity of the region. For example, in Punjab, Haryana, Uttar Pradesh, Bihar, and West Bengal average productivity of cereals has stagnated at 7.5–9.5 t ha^{-1} year^{-1} (Ladha, 2003). It seems there is still a large yield gap that needs to be rectified and grain productivity enhanced in many locations of the plains. It seems fertilizer inputs needed to enhance yield is ever higher, because "partial factor productivity" has decreased in this region. Partial factor productivity of fertilizer is defined as grain output divided by fertilizer input (Snyder and Bruulsema, 2007). For example, in case of fertilizer (N, P, and K) supply the efficiency (or partial factor productivity) has steadily declined. Hence, fertilizer input has to be increased to achieve higher yield goals. Following is an example from eastern Indo-Gangetic plain that depicts decline in fertilizer efficiency in the rice-wheat belt:

1960–1970 fertilizer supply increased to 1.47 m t year^{-1}; increase in productivity was 26.4 m t grains at an efficiency ratio of 17.9 kg grains kg fertilizer (N, P, and K) applied)

1970–1980 fertilizer supply increased to 2.44 m t year^{-1}; increase in productivity was 31.09 m t grains at an efficiency ratio of 12.7 kg grains kg fertilizer (N, P, and K) applied

1980–1990 fertilizer supply increased to 5.28 m t year^{-1}; increase in productivity was 46.80 m t grains at an efficiency ratio of 8.9 kg grains kg fertilizer (N, P, and K) applied

1990–2000 fertilizer supply increased to 3.18 m t year^{-1}; increase in productivity was 19.53 m t grains at an efficiency ratio of 6.3 kg grains kg fertilizer (N, P, and K) applied

---

Source: Singh and Bansal (2010)

Due to decline in fertilizer use efficiency, it seems farmers in the rice-wheat belt are required to apply 10–60 kg N ha^{-1} in excess to achieve the same grain production level (Ladha, 2000). No doubt, higher supply of fertilizer has detrimental consequences on soil quality. Nutrient loss through percolation and seepage may increase and contamination of groundwater is a clear possibility.

In the Eastern Indo-Gangetic belt, crops are fertilized to achieve higher yield goals. Fertilizer supply recommended by Agricultural Ministry is 141–269 kg N ha^{-1}, 101–131 kg TSP ha^{-1}, and 69–121 kg K ha^{-1}. However, actual use is 135–192 kg N, 28–47 kg P, and 17–37 kg K ha^{-1}. It leaves a deficit of 4–28 kg N, 64 kg P, and 60 kg K ha^{-1} (Ministry of Agriculture, Bangladesh, 2004). Farmers in the Delta seem to apply N in quantities as recommended or required by the crop. However, it falls short with regard to other two major nutrients, P and K. Clearly, if there is growth limitation it is perhaps caused by lack of P and K. It may create imbalance in the ratios of major nutrients required by crops.

6.2.4   NITROGEN IN PRAIRIES

Nitrogen dynamics in the South Asian prairies has immediate impact on growth and productivity of agricultural prairies. Nitrogen supply, its exploration and utilization by crops, loss via natural and anthropogenic processes and removal via grains/forage are some aspects that need greater attention. Soil-N inherent in the profile, especially its chemical forms are of significance to development of natural vegetation/crops. As stated above, soil-N is relatively low/moderate in the South Asian region. Natural vegetation and cropland both receive N from different sources. Nitrogen addition to soil profile occurs through atmospheric deposits, precipitation, and dust storms. This fraction is generally small in the agricultural prairies. Farmers neglect it while devising fertilizer-N schedules. However, in locations with subsistence farming practice, even small quantities of N derived from atmosphere may be of value to crop growth. Farming belts in South Asia receive N through other sources such as organic manure application, crop residue recycling, green manures, and biological nitrogen fixation (BNF). Inorganic fertilizer inputs are the major form of N supply to crop fields.

Nitrogen supply to South Asian agricultural prairies is crucial since crop productivity is directly dependent on soil-N status. Fertilizer-N consumption in India, for example, has increased enormously since mid-1900s. In 1950, it was 0.6 m t year^{-1}. During 2000–2002, it averaged 11.3 m t year^{-1} (FAI, 2003). Regarding Indo-Gangetic plains, N is supposed to be the most important soil fertility-related input to the biome. Fertilizer-N usage actually increased with the release of fertilizer-responsive genotypes in 1960s. According to Pathak et al. (2006), about 98 kg N ha^{-1} is supplied through inorganic chemical fertilizers, 37 kg N ha^{-1} via FYM and organic manures, 17 kg N ha^{-1} through biological N fixation and 7 kg N ha^{-1} is derived through atmospheric deposits. As a consequence of rice-wheat cropping, about 175 kg N ha^{-1} is removed from soils, 14 kg N ha^{-1} is lost via volatilization, 12 kg N ha^{-1} is lost via leaching and seepage, and 4 kg gets emitted as N_2O or N_2. It seems NH_3 volatilization is the most conspicuous loss from the rice-wheat belt. During recent years, farmers in the Indo-Gangetic plains and in other regions are guided, based on projections from several computer models and simulations that consider several details about soil fertility, physicochemical traits, crops, genotypes, and agroclimate.

In nature, soil-N is emitted into atmosphere in different forms. Mechanisms such as volatilization, gaseous emission, dusts, and pollutants from industrial process aid transfer of N to atmosphere. However, a large portion of N emitted is returned to earth via natural process. Study of regional precipitation pattern and N deposits through it; suggest that average concentration of NH_4^+ in the rainfall is about 9 μg eq l^{-1} with a range of 4–16 μg eq l^{-1}. Lowest NH_4^+-N was traced in samples from interior farming zones of India. Precipitation samples near cropland located closer to suburban had over 14 μg eq l^{-1} NH_4^+-N (Kulshrestha et al., 2005; Sharma et al., 2008). Precipitation sample from Industrial location has high concentration of NH_4^+-N. It ranged from 18 to 37 μ eq l^{-1}. Regarding NO_3–N in precipitation, samples from rural, suburban, and industrial zones in India contained 22, 11, and 21 μ eq l^{-1} respectively. Following is the annual NH_4^+-N and NO_3–N fluxes recorded in India on an area basis, during past 5 years (in μ eq^{-1} m^{-2} year^{-1}):

NH_4^+–N: Rural farming zones—10; Suburban locations—13; Urban—18; and Industrial—26

NO_3–N: Rural farming zones—18; Suburban locations—14; Urban—26; and Industrial—29

Source: Sharma et al. (2008)

Computer-based weather models and those treating N dynamics in the ecosystem in detail have forecasted that, on an average, cereal belts in India may receive about 350 mg N m^{-2} year^{-1} as NH$_4^+$N and 130 mg N m^{-2} year^{-1} as NO$_3$–N deposits. Now, if we consider that large stretches of subsistence or low input crop production trends in drier tracts of Pakistan, India, Terai in Nepal and Bangladesh, then atmospheric inputs of N are valuable. Since, inorganic fertilizers are not used to any great extent, crops or natural vegetation depend on atmospheric-N deposits. There is no doubt that atmospheric N deposits are essential to natural pastures and shrub land vegetation common in plains.

6.2.5   FERTILIZER SUPPLY AND NUTRIENT DYNAMICS

During past 5 decades, fertilizer supply to prairies in South Asia has played a key role in generating food grains. Fertilizer supply has increased steadily since 1960s. It has led to intensification of crop production. Large share of fertilizer has been garnered by cereals such as wheat, rice, and maize. Following data depicts use of fertilizers in South Asia:

Fertilizer Supply and Productivity of Rice and Wheat in Indo-Gangetic Plains				
	Pakistan	India	Nepal (Terai)	Bangladesh
Fertilizer supply (kg ha^{-1})	131	102	122	160
Paddy yield (kg ha^{-1})	2,754	3,131	2,754	3,510
Wheat (kg ha^{-1})	2,325	2,708	1,806	2,150

Source: Ministry of Agriculture, Bangladesh (2004)

Fertilizer supply in the Eastern Gangetic plains is trifle low and insufficient to support crop production and reap yield goals stipulated. Most crops grown in the Delta actually extract nutrients in quantities higher than that supplied through fertilizers/organic manures. Nutrients removed in crops, that lost through leaching, erosion, and emissions are higher than that supplied. Hence, crop fields in Eastern Gangetic plains and Bangladesh experience depletion of nutrients from soil profile. Depletion of soil nutrients experienced due to adoption of various crop rotations in Bangladesh is as follows (input–output) (in kg ha^{-1}):

Boro-fallow-Taman = −175; Boro-T.Aus-Taman = 335; Boro-GM-Taman = 121; Mustard-Boro-T.Aman = 191; Potato-T.Aus-T.Aman = 245; Potato-Jute-T.Aman = 246; Mustard-Jute-T.Aman = 318; Wheat-T.Aus-T.Aman = Wheat-Bean-T.Aman = 210; Sugarcane-Potato = 112

Source: BARC (2012)

During past decade, farmers in Rice-Wheat belt of North India have been exposed to Precision farming or Site Specific Nutrient Management (SSNM) procedures. It involves series of soil nutrient analysis, periodic monitoring of plant nutrient status and application of nutrients based on decision support systems guided by computer models/simulations and yield goals. Precision techniques often require lesser quantities of fertilizers to achieve the same yield goals. Most importantly, precision techniques allow farmers to attain uniformity in soil fertility and avoid undue accumulation of nutrients in the soil profile. Therefore, loss of nutrients through emissions and groundwater contamination is minimized. Report by Singh and Bansal (2010) suggests that wheat productivity under "Farmer's practice" is 2.56–5.45 t grain ha^{-1}. State recommendation resulted in 3.76–6.28 grain ha^{-1} and Site-Specific Nutrient management produced 3.87–6.55 t grains ha^{-1}. Precision techniques resulted in 3–25.3 percent higher grain yield over state recommendation.

6.2.6 SOIL ORGANIC MATTER, ORGANIC MANURES, AND CROP RESIDUE RECYCLING

Soil types used for crop production should be sufficiently endowed with organic matter. However, soil types encountered in the South Asian prairies differ with regard to SOM content. Soil types traced in the Indus valley are well supplied with organic matter, yet incessant depletion necessitates its replenishment. Similarly, Mollisols and Inceptisols traced in the Gangetic Plains and Delta are moderately well supplied with organic matter. The SOM affects several physicochemical properties of soil relevant to crop production. The major soil types traced in South India, mainly Vertisols and Alfisols are relatively poor in organic matter and need periodic addition of FYM and recycling of crop residues.

In Pakistan, soil types used for crop production in Punjab and Sind possess less than sufficient levels of SOM. A field with less than 1.3 percent SOC is considered as poor in SOM and it requires FYM and/or organic manure supply. General soil surveys conducted by agricultural agencies suggest that C content of soil in Indus valley ranges from 0.51 to 1.38 percent. Most regions within the Indus valley have <1.0 percent organic-C. There are innumerable reasons for low organic-C in the cropping zones of Pakistan. They are:

(a) High annual mean temperature in the cropping belt induces rapid decomposition of organic matter. Warmer soils possess high microbial activity that respires off a sizeable fraction of SOC as CO_2. Conventional tillage induces oxidization of soil and evolution of soil-C as CO_2
(b) There are several orders of soils found in the Indus plains that are inherently low in organic matter. They have to be corrected using FYM
(c) Soil types such as Aridisols and Entisols found in the Western regions of Pakistan are inherently low in organic matter (0.8%); Agricultural cropping in the entire Indus plains was sustained by recycling crop residues and application of FYM /organic manures. However, with advent of inorganic fertilizers, replenishment of SOM has markedly reduced

(d) Subsistence farming conditions in some parts of Pakistan does not allow farmers to apply extraneous FYM and fertilizer-N. It leads to low organic matter content. Intensive tillage using heavy disks and chisel plow often leads to loss of SOC. Deep tillage exposes soils to loss of SOC. According to Ijaz (2013), several different types of organic matter sources are available for farmers in the Indus valley to use in the fields. A few examples are crop residues, green manures, cover crops, animal manures, compost, biogas compost industrial wastes, etc. Adoption of zero tillage along with supply of different organic matter sources helps in restoring SOC content

6.2.7 CARBON SEQUESTRATION IN THE PRAIRIES

According to Flint and Richards (1991), decline in C sequestration in the Indo-Gangetic belt was about 2.62 Gt during past century. Evolution of CO_2 during 1880–1920 was 911 m t, during 1920–1950 it was 750 m t and 964 m t during 1959–1980. Forest land and shrubs emanated maximum amount of CO_2 and accounted for 90 percent of C loss from the vegetation. The conversion of forest to cropland was the main cause for reduction of C-sequestration.

Indo-Gangetic plain is among most productive areas of South Asia. It provides lively-hood and gainful occupation to farmers and serves with food grains to a large population over 650 million. Soil fertility is one of the most important aspects that stabilize grain harvests. The wetland rice and arable wheat are the two main crops cultivated by the farmers. Hence, soil experiences periodic alteration between sub-merged wet land conditions and dry arable conditions. Soil nutrient deficiencies are common. Fertilizer inputs are almost mandatory in all the sub regions of the plains from Pakistan, throughout Gangetic zone right up to Delta areas of Bangladesh. Soil fertility is affected by the organic matter content to great extent. Agronomic measures that reduce SOM loss due to erosion, emissions, and removal of crop residues are necessary. Sequestration of C in soil is almost essential if soil quality of the prairies is to be sustained. Soil organic carbon forms a large share of total carbon in the top 30-cm layer of profile. However, SOC decreases from 83 percent of total C at 30 cm to 30 percent at 150 cm depth. Therefore, preserving SOC in the top layers is crucial if crop productivity is to be sustained (Bhattacharyya et al., 2001; Bhattacharyya and Pal, 2002). Mollisols and Vertic Inceptisols traced in the Indian plains may possess relatively higher SOM. Yet, application of farm yard manure (FYM) and green manure helps in improving SOC in the Rice-Wheat cropping zones. Cultivation of cover crops during short summer season is another proposition to improve SOC of prairies.

Studies on C-sequestration achieved due to interaction between intensity of tillage and crop residue application suggest that conventional tillage does induce a certain degree of loss of SOC. According to, Ghimire (2013), loss of SOC is attributable to degradation of soil aggregation, induction of soil oxidative reactions and microbial respiration. Higher SOC content in the upper layers of fields with zero tillage is attributable to crop residue that reduces loss of soil moisture, and protects

SOC from oxidation and mineralization processes. The SOC under no-tillage was consistently higher in upper 0–15 cm layer by 29 percent over conventional chisel plow (Ghimire, 2013). Overall, rice-wheat cropping system practiced in the Terai region of Nepal, sequestered C, at rates equivalent to other tropical cropping systems. It seems, anaerobic conditions during rice phase of the cropping system retarded mineralization and held C in the soil profile for a longer stretch of time.

Soil organic carbon depletion is rampant in many regions within the Alfisol plains of South India. Anthropogenic activities are among the major causes of SOM depletion. Farmers are hence advised to adopt agronomic methods that sequester C in the profile and restore SOC. Nishant et al. (2013) have suggested that it is worthwhile to grow as many different forage species during the summer months instead of a fallow without crops. Evaluations with forage species suggest that if un-cropped fields had 1.03 percent SOC, fields planted with forage maize possessed 1.35 percent SOC, fodder cowpea plots had 1.33 percent SOC, lucerne fields had 2.25 percent SOC and hybrid napier plots had 1.45 percent SOC. Clearly, selecting most appropriate forage species is essential if SOC contents are to be maximized. It is believed that interactive effects of soil types, forage species, and soil properties governs C-sequestration trend to a certain extent.

6.3   SOIL FERTILITY AND ENVIRONMENTAL CONCERNS IN SOUTH ASIAN PRAIRIES

6.3.1   SOIL EROSION

Soil erosion due to intense precipitation events is rampant in the Indus plains and hill country areas of Pakistan. About 600 thousand ha of land is said to be afflicted by water mediated erosion. According to Mian and Mirza (1993), water erosion is marked in Indus plains, Potowar plateau, and slopes of mountains. They are primarily attributable to geological processes. Soil erosion due to water has undoubtedly increased during past few decades due to anthropogenic activities. Destruction of natural vegetation, grasslands, lack of programs to curb soil erosion and reestablish vegetation has enhanced water mediated erosion. Soil management operations such as clearing the shrub vegetation, loosening soil with diggers, disks and chisel plows, repeated plowing of topsoil have caused enhanced loss of topsoil and nutrients from agricultural landscapes. Shallow tillage and formation of plow pans that affect water infiltration are other factors that induce soil erosion. Crop production on hill slopes with rapid movement of water also causes erosion and loss of fertility. Inadequate terracing accentuates the soil erosion. Loss of soil structure (% aggregation) and soil organic matter makes it more vulnerable to erosion. Pakistan has 11,171 thousand ha of land vulnerable to repeated soil erosion. The extent of water mediated erosion of soil varies in different provinces of Pakistan. Soil erosion due to water is rampant in Northwest Frontier Province (4,292,000 ha), Baluchistan (2,634,000 ha), Northern areas (2,282,000 ha), and Punjab (1,904,000 ha). Following is the severity with which water-induced erosion afflicts the agricultural cropland in Pakistan:

Degree of Erosion	Type of Water Erosion	Area
		(x 1000 ha)
Slight	Sheet and Rill	398
Moderate	Sheet and Rill	3,581
Severe	Rill, Gully, Stream Bank	3,745
Very Severe	Gully, Pipe, and Pinnacle	3,446

Source: Mian and Mirza (1993)

There are several suggestions made to control water-induced soil erosion, so that natural vegetation and crops are protected. In the Mountain regions and Potowar area, protection of natural vegetation from surface soil erosion is important. Illegal cutting, burning, and uprooting of grass/shrubs should be avoided. Replanting of slopes with grass, herbs, and shrubs may be helpful. Restricting crops in the slopes will help in avoiding loosening of soils. Formulating cropping systems that help in avoiding sheet erosion and loss of surface soil is important. Growing green manure and cover crops during fallow period is useful in restricting water erosion. Growing trees with deep and strong roots in areas prone to gully erosion is important. Shaheen et al. (2011) state that restoring land productivity by curbing soil erosion and adopting an array of soil conservation systems such as contour bunding, mulching, organic matter application, and others are needed if the productivity of agricultural prairies has to be sustained. According to Nizami et al. (2004), Potwar plateau in Punjab has fairly flat agricultural zone with some undulating zones. Yet, soil erosion and loss of fertility due to sheet and rill erosion seems rampant. Loss of clay fraction and organic matter induces soil degradation. Therefore, soil conservation systems are almost mandatory. Field evaluations suggest that, if farmer's practice results in 2.98 t wheat grain ha^{-1}, improved fertilizer application provides 3.48 t grain ha^{-1}. Soil conservation procedures adopted along with higher fertilizer dosages provides 3.58 t gain ha^{-1}. In addition, this treatment restores soil fertility and reduces soil erosion (Shaheen et al., 2011).

Wind erosion of topsoil and loss of fertility is common in drier belts. Wind erosion occurs in summer months when fields are left unprotected. Droughts, famines, and windy conditions also create dust bowl. The soil loss via wind may vary with speed, height, and characteristics of sand particle that gets lifted and deposited in a different location. The degree of wind erosion varies. In Pakistan, about 2,595 thousand ha of land is exposed to slight wind erosion, 486 thousand ha to moderate level of erosion, 1,668 thousand ha is affected by severe to very severe wind erosion (Mian and Mirza, 1993; Khan, 2012). In Pakistan, wind erosion of agricultural zones is a common feature in sandy and arid belts of Thal and Cholisthan deserts of

Punjab, Thar regions of Sind, and Kharan desert region in Baluchistan. The extent of devastation caused by wind erosion depends on several factors related to soil characteristics, nutrient carrying capacity of sand or dust particles, soil moisture content, wind velocity, height of the wind, and topography. Sand drifts due to creeping soil particles are also a common occurrence near the deserts. Shifting sands may affect seed germination, seedling establishment, and may affect crop growth and grain formation. There are several causes attributed to wind aided erosion. They are degradation of topography, loss of contour ridges, over grazing of surface vegetation, uprooting and burning of natural herbaceous cover and shrubs, and so on. Extension of cultivation into natural semistabilized sandy dunes, loosening and creation of crop fields on partially stabilized sand dunes, fallowing without cover crops may induce loss of topsoil from the plains region. Untimely tillage, excessive tillage, and frequent deep tillage are procedures that accentuate loss of topsoil via wind and water. Agronomic measures that control wind erosion are avoiding uprooting of shrubs, avoiding excessive grazing, restricting cultivation on sandy ridges, stabilization of sand dunes, growing shelter belts and wind breaks and growing cover crops.

Soil degradation in the Indo-Gangetic plains has direct impact on the agricultural prairies, especially the cereal grain harvests. Almost all subregions of Indo-Gangetic plains that is, Trans, Upper, Western, Middle and Lower Gangetic plains are prone to soil degradation of varying intensities. States such as Punjab, Haryana, Western Uttar Pradesh, and Bihar in India are among most intensely cultivated zones both during kharif (rainy season) and rabi (post rainy season). Therefore, soil degradation or any other maladies will affect grain yield perceptibly. For example, in Punjab, Sidhu et al. (2002) evaluated different forms of soil degradation in the plains. They found that soil was affected by water logging in 3 percent of rice-wheat cropping zone, by salinity/sodicity in 3 percent of state's area and by water erosion in 13.3 percent of state's cereal belt. Soil erosion due to water is supposedly among important factors that reduce productivity. It reduces soil fertility markedly. In a single state, for example, Punjab profits from cereal grains could reduce by 18 percent, if unchecked.

6.3.2 GASEOUS EMISSIONS

Greenhouse gas emissions occur due to natural factors and several different anthropogenic activities, including agriculture. Agricultural activity is among important factors that generate greenhouse gas. South Asian prairies are no exceptions. They generate large quantities of methane and nitrous oxide that needs to be minimized. Following is an estimate of greenhouse gas emissions in South Asian nations:

Greenhouse Gas Emission	Methane Emissions		Nitrous Oxide Emissions	
	Share of Agriculture	Total	By Agri-culture	Total	By Agriculture
	(%)	(kt of CO_2 eq)	(%)	(kt of CO_2 eq)	(%)
Pakistan	38.6	1,37,000	63.5	26,800	74.2
India	28.4	5,84,000	64.4	2,13,000	73.4
Bangladesh	61.2	92,400	70.5	21,400	83.1
Nepal	87.2	22,100	82.9	4,520	76.8

Source: FAOSTAT (2013); http://www.fao.org/docrep/docrep/015/i2490e/i2490e.04.pdf pp 320–335 (September 12, 2013)

Anthropogenic activities such as farming, pastures, dairy cattle, fuel usage for transport and industries all generate N-emissions from land surface and release it to atmosphere. Nitrogen emissions occur most commonly as reactive-N (Nr) species such as nitric oxide (N_2O), nitrous oxide (NO_2), and NH_3. Emission of Nr is rampant across farming locations that adopt high-intensity crop production. Nitrogen emissions are generally higher in regions supporting high input. A few examples are wet land rice production, intensively cultivated arable crops such as wheat in the plains, sugarcane, cotton, oilseeds, and other cash crops. Nitrogen emission from India was calculated as 75.94 g year^{-1} during 2005 (Sharma et al., 2008). Nitrogen emissions from rice belts in Uttar Pradesh (15.5 Gg y^{-1}), Andhra Pradesh (9.2 g year^{-1}) and Cauvery delta of Tamil Nadu (4.2 g g year^{-1}) in India, Punjab plains in Pakistan, and Delta zones of Ganges in Bangladesh are relatively high. Nitrogen emissions depend on several factors related to geographic location, soil type, fertilizer supply, crop and atmospheric conditions. For example, in the North Indian plains, rice provided with 140 kg N ha^{-1} emitted 0.15–0.23 kg N ha^{-1} during a crop season. Whereas, wheat grown in the same region with 120 kg N ha^{-1} emitted 0.77 kg N ha^{-1} (Kumar et al., 2000; Majumdar et al., 2000, 2002; Pathak et al., 2002).

Nitrogen Emission as N_2O (Gg year^{-1}) from Agrarian Regions of India—Examples

Northern Plains: Bihar—3.90; Delhi—0.10; Uttar Pradesh—15.53; Punjab—7.69; Haryana—4.38

Eastern India: Assam—0.21; Nagaland—0.53; Orissa—1.34; West Bengal—3.38

Western India: Gujarat—4.49; Maharashtra—7.50; Rajasthan—4.3

Southern India: Andhra Pradesh—9.5; Karnataka—4.35; Tamil Nadu—4.3; Kerala—0.71

Nitrogen Emission from India Compared to World

N-emissions from world—9.96 Tg year^{-1}; from India—0.26 Tg year^{-1}

N-emissions from World Agricultural belts—3.5 Tg year^{-1}; Indian Agriculure—0.24 Tg year^{-1}; India Agricultural Soils—0.08 Tg

---

Source: Sharma et al. (2008); Bhatia et al. (2004)

The rice-wheat cropping belt of South Asia is supplied with largest share of total fertilizer-N produced. Nitrogen supply to soil is high in order to support intensive cultivation trends and high yield goals stipulated by farmers. About 180–240 kg N ha^{-1} is applied through fertilizers. The arable/wet prairies also receive N from several other sources such as FYM, crop residue recycling, and atmospheric sources. Fertilizer-N is not utilized efficiently. About 30–42 percent of fertilizer-N is all that is traceable in the crop phase of the ecosystem. A sizeable portion of soil-N and that applied as fertilizer-N is lost to atmosphere as N-emissions, such as N_2O, NO_2, and N_2. Soil-N transformation processes such as mineralization, nitrification, and de-nitrification aid the process of N- emissions. Nitrogen emissions are accentuated in field supplied with higher quantities of fertilizers such as urea, KNO_3, Di-ammonium phosphate, and FYM. Pathak et al. (2002) state that Indo-Gangetic plain that supports rice-wheat rotation is among most important N-emitting biomes. It spreads across 26 million ha in Pakistan, India, and parts of Bangladesh. It also consumes relatively a very large amount of fertilizer-N source. The rice-wheat belt is known to emit 654 g N ha^{-1}. However, if fertilizer-N is applied, the N_2O emission gets accentuated and emits 1,520 g N_2O ha^{-1}. Supply of organic manures and use of nitrification inhibitors helps in reducing NO_3 accumulation and N_2O emissions. It seems periodic alternation of arable and wet conditions in the rice belt aids N-emissions. Rice fields supplied with 240 kg N ha^{-1} emitted about 1,570 g N ha^{-1} equivalent to 0.4 percent of fertilizer-N supply. If nitrification inhibitors are used, then N-emissions get reduced to 1,096 g N ha^{-1} (Pathak et al., 2002).

Fertilizer-N supply to rice/wheat crops grown in the plains, hills and coasts of South Asia is relatively higher. It helps in restoring soil-N that was depleted by previous crops that lost through natural processes such as emissions, percolation and seepage. It helps mainly to satisfy the needs of present crop and achieve yield goals aimed by farmers. Soil-N unused that accumulates in the profile is obviously vulnerable to loss via emissions. Nitrous oxide emission, for example, is high in the Indo-Gangetic plains. Pathak et al. (2013) believe that several soil processes, especially alternation between aerobic and anaerobic condition, changes in redox conditions and soil microflora accentuate soil-N loss through emissions. Soil drying and higher temperature during rice cultivation also induces emissions. Hence, Pathak et al. (2013) suggest that use of nitrifying inhibitors that reduce soil-N accumulation and loss via seepage too is required. It is believed that over 13.5 million ha in the Indo-Gangetic belt is prone to soil-N loss via emissions. Evaluations using nitrification inhibitors suggest that reduction in N emission varies based on type of nitrification inhibitor applied. Application of hydroquinones reduced N-emissions by 5 percent, thiosulfates reduced it by 31 percent, and DCD reduced it by 29 percent.

6.3.3   AMMONIUM (NH₃⁺) EMISSIONS

South Asian agricultural prairies are among the major NH_3 emitters into atmosphere. Fertilizer supply to fields is the main generator of NH_3. However, estimates of NH_3

emissions are said to be still uncertain in many regions within the prairies because emission factors are not yet standardized (Sharma et al., 2008). The loss of inherent soil-N and that applied as fertilizer as NH_3 emissions vary based on location, agro-climate, soil type, crop, fertilizer formulation, and rates. Soil-N loss as NH_3 emission ranges from 1 to 22 kg N ha^{-1}. There are reports that loss of N as NH_3 is greater at 20–30 kg N ha^{-1} in the Indo-Gangetic plains. Volatilization of N as NH_3 in the Eastern Gangetic plains was estimated at 36 kg N ha^{-1} if the soil was unfertilized and 60.1 kg N ha^{-1} if fertilized with Urea. Daily rate of NH_3 emission was high if the tropical warm conditions prevailed. It was 0.45 kg N ha^{-1} for rice and 0.5 kg N ha^{-1} for wheat. Urea inputs are vulnerable to loss as NH_3 emissions. Incidentally, more than 90 percent of N supply in the major agrarian zones of South Asia is achieved using urea. Estimations by Pathak et al. (2006) suggest that NH_3 volatilization results in 16–62 kg N ha^{-1} loss in Indo-Gangetic plains. The average loss of N as NH_3 in India is 30 kg N ha^{-1} if average fertilizer-N supply is taken as 90 kg N ha^{-1}.

The livestock in India and Pakistan are other major emitters of NH_3. India is said to contribute about 1.43 Tg NH_3 $year^{-1}$. North Indian plain is among top emitters of N in different forms (Aneja et al., 2010). The emission of gaseous reactive N is caused by both sprawling cereal belt and high number of livestock found in the region. It is worthwhile to note that each cattle head emits 4.4 kg NH_3-N $year^{-1}$ and 0.32 kg N_2O-N $year^{-1}$, Buffalo emits 3.4 kg NH_3-N $head^{-1}year^{-1}$ and 0.39 kg N_2O-N $head^{-1}$ $year^{-1}$; and sheep emit 1.2 kg NH_3-N $head^{-1}$ $year^{-1}$ and 0.21 kg N_2O-N $head^{-1}$ $year^{-1}$. Consider the vast number of livestock traced in the Gangetic plains; the total emissions could be very high. Livestock caused release of 1,392 Gg NH_3-N $year^{-1}$ and 136 Gg N_2O $year^{-1}$ during 2010. Fertilizer supply to cropland induced emissions equivalent to 2,221 Gg NH_3-N and 126 Gg$N_2$0-N during 2010. Fertilizers left unused in the soil profile is the most vulnerable fraction that gets lost as emissions. States such as Punjab, Haryana, Uttarakand, Uttar Pradesh, Bihar, and West Bengal in the Indo-Gangetic plains; Maharashtra and Gujarat in the Western India; Andhra Pradesh, Karnataka, and Tamil Nadu in the South are among major fertilizer users (Aneja et al., 2010). Overall, N-emissions from Indian cropland almost match that caused by agrarian regions of China, USA, and European Union. There is need to reduce fertilizer-usage, improve fertilizer-use efficiency and adopt agronomic procedures such as Site-Specific Nutrient Management and Precision technique that minimizes undue accumulation of fertilizer-N in the soil profile.

6.3.4   CARBON DI-OXIDE (CO_2) EMISSIONS

Soil microbial respiration is supposedly a major contributor to CO_2 efflux from agrarian regions of South Asia. Other important sources are fossil fuel burning and industrial emissions. Studies on influence of soil respiration rates on CO_2 efflux in India are feeble. Excessive efflux of CO_2 is detrimental, since it can reduce C-storage in soils. It leads to loss of soil quality. The CO_2 efflux in the agricultural prairies depends on several factors related to soil, crop and environment. Farmers practice wide range of crop rotations that include wet land rice, arable wheat, legumes, oilseeds, vegetables, or fallows. For most crop rotations that are popular in Indo-Gangetic plains, CO_2 emission ranged from 0.52 to 0.70 g CO_2 m^{-2} h^{-1} (Srivastava et al., 2012). Crop rotations that included fodder jowar and berseem emitted higher amounts of CO_2 (0.75 g CO_2 m^{-2} h^{-1}). Similarly, those with vegetables after cereal crop emitted 0.70 g m^{-2} h^{-1}. Rotation with arable crops such as wheat and mustard also emitted higher levels of CO_2 about 0.71–0.72 g CO_2 m^{-2} h^{-1}. Crop rotations with fallow, for example, wheat-fallow or rice-fallow emitted 0.56–0.60 g CO_2 m^{-2} h^{-1}. Srivastava et al. (2012) state that, in the Indo-Gangetic plain, quantity of CO_2 emission was highest from field with wheat-fallow sequence, followed by wheat-maize > wheat-pulse > wheat-sugarcane > wheat-jowar > wheat-vegetable in that order. The CO_2 emission in the plains is affected by season. Monsoon period from June to end of September generated highest amount of CO_2 ranging from 0.75 to 0.95 g CO_2 m^{-2} h^{-1}. It is attributable to warm tropical conditions in soil. Winter months from November to next March, with low temperature induced lower levels of CO_2 emissions, ranging from 0.39 to 0.65 g m^{-2} h^{-1}. Soil tillage is among important factors that induces oxidative conditions in soil, improves soil microbial activity and respiration leading to higher levels of CO_2 emissions. As CO_2 emissions increase, ability of soil to sequester C reduces leading to deterioration of soil quality.

Methane emission is an important environmental concern that generates greenhouse effects. Methane emission is predominant in the flooded ecosystems that support paddy cultivation. The anaerobic condition found in the submerged lowland rice field allows organic matter decomposition leading to methane generation. Upland paddy fields that account for only 10 percent of rice production and < 15 percent rice area, does not cause methane emission to any perceptible level. Globally, lowland paddy fields are known to generate about 20–100 t g CH_4 $year^{-1}$ and accounts for 15–20 percent of total methane emission to atmosphere (IPCC, 2012). Methane fluxes from wet lands differ markedly based on geographic location, soil type, fertilizer supply trends, organic matter content, and weather pattern. Overall, in India, rice production zones emit 10 (5–15) g CH_4 m^{-2} (Mitra, 1992; 1996; Parashar, 1997). Following is an example that demonstrates difference in total methane emission per rice crop season and fluctuations within short period of 1 hr.

Central Gangetic Plain: Allahabad—0.2 mg m^{-2} hr^{-1}, 0.5 g m^{-2} $season^{-1}$; Faizabad—0.8 mg m^{-2} hr^{-1}, 2.0 g m^{-2} $season^{-1}$

Eastern Gangetic Plain: Barrackpore—0.7–20.2 mg m^{-2} hr^{-1}, 1.8–6.3 g m^{-2} $season^{-1}$; Kalyani—4.1 mg m^{-2} hr^{-1}, 10.8 g m^{-2} $season^{-1}$; Purulia—4.2 mg m^{-2} hr^{-1}, 11.0 g m^{-2} $season^{-1}$; Garagacha—11.0 mg m^{-2} hr^{-1}, 29.0 g m^{-2} $season^{-1}$

South India: Chennai—5.8 mg m^{-2} hr^{-1}, 11.0 g m^{-2} $season^{-1}$; Trivandrum—5.1 mg m^{-2} hr^{-1}, 9.0 g m^{-2} $season^{-1}$

East Coast: Cuttack—2.7–7.1 mg m^{-2} $hr^{-1,}$ 7–19 g m^{-2} $season^{-1}$

Source: IPCC (2012); Mitra (1992, 1996); Parashar (1997)

Methane has high greenhouse effect potential compared with several other gaseous emissions. During recent years, paddy farming has been modified several times. New agronomic procedures have been consistently tried, tested, standardized and utilized in larger areas of the wetlands. "System of Rice Intensification (SRI)" is one such procedure that is gaining ground. Field trials that compared SRI with conventional traditional puddled rice system indicates that SRI emits lesser quantity of CH_4 at 8.61 kg ha^{-1} season^{-1} compared to conventional system that emits 22.59 kg CH_4 ha^{-1}season^{-1} (Jain et al., 2013).

6.3.5 FIRE IN THE AGRICULTURAL PRAIRIES AND CO_2 EMISSIONS

Fire in agricultural farms to clear the stubbles and crop residues is a common practice. Natural prairie grasses that occur in the farming zones and in the reserves are also subjected to burning periodically. It adds ash that is rich in minerals and recycles plant nutrients *in situ*. However, it emits large amounts of carbonaceous particles and CO_2 into the atmosphere. These emissions potentially generate greenhouse effect and warming. Burning crop residues/stubbles *in situ* is a very common occurrence in agrarian zones. In fact, rice-wheat cropping system that spans across Indus and Gangetic valleys is a prime generator of carbonaceous soot, CO_2, and methane. Burning residues during rice-wheat rotation is a popular agronomic procedure in Indo-Gangetic plains. At present, "Atmospheric Brown Cloud" that looms all across regions that practice crop residue burning is said to be a prominent environmental concern in South Asia (Gupta et al., 2004, 2012a, 2012b; Gustafsson et al., 2009). Reports suggest that farmers in Gangetic plains and Deccan generated over 116 m t year^{-1} crop residues during 2001–2005 that was burnt, causing atmospheric pollution with CO_2. Crop residue burning contributes 14 percent of black carbon emission, 10 percent of CO, 6 percent of CO_2 and 1 percent SO_2. According to Venkatraman et al. (2006), in India, crop residue burning is prominent during off-season or summer spanning from April to June. Crop residue burning is more rampant in the Western Indo-Gangetic plains, followed by Central India > Northeast > South Indian plains (Table 6.3). Overall, in India, it seems open burning of crops/natural grasses generates and contributes about 25 percent of black carbon, CO_2 and CO, and about 9 percent of particulate matter traceable in atmosphere. Major suggestion to thwart atmospheric pollution is to reduce burning of crop residues. Instead, crop residue could be recycled *in situ* to enhance C-sequestration in the soil phase of biome. In the wet prairies of Bangladesh, farmers are known to burn a portion of crop residue and recycle the rest *in situ* into soil (See Haider, 2012). Prevention of crop residue burning can avoid undue increase in soil/atmospheric temperature. Further, it avoids outdoor pollution and related ill effects to human beings and fauna.

TABLE 6.3 Crop waste burnt in different parts of India and Pakistan

Region in India	Crop Waste Bunt (T g y^{-1})		
	Cereals	Sugarcane	Other Species
Indo-Gangetic Plains	43.8	12.1	11.8
Central India	4.5	10.3	0.1
North-eastern India	13.9	1.6	0.1
South India	2.9	6.1	5.0

Source: Venkatraman et al. (2006)

6.3.6 DROUGHTS

Droughts are almost routine events in South Asian prairies. Droughts ravage parts of one or the other ecoregions of South Asia each year. Theoretically, droughts are defined as just negative deviation in the normal frequency and quantity of precipitation and resultant soil moisture levels. Droughts occur due to acute water shortage with disastrous consequences to crops and livestock. Drought affects large section of human population than any other environmental disaster (SAARC, 2012a, b, c). Its consequences such as lowering groundwater level, devastation of crops, reduction in livestock, deficit human food supply, and economic loss are dependent on intensity. Hence, drought preparedness and ability to buffer the detriments of water scarcity are a necessity (Singh, 2011). The causes of droughts in the Indian subcontinent could be inadequate monsoon rainfall, high temperature leading to loss of moisture via evapotranspiration, excessive groundwater depletion, inadequate water distribution, shift in agricultural practices and high water usage by crops and human dwellings without adequate storage (SAARC, 2012a). Meteorologically, sea surface temperature anomaly around Indian subcontinent in relation to atmospheric circulation and large-scale pressure oscillation in atmosphere in the Pacific Ocean may cause distortions in normal monsoon rains (SAARC, 2012a, c). El Nino is one such event that has drastic effects on monsoon precipitation pattern.

Drought effects are generally severe in any agroecosystem. In South Asia, its effects get accentuated because a large portion of agricultural prairies in Pakistan and India are rain fed. Farms occupy dry lands with ordinarily low rainfall. In India, about 57 percent of cropping region is drought prone. About 51.2 million ha of agricultural cropland is said to be vulnerable to drought stress (Kumar et al., 2005; NRSC, 2013a, b). Droughts occur in cycles and are periodic in occurrence in India. North-western region of India encompassing Rajasthan, Haryana, Punjab are prone to drought every 2–3 years, regions such as Gujarat, Uttar Pradesh perceive drought every 5 years. Similarly, Southern Indian states such as Karnataka, Andhra Pradesh, and Tamil Nadu suffer droughts every 5 years. Droughts are also recorded in tropical wet regions such as Assam and and other parts of Northeast of India. In Pakistan, large section of Punjab plains is drought prone and crop loss could be severe, if irrigation support is lacking during crucial stages of crop production. Gangetic and Padma Delta in Bangladesh too are vulnerable to droughts. The natural grassland and cropping zones of Terai in Nepal also suffers drought.

The impact of droughts on agricultural prairies has generally been drastic. For example, repeated droughts that occurred in the Eastern Gangetic plains during mid-1930s, reduced crop production entirely and induced farmers to migrate in different directions. Agricultural prairies were almost decimated due to summer temperatures and lack of precipitation. The predicted loss was over 50 percent due to droughts that occurred during 1957–58. It is interesting to note that, depreciation of pearl millet yield in Northwest India was directly related to drought intensity. During 1987, pearl millet yield reduced by 78 percent in areas that received 300 mm precipitation, by 74 percent if precipitation was 300–400 mm and by 43 percent if precipitation was > 400 mm (Singh, 2011; NRSC, 2012a, b). Droughts that occurred during 2002/2003 depreciated grain harvest by 29 million tons. Production of rice and wheat reduced by 25 percent, compared to normal year. Food grain production reduced by 7 percent of normal, due to drought, that occurred during 2008/2009, in India. Droughts are severe environmental constraints in India. Droughts affect exchequer of individual farmer as well as region. For example, it is believed that during 1976–1996, India's farm exchequer reduced by 1.8 billion INR each year due to drought. Drought management is an important aspect in all agrarian zones of India. Several measures are adopted by the Government agencies that are supposed to help crop production in large areas. Farmers adopt different measures such as water storage, water harvesting, ponding, mulching, devising efficient irrigation schedules, deficit irrigation, and so on. (Sharma, 2002a, b)

Indian plains have experienced droughts of varying intensity, but there were few that were devastating on the prairie crop production. During past two centuries, severe droughts occurred on18 different years during 1800–1899; and on 23 different years between 1900 and 2003. These were severe and affected water resources and crop production. The percentage area affected by drought in India ranged from 13.1 (in 1971) to a high of 49.2 (in 1987). Droughts were more common in states with large tracts of semiarid and arid region. Similarly, drier tracts of Southern Pakistan, Terai in Nepal are prone to drought.

Droughts occur in the Indus valley fairly frequently. Droughts are actually controlled to a great extent by a massive canal network emanating from Indus and its tributaries. Yet, drought affects the drier regions of Baluchistan where natural precipitation is low ranging from 125 to 250 mm annually. Erratic distribution of precipitation adds to drought-related problems. The agricultural regions close to Thar desert on the left bank of Sutlej and Indus are again exposed to severe drought, despite several types of control measures (SAARC, 2012b). It is said that among consequences of droughts, seminomadic pastoralism and migration is common in the drought prone regions of Baluchistan and Northwest Frontier province. Drought management agencies are in place in several locations within Pakistan. It helps farmers to overcome drought effects on crop productivity (SAARC, 2012b; NSPA, 2013). For example, Sindh Arid Zone Development Authority monitors and provides resources to overcome drought effects on wheat/rice crop production in Sindh region of Pakistan. Agrarian region in Pakistan has experienced several severe droughts during recent history. Some of them affected the cropping zones in Punjab and Sindh rather severely. For example, drought of 1899, 1920, and 1935 in Punjab, during 1902 and 1951 in Northwest region, during 1871, 1881, 1931, 1947, and 1999 in Sindh were devastating. Most recent drought in Pakistan occurred during 2000–2001 that lasted for nearly 2 years. It reduced the crop productivity by 10 percent. The water shortage during drought period was 51 percent of the normal years (Ahmad et al., 2004; NSPA, 2013). The total flow of water in the major rivers declined from 169 billion m^3 to 109 million m^3 because of drought in 2001. There are several measures that government agencies and farmers adopt to overcome drought period. Traditional systems such as *Sailaba* (spate irrigation) and *Kushkaba* (runoff irrigation) are still in vogue in many regions (Ahmad et al., 2004).

Bangladesh too is exposed to seasonal aridity. Crops in Northeast of the nation may suffer due to drought, if scarcity of water perpetuates for more than 4–6 months. The wet prairies that support rice production and the succeeding legume crop may suffer. Droughts are severe only if it affects more than 20 percent of agrarian regions of Bangladesh. Severe droughts were identified during 1957, 1965–66, 72–73, and 82–83 (SAARC, 2012a, b). In Bangladesh, pre-monsoon drought affects normal growth and yield formation by HYV Boro, Aus, wheat, pulses, sugarcane, and potatoes. Droughts in the Delta zone vary with regard to spatial effect, intensity and loss of crop yield. Nearly 0.4 million ha are affected by severe drought during Kharif and 0.35 million ha during Rabi season (Ahmed, 2006).

In the Terai region of Nepal, agricultural agencies monitor the weather pattern and drought cycles regularly (SAARC, 2012a, b, c). Dry lands of Terai produce large quantities of cereals and legumes needed by the local population. A severe drought will be detrimental to productivity of this important agrarian region. However, droughts in the prairies of Terai are overcome by augmenting water from network of rivers that flow from Himalayas. The groundwater is well replenished by the flowing rivers and torrential rains that occur in the foothills. Farmers, therefore try to store water or arrange for irrigation during critical period of crop growth. It suffices to support optimum crop productivity. Droughts are also common in the Sri Lankan rice producing zones. Droughts that occur for short or longer duration both are experienced in the drier regions. Insufficient precipitation causes reduction in grain yield. In most areas, irrigated paddy and dry land crops are dependent on stored water. We may note that during 1988–89, severe drought had affected crop production in large areas of North-Central and Southern Provinces of Sri Lanka.

6.3.7 FLOODS

An agricultural zone is considered as water logged if water table occurs within 3 m depth from surface. Water logging may not be a problem in Indus valley, since in most areas groundwater level is well below crop's rooting zone. Water logging may affect about 5–10 percent of cropland and small regions of natural prairies of Pakistan. Some of the reasons for water logging are excessive percolation of water from canal system; highly permeable soils; obstruction of natural drainage systems that restricts free water flow; improper alignment of canals and embankments and mainly inefficient disposal of rain water (Mian and Mirza, 1993). Flooding and ponding is actually a problem seen more due to rivers flowing out of embankments and whenever torrential rains occur. About 2,557 thousand ha of cropland is exposed to river flooding and torrent flooding in Pakistan. Ponding may affect 9,36,000 ha of agricultural land in Pakistan. Ponding-related problems can be grouped into frequent ponding, shallow ponding, and occasional shallow ponding (Mian and Mirza, 1993).

Floods are frequent in the Gangetic plains of North India. They occur during monsoon season if precipitation events are intense. Major rivers tend to over flow the banks and inundate large tracts of crop fields. Gangetic delta exists almost at sea level. Its low-lying topography makes it highly vulnerable to floods by all three major rivers and their tributaries. Food control systems have been in vogue through centuries. Flood control and drainage is essential to reduce inundation of crop fields. It seems about 20 percent of Bangladesh is prone to flood damage. Major flood control programs were initiated in Bangladesh during 1958. Surface drain systems are also in place in many regions. They are known to help wet land farmers in 1,383 thousand ha (ICID, 2012). Cyclones and storms surges are more frequent during monsoon. They do cause loss of topsoil in great proportions. They even affect a standing crop sometimes resulting in massive loss to farmers (Ahmed, 2006).

6.3.8 SALINITY AND ALKALINITY OF SOILS

South Asian prairies and cropped land thrives on soils of different fertility levels. They are also afflicted by several different types of maladies that inhibit crop growth and productivity. The maladies may be perceived for different lengths of period and intensities. Soil salinity/alkalinity is a problem that is scattered across Indo-Gangetic plains, Peninsular India, and Coasts. Salinity, no doubt reduces crop yield. It destroys soil structure, affects soil aggregation properties, and nutrient availability. Salinity affects rooting and retards biomass accumulation by natural vegetation/crops.

Indo-Gangetic belt that spans across Pakistan, India, and Bangladesh is a vast stretch of fertile plains that predominantly supports rice-wheat production. In Pakistan, according to Hassan (2013) about 2.67 m ha of rice-wheat cultivation zones in Punjab are afflicted by salinity/alkalinity. The salinity problem gets accentuated if droughts too are frequent. Fresh water supplies are needed to drain out the salts. This commodity too becomes scarce at crucial periods. About 2,667 thousand ha in Punjab, 2,109 thousand ha in Sindh are afflicted with various types of salinity/sodicity. In all about 5,327 thousand ha of crop belt in Pakistan suffers due to salinity (Mian and Mirza, 1993). There are at least three types of salinity affected soils that reduce crop productivity. They are Patchy/surface salinity, gypsiferrous salinity, porous salinity, and dense salinity. Large areas of Mollisols and Inceptisols are still reclaimed using fresh water from riverine source to drain out salts. One of the propositions is to cultivate salt-tolerant crops and their genotypes. Crop genotypes that are tolerant to salinity have been screened, selected for optimum grain formation, and introduced into Indus plains. For example, wheat varieties such as Sarsabz, cotton variety NIAB-999, and mung bean mutants NMST-2 or NM-9800 are salt-tolerant. Their introduction into salinity affected zones will affect crop growth favorably. Rice is fairly more tolerant to salinity than wheat or legumes. Salinity-tolerant agroforestry species and grasses have also been grown in the soils afflicted with salinity. Among the agronomic procedures, testing soils is necessary. The Electrical conductivity (Ec) should be below 8 dS m^{-1}, if not the fields should be flooded with fresh water prior to seeding. Seed germination will be erratic and low if soils are saline. The imbibition of water by seeds is hampered in saline conditions. Further, seedling growth and crop stand will be affected. Hassan (2013) states that growing salt-tolerant fruit trees such as guava, date palm, ber, jam should be preferred. The prairie crops should be allowed to establish as understory crop. In addition to crops, natural vegetation too is affected if salinity beyond threshold and prairie vegetation will be scattered and irregular. The permissible upper limits of characteristics of irrigation water used on Loamy soils of Indus Plains of Pakistan is as follows: Electrical conductivity- 0.75 m mho cm^{-1}; salt concentration- 7.5 meq l^{-1} or 500 ppm; sodium absorption ratio- 13 m mol l^{-1}; Residual Sodium Carbonate- 2.5 m eq l^{-1} (Mian and Mirza, 1993).

In the Gangetic plains of India, grain yields are moderately high due to inherent soil fertility and high fertilizer inputs. Together, rice and wheat yield about 9.5 t grain ha^{-1} rotation^{-1}. However, this biome is affected by salinity/sodicity to different intensities, and in patches all across the plains. Fertile Mollisols and Inceptisols are rendered unfit for crop production in some places. Digitized soil maps that depict extent of salinity/sodicity in different sub regions of Indo-Gangetic plains are available (NRSA, 1997; Mandal and Sharma, 2001, 2006). Soil maps depicting salt affected areas and GPS guided location of such problems is also possible (Sharma, 2002b). Such maps are useful in devising procedures to drain out salts and rectify salinity of soil. In general, soil affected by salinity, salinity/sodicity and sodicity alone range from 4.24 to 10.74 percent of different sub-ecoregions of Indo-Gangetic plains. Salinity is a severe problem in some zones. For example, in Trans-Gangetic zone comprising Punjab and Haryana that receive about 550–600 mm precipitation, salt affected fields occurred in 73 percent area. In Eastern Gangetic belt, 64 percent area had mild problem related to salinity/sodicity. In the Upper-Gangetic plains

(Meerat, Ghaziabad and Bulandhshahar), about 3 percent of area showed up salinity/sodicity. In some parts of Middle-Gangetic belt salinity/sodicity is felt severely affecting the crop yield. In Bihar and West Bengal, salinity affects 12 and 7 percent of cropping zones respectively. The coastal/delta region of Bangladesh is affected by salinity to different intensities and it affects crop production.

In Bangladesh, Gangetic belt has large areas filled with alluvial silts deposited by rivers and coastal waves. At the same time, salinity/alkalinity is a major soil-related malady that affects natural vegetation and crops alike (Ahmed, 2006). However, delta zone is also host to large areas of salty mangrove vegetation. Mangroves tolerate salty conditions. Here, natural vegetation that is well adapted to salinity/alkalinity thrives better and outcompetes crops. Still, there are certain rice varieties that are grown on soil inundated periodically by tidal waves. They tolerate both flooding and saline conditions. About 35 percent of Shatkira, 29 percent of Khulna and 36 percent of Bagherat areas are classified as salinity affected zones. They need amendments and specific management procedures. The very strong saline areas (> 8 d S m^{-1}) do not support cropping and natural vegetation too is sparse. Even in soils with slightly saline conditions (4 dS m^{-1}) crops grow with reduced growth rate. Rice is fairly tolerant to salinity, but wheat and legumes are highly susceptible (IRRI, 2000).

Eastern Gangetic plains actually experiences periodic increase in saline zones due to tidal waves and sea water inundation. Salinity ingression due to low flow sea water intrusion is a severe problem. It affects soil productivity (Ahmed, 2006). The area experiences cycles of saline sea water intrusions and out flow during dry season. Fresh water drainage through riverine floods dilutes and reduces salinity. Salt water evaporates and leaves precipitated or concentrated salt solution in the soil profile. It can be highly detrimental to crop root growth and nutrition. Climate change effects on sea water level and riverine fresh water flows literally dictate the extent of salinity that natural vegetation/crops will have to experience. Timing of monsoon, accumulation of fresh water in paddy fields and drainage too affects salinity. Construction of bunds to keep tidal waters away from crop fields is a common measure. During recent years, one of the trends with farmers in Delta zone is to grow salt-tolerant genotypes of crops, so that reasonably good grain/forage yield could be realized.

6.4 WATER RESOURCES AND PRODUCTIVITY OF PRAIRIES

South Asia is well endowed with water resources. Several rivers, tributaries, lakes, dams, and irrigation channels contribute water to crops and natural vegetation. Following data depicts extent of total land and that irrigated in South Asian nations:

Total cultivable land and irrigated land in South Asian region (in thousand ha, in 2001):

	Cultivable land	Irrigated land	% Irrigated land
Bangladesh	8,485	4,421	53
India	1,69,900	54,800	32
Pakistan	22,160	17,820	80
Sri Lanka	18,300	4,924	27

Source: FAOSTAT (2010); Bangladesh Bureau of Statistics (2005); Bhatti et al. (2009)
Note: About 30 percent of total irrigated cropland exists in south Asian nations

Pakistan's irrigation system is vast and comprises of three major reservoirs, 19 barrages and 43 main canal networks with a conveyance length of 57,000 km. It serves over 17.8 m ha. The Indus plain in Pakistan is served by several different resources of water (Nizamani et al. 1998; Kazi, 2013). They are keys to sustenance, productivity, and perpetuation of agricultural prairies of Pakistan and natural vegetation. Water resources available for use by farmers are:

(a) Precipitation: Precipitation ranges from 100 mm in dry/arid regions of Baluchistan to 1,500 mm in the foothills of northern mountain ranges. Torrential rains are also experienced in many areas of Pakistan. Totally, rain water contributes about 16.5 billion km^3 to the Indus basin. A share of it is stored using barrages and supplied as required.

(b) Surface water: Major river systems are supplied water through glacier melt, snow melt, rainfall and runoff. River water is collected by various dams at Tarbela, Attock, Mangla, and Maral for rivers such as Indus, Kabul, Jhelum, and Chenab. In Sindh, Sukkur barrage collects water from eastern rivers. River flows are smaller in quantity during rabi.

(c) Groundwater sources: The Indus basin has extensive groundwater aquifer covering an area of 16.2 M ha. The aquifer is recharged by river and rainfall. The development of canal system increased water percolation into aquifers. The total contribution of irrigation water by groundwater currently is 12 billion km^3 and about 11 percent of aquifer water is made available to crop production in the plain (Randhawa, 2005). Some of the concerns connected with irrigation relates to quality of water, irrigation use efficiency and availability to farmers at critical periods. Optimization of canal flows and improving per capita availability of water to crops is important (Qureshi et al., 2012).

Hydrologically, Pakistan can be demarcated into three major regions. They are (a) Indus Basin that covers large tracts of cereal cultivation. It occupies about 5,66,000 km^2 and accounts for 71 percent of Pakistan's territory (IGES-GWP South Asia, 2012). It comprises entire Punjab, Sindh, and Northwest Frontier Province, plus a small portion of Baluchistan; (b) The Karan desert in Baluchistan's dry region comprising 15 percent of Pakistan's area; and (c) The arid Makran coast along the Arabian sea spreads into rest 14 percent of Pakistan's area. It is interesting to note that total water withdrawal is about 166 km³, of which 95.6 percent is utilized for agricultural cropping and rest for urban drinking water programs (FAO, 2007f). Agricultural prairies in the Indus plains are also nurtured using groundwater sources that accounts for 63 km³ annually. There are over 500,000 tube wells that operate in the Indus cropping zones. The irrigation system in Pakistan can be classified as follows: (a) Government canals that support crop production in 6.3 million ha. These canals account for water supply to 58 percent of cropping zones in Punjab and 28 percent in Sindh; (b) private canals serve 0.43 million ha in Northwest Frontier province and covers about 83 percent of cropping area in that state; (c) tube wells supply water to 3.45 million ha of which 82 percent is found in Punjab plains; (d) open wells serve about 0.2 million ha of cropland with water. About half (55%) of open wells occur in Punjab; (e) Canals and tube wells together supply 7.28 million ha of cropland with water in Punjab, (f) other sources cover about 0.18 million ha of agricultural zone in Pakistan. Overall, the total irrigated land in Pakistan is 18.2 million ha and that under rain-fed condition is 4.2 million ha (FAO, 2007f).

Irrigation systems are utilized to support wide array of crops that are cultivated under diverse cropping systems. In Punjab plains, wheat and rice based cropping systems that spread into 15.5 million ha are supplied with water at critical stages using canals and tube wells. Maize, wheat, and oilseed rotations in many regions are grown under rain-fed conditions (FAO, 2007f). In Sindh, about 1.1 million ha of cropped area is under rice-wheat rotation. This zone is supplied water through canals. The arable crops such as cotton-wheat rotations, maize monocrops, and maize/wheat intercrops are grown under rain-fed conditions. Irrigation is restricted to only critical stages of the crop (FAO, 2007f). Agricultural prairies in Northwest Frontier Province cove about 0.89 million ha that is propped using canals and open wells. In Baluchistan, about 1.07 million ha of land under fruit trees, field cops, and vegetables is supplied with water through tube wells, canals and karez (FAO, 2007f). Overall, Pakistan is among nations whose agrarian regions are mostly irrigated. About 82 percent of cropped area is under assured irrigation. It seems about 2.3 million ha of irrigated cropped land is afflicted with soil maladies such as salinity and sodicity (FAO, 2007f).

The total utilizable water resource of India is assessed at 1,086 km³. Regarding surface water resources, it is estimated that river basins constitute major source and average annual flow is 1,953 km³. It seems there is still scope to improve utilization of water in the Ganga-Brahmaputra basins. Average annual flow of major river systems in India are as follows (in km³): Indus- 73.3; Brahmaputra- 525; Meghna-625, Mahanadi-67, Godavari-110; Krishna- 69; Pennar-6.32; Cauvery-21; Tapati-15; Narmada-46; Sabarmati-3.8; (Kumar et al., 2005).

Regarding groundwater resources, annual groundwater recharge from rainfall is about 343 km³, which is equivalent to 8 percent of annual precipitation received on the surface. Annual groundwater recharge attained by canals is said to be 89.46 m³. Therefore, total replenishable groundwater resource is 431 km³. However, total groundwater available for irrigation is 361 km³, but that utilized actually is only

325 km³ (Kumar et al., 2005). The groundwater potential of major rivers in India such as Ganges in the Northern plains and Godavari, Krishna and Cauvery in South, Narmada and Tapti in the West and Mahanadi in the east is important. Groundwater serves large areas of natural vegetation and agrarian regions of India. In India, about 32 percent of groundwater potential has been developed and it serves in irrigating crops, industries, and urban settlements.

During 1950–51, about 22.6 million ha of agrarian prairies were irrigated in the Indian Plains. The irrigation potential of India is estimated at 140 m ha. About 76 m ha could be supported by surface irrigation schemes and 64 M ha through groundwater (Kumar et al., 2005). Currently, over 59 m ha of cultivated land is under irrigation in India.

Estimates of Water Resources of India	
Precipitation mm year⁻¹	**1,083 (1,129)**
Precipitation volume km³	**4,000 (3,560)**
Surface water entering the country km³ year⁻¹	**500 (635)**
Total internal renewable water resources km³ year⁻¹	**1,446 (1,662)**
Total renewable water resources/km³ year⁻¹	**1,869 (1,911)**
Annual TRWR/per capita	**1,581 (1,616)**

Source: IGES-GWP South Asia (2012); Ministry of Water Resources (MWR) (2008); AQUASTAT (2013)
Note: Values in the parenthesis are derived from AQUASTAT

South India supports large posse of crop species that differ in water requirements. Precipitation is the major water resource, in addition to rivers, canals, groundwater, open wells and lakes (Guilmoto, 2003). Totally about 9,329,672 ha of cropland in South India are irrigated. Precipitation received during monsoon period is in fact the main stay for several arable crop species and low land rice. Agricultural prairies in the Deccan region and plains receive about 830–1,120 mm precipitation annually (Kubo, 2005; Krishna, 2010a). Southern India receives water through both southeast and northeast monsoon. Regions in Western Ghats and Coasts may receive high amounts of precipitation ranging from 1,800 to 2,800 mm year⁻¹. The dry land region is supplied with scanty water ranging from 450 to 550 mm year⁻¹. South Indian farmers try to match crop's water needs with supply, especially during critical stages of crops. Generally, dry land cereals such as sorghum, maize, finger millet, setaria, and panicum need 450–500 mm, legumes require 350–600 mm depending on species; oilseeds like groundnut sunflower and sesamum need 500–700 mm; cash crops like cotton, chillies, tobacco need 700 mm ; vegetable like tomato, potato, onion, beans, and cabbage need 350–700 mm based on species. In addition to various natural factors, yield goals set by farmers and water use efficiency (WUE) decides water needs of crops grown in South Indian prairies. Wet land crops such as rice needs 900–1,200 mm and sugarcane 1,300 mm (Ikisan, 2007a; Krishna, 2010a).

Following is the water needs, grain productivity, and water use efficiency (WUE) of major cereals of South India:

	Rice	Sorghum	Pearl Millet	Maize	Finger Millet
Water requirement (mm)	2,000	500	500	625	410
Grain yield (kg ha^{-1})	6,000	4,500	4,000	5,000	4,137
WUE (kg ha^{-1} mm^{-1})	3.0	9.0	8.0	8.0	11.4

Source: Krishna (2010a)

The average annual water requirement in India is forecasted as follows: by 2010 water requirement for irrigation reached 524 km^3 (surface water 318 km^3 + groundwater 206 km^3) but by 2025 it increases to 611 km^3 (surface water 366 km^3 + groundwater 236 km^3). Considering that agricultural productivity has to increase further and irrigation requirements would increase, it is generally accepted that management of water resources and improving irrigation efficiency is of utmost importance both in the Indo-Gangetic plains and Peninsular India. Agronomic procedures that conserve soil moisture and irrigation water are essential. Since, 70 percent of water potential is utilized to support crop production, it is said even a 10 percent saving through efficiency measures is of great utility. It seems, just use of sprinkler irrigation reduces water usage by 56 percent for crops such as bajra and jowar (Kumar et al., 2005). Drip irrigation of cash crops is highly efficient. Methods that allow reuse and recycling of water are also required in areas with scarcity. In future, perhaps as precision agriculture becomes more amenable and popular, dispensation of water will be more efficient in each field. The water resource utilization by a larger belt becomes efficient.

Bangladesh is well endowed with water resource. Annual precipitation is 2,666 mm equivalent to 384 km^3 water. Total internal renewable resource is 105 km^3. The per capita annual water availability is 8,370 m^3 (IGES-GWP South Asia, 2012).

Water resource utilization and regular irrigation of crops were in vogue in Eastern Gangetic plains since several centuries. Canals were regulated for water flow into crop fields using temporary mud structures. During monsoon, water flooded into entire region, but in summer it was well controlled through irrigation canals. Water resources were controlled by Zamindars using a system called "pulbandi." This system of water resource allocation continued till 1947. Land reforms and development of small irrigation projects since 1960s has helped Bangladesh farmers in growing winter/summer crops with yield goals. Irrigation through canals and portable pumps were highly successful in the delta zone. Currently, irrigation potential of Bangladesh is estimated at 7.5 million ha out of which about 3.6 million ha is already under regular irrigation.

Eastern Gangetic plain encompasses flood plains of three major rivers, Ganges, Brahmaputra, and Meghna. Tributaries such as Teesta, Dharla, Dudkumar, Surma, and Kshiyara are other riverine sources of water. Brahmaputra flows for 220 km, Ganges for 21 km and Meghna 160 km within Bangladesh territories. These rivers irrigate crops and provide water along their flow path (ICID, 2012). The agricultural prairie in Bangladesh is predominately rice-based and irrigation is almost mandatory.

Groundwater is a major source of water that helps rice and other crops to thrive in the Delta zone. There are two aquifers in Bangladesh. They are upper aquifer and main aquifer. These two aquifers, it seems are hydrologically interconnected. They occur below ground at 5–75 m depth depending on location. Groundwater gets recharged during rainy season, beginning in May. Recharge of groundwater occurs through infiltration of rainfall. Areas with highest recharge are in Dinajpur, Mymensingh, Sylhet, Naokali, and Chittagong. Rajshahi and Pabna are among lowest recharge potential zones. Groundwater levels are high during August through October.

Agricultural crops in Bangladesh are irrigated through various sources and using different methods. The surface irrigation that includes riverine zones, canals and low lift pumps caters to 1,209 thousand ha. Groundwater is the major source of irrigation to crops in the Delta. It includes Deep tube well, Shallow tube well, and Hand tube well. These sources irrigate over 2,482 thousand ha. About 71 percent of irrigation potential is utilized to support rice and 9 percent to grow winter wheat in Bangladesh. Irrigation is mostly used during dry season for aman rice production. Irrigated Boro paddy yields about 4–4.5 t grains ha^{-1} (ICID, 2012). According to Water Development Board of Bangladesh, surface irrigation facility increased rather marginally from 1,221 thousand ha in 1988 to 1,435 thousand ha in 2005. Groundwater facility improved from 3,129 thousand ha in 1988 to 3,373 ha in 2005. Clearly, irrigation support to agricultural prairies in the Delta region has been constant for past 2 decades. Following is the area of each major crop grown in Bangladesh during 2005 (in thousand ha): Aus-112; Aman-339; Boro-3661; wheat-364; Potato-199; Vegetables-152; others (oilseeds, sugarcane, cotton) - 318 (Ministry of Agriculture, Bangladesh, 2005).

Nepal receives water through precipitation and is also endowed with perennial rivers and lakes. Annual precipitation is 1,500 mm equivalent to 150 km^3. (IGES-GWP South Asia, 2012). A part of precipitation occurs as snow. Snow melt is an important water resource that helps in irrigating crops in the Terai plains. About 40 percent of irrigation potential has been used to produce crops.

6.4.1 CROPS, CROPPING SYSTEMS AND PRODUCTIVITY

The Hill country in the North, semiarid and arid belts in western provinces and fertile Indus plains that extend throughout Pakistan, supports variety of crops. They together make up the highly productive agricultural prairies that support a large population of human beings and farm animals. Major food crops of Pakistan are wheat, rice, jowar, maize, bajra, and barley. Pulses grown are chickpea, mungbean, mash, masoor, mattari, and pigeonpea. Major cash crops are Sugarcane, cotton, tobacco, sugar beet, and jute. Oilseeds such as brassica, rapeseed, mustard, sesamum, linseed, sunflower, groundnut, and castor are spread all through the hill country and Indus valley. Vegetables such as chillies, onion, garlic, coriander, turmeric, and ginger are found scattered in all regions of Pakistan (Amjad, 2003)

Major agricultural crops grown in the Indus plains, in particular, are wheat, cotton, rice and sugarcane. In addition, prairies support a variety of vegetables and horticultural crops in the riverine zones. Wheat production zones are the largest and occupy over 8.3 m ha of the fertile Mollisol plains. Wheat-based prairies are dominant in Punjab and Sind. The productivity of wheat is relatively high in the plains due to application of fertilizers and irrigation. Total production of wheat is about 191,183 thousand t grain ha^{-1} at an average productivity of 2,388 kg grain ha^{-1}. Cotton is an important cash crop that spreads all across the Indus valley. It occupies about 2.8 m ha and contributes about 10,211 thousand bales annually at an average productivity of 621 bales ha^{-1}. Rice is an important wetland crop and a staple cereal to large section of population in Pakistan. Rice is cultivated on 2,335 thousand ha annually (Figure 6.4; IRRI, 2008). Productivity is moderate at 2.3 t grain ha^{-1} and total contribution of rice grains is 4,438 thousand t annually. Sugarcane forms the tall prairies in Punjab, Sind and the Delta zone. It spreads into areas with facility for irrigation. Sugarcane based prairies occupies 1,100,000 ha, and offers 52,056 t cane at an average productivity of 47,324 kg cane ha^{-1} (Amjad, 2004; IRRI, 2008).

FIGURE 6.4   Rice growing regions of Indus valley (Dotted zone along Indus and tributaries)
Note: The Indus plains support diverse natural vegetation and intensely cultivated agricultural prairies that includes major cereals such as rice, wheat, maize wheat, legumes, and oilseed crops. Rice-wheat and wheat-cotton cropping systems dominate the plains of Pakistan.
Source: IRRI (2013a)

Major crops grown in the Indo-Gangetic Belt are:

Cereals such as Wheat (*Triticum aestivum; Triticum durum*), Rice (*Oryza sativa*), Maize (*Zea mays*), Pearl Millet (*Pennesitum glaucum*), Sorghum (*Sorghum bicolor*), Proso millet (*Panicum milleaceum*), Foxtail millet (*Setaria italica*), Kodo millet (*Paspalum scrobiculatum*);

Legumes and Oilseeds such as Arhar (*Cajanus cajan*), Black gram (*Vigna mungo*), chickpea (*Cicer arietinum*), Cluster bean (*Cyamopsis tetragonolobosa*), Peas (*Pisum sativum*), Moth bean (*Vigna aconitafolia*), Rapeseed (*Brassica juncea*), Safflower (*Carthamus tinctorious*), Groundnut (*Arachis hypogaea*), Sunflower (*Helianthus anuus*);

Fodder crops such as Deenanath grass (*Pennesitum pediculatum*), Guinea grass (*Panicum maximum*), Rhodes grass (*Chloris guyana*), Oat (*Avena sativa*), Cowpea (*Vigna unguiculata*), Guar (*Cyamopsis tetragonolobosa*), *Leucana leucocephala*, Velvet bean (*Macuna deeringiana*);

Fibre crops such as Cotton (*Gossypium herbaceoum; G. arboretum, G. hirsutum*), jute (*Chorcorus spp*);

Vegetables such as Ash gourd (*Benincasa hispida*), Bhendi (*Abelmoschus esculentus*), Bottle gourd(*Lagenaria sicararia*), Bitter gourd (*Momordica charantia*), Cauliflower (*Brassica oleracea*), Cucumber (*Cucumis sativa, Cucumis melo*), Pumpkin (*Cucurbita pepo*), Ras dhana (*Amaranthus caudatus*), Ridge gourd (*Luffa acutungula*), Tomato (*Lycopersicum esculentum*), Brinjal (*Solanum melangina*), Potato (*Solanum tuberosum*), Satawar (*Asparagus sp*), Bathua (*Chenopodium album*), Lettuce (*Lactuca sativa*), Methi (*Trigonella foenum-graecum*), Spinach (*Spinacea oleracea*), Sweet potato (*Ipomea batatus*);

In addition to the above listed crops that dominate the Indo-Gangetic prairie, farmers here cultivate several others that are spices, medicinal plants, timber species, fruit trees, and multipurpose crops. (Singh 2012; see ICAR, 1975). In the Indus plains, salinity is being negotiated by planting agroforestry trees and grass species that tolerate salinity better. For example, salt-tolerant fodder species such as Kallar grass (*Leptoticha fusca*), Para grass (*Sporobolus arabicus*), Swank (*Echinochloa sp*), *Sesbania sp* and *Atriplex* have been planted in regions afflicted with salinity (NIAB, 2013).

South Asian prairies traced in India comprises of several cereals, legumes, oilseed crops, cotton, sugarcane, and so on. They form prairies of short or tall stature depending on crop species and genotype. Many of them are cultivated as monocrops, but most of them are found intercropped with other crop species of value to farmers of the location. These prairies negotiate vagaries of agroclimatic parameters, soil fertility variations, different levels of fertilizer and irrigation supply. Let us consider a few examples of cereals and legumes.

Wheat is a crop native to "Fertile Crescent" region of West Asia, where it was domesticated. Wheat reached the Indo-Gangetic plains during 2nd millennium B.C. Since then, wheat cultivation has been in vogue in entire Indo-Gangetic belt. Wheat is predominantly a winter crop sown during early November in the entire Indo-Gangetic plains. Wheat is grown on Mollisols rich in organic matter in Pakistan and Punjab plains of India. Wheat based prairies thrive excellently on Inceptisols and Alfisols traced in the Gangetic plains. Wheat crop thrives on residual moisture and winter rains that may range from 200 to 300 mm depending on location in the plains. Wheat crop requires 500–600 mm precipitation in order to grow and form grains. Land races of wheat grown during early 1900s till 1950s were replaced with high-yielding semidwarfs that were tolerant to diseases such as brown rust and smut. Wheat hybrids generated moderately high grain yield at 2–3 t grain ha^{-1}, if the crop is supplied with fertilizers, organic manures and irrigated timely. Irrigation at crucial stages such as tillering and grain fill is important. Wheat based sequence practiced until 1970s included Oilseed Brassica-wheat and Pulse-wheat. However, since 1970s, Rice-wheat cropping system has evolved into dominant cropping system in the prairies. Wheat grain generated in the Indo-Gangetic plains is predominantly consumed *in situ* and it supports over 800 million people residing in the Pakistan, India, and Bangladesh. Currently, average productivity of hexaploid wheat (*T. aestivum*) is about 3.1–3.5 t ha^{-1}. Durum wheat also thrives in patches within the Indo-Gangetic plains (See Biswas and Narayansamy, 1997; Velayutham, 1997; Pasricha and Brar, 1999; Pathak et al., 2002; Krishna, 2003, 2014).

Rice is a staple cereal in most areas of South Asia. Rice-based wetland prairies dominate the landscape throughout South Asia, except in dry regions without irrigation facility. Rice ecosystem extends into 45 million ha and offers 145 million t grains in India. Rice thrives excellently on Inceptisol, Mollisols, and Alluvial soils found in the Northern Indian plains. It is grown on Alluvial soils in the Napalese Terai and Bangladesh. In South India, rice thrives on Alfisols, Vertisols, and Alluvial soils. Rice is grown during kharif in the Indo-Gangetic plains. It needs a minimum of 900–1,200 mm water to grow and form grains. Rice is grown intensively in the Coastal plains and River Delta region of Peninsular India. Fertilizer supply is high at 180 kg N, 60 P and 100 kg ha^{-1} and is commensurate with yield goals of 6–8 t ha^{-1}. Low land rice grown under rain-fed conditions may offer 2–3.5 t grain ha^{-1}. Upland Paddy is also grown in some pockets of hilly regions of South Asia. The rice based wet prairies alternate with arable crops in many regions. However, there are several locations where intensive rice monocrops are preferred. Water use efficiency of rice is low and therefore a monocrop during all three seasons needs commensurately higher water supply. In South Asia, rice-based cropping systems in plains include crops such as pigeonpea, chickpea, cowpea, oilseeds such as groundnut, sunflower, or vegetable (Sharma et al., 2006; Krishna, 2010b; IRRI, 2012; IRRI, 2013b).

Sorghum reached peninsular Indian coasts sometime during 3rd millennium B.C. from Eastern African region of Mozambique. The Neolithic community spread the sorghum cultivation techniques all across peninsula through migration (Fuller et al., 2004). Sorghum is an arable crop that thrives on Vertisol and Alfisol plains of South India. It is also grown in Indo-Gangetic plains, but under low fertility conditions. Sorghum belt in India has depreciated in area from 18.1 m ha in 1960 to 9.3 m ha in 2006. Yet, total grain harvest has not declined sharply, because it has been offset by enhanced productivity. Productivity of sorghum has increased from 528 kg ha^{-1} in 1960 to 1.1 t ha^{-1} in 2006 (FAO, 2006). Sorghum thrives luxuriantly on Vertisol plains of Maharashtra, North Karnataka, and Andhra Pradesh. It is grown as rotation crop on Red Alfisols found all over in South India. Sorghum is an arable cereal in some locations within Sindh region of Pakistan. Sorghum crop needs 550–600 mm per season to mature and form seeds. Fertilizer supply includes all three major nutrients and micronutrients. About 80 kg N, 40 kg P, and 60 kg K is applied for a crop that offers 2–2.5 t grain ha^{-1}.On an average, sorghum crop removes 22 kg N, 13 kg P$_2$O$_5$, and 34 kg K$_2$O to produce 1 t grain ha. Sorghum based taller prairies are usually intermixed with other cereals such as maize or pearl millet. It is intercropped most frequently with pigeonpea, cowpea, or groundnut in Peninsular India (Krishna, 2010c; Krishna, 2014).

Maize was introduced into South Asia by the European Travellers during Medieval period. Maize based tall prairies are predominant in dry arable cropping zones of South Asia. Maize is preferred because it offers high grain yield and relatively larger quantity of forage compared to other cereals. During recent years, maize has gained in area and replaced cereals such as sorghum and pearl millet in India. Maize belt is pronounced in Rajasthan, Gujarat, Bihar, Maharashtra, Karnataka, Andhra Pradesh, and Tamil Nadu. Moderately fertile Vertisols and Alfisols regions of South India support vast tracts of maize intercrops with legumes or oilseeds such as sunflower and groundnut. Maize requires 550–600 mm precipitation per season to grow and form mature cobs. Maize is a preferred cereal in the Terai region of Nepal. It thrives well on Alluvial soils of Terai. The productivity of maize depends on inherent soil fertility and fertilizer supply. Maize crop produces 0.5–2 t grain ha^{-1} under low fertility and low input conditions. Moderately intense cropping zones offer 3.5–5 t grain ha^{-1}. Maize is also produced intensively with high fertilize inputs and irrigation. It offers 7–8 t grain ha^{-1} plus 12–15 t forage (Krishna, 2013; Krishna 2010d; Ikisan, 2007b). In India, maize crop extends into 7.5 million ha and its average productivity is 2.1 t grain ha^{-1} (Krishna, 2013; 1014).

Pearl millet has a secondary center of genetic diversity in Northwest India and Dry tracts of Pakistan. It has been cultivated in this region since Neolithic period (2nd and 3rd millennium B.C.) (See Krishna, 2014). Pearl millet based prairies here are moderate in height at 3–3.5 m, but shorter than landraces traced in West Africa (4–5 m). Pearl millet thrives well in the arid regions of Pakistan found in the eastern part of Sindh. Pearl millet forms moderately tall prairie in the dry and arid regions of Rajasthan. Pearl millet thrives on dry sandy soils found all through the fringes of Thar Desert in Northwest India. Pearl millet prairies are interspersed with small trees and shrubs. In Gujarat, Maharashtra, and Andhra Pradesh, pearl millet is an important cereal crop. It also offers forage. Pearl millet is mostly preferred on low fertility Alfisols. It produces about 2 t grains ha^{-1} plus forage. Pearl millet grown as rain-fed crop requires 450–500 mm precipitation. Pearl millet is most often a subsistence crop in Rajasthan, hence extraneous supply of nutrients is restricted to crop residue recycling and FYM. Pearl millet is intercropped with hardy legumes such as guar or cowpea (Gill, 1991; Andrews and Kumar, 1992; Krishna, 2014). Since 1980s, Pearl millet hybrids with tolerance to diseases have been preferred in India.

Finger millet is native to Southern Indian plains. Archaeological evidences from Southern Indian locations such as Hullur suggest that finger millet (*Eleucine coracana*) was cultivated during late Neolithic phase (Fuller et al., 2004). Finger millet is a dry land cereal more frequently traced in the Alfisol plains of South India. Finger millet based prairies are predominant in Karnataka, southern Andhra Pradesh, and Western Tamil Nadu. Finger millet is a staple cereal to a certain portion of human population in South India. Finger millet varieties were generally grown during Kharif season. Finger millet genotypes called as Cary, Kempu and Hulupuria were common during 19th and 20th century. Finger millet varieties such as Poorna and Hamsa have dominated the landscape for over 40 years now in the South Indian plains. Since, 1975, several hybrids derived from African and Indian finger millet land race are dominating the dry lands. For example, Indaf series and more recently GBP series are common in the Alfisols zones (UAS, 2006). The crop is sown during July/August and harvested by late November. Finger millet is grown as a subsistence crop on low fertility Alfisols of Karnataka and Andhra Pradesh. Finger millet is also cultivated with moderate supply of fertilizers say 80 kg N, 40 kg P, and 40 kg K ha^{-1} depending on yield goals. It is a drought-tolerant crop. Finger millet is rotated with legumes such as groundnut, cowpea, and pigeon pea. Finger millet is grown in sequence with maize, legumes such as groundnut or cowpea in Tamil Nadu. Average productivity of Finger millet is 1.5–2.5 t grain ha^{-1} plus 5–10 t forage, if grown with supply of fertilizers and irrigation during critical stages (Rachie and Peters, 1977; Coleman, 1920; UAS, 2006; Krishna, 2010e, 2014).

Sugarcane (*Saccharum spp*) is native to swampy region of Northeast India, Bangladesh, and Yunan region of China. Sugarcane was domesticated some 6–8 thousand years ago. South Asian sugarcane belt is well distributed into Sindh and Punjab provinces of Pakistan, entire North Indian plains, swamps, and riverine regions of Bangladesh. Sugarcane production in Peninsular India is predominant in Maharashtra, Karnataka, Andhra Pradesh, and Tamil Nadu. Its cultivation is intense in the Indo-Gangetic plains. Sugarcane stretches in the river valleys and plains form tall wet prairies of high productivity. Sugarcane is a cash crop grown in regions with assured irrigation. It is grown mainly to produce sugar, gur, and alcohol. Sugarcane is a deep-rooted crop and needs elaborate land preparation. Sugarcane region in Sindh and Punjab supports genotypes that are early, medium, or late maturing types. Tissue cultured varieties are also traced in the sugarcane belt of Pakistan and India. Sugarcane requires relatively higher quantities of water. It does not demand stagnating water, but fields are kept wet to maximum water folding capacity. Sugarcane needs 1,500 mm water per season. Sugarcane is supplied with all major nutrients and micronutrients based on soil tests. Fertilizer prescription is high at 275 kg N, 62 kg P, and 112 kg K ha⁻¹ in Peninsular India. Sugarcane crop stays in field for longer duration of 3–5 years once established and hence nutrient needs are calculated proportionately. The productivity of sugarcane ranges from 30 to 60 t cane ha⁻¹. Sugarcane prairies are well entrenched in the wetter regions with irrigation facility. Sugarcane based prairies of South Asia thrive on consistent and inelastic demand for sugar and related products by over 1.5 billion people plus biofuel (Gill et al., 1998; Sharpe, 1998; Verma, 2004; Solomon and Singh, 2005; TNAU, 2008; Ikisan, 2012).

Cotton cultivation is in vogue in the Indus valley and Gangetic terrain since 3,000 B.C. It seems cotton cultivation spread into entire South Asian region during 2nd millennium B.C. (Fuller, 2008). Currently, cotton belt in Pakistan thrives on Mollisols found in Punjab and Alluvial zones of Sindh. Cotton is predominantly a rain-fed crop, but it is also grown with irrigation support in Multan. In addition, cotton cropping extends into Bahawalpur, Dera Ghazi Khan, Faisalabad, Lahore, and Sargoda divisions. Cotton varieties such as Nayab, MHH93, BT-cotton are popular in Pakistan (Cotton Research Centre, 2011). In India, cotton production flourishes on Vertisols of Central and Southern India that are known as "Black Cotton Soils." Cotton based prairies are supplied with fertilizers, irrigated, and sprayed with pesticides to achieve the yield goals. Cotton species such as *Gossypium arboretum, G. Hirstum*, and *G. herbaceaum* are common in South Asian region. Cotton belt in India produces 3 m t year⁻¹, Pakistan 2 m t year⁻¹, and Bangladesh 0.25 m t year⁻¹. The average productivity of cotton is 590–930 kg ha⁻¹ depending on soil fertility, irrigation, and yield goals. During recent years, transgenic cotton known as BT-cotton is being gradually introduced into the Vertisol prairies. Cotton is often rotated with sorghum, maize, or groundnuts. Intercrops of cotton-maize, cotton-sorghum, and cotton- pigeonpea flourish in the dry lands of South India (Krishna, 2014; Pal, 2011).

Sunflower is an important oilseed crop in Central and South India. It is a prominent rotation crop in dry land zones of India. Sunflower thrives on wide range of soil types such as Mollisols and Inceptisols in the Indo-Gangetic plains, on Vertisols of Central India, and Alfisols of South India. Sunflower is a versatile crop and fits into several variations of crop sequences. It is grown in all three seasons depending on precipitation pattern and irrigation facilities. Sunflower thrives well in regions with average temperate of 23°C–30°C. It tolerates drought. High humidity and rainfall during seed set is detrimental. Sunflower requires 600–750 mm per season. Productivity of sunflower composites and hybrids such as KBSH 1 is about 1.2 t grain ha⁻¹. Sunflower genotypes that occupy large areas in the Indus Valley are Hysun, SF-177, Pioneer-6435, SunCross-42, and Parsun-1. Sunflower is sown in December/January in Punjab and Sindh region. Sunflower is also grown in Autumn. Autumn crop begins during July 3rd/4th week. Sunflower crop in the sandy zones of Sindh may produce 800–1,100 kg gran ha⁻¹ (Maiti et al., 2007; Ikisan, 2008a, 2008b, 2008c, 2008d, 2008e, 2008f; Krishna, 2014).

6.4.2 GENETICALLY MODIFIED CROPS IN SOUTH ASIAN PRAIRIES

South Asian prairies have been exposed to variety of incursions, introductions, on-slaught and slow acclimatization of new plant species, crops and their genotypes, through several millennia. Natural processes such as wind/water mediated dispersal; human migration, animal agents, birds, and so on have all facilitated transmission of seeds into the Asian region from both nearby and farther locations. Human migration, conquests, travelers, and traders have introduced a variety of new crops into Asian plains and hill country. Such events might have affected native vegetation and crop species in different ways and at different intensities. For example, a weed species intermixed with wheat imports from different continents might have caused havoc with farmers. A hybrid maize or sorghum introduced into Asian prairies might have enormously improved food grain production in Peninsular India. Introduction of sunflower in 1970 has improved oilseed production in the dry lands of Western and Southern India, rather perceptibly. Hybrid rice is gaining in area and has potential to improve grain yield. Hybrid rice is slowly but surely replacing cultivars that hitherto were well entrenched into the prairie ecosystem.

During the past decade, Asian prairies have been under pressure to accept introduction of "Genetically Modified crops," commonly referred as GM crops. It is a new type of invasion into agrarian regions of South Asia. These crop genotypes are modified using molecular techniques. Genes known to suppress diseases or insect attack are introduced using Molecular cloning methods. We need to evaluate, forecast and simulate the effects of introducing GM crops on the agricultural prairies of Asia mainly, Indo-Gangetic region, Deccan, Southern Plains, and Coasts. A wrong genotype potentially can create havoc with seed production of even related species.

In the Indus plains, BT-cotton, a genotype with genes that resist Lepidopteron pests is being tested. Similarly, a maize genotype known as "VT Double Pro" that resists pest attack is being evaluated for environmental effects. The flow of BT or other introduced genes in the plant community needs verification (Shaikh, 2012b). We should realize that many of the crop species are highly cross pollinated in nature and as such may transmit pollen/genes to variety of related species. Plant scientists at various centers in Pakistan have been evaluating the pros and cons of introducing GM crops. Zafar (2007) states that despite knowledge and trained personnel who could produce BT crops and ability to import as many different BT or other GM crops, so far, GM crops have not been given permission by the Agricultural Department of Pakistan. It has been suspected that a few illegal intrusions of GM crops have occurred with regard to culture of BT-Cotton. Further, Zafar (2007) suggests that farmers in Indus valley could wait and watch for the effects of GM crops on other biomes in India and China. Currently, Agricultural agencies in Pakistan have delayed introduction of GM crops, pending careful evaluation in the plains. Ebrahim (2013) states that GM crops have faced set-back in the Indian agrarian region. Now, Pakistan seems to be the next testing round. GM crops were banned for use in agricultural farms of India. Agricultural states in North India and Central plateau have already rejected use of GM crops. Pakistan too has indefinitely delayed introduction of BT cotton and BT maize in its agricultural belts. It is believed that GM crops may necessitate repeated purchasing of seeds from other countries. They also believe that seed companies from other regions may monopolize GM seed production.

Let us consider some field reports from Indus valley. Shaikh (2012b) states that BT cotton has found its way into some regions of Indus plains. It is being cultivated since 4–5 years. Reports suggest that in the Sindh region, boll worm attack has reduced due to use of BT-Cotton. BT-cotton has provided control of insects in the humid regions of Pakistan. Sindh Grower's Board has stated that BT cotton may reduce insect attack, still pesticide spray is required. BT cotton crop needs only 4–5 sprays if a non-BT crop needs 9–10 sprays. Farmers in Sindh have stated that BT cotton offers higher yield during first 3–4 cycles but its advantage subsides later. The productivity, in fact, has plunged from 5.5 t ha^{-1} to 3.2 t ha^{-1} in 5 years of continuous use of BT cotton (Shaikh, 2012a). Now, let us consider some practical notes about BT cotton grown around Bahawalpur, in Pakistan. Ahmad Rao (2013) states that, cotton is an important cash crop and it is known as "White Gold." It offers 8.2 percent of agricultural exchequer and accounts for 3.25 GDP of Pakistan. Therefore, need to enhance productivity is utmost. Farmers are prone to use BT-Cotton, since it requires lower investment on pesticides and offers greater profits. About 40,000 kg of BT cotton seeds (var. IR-FH901, IR-NIBGE-@, IR-CIM 448 and IR-CIM) has been sown in about 8,000 ha. The seed germination is moderate at 65–85 percent. Pest incidence decreases but it still needed 4–5 sprays of pesticides. BT-Cotton yielded 23–28 maund (1 mound = 40 kg), but traditional non-BT cotton yielded only 17–23 mound ac^{-1} (Ahmad Rao, 2013). We may also have to understand the influence of BT crops on nontarget insects and other forage eating species.

6.4.3 CROPPING SYSTEMS

Indus region in Pakistan supports several different cropping systems that match the agroclimate, soils, and economic goals. Yet, most cropping systems are cereal based.

Cropping Pattern in Pakistan

Punjab state: Cotton-wheat 5.5 m ha; Rice-wheat 2.8 m ha; Pulse-wheat 1.9 m ha; and Mixed crops-4.1 m ha

Sindh: Cotton-wheat 1.6 m ha; Rice-wheat 1.1 m ha; Maize-wheat 0.9 m ha; and Mixed crops 1.3 m ha

Northwest Frontier Province: Maize-wheat 0.9 m ha; Pulse-wheat 0.36 m ha; and Mixed crops 0.53 m ha

Baluchistan: rice-wheat 0.35 m ha; Vegetables-wheat 0.30 m ha; and Mixed crops 0.40 m ha

Source: FAO (2007f)

Wheat-based cropping systems are perhaps most dominant in entire Indus valley. Wheat is intercropped or rotated with variety of different crop species. Major cropping systems that include wheat are wheat-cotton, wheat-rice, wheat-pulse, and wheat-summer fodders. During recent years, green manure crops are grown during short fallow period. In the Indus valley, about 73 percent of wheat is rotated with rice, wheat-cotton, and wheat-pulse are other rotations. Wheat-based cropping systems are exhaustive and need fertilizer and organic matter supply. Green manure crop such as Sesbania is grown during fallow. It helps in recycling nutrients and adds organic-C to soil (Kamal, 2002). Productivity of rice-wheat is 4.77 t ha^{-1}. Inclusion of green manure Jantar in the sequence increased grain productivity to 5.18–5.34 t grain ha^{-1} (Kamal, 2002)

6.4.4   RICE-WHEAT CROPPING BELT OF INDO-GANGETIC PLAINS

The rice wheat cropping system that evolved during early 1970s in the Gangetic belt is an important food grain generating system. These agricultural prairies are intensely cultivated. They experience alternation of wetland and dry arable conditions (Biswas and Narayansamy, 1997; Velayutham, 1997). The rice-wheat belt of Indo-Gangetic plains can be demarcated using different criteria such as soil types, precipitation pattern and irrigation, crop genotypes preferred and cropping systems adopted. Generally, based on physiographic region, rice-wheat belt can be grouped into (a) Indus plains covering Punjab and Sindh region of Pakistan; (b) Trans Gangetic Plains (TGP) comprising Punjab in Pakistan and India, Haryana, Rajasthan, Delhi and Uttarakand region; (c) Upper Gangetic plains (UGP) covering regions between Ganges and Yamuna rivers; (d) Middlle Gangetic Plains (MGP); and (e) Lower Gangetic Plains (LGP) (Pasricha and Brar, 1999). Soils traced in Indo-Gangetic plains are generally branded as moderately fertile and with organic matter at optimum levels of 1–3 percent. Yet, FYM and organic manures are supplied each season in order to replenish SOM lost during incessant cropping. Major soil types encountered in the Indo-Gangetic plains are Mollisols, Inceptisols (Aquepts), and Alfisols (Ustalfs). To a certain extent, Aridisols are also traced. The basic cropping sequence practiced is rice during kharif season, wheat in winter and a fallow, or green manure, or a short season crop. Rice-wheat cropping pattern is actually interspersed with farmers preferring rice monocrops, rice-fallow, rice-pulse, rice-oilseeds. The rice-wheat rotation is advantageous due to various reasons such as shorter duration and better land use efficiency. Water use efficiency of rice-wheat cropping system is higher at 82 kg ha^{-1} cm^{-1} compared to rice-rice monocrop (46 kg ha^{-1} cm^{-1}). Production efficiency of rice-wheat is again higher at 43 kg day^{-1} ha^{-1} compared to rice-rice at 42 and rice-mustard at 36 kg day^{-1} cm^{-1}. Total grain harvest and benefit cost ratio is also higher with rice-wheat system compared to other sequences. Fertilizer supply, especially urea and di-ammonium phosphate is very high and is based on the yield goals. A single rice-wheat cycle may receive 180–240 kg N ha^{-1} depending on soil test values. High fertilizer-N inputs may result in proportionately higher N-emissions such as NH_3, N_2O, and NO_2. It seems, in the Central Gangetic plains, despite split application of fertilizer-N about 10–15 kg N is lost as NH_3 emissions during rice-phase of the rotation. Then, about 7–10 kg N ha^{-1} is then lost during wheat season. Overall, about 8–10 percent of fertilizer-N could be lost each year (Mandal and Kar, 1995; Krishna, 2003). Farmers are advised to apply urease inhibitors, in order to reduce loss of soil-N. In addition to fertilizer-N, supply of P, K and micronutrients are almost mandatory, since incessant cropping during past 5 decades has depleted soil nutrients. Maintaining proper ratios of soil nutrients is important. Supply of Zn as $ZnSO_4$ once in 3 years along with FYM reduces micronutrient deficiency. During recent years productivity of rice has ranged from 4 to 5 t ha^{-1} and that wheat at 3–3.5 t ha^{-1}.

6.4.5 PEARL MILLET-FALLOW

Pearl millet-fallow systems are dominant in the semiarid and drier regions of Northwest India. Sandy Aridisol in the fringes of Thar Desert support pearl millet cultivation. In Northwest India, pearl millet expanses create taller prairies (2–3 m in height), with intercrops of legumes such as guar, cowpea, or groundnut. The region is interspersed with Kejri trees that are an important source of foliage even during dry summer. They add to soil fertility and preserve cool microclimate. In this region, agricultural prairies are comparable to "Parkland" regions of Senegal and other countries of West Africa. Pearl millet is sown with the onset of monsoon rains during July/August and harvested in mid-October. Several different cultivars of Pearl millet are cultivated and all of them tolerate drought. Productivity is constrained by low precipitation pattern, intermittent droughts, poor soil fertility, and reduced supply of nutrients through inorganic and organic manures. As such, soils are low in organic matter content. Productivity is low at 850–1,200 kg ha^{-1}, since pearl millet is grown mostly as subsistence crop in Northwest plains of India. In Maharashtra, pearl millet is grown both under rain-fed and irrigated conditions. Productivity ranges from 1.5 to 3 t grain ha^{-1}, depending on soil fertility and cultivar (Gill, 1991; Andrews and Kumar, 1992; Krishna, 2014)

The Southern Indian plains and hills support variety of different cropping systems. It generates vegetation with a mosaic of crop species. The crop combination of prairies is dynamic and keeps changing each season. Rice monocrops in all three seasons, rice-fallow and rice-arable crop are dominant in the wet land areas. Rice monocrops are intensively nurtured in Deltaic regions of rivers Godavari, Krishna and Cauvery. Productivity per season is above 5 t grain ha^{-1}. This is because rice a preferred cereal in this region. However, farmers in dry land zones adopt cropping systems such as maize continuous; maize-legume-fallow; maize-legume, maize-cotton; maize-sorghum and maize-vegetables. Sorghum is a dominant cereal in the

Vertisol areas of South India. It is a versatile cereal that gets intercropped or rotated with several other species. Productivity of sorghum has increased from 0.8 to 1.2 t grain ha^{-1} in 1970s to 2–3 t grain ha^{-1} in 2008. However, during recent years, its area of expanse has decreased perceptibly. Currently, sorghum-cotton, sorghum-sunflower, sorghum-legumes are dominant in Peninsular India. Cotton-based rotations are most conspicuous in the Vertisol plains of Karnataka, Maharashtra, and Andhra Pradesh. Finger millet cropping zones are confined mostly to Red Alfisols zones and dry areas of Karnataka and Andhra Pradesh. Finger millet is intercropped with legumes such as groundnut, dolichos, horse gram and mustard. Finger millet is a component of several combinations of rotations such as finger millet-groundnut, finger millet-cowpea-maize; finger millet-sorghum- cotton; finger millet-sunflower-groundnut; finger millet-vegetable, and so on (See Krishna, 2010a; Plate 6.1).

PLATE 6.1   Finger millet based prairies are frequently encountered in the Southern India Alfisol Plains.
Source: Krishna, K. R. (2008)

6.4.6   RICE-BASED CROPPING SYSTEMS OF SOUTH INDIA

Low land rice based cropping systems are dominant in entire South India, wherever precipitation is congenial and irrigation facilities are available. The riverine belts and deltas are almost occupied all through the year with wet land paddy. Deltas of rivers Godavari, Krishna, and Cauvery are among world's most intensely cultivated regions and they specialize in wet land rice. Rice monocrops are preferred by 31 percent of farmers residing in South India. Rice monocrops are also common in West Coast of Peninsular India. Rice-based cropping systems have been dynamic with regard to genotypes and species that follow in sequence (Janaiah, 2003; Joshi

2007; and ICAR, 2008). For example, during past 5–8 years, rice hybrids have slowly replaced hitherto semidwarf high-yielding varieties (See Krishna, 2010b). The wetland prairies are used to grow variety of arable crops after the harvest of rice. Major rice-based cropping systems encountered in agrarian regions of South India are rice-rice (monocrops), rice-groundnut, rice-pulses, rice-fallow, rice-vegetables, rice-maize, and rice-finger millet. The area covered by each rice-based rotation has changed each year. For example, between mid-1980s and 2000, area occupied by rice monocrops decreased by 7.6 percent, rice-pulse area increased by 11.3 percent, rice-vegetable increased by 12.8 percent. Overall, rice-based crop rotations increased by 22 percent. Rice-arable crop rotations are gaining in area. This procedure causes upheaval in soil physicochemical conditions, soil nutrient transformations, nutrient availability and microbial load, diversity and activity. During recent years agronomic procedures such as Integrated Nutrient Management, "Maximum Yield" procedures, Intensification and "Site-Specific Nutrient Management (Precision Farming)" have influenced the rice belt of South India (See Krishna, 2010b).

6.4.7 LEGUMES IN CROPPING SYSTEMS OF SOUTH ASIA

Through the ages, Indo-Gangetic plain has been subjected to a variety of crops and cropping systems that included cereals, legumes, oilseeds, and cash crops. Cropping systems have been dynamic based on natural and man-made factors. Indo-Gangetic plain is now a vast stretch a natural grassland vegetation and a dominant cereal belt. Rice-Wheat cropping sequence occupies most part of the plains and for most part of the year. Rice stays in the field until end of rainy season (Kharif) and wheat lasts entire winter plus a short period in summer too. This agricultural prairie is therefore cereal dominated. However, farmers in the plains do grow important legumes and oilseed crops as rotation crops or intercrops. The current trend is to utilize short summer season to grow legumes. Specific genotypes of legumes that mature early within 60–70 days are preferred. A long duration pigeon pea usually takes the place of cereal in the sequence. However, there are short season pigeonpea that can be relay sown. This way, farmers in Indo-Gangetic plain produce cereals plus legumes that are essential as protein source. Regarding soil conditions, we must appreciate that since past 4–5 decades, when rice-wheat sequence became common, Indo-Gangetic belt has experienced periodic alterations and drastic changes to soil physicochemical properties. Of course, this is attributable to the wet land rice with anaerobic soil conditions that alternates with arable wheat when soil is under oxidized aerobic state. Legume-wheat rotations do not experience such upheavals in soil physicochemical conditions. Legumes cultivated in the plains may yield moderate or low levels of grains. Yet they are grown as intercrops and sole crops (Table 6.4). Farmers tend to allocate legumes to dry land areas and those inherently low in fertility. Hence, legume productivity is commensurately low. Some of the most common rotations practiced in the Indo-Gangetic plains are Pigeonpea-wheat, rice-wheat-legume, rice-chickpea, rice-wheat-legume pasture in India; Groundnut/ Mungbean-wheat, rice-wheat-legume cover crop, pigeonpea-wheat in Pakistan; Rice-wheat-mungbean in Terai region of Nepal; and rice-jute-fallow-legumes in Bangladesh (Ramakrishna, et al., 2000).

TABLE 6.4   Legume cultivation in South Asia compared with that in Rice-Wheat belt in Indo-Gangetic Plain (1,000 ha)

Country	Total Legumes Area	Legumes in Rice-Wheat Belt
Pakistan	1,576	262
India	18,790	1,990
Nepal	253	197
Bangladesh	785	750

Source: IRRI (2000); Ramakrishna (2000)
Note: In India and Pakistan legumes are grown more outside the rice-wheat belt.

Farmers in the plains are advised to select short duration legumes such as cowpea, mung bean, or French bean. Specific short duration varieties of legumes are available that fit the rice-wheat time table. Aspects such as no-tillage, soil fertility measures, irrigation during crucial stages and pest control during summer are emphasized. We may realize that cereals that dominate the Indo-Gangetic plains are supplied with relatively high dosages of fertilizers, especially N. Soil-N extracted away by cereals needs to be replenished to sustain the fertility and yield goals. *Farmers do not derive advantage from atmospheric-N fixation that otherwise gets added, if legumes are grown as major crops. On a large scale, yearly loss of N input due to nonutilization of BNF is formidably high. This is not a good situation to continue, perhaps legumes need to be interjected in larger areas within Indo-Gangetic plains. It is striking to note that in Brazilian Cerrados and Pampas of Argentina, soybean cultivation in vast expanses literally replaces inorganic fertilizer-N requirements because of BNF. It adds 30–40 kg N ha⁻¹ per season. These regions need less number of factories that produce chemical fertilizer-N through Haber's process. Deriving N from legume-Bradyrhizobium symbiosis is ecologically a more harmonious venture.* No doubt, farmers in Indo-Gangetic plains are glaringly less efficient in utilizing natural process that adds to soil-N. We may agree that cereal root exudates too support N fixation in the root zone to a certain extent. Although not well characterized, legumes too support asymbiotic N-fixation in the root zone soil. This is in addition to that achieved in root nodules.

Groundnut is an important oilseed crop in the semiarid plains of Peninsular India. It is a short statured legume that offers pod underground. Seeds are rich in fat and protein. Groundnut expanses form short prairie vegetation. It is predominantly grown during rainy season, but in regions with irrigation, it is grown during all three seasons. Groundnut-based cropping systems thrive on variety of soil types. This crop is preferred in dry land zones with low fertility gravely/sandy soils (Krishna, 2008, 2010a). In Gujarat, groundnut expanses are maintained under intensive cropping conditions or mixed in different ratios with sorghum, pearl millet or maize

Groundnut dominates landscapes of several districts of Gujarat like Junagad and others, where it is grown as monocrop. Groundnut pod yield is relatively low in Junagad and Saurashtra because the crop is cultivated incessantly resulting in depletion of soil nutrients. Groundnut intercrop with pigeonpea is common in Maharashtra, Andhra Pradesh, and North Karnataka (Adhikary and Sarkar, 2000). Groundnuts are also intercropped with sunflower. Groundnut/maize and Groundnut/sorghum intercrops are preferred in the Alfisol plains of Karnataka and Tamil Nadu. In the hilly tracts, groundnut is intercropped with cassava. Groundnuts are strip cropped with agroforestry trees (Manickam et al., 2001; Singh et al., 2001; and Krishna, 2010d). Groundnut is rotated with several different cereals, legumes and vegetables. Groundnut-wheat rotations that offer better land use efficiency and higher grain yield are common on loamy soils of North-western Maharashtra. In the Gangetic plains, wheat grown after groundnut is said to derive up to 40 kg N ha^{-1} advantage due to residual-N available after groundnut. Fertilizer-N is usually curtailed during groundnut phase of the rotation, but other nutrients such as P, K and micronutrients are supplied. Generally, during groundnut-cereal rotations, groundnuts are supplied with 20–30 kg S ha^{-1} because it aids better pod formation and accumulation of oil (Aulakh and Pashricha, 1997).

Groundnut-rice rotation dominates the intensive cropping zones of Karnataka and Andhra Pradesh. Rice-groundnut is also preferred in the Eastern plains of India (Pal and Gangwar, 2004). Rice-groundnut rotations in the Northern plains offers 6.3 t rice grain ha^{-1}and 3.2 t groundnut pod ha$^-$ (Pal and Gangwar, 2004). Rice-Groundnut rotation involves alternation of wetland and dry arable conditions in the soil. Prior to sowing paddy it requires deep tillage, fine tilth, and puddling to flood the fields. This creates anaerobic conditions, in addition to loss of soil aggregation. Groundnut that succeeds needs gravely or sandy, loose soil conditions in order to germinate, grow, and form pods below ground. Rice-groundnut cropping system is also common in Coastal districts of Tamil Nadu and Andhra Pradesh. Rice-groundnut based prairies need relatively lower quantities of fertilizer-N supply because groundnut adds to soil-N through biological nitrogen fixation, that occcurs in symbiosis with Bradyrhizobium. Soil-N may improve by 20–60 kg N ha^{-1} season during groundnut phase of the crop rotation (See Krishna, 2008, 2010a). During recent years, farmers are guided to use integrated nutrient management (INM) practice. The INM involves application of nutrients and water using as many different sources, without affecting the ecosytematic functions of the prairies. Rice-groundnut-green manure rotation helps in restoring soil organic matter and nutrients

Sunflower is a preferred crop on sandy well drained Red Alfisols. It is a versatile crop; hence it is grown as intercrop or in rotation with several different crop species. In the plains, sunflower sequences in irrigated areas are cotton-sunflower-cotton-wheat; rice-sunflower-rice-wheat; potato-sunflower-potato; Potato-sunflower-maize-wheat. In the rain-fed regions, sunflower-fodder- wheat is a preferred.

6.4.8 LAND USE AND CHANGE IN SOUTH ASIAN PRAIRIES ZONE

During past two centuries, major land use pattern and changes if any, in the vast Indo-Gangetic plains has been caused by deforestation, agricultural expansion, and wetland clearance, especially in the Deltas of Indus, Ganges, Brahmaputra, and Meghna (Flint and Richards, 1991). The land use pattern in Pakistan is almost like any other developing country that is predominantly agrarian. Pakistan has 79,610 ha of land that could be utilized to support crops and natural vegetation. Agricultural crop production and Rangelands occupy about 60 percent of total land area. Agriculture engulfs 21,733 thousand ha and range lands 25,475 thousand ha. Coniferous forests that flourish in the Himalayan zone occupies 1,353 thousand ha. Pakistan has large expanse of waste land in the Delta zone and in the snow belt of Northeast (Khan, 2012). Wastelands are not cultivable, because of soil degradation, drought, floods and salty swampy conditions in the Indus delta. Irrigated plantations occupy 80 thousand ha. Scrub vegetation and riverine forests are other ways that land has been used.

Regarding Northwest India, land use analysis by Singh (2013) suggests that in the four major states 84 percent of land is under intense farming of crops, 5 percent is under forests, 8 percent is under urban and infrastructure. During 1960s, Northwest plains supported diverse crops such as wheat, barley, legumes, brassicas, cotton, and vegetable. However, since, 1970s, rice-wheat agroecosystem has engulfed entire Indo-Gangetic region. Rice-wheat cropping system has been intensified and it has replaced crops such as chickpea, bajra, barley, millet, and pulses. Sugarcane farming belts have not experienced severe alteration. Land use has changed predominantly towords production of cereals. High demand for cereal grains has induced farmers to intensify using higher quantities of fertilizer based nutrients and water. Agricultural cropping intensity in India is high at 170–194 in Punjab, Haryana and West Bengal. It is moderately high ranging from 124 to 155 in Southern Indian region including Andhra Pradesh, Karnataka, Tamil Nadu, and Kerala (Mohanty, 2013).

During 50 years from 1901 to 1950, the British India experienced changes in the way fertile land was utilized by farmers. Forested zones increased marginally from 22.3 m ha to 27.5 m ha in 50 years. During the same period, net cropped area increased from 76.2 m ha to 86 m ha. Fallows or land unused for cropping also increased. The waste land that could be brought under cultivation with small amendments increased from 33.7 m ha to 37 m ha. In other words, commensurate with population status and demand for food grains, the cropping area increased marginally and did not affect the forested zones. Following is the cumulative land use change pattern in India, Pakistan, and Bangladesh during 1951–2001, that is, in 50 years:

	Year	Forests	Cropland	Fallows	Cultivable Wasteland
India	1951	48.9	119	28.96	40.4
	2001	69.5	141	24.91	27.3
Pakistan	1951	1.39	11.6	3.54	9.16
	2001	3.61	16.3	5.67	9.13
Bangladesh	1951	1.27	8.38	0.63	1.81
	2001	2.32	8.08	0.40	1.93

Source: Kurosaki (2007)

Agrarian zone in India has expanded rather briskly from 1950 to 2001 by 40 m ha. Perhaps, there is small possibility for further increase in cropland. Hence, land use efficiency has to increase and productivity of grains has to be higher. Basically, intensity of crop production has to increase. Forested zones have been cleared in large scale during early 1900s and until 1970s to expand farming zones. Yet, forested zones have shown marginal increase in area by 20 m ha in 50 years (i.e., tendency of farmers to allow fields unutilized, has decreased). Vast stretches of wasteland has been developed gradually.

In Pakistan, both forest area and cropland have shown increase in area during past 5 decades. Cropland has increased by over 5.5 m ha in 50 years. In Bangladesh, cropland and fallow areas have decreased marginally. Forest area has fluctuated marginally around 1.27 m ha and waste land has not altered much in area. Clearly, population size and rapid increase in demand for food grains have necessitated conversion of large areas of forests, shrub land and waste land in the Indo-Gangetic plains and Peninsula into agrarian zones. This aspect affects the general agroclimate, nutrient dynamics, C-sequestration, biomass accumulation, and residue recycling patterns.

6.5   CROP PRODUCTIVITY IN SOUTH ASIAN PRAIRIE ZONES

Pakistan currently has 33.63 million ha of agricultural land. The growth rate of area has been low during past decade. In fact, it has depreciated in few years. It is believed that expansion of cropland is associated with population level and its growth rate, export opportunities, market value of grain, technical and institutional infrastructure, innovation and need for rural development (Khan and Hye, 2010). Further analyses have suggested that on a long-term basis, expansion of agricultural cropping zones in Pakistan has been affected more by human population, agricultural prices, income, agricultural technology, precipitation pattern, and soil fertility.

Wheat is the main cereal food source for Pakistanis. Wheat production and exchequer derived from it accounts for 3 percent of nation's GDP. Wheat-based prairies are therefore common and they occupy about 8,494 thousand ha, equivalent to 40 percent of total cropped area. Wheat grain production is about 23.5 m t year^{-1}.

Modern techniques involving fertilizers, improved semidwarfs and hybrids, irriga-tion, and disease/pest tolerance have been highly beneficial in improving wheat productivity. The total wheat production in 1960 was 4 million t and now it is 23 million t. It helped the Pakistan gain self-sufficiency in wheat grain requirements. The productivity of wheat in Pakistan increased steadily from 750 kg grain ha^{-1} in 1960 to 2,750 kg ha^{-1} in 2005 (Mahmood, 2013). Productivity of wheat in the Indus plains is relatively low at 2,451 kg grain ha^{-1} plus 6 t forage compared with world average of 3,086 kg grain ha^{-1} plus 7–8 t forage ha^{-1} (Khan et al., 2012) or 2,820 kg ha^{-1} in the Gangetic plains or 4,568 kg ha^{-1} in Northeast China (MINFA, 2010). Productivity of wheat fields given irrigation has steadily increased by 30 percent over past 3 decades and is currently 2,170 kg ha^{-1} (Nazir, 2012; FAOSTAT, 2013; and Saif, 2013). Wheat crop without irrigation yields 1,145 kg grain ha^{-1}. It seems grain yield in Indus valley is proportionate to the extent of manures and irrigation supply. Intensity of cropping and yield goals has to be revised if land use efficiency is to be enhanced. Some of the other causes suggested for reduced productivity are periodic drought, unbalanced nutrient ratios in soil, limitation of micronutrients, and rain-fed conditions. Some of the ambitious projections indicate that during 2010 till 2050s, wheat grain yield may range from 3,500 to 4,500 kg ha^{-1} in the Indus Plains (Mahmood, 2013).

Rice grown under low land flooded condition is the other major cereal grain crop of Pakistan. Punjab and Sindh are important rice producing zones. They account for 90 percent of rice belt of Pakistan. Together with wheat as rotation crop during win-ter, it has generated the rice-wheat agroecosystem of the vast Indus plains (Ahmad and Iram, 2013). Rice is an important export crop and it constitutes 6.1 percent of Pakistan's GDP. The total production of rice has experienced fluctuations because of several reasons such as demand for export, droughts, availability of irrigation, and so on. (Saif, 2013). Total rice production in the Indus valley was 5,438 thousand t during 2012.The scented rice grains, known as Basmati is among the major export-able rice varieties of Pakistan. Long grain white rice is the most common variety preferred (IRRI, 2013a; Pakissan, 2013). Rice-based prairies extended into 2.04 million ha and produced 1.5 t grain ha^{-1}. Currently, the average productivity of rice fields is 1,754 kg grain ha^{-1} but experimental fields produce up to 2.75 t grain ha^{-1} (Nazir, 2012; IRRI, 2013a). During past 5 decades, rice grain yield has increased rather gradually from 850 kg grain ha^{-1} in 1960 to 2,085 kg ha^{-1} in 2006. Climate change effects on rice yield could be marked during the next 5 decades. Enhanced ambient temperature may depreciate rice grain yield.

Maize is an important grain and forage crop. It is also used to produce vegetable oil. Maize cropping zones occupy 1,056,000 ha. Maize production is concentrated in Northwest Frontier Province and Punjab. Maize productivity is about 1,507 kg grain ha^{-1} plus forage (Nazir, 2012).

Indus valley supports a posse of small grain cereals/millet and pulses. Cereals such as barley, sorghum, pearl millet are common in less fertile zones. Their produc-tivity is < 1 t grain ha^{-1}. Major legume crops of Indus plains are chickpea, pigeonpea, and black gram. Legumes are grown more in Sindh and Northwest Frontier province. Legumes are mostly grown as intercrops or rotation crops with cereals. During 2007, pulse crops occupied 1,056,000 ha and total production was 8,42,000 t grains (Saif, 2013).

Brassicas such as rape, lahi, and mustard are important oilseed crops. Sunflower and groundnut are other major oilseed crops traced in Indus plains. They are adapted well to dry land conditions. Total area under oilseed crops has experienced a decline from 4,35,000 ha in 1975 to 2,19,000 ha in 1992. The decrease in expanse was compensated by enhanced productivity (Nazir, 2012). Currently, groundnut is an important rainy season crop grown in barani and irrigated regions. It offers seeds with high oil and protein content. Its cultivation is gaining ground in Sindh (Sukkur, Ghotki, Khairpur, and Sanghar) and Northwest Frontier (Kohot and Khurran) regions (Ahmed, 2013).

Cotton is an important cash crop of Pakistan and it contributes about 1.8 percent of GDP of the Nation (Saif, 2013). Cotton production zones extend into 2.83 m ha in the Indus plains. The cotton cropping zone has increased by 53 percent during past 3 decades. Pakistan is said to be among efficient producers of cotton. During past 3 decades, productivity of cotton in Punjab plains increased from 249 kg ha^{-1} to 865 kg ha^{-1} and in Sindh it increased from 343 kg ha^{-1} to 835 kg ha^{-1} (Nazir, 2012). Annually, Pakistan produces about 15.2 million bales.

Sugar cane is an important wet prairie crop of Indus plains. It creates a tall prairie zone that lasts for longer duration on the field. Sugar cane based prairies extend into 8,96,100 ha. Punjab has 5,36,000 ha, Sindh 2,55,000 ha, and Northwest Frontier zone has 1,04,000 ha. Sugarcane productivity has increased from 36.5 t ha^{-1} in 1975 to 48.4 t ha^{-1} in 2010 (Nazir, 2012, FAOSTAT, 2011). Fertile zones in Sindh produce 58.5t ha^{-1}. Sugarcane contributes about 0.7 to Pakistan's GDP and is a major raw material for sugar, gur, and alcohol.

Pakistan has large expanses of range land and pastures. Pastures are common in regions with <200 mm annual precipitation. Natural pastures have suffered from centuries of heavy grazing and insufficient soil management procedures. During recent years, several forage crop species have been grown to supply fodder to cattle and other farm animals. The pastures cover an area of 2.75 m ha and their productivity is 19.4 t ha^{-1}. Mixed pastures that support cereals and legumes are also common. Major forage species are sorghum that occupies 5,15,000 ha and yields 14.8 t forge ha^{-1}; berseem occupies 8,20,000 ha and produces 27.7 t forage ha^{-1}; guar 3,11,000 ha and 11.4 t forage ha^{-1}; millets 1,05,000 ha and 7.2 t ha^{-1}; Lucerne 1,52,000 ha and 32 t forage ha^{-1} (Muhammad, 2002).

The average productivity of major crops grown in India has increased during past 5 decades. Productivity increase has been marked in case of crops such as wheat, rice, moderate with oilseeds and marginal with pulses. Following is the productivity (kg ha^{-1}) trend for different crops cultivated in India:

	Rice	Wheat	Pulses	Oilseeds	Sugarcane	Tea	Cotton
1970	1,123	1,307	524	579	48,322	1,182	106
1990	1,740	2,281	578	771	65,395	1,652	225
2010	2,240	2,938	689	1,325	68,596	1,669	510

Source: FAOSTAT (2011)

KEYWORDS

- **Agroecological zones**
- **Alluvial soil**
- **Cash crops**
- **Gangetic plains**
- **Precipitation pattern**
- **Soil erosion**
- **Western Ghats**

REFERENCES

1. Abrol, Y. P.; Sangwan, S.; and Tiwari, M. K.; Land Use: Historical Perspectives. Focus on Indo-Gangetic Plains. New Delhi: Allied Publishers; **2002**, 667 p.
2. ADB. Climate Change and Migration in Asia and the Pacific. Bangkok, Thailand; Asian Development Bank; **2010**, 1–42 p.
3. ADB. Key Indicators for Asia and Pacific. Mandaluyong City, Philippines: Asian; Development Bank; **2011**, 10 December 2011.
4. Adhikary, S.; and Sarkar, B. K.; Pigeonpea (*Cajanus cajan*) intercropping with legumes in Bihar plateau at different levels of phosphate and cropping patterns. *Ind. J. Agron.* **2000**, 45, 279–283.
5. Aggarwal, P. K.; A Rice-Wheat Consortium for the Indo-Gangetic Plains. Rice-Wheat Consortium. Paper Series No 10, **2000**, 16 p.
6. Ahmad, M.; Afzal, M.; Ahmad, A.; Ahmad, Au. H.; and Azeem, M. I.; Role of Organic and Inorganic nutrient sources in improving wheat crop production. Cercetari Agronomice in Maldova. **2013**, *43*, 15–122.
7. Ahmad Rao; First BT Cotton grown in Pakistan. Pakissan. **2013**, 1–4, September 12, 2013, http://www.pakissan.com/english/ advisorybiotechnology/ first.bt.cotton.grown.in.pakistan.shtml
8. Ahmad, S.; Hussain, Z.; Qureshi, A. S.; Majeed, R.; and Saleem, S.; Drought Mitigation in Pakistan: Current Status and Options for Future Strategies. Colombo, Sri Lanka: International Water Management Institute; Working Paper No, **2004**, *85*, 1–23.
9. Ahmad, S.; and Iram, S.; Rice-Wheat Cropping Pattern and Resource Conservation technologies. Pakissan. **2013**, 1–3, September 8, 2013, http://www.pakissan.com/English/agri.overveiw/rice.wheat.cropping.pattern.shtml

10. Ahmed, A. U.; Bangladesh: Climate Change Impacts and Vulnerability-A Synthesis. Climate Change Cell, Government of the People's Republic of Bangladesh, Dhaka; **2006**, 1–48.

11. Ref>Ahmed, S.; Groundnut: A Potential Crop. Pakissan. **2013**, 1–3, September 8, 2013, http://www.pakissan.com/english/agri.overveiw/ groundnut.a.potential.crop.shtml

12. Amjad, D.; Country Report: Pakistan. Proceedings of TAC of the Asian and Pacific Centre for Agricultural Engineering and Machinery (APCAEM) Held at Hanoi, Vietnam. **2004**, 1–28.

13. Andews, D. J.; and Kumar, A.; Pearl millet for food, feed and forage. *Adv. Agron.* **1992**, *48*, 90–139.

14. Aneja, V. P.; Schlesinger, W. H.; Erisman, J. W.; Harma, M.; Behera, S. N.; and Battye, W.; Reactive Nitrogen Emissions from Crop and Livestock Farming in India. ING Bulletins on Regional Assessment of Reactive Nitrogen Bulletin No 3 1018, **2010**.

15. AQUASTAT. Water resources: Review of water resources statistics by country. Water Resources of India. Rome, Italy: Food and Agricultural Organization of the United Nations; **2013**, 1–3, October 2, 2013, www.fao.org/nr/aquastat

16. Ashalatha, K. V.; Gopinath, M.; and Bhat, R. S.; Impact of climate change on rain fed agriculture in India: A case study of Dhār wad. *Int. J. Environ. Sci. Dev.* **2012**, *3*, 68–371.

17. Asouti, E.; The origins of Agricultural in South india and contribution of Charcoal analysis. Liverhume Trust, United Kingdom. **2006**, 1–3, November 30, 2013, http://pcwww..liv.ac.uk/~easouti/southindia_prelimnaryrsults.htm

18. Aulakh, M. S.; and Pasricha, N. S.; Role of Balanced fertilizers in oilseed based cropping systems. Fertilizer News. **1997**, 42, 101–111.

19. Baber, Z.; The Science of Empire: Scientific Knowledge, Civilization and Colonial Rule in India. State University of New York Press. **1996**, 439.

20. Bangladesh Bureau of Statistics. Irrigated Land as Proportion of Cultivable Land in South Asian Countries. Dhaka, Bangladesh: Ministry of Agriculture; **2002**, 1, http://www.moa.gov.bd/statistics/Table5.12_ILAC.htm

21. BARC. Estimation of Nutrient Depletion in Major cropping pattern in Bangladesh. Bangladesh Agricultural Research Council, Dhaka. **2012**, 1, September 22, 2013, http://www. moa.gov.bd/statistics/Table4.04a.htm

22. Bhatia, A.; Pathak, H.; and Aggarwal, P. K.; Methane and Nitrous oxide emissions from agricultural soils of India and their global warming potential. *Current Sci.* **2004**, *87*, 317–324.

23. Bhattacharyya, T.; and Pal, D. K.; Carbon Sequestration in Soils of the Indo-Gangetic plain. **2002**, 1–5, September 14, 2013, http://www. knowledgebank.irri.org/ckb/PDFs/croppingsystem/environmental/carbon%20sequestration%20in%20soils%20of%20the%20indo-gangetic%20plains.pdf

24. Bhattacharyya, T.; Pal, D. K.; Velayutham, M.; Chandran, P.; and Mandal, C.; Total carbon stock in Indian soils: Issues, priorities and management. In: Land Resource Management for Food and Environment Security. New Delhi; Soil Conservation Society of India; **2001**, 1–46.

25. Bhattacharyya, A.; and Werz, M.; Climate change, migration and conflict in South Asia. Centre for American progress. **2012**, 1–93, September 14, 2013, http://www.org/downloads/bhattacharyya-Werz-Climatechange-Migration-Conflict-in-South-Asia.pdf

26. Bhatti, A. M.; Suttinon, P.; and Nasu, S.; Agriculture water demand management in Pakistan: A review and perspective. Society for Social Management Systems. **2009**, 1–7, October 5, 2013, http://academic.research.microsoft.com/Paper/11535246.aspx

27. Biswas, T. D.; and Narayanaswamy, G.; Sustainable Soil Productivity Under Rice-Wheat System. New Delhi: Indian Society of Soil Science; **1997**, 18, 88 p.

28. Brady, N. C.; Nature and Properties of Soils. New Delhi, India: Prentice Hall of India; **1975**, 567 p.

29. Brilliant, R.; Varghese, V. M.; Paul, J.; and Pradeepkumar, A. P.; Vegetation analysis of Montane forest of Western Ghats with special emphasis to RET species. *Int. J. Biodiv. Conserv.* **2012**, *4*, 652–664.

30. Buchanan, F.; A Journey from Madras through the Countries of Mysore, Canara and Malabar. London: W. Bulmer and Company, Cleveland Row, St James; (in three volumes) **1807**, 1, 1–370; 2, 1–510; 3, 1–440.

31. Chaudhry, Q.; and Rasul, G.; Agroclimatic classification of Pakistan. *Q. Sci. Vis.* **2004**, *9*, 59–66.

32. Clift; A Brief history of Indus River. In: The Tectonic and climatic evolution of the Arabian region. *Geol. Soc. Special Publ.* **2002**, 195, 237–258.

33. Coleman, L. C.; The Cultivation of Ragi in Mysore. Mysore: Bulletin of the Department of Agriculture; **1920**, 1–152.

34. Constatntini, I.; The Beginings of Agriculture in Kachi Plains: The Evidence of Mehrgarh in Pakistan. South Asain Archaeology. Allchin, E.; ed. Cambridge University Press; **1983**, 29–33.

35. Cookman, C.; Climate change and the impact of Pakistan's floods. In: Climate Change, Migration and Conflict in South Asia. Bhattacharyya, A.; and Werz, M.; eds. Centre for American Progress. **2012**, 20–27 September 14, 2013, http://www.org/downloads/ bhattacharyya-Werz-Climatechange-Migration-Conflict-in-South-Asia.pdf

36. Cotton Research Centre. Cotton varieties. Central Cotton Research Institute, Multan, Pakistan. **2011**, 1–3, November 8, 2013, http://www.icpsr.org.ma/?Page=showInstitute&InstituteID=ccri565&CountryID=pakistan

37. Deshingkar, P.; and Akter, S.; Migration and Human Development in India. United Nations Development Programme. Research Paper 1–38, 2009.

38. Deshingkar, P.; and Farrington, J.; Rural Labour Markets and Migration in South Asia: Evidence from India and Bangladesh. **2006**, 1–31, September 14, 2013, http;//openknowledge.worldbank.org/bitstream/handle/10986/9199/wdr2008_0011.pdf?sequence=1.htm

39. Ebrahim, Z.; GM Seeks New Pasture in Pakistan. **2013**, 1–4, September 12, 201, http://www.ourworld.unu.edu/en/gm-seeks-new-pastures-in-pakistan.htm

40. Erenstein, O.; Leaving the Plow Behind: Zero-tillage Rice-wheat Cultivation in the Indo-Gangetic Plains. Washington, DC: International Food Policy Research Institute; 2009a , 65–70, http://www.ifpri.org/book-5826/millionsfed/cases/ricewheat

41. Erenstein, O.; Zero-tillage in the Rice-Wheat Systems of the Indo-Gangetic Plains: A Review of Impacts and Sustainability Implications. Washington, DC: International Food Policy Research Institute; IFPRI Discussion Paper No 00916, **2009b**, 1–86, September 24th, 2013, http://www.ifpri.cgiar.org/sites/default/files/publications/ifpridp00916.pdf

42. FAI. Fertilizer Statistics-2002/2003. New Delhi; Fertilizer Association of India; **2003**, 56–89.

43. FAO. Zero Tillage: When Less Means More. Rome, Italy: Food and Agricultural Organization of the United Nations; **2001**, 1–3, September 11, 2013, http://www.fao.org/ag/magazine/0101sp1.htm

44. FAO. FAOSTAT Agricultural. Rome, Italy; **2006**, 1–8, data.http://faostat.fao.org

45. FAO. The Fertilizer Sector. Food and Agricultural Organization of the United Nations, Rome, Italy; **2007a**, 1–5, September 12th, 2013, http://www.fao.org/docrep/007/y5460e/y5460e07.htm

46. FAO. The Fertilizer Sector. Food and Agricultural Organization of the United Nations, Rome, Italy; **2007b**, 1–7, September, 12, 2013, http://www. fao.org/docrep/007/y5460e/y5460e07.htm

47. FAO. Fertilizer use by crops. Food and Agricultural Organization of the United Nations, Rome, Italy; **2007c**, 1–6, September, 12, 2013, http://www. fao.org/docrep/007/y5460e/y5460e0a.htm

48. FAO. Fertilizer recommendation. Food and Agricultural Organization of the United Nations. Rome, Italy; **2007d,** 1–9, September, 12, 2013, http://www. fao.org/docrep/007/y5460e/y5460e08.htm

49. FAO. Organic and biological sources of plant nutrients. Food and Agricultural Organization of the United Nations. Rome, Italy; **2007e,** 1–4, September, 12, 2013, http://www. fao.org/docrep/007/y5460e/y5460e09.htm

50. FAO. Agro-ecological zones and crop production regions. Food and Agricultural Organization of the United Nations. Rome, Italy; **2007f,** 1–13, September 10, 2013, http://www.fao.org/docrep/007/y5460e/5460e06.htm

51. FAOSTAT. Statistical year book. Food and Agricultural Organization of the United Nations. Rome, Italy. **2010,** 9, December 20, 2013, http://www.faostat.org

52. FAOSTAT. FAO Statistical Yearbook. 20 December 2011. **2011,** October 14, 2013, faostat.fao. org

53. FAOSTAT. FAO-irrigation statistics, Food and Agricultural Organization of the United Nations. Rome, Italy; **2013,** September 13, 2013, http://www.faostat.org

54. Flint, E. P.; and Richards, J. R.; Historical analysis of changes in land use and carbon stock of vegetation in South and Southeast Asia. *Canad. J. Forest Res.* **1991,** 21, 91–110.

55. Fuller, D. Q.; Archeobotanical and Settlement Survey, South Indian Neolithic Period. London, United Kingdom; Institute of Archaeology. University College of London; 2005a, 1–3, http://www. ucl.as. uk/ archeology/ staff/profiles/fuller/India.html

56. Fuller, D. Q.; Archaeobotany of Early Historic Sites in Southern India. London, United Kingdom: Institute of Archaeology, University College of London; 2005b, 1–3, http://www.ucl.ac.uk /archaeology /staff/profiles /fuller/ tamil.htm

57. Fuller, D. Q.; Ceramics, Seeds and Culinary Change in Prehistoric India. **2006,** 21, http://www.antiquity.ac.uk/ ant/079/0761/ant0790761.pdf

58. Fuller, D. Q.; The spread of textile production and textile crops in India beyond the Harappan zone; An aspect of craft specialization and systematic trade. In: Linguistics, Archaeology and the Human Past. Indus Project Occasional Papers 3 Series. Osada, T.; and Uesugi, A.; eds. **2008,** 1–18, http://www.ucl.ac.uk/ archaeology/people/ staff/fuller/usercontent_profile/ Textilesbeyondindus.pdf

59. Fuller, D. Q.; Korisettar, R.; Venkatasubbiah, P. C.; and Jones, M. K.; Early plant domestications in Southern India; some preliminary archaeobotanical results. *Veg. History Archaeobot.* **2004,** *13,* 115–129.

60. Gajbiye, K. S.; and Mandal, C.; Agroecological Zones, their Soil Resources and Cropping Systems. Nagpur, India: National Bureau of Soil Survey and Land Use Planning; **2007,** 142, September 20, 2013, http://agricrop.nic.in/Farm%20Mech.%20.PDF/05024-01.pdf

61. Ghimire, R.; Soil Organic Carbon Sequestration as Affected by Tillage, Crop Residue and Nitrogen Application to Rice-Wheat System. **2013,** 1–7, September 11, 2013, http://www.academia.edu/601355/ soil_organic_carbon_sequestration_as_affected_by_tillage, _crop_residue_and_nitrogen_application_to_RiceWheat_System.htm

62. Gianessi, L.; Herbicides essential for zero-till wheat on the Indo-Gangetic plains. *Crop Life Found. Int. Case Study.* **2013,** *84,* 1–4.

63. Gill, K. S.; Pearl Millet and Its Improvement. New Delhi: Indian Council of Agricultural Research; **1991,** 483 p.

64. Gill, P. S.; Srivastava, S. C.; and Johri, D. P.; Manual of Sugarcane Production in India. National Institute of Sugarcane. New Delhi: Council of Scientific and Industrial Research; **1998,** 78 p.

65. Gregory. P.; "Mehrgarh" Oxford Companion to Archaeology. England: Oxford University Press; **1996,** 329 p.

66. Guha, S.; Growth, stagnation or decline? Agriculture Productivity in British India. New Delhi: Oxford University Press; **1992,** 48–168 p.

67. Guilmoto, C. Z.; Irrigation and the Great Indian Rural Database. Vignettes from South India. **2003,** 1–8, http://demographie.net/ sifp/output/epw%20irrigation%20in%20South%20India. pdf

68. Gupta, R.; Causes of Emissions from Agricultural Residue Burning in North-West India: Evaluation of a Technology Policy Response. Kathmandu, Nepal: South Asian Network for Development and Environmental Economics (SANDEE) Working Paper, WP 66–12, **2012a,** 1–87 p.

69. Gupta, R.; Tackling Black Carbon Pollution from Agricultural Fires: The Case of Punjab. New Delhi: Indian Statistical Institute; 2012b, 1–18, September 13, **2013,** http://www.cere.se/documents/Ulvon/presentation_2012/Ridhima.pdf

70. Gupta, R. K.; Naresh, R. K.; Hobbs, P. R.; Jiaguo, Z.; and Ladha, J. K.; Sustainability of post-green revolution Agriculture: The rice-wheat cropping systems of the Indo-Gangetic Plains and China. In: Improving the Productivity and Sustainability of Rice-Wheat Systems: Issues and Impacts. Ladha, J. K.; Hill, J. E.; Duxbury, J. M.; Gupta, R. K.; and Buresh, R. J. eds. Madison, Wisconsin, USA: American Society of Agronomy; **2003,** 76–98.

71. Gupta, P. K.; et al. "Residue burning in rice-wheat cropping system: Causes and implications." *Res. Commun. Current Sci.* **2004,** *87,* 1713–1717.

72. Gupta, R.; and Seth, A.; A review of resource conserving technologies for sustainable management of the rice-wheat cropping systems of the Indo-Gangetic plains (IGP). *Crop Protect.* **2007,** *26,* 436–447.

73. Gustafsson, Ö.; et al. "Brown clouds over South Asia: Biomass or fossil fuel combustion?" *Science.* **2009,** *323,* 495–498.

74. Haider, M. Z.; Options and Determinants of Rice Residue Management Practices in the South-West Region of Bangladesh. Kathmandu, Nepal: South Asian Network for Development and Environmental Economics (SANDEE); SANDEE Working papers No 71–112, **2012,** 1–32.

75. Hasan, A.; and Raza, M.; Migration and small towns in Pakistan. International Institute for Environment and Development. Working Paper No 15: **2009,** 1–129.

76. Hassan, G.; Agri Overveiw: Saline agriculture and reclamation approach. **2013,** 1–3 September 8, 2013, Pakkisan.com http://www.pakissan.com/ english/agri.overveiw/saline.agriculture.and.reclamation.approach.shtml

77. Hobbs, P. H.; Sayre, K.; and Gupta, R.; The role of conservation agriculture in sustainable agriculture. Philosophical Transactions of the Royal Society B. Biological Sciences. **2008,** *363,* 543–555.

78. Humayun, S.; Collison of the Ganges-Brahmaputra Delta with Burma arc. **2013,** 1–45, September 12, 2013, http://www.univdhaka.academia.edu/ SyedHumayunAkther/papers/274196/ Collison_of_Ganges-Brahmaputra_Delta_with_the_Burma_Arc

79. Hussain, T.; Yaseen, M.; Jilani, G.; and Abbas, M. A.; Prospects of Using Biofertilizers for Crop Production in Pakistan. University of Agriculture, Faisalabad, Pakistan, **2010,** 1–12, http://www.infrc.or.jp/english/KNF_Data_BaseWeb/ 2nd_Conf _S_6_7. html

80. ICAR. Hand Book of Agriculture. New Delhi, India: Indian Council of Agricultural Research; **1975,** 1580.

81. ICAR. Rice. **2008,** 1–32, September 20, 2013, http://dacnet.nic.in.rice

82. ICID. Bangladesh. International Commission on Irrigation and Development. New Delhi; **2012,** 1–8, September 13, 2013, http://www.icid.org /i_dbangladesh.pdf

83. IGES-GWP South Asia. Technical Report on Issues Related to Water and Agriculture in South Asia. Kanagawa, Japan: Hayama: Institute for Global Environmental Strategies; **2012,** 33, http://www.gwp.org/Global/GWP-SAs_Files/APAN/ gwpapantechnicalreport.pdf

84. Ijaz, S. S.; Organic Matter Status of Pakistan Soils and its Management. Pakkissan. **2013,** 1–3, August 9, 2013, http://www.pakissan.com/ english/advisory/organic.farming.matter.status.of-pakistan.shtml

85. Ikisan, Irrigation. **2007a,** 1–4 September 23, 2013, http://www.ikisan.com/links/ap_irigation. shtml
86. Ikisan, Maize: Crop Technologies. **2007b,** 1–9, http://www.ikisan.com/lomks/ap_maizeCrop-Technologies.shtml
87. Ikisan, Sunflower: History. **2008a,** 1–7, March 20, 2012, http;//www.ikisan…. check
88. Ikisan, Sunflower: Seed Varieties. **2008b,** 1–4, March 20, **2012,** http://www.ikisan.com/links/ ap_sunflowerSeed%20 Varieties.shtml
89. Ikisan; Sunflower: Soils and Climate. **2008c,** 1–5, http://www.ikisan.com/links/ ap_sunflowerSoils %20 And%20Climate.shtml
90. Ikisan; Sunflower: Crop Establishment. 2008d, 7, http://www.ikisan.com/links/ap_sunflower-Crop %20Establishment.shtml
91. Ikisan; Sunflower: Nutrient Management. **2008e,** 1–8, August 20, 2012, http://www.ikisan. com/links/ap_sunflower Nutrient%20Management.shtml
92. Ikisan; Sunflower: Hybrid Sunflower. **2008f,** 1–6, May 20, 2012, http://www.ikisan.com/links /ap_sunflowerHybrid %20Sunflower.shtml
93. Ikisan; Nutrient Management in Sugarcane. **2012,** 1–5, October 20, 2013, http://www.ikisan. com/crop%20Specific/Eng/links/in-Sugarcane% 20NutrientManagement
94. Ikisan; Cultivation Practices. **2013,** 1–3, September 10, 2013, http://www.ikisan.com/basicsof agriculture/CultivationPractices.htm
95. Inam, A.; Clift, P. D.; Giosan, L.; Tabrez, A. R.; Tahir, M.; Rabbani, M. M.; and Danish, M.; The geographic and Oceanographic setting of the Indus River. In: Large Rivers: Geomorphology and Management. Gupta, A.; ed. New York: John Wiley and Sons Ltd; **2007,** 333–346.
96. IPCC. IPCC Guidelines for National Greenhouse Gas Inventories-Reference Manual. International Panel on Climate Change. **2012,** 1–28, September 14, 2013, http://www.ipcc-nggip. iges.jp/public/gl/guidelines/ch4ref5.pdf
97. Iqbal, M. M.; and Khan, A. M.; Climate Change Threats to Biodiversity in Pakistan. Global change impact studies centre, Quaid-i-Azam University Campus, Islamabad, Pakistan, World Environment Day Series. **2010,** 119–129.
98. IRRI. Management Factors Affecting Legumes Production in the Indo-Gangetic Plains. **2000,** 1–7, September 13, 2013, http://www.knowledgebank. irri.org/ckb/pdfs/cropping systems/agronomical/managment.htm
99. IRRI. Rice in Pakistan. International Rice Research Institute, Manila, Philippines, **2008,** 1–3, September 9, 2013, http://www.irri.org/index.php? option=com_k2&view=item&id=12077:rice-in-pakistan&lang=en
100. IRRI. Rice in India. Manila, Philippines: International Rice Research Institute; **2012,** 14, October 12, 2012, http://www.irri.og/partnrships/ contents_profiles/asia_oceania/india/rice_ ch_china.
101. IRRI. Rice in Pakistan. Manila, Philipines: International Rice Research Institute; **2013a,** 1–3, October 12, 2013, http://www.irri.org/index.php?option =com_k2&view=item&id=12077:rice-in-pakistan&lang=en.htm
102. IRRI. Rice in India. International Rice Research Institute, Manila, Philippines. **2013b,** 1–3, October 15, 2013, http://irri.org/index.php?option =com_k2&view=item&id=8744:rice-in-india&lang=en
103. Jain, N.; Dubey, R.; Dubey, D. S.; Singh, J.; Khanna, M.; Pathak, H.; and Bhatia, A.; Paddy and Water Environment. **2013,** 1–12, October 5, 2013, http://www.linkspringer.com/article/10.1007/s10333-013-0390-2
104. Janaiah, A.; Hybrid rice in Andhra Pradesh: Findings of a survey. *Econ. Polit. Wkly.* **2003,** *38,* 2513–2516.
105. Joshi, M.; Dynamics of Rice-Based Cropping Systems in the Southern Transitional Zone of Karnataka. **2007,** 1–5, September 24, 2013, http://www. irri.orgpublications/irrn/pdfs/ vol27no2/irrn27-2socecono.pdf

106. Kamal, J.; Effect of green manuring Jantar (*Sesbania acculata*) on the growth and yield of crops grown in wheat-based cropping systems. Proceedings of the International Technology, Education and Environment Conference. **2002**, 356–376.

107. Kazi, A. M.; Overview of Water Resources in Pakistan. Pakissan.com. **2013**, 1–15, August 9, 2013, http://www.pakissan.con/english/watercrisis/ overview.of.water.resources.in .pakistan. shmtl

108. Kazmi, A. H.; and Qasimjan, M.; Geology and Tectonics of Pakistan. Nazimabad, Pakistan: Geographic Publishers; **2011**, 41–65 p.

109. Kerr, J.; Sustainable Development of Rain Fed Agriculture in India. Washington DC USA: International Food Policy Research Institute; **1996**, 87 p.

110. Khan, J. A.; Soil survey of Pakistan: History, Achievement and Impact on Agriculture. Soil Survey of Punjab, Pakistan. **2012**, 1–39, September 13, 2013, http://www.fao.org/fileadmin/ user_upload/GSP/docs/presentation_China.feb2012/khan.pdf

111. Khan, R, E.; and Hye, Q. M. A.; Agricultural land expansion in Pakistan: An empirical analysis. *Ind. J. Agricu. Res.* **2010**, *44*, 235–241.

112. Khan, M. D.; Iqbal, M.; and Ullah, R.; Effect of Phosphatic fertilizers on chemical composition and total Phosphorus uptake by Wheat (*Triticum aestivum*). *Life Sci. J.* **2012**, *9*, 1245–1249.

113. Krishisewa; Agroclimatic zones of India. **2007**, 1–4, http://www.krishisewa.com/krishi/ azone.html

114. Krishna, K. R.; Agrosphere: Nutrient Dynamics, Ecology and Productivity. Enfield, New Delhi: Science Publishers Inc.; **2003**, 142–178 p.

115. Krishna, K. R.; Peanut Agroecosystem: Nutrient Dynamics and Productivity. Oxford, England: Alpha Science International Inc.; **2008**, 145–207 p.

116. Krishna, K. R.; Agroecosystems of South India. Boca Raton, Florida, USA: Brown Walker Press Inc.; 2010a, 51–102 p.

117. Krishna, K. R.; Rice in South India. In: Agroecosystems of south India: Nutrient Dynamics and Productivity. Krishna. K. R.; ed. Boca Raton, Florida, USA: Brown Walker Press Inc.; 2010b, 103–174 p.

118. Krishna, K.; Sorghum Agroecosystem of South India. In: Agroecosystems of south India. Krishna, K.R. (Ed.) BrownWalker Press Inc. Boca Raton, Florida, USA. 2010c, 175–239 p.

119. Krishna, K.R. 2010 d Maize Agroecosystem of South India. In: Agroecosystems of South India. Krishna, K. R.; ed. Boca Raton, Florida, USA: Brown Walker Press Inc.; 240–278 p.

120. Krishna, K. R.; Finger millet cropping zones of South India. In: Agroecosystems of South India. Krishna, K. R.; ed. Boca Raton, Florida, USA: Brown Walker Press Inc.; 2010e, 279–311 p.

121. Krishna, K. R.; Maize Agroecosystem: Nutrient Dynamics and Productivity. Waretown, New Jersey, USA: Apple Academic Press Inc.; **2013**, 343 p.

122. Krishna, K. R.; Agroecosystem: Soils, Climate, Crops, Nutrient Dynamics and Productivity. New Jersey, USA: Apple Academic Press Inc.; **2014**, 553 p.

123. Krishna, K. R.; and Morrison, K. D.; History of southern Indian Agriculture and Agroecosystems. In: Agroecosystems of South India. Krishna, K. R.; ed. Boca Raton, Florida, USA: Brown Walker Press Inc.; **2010**, 1–51 p.

124. Kubo, K.; the cropping pattern changes in Andhra Pradesh during 1990s. Implications for Micro Level Studies. **2005**, 1–28, http://www.ide.go.jp/English/publish/jrp/pdf/jrp_135_05.pdf

125. Kulshrestha, U. C.; Grenat, L.; Engardt, M.; and Roshe, H.; Review of precipitation monitoring studies in India. A search for regional patterns. Atmosphere and Environment **2005**, *39*, 7403–741.

126. Kumar, P. D.; et al. Economic analysis of total factor productivity of crop sector in Indo-Gangetic Plain of India by district and region. Agricultural Economics Research Report 2. New Delhi: Indian Agricultural Research Institute; 2002, 78–89 p.

127. Kumar, R.; Singh, R. D.; and Sharma, R. D.; Water resources of India. *Current Sci.* 2005, *89*, 794–811.

128. Kumar, U.; Jain, M. C.; Pathak, H.; Kumar, S.; and Majumdar, D.; Nitrous oxide emission from different fertilizers and its mitigation by Nitrification inhibitors in irrigated rice. *Biol. Fert. Soils.* 2000, 32, 474–478.

129. Kumbhar, M. I.; Sheikh, S. A.; Soomre, A. H.; and Khoohar, A. A.; Sustainable agricultural practices as perceived by farmers in Sindh Province of Pakistan. *J. Basic Appl. Sci.* 2012, *8*, 325–327.

130. Kurosaki, T.; Land Use Changes and Agricultural Growth in India, Pakistan and Bangladesh-1901 to 2004. 2007, 1–35, September 19, 2013, http;//www.isid.ac.in/planning/confrenceDec07/papers/Takashikurosaki.pdf

131. Ladha, J. K.; Improving the Productivity and Sustainability of Rice-Wheat Systems of the Indo-Gangetic Plains. A Synthesis of NARS_IRRI Partnership Research. IRRI Discussion Paper No 40, 2000, 31 p.

132. Ladha, J. K.; Improving the Productivity and Sustainability of Rice-Wheat Systems: Issues and Impacts. American Society of Agronomy Special Publication; 2003, 65, 76 p.

133. Ladha, J. K. J. E.; Hill, J. M.; Duxbury, R. K.; Gupta, and Buresh, R. J.; eds. Improving the Productivity and Sustainability of Rice-Wheat Systems: Issues and Impacts. Madison, Wisconsin, USA: American Society of Agronomy, Crop Science Society of America, Soil Science Society of America; 2003.

134. Library of Congress, USA. Agriculture. Washington, DC: United States Library of Congress, 2013, 1–7, September 10, 2013, http://www. countrystudies.us/india/102.htm

135. Maiti, R. K.; Singh, V. P.; Purohit, S. S.; and Vidyasagar, P.; Research Advances in Sunflower (*Helianthus annuus*). Jodhpur: Agrobios Publishers; 2007, 512.

136. Mohammad, W.; Shah, S. M.; Shehzadi, S.; and Shah, S. A.; Effect of tillage, rotation and crop residues on wheat crop productivity, fertilizer nitrogen and water use efficiency and soil organic carbon status in dry area (rain fed) of North-west Pakistan. *J. Soil Sci. Plant Nutr.* 2012, *12*, 715–727.

137. Mahmood A.; Climate change and food security in Pakistan. *Pakistan Meteorol. Dept.* 2013, 1074, September 20, 2013, http://www.pakmet. com.pk

138. Majumdar, D.; Kumar, S.; Pathak, H.; Jain, M. C.; and Kumar, U.; Reducing nitrous oxide emission from an irrigated rice field of North India with nitrification inhibitors. *Agricu. Ecosyst. Environ.* 2000, *81*, 163–169.

139. Majumdar, D.; Pathak, H.; Kumar, S.; and Jain, M. C.; Nitrous oxide emission from a sandy loam Inceptisol under irrigated wheat in India as influenced by different nitrification inhibitors. *Agricu. Ecosyst. Environ.* 2002, *91*, 283–293.

140. Malik, S. M.; Awan, H.; and Khan, N.; Mapping vulnerability to climate change and its repercussions on human health in Pakistan. *Global Health.* 2012, *8*, 31–35.

141. Mandal, D. K.; and Kar, S.; Water and nitrogen use by rice-wheat on Ultic Haplustalf. *J. Ind. Soc. Soil Sci.* 1995, 43, 9–13.

142. Mandal, A. K.; and Sharma, R. C.; Mapping of water logged and salt affected soils in IGPN command area. *J. Ind. Soc. Rem. Sen. Agency.* 2001, *29*, 229–235

143. Mandal, A. K.; and Sharma, R. C.; Computerized data base of Salt affected soils for Agro-Climatic regions in the Indo-Gangetic Plain of India using GIS. *Geocarto Int.* 2006, 21, 47–57.

144. Manickam, G.; Durai, R.; and Gnanmurthi, P.; Production potential and economic returns of different integrated weed management practices on groundnut-based cropping systems. *Crops Res. (Hissar).* 2001. *21,* 49–52.

145. Manickam, V.; Shanthisree, K.; and Muralikrishna, I. V.; Study on the vulnerability of Agricultural Productivity to climate change in Mahabubnagar District of Andhra Pradesh. *Int. J. Environ. Sci. Dev.* **2012,** *3,* 528–532.

146. Mian, M. A.; Aslam, M.; and Khalid, M.; Fertilizer use efficiency in Rice as influenced by Organic manure. Proceedings of National Soil Science Congress, Lahore, Pakistan. **1988,** 285–293 p.

147. Mian, M. A.; and Mirza, Y. J.; Pakistan's Soil Resources. Environment and Urban Affairs Division. Government of Pakistan. IUCN World Conservation Union, UNDDP Report. **1993,** 1–32 p.

148. MINFA. Agricultural Statistic of Pakistan 2009–10. Islamabad, Pakistan: Ministry of Food and Agriculture. Government of Pakistan; **2010,** 1–245.

149. Ministry of Agriculture, Bangladesh. Use of Fertilizer in Bangladesh. Ministry of Agriculture, Bangladesh. **2004,** 1, September, 22, 2013, http://www.moa.gov.bd/statistics/Table4.07. UFB.htm

150. Ministry of Agriculture, Bangladesh. Irrigated area under different crops-2005. Ministry of Agriculture, Bangladesh. **2005,** 1, September, 22, 2013, http://www.moa.gov.bd/statistics/ Table5.05IAC.htm

151. Ministry of Water Resources (MWR). National water mission under National action plan on climate change. Comprehensive Mission Document Volume—ii. New Delhi. 15 January 2012. **2008,** 1–30, October 2, 2013, http://www.india.gov.in/allimpfrms/ alldocs/ 15658.pdf

152. Mitra, A. P.; Greenhouse Gas Emission in India: 1991 Methane Campaign. Science Report No 2. New Delhi, India: Council of Scientific and Industrial Research and Ministry of Environment and Forests; **1992,** 1–23.

153. Mitra, A. P.; Greenhouse Gas Inventory for India; 1995 Update, Global Change Series, Council of Scientific and Industrial research, and Ministry of Environment and Forest, NPL, New Delhi, India; **1996,** 1–89.

154. Mitra, A.; and Murayama, M.; Rural to urban migration: A district level analysis of India. IDE Discussion Paper **2008,** *137,* 1–65.

155. Mohammed, W.; Shah, S. M.; Shehzadi, S.; and Shah, S. A.; Effect of tillage, rotation and crop residue on wheat crop productivity, fertilizer Nitrogen and water use efficiency and soil organic carbon status in dry areas of Northwest Pakistan. *J. Soil Sci. Plant Nutr.* **2012,** *12,* 715–727.

156. Mohanty, S.; Population Growth, Changes in Land Use and Environmental Degradation in India. **2013,** 1–23, September 12, 2013, http://www.iussp2009princepton.edu/papers/91994

157. Morrison, K. D.; Daroji Valley: Landscape History, Place, and the Making of a Dryland Reservoir System. Delhi: Manohar Press, Vijayanagara Research Project Monograph Series 18; 2008.

158. Muhammad, D.; Country Pasture/forage resource profiles: Pakistan. Grassland and Pasture Crops. Rome, Italy: Food and Agricultural Organization of the United Nations; **2002,** 1–14, September 11, 2013, http://www.fao.org/ag/AGP/AGPC/docCounprof/ Pakistan.htm

159. Naeem, M.; Integrated Nutrient Management for Soil Fertility and Crop Productivity of Water Eroded Lands District Swat. NWFP Agricultural University, Peshawar, Pakistan, PhD Thesis, **2009,** 186 p.

160. Naidu, B. V. N.; Groundnut. Annamalai Economic Series. Annamalai University Bulletin No 7: 148 National Remote Sensing Agency1997 Salt Affected Soils of India. National Remote Sensing Agency, Hyderabad, India: Department of Space; **1941,** 1–32 p.

161. Nazir, S.; Major crops of Pakistan. Agrihunt. **2012,** 1–3, September 8, 2013, http;//www. agrihunt.com/major-crops/39-major-crops-of-pakistan.html

162. NIAB. Research Progress of Soil Science Division. Faisalabad, Pakistan: Nuclear Institute for Agriculture and Biology; **2013,** 1–10, September 30, 2013, http://www.niab.org.pk/soil. htm

163. Nishant, B.; Rajkumar, J. S. I.; Meenakshi Sundaram, S.; Shivakumar, T.; Sanakran, V. M.; and Vanam, T. T.; Sequestration of Atmospheric carbon through forage crops cultivated in Ramayanpatti, Tirunalveli district, India. *Res. J. Agricu. Forest. Sci.* **2013,** *1,* 11–14.

164. Nizamani, A.; Rauf, F.; and Khoso, A. H.; Case Study: Pakistan-Population and water resources. **1998,** 1–14, August 9, 2013, http://www.aaas.org/ international/ehn/waterpop/paki. htm

165. Nizami, M. M. J.; Shafiq, M.; Rashid, A.; and Aslam, M.; The Soils and their Agricultural Development Potential in Pothwar. WRRI-LRRP National Agricultural Research Centre, Islamabad, Pakistan **2004,** 1–158.

166. NRSA. Salt affected soils of India. National Remote Sensing Agency. Hyderabad, India: Department of Space; Internal Report, **1997,** 1–5 p.

167. NRSC. Agricultural drought scenario in India. National Remote Sensing Agency. New Delhi: Government of India; **2013a,** 1–5, October 3, 2013, http://www.dsc.nrsc.gov.in/DSC/ Drought/indexjsp?include1=homelink1_b1.jsp&&include2=homelink1_b2.jsp

168. NRSC. Space Technology for Agricultural Drought Monitoring, National Remote Sensing Agency. New Delhi: Government of India; **2013b,** 1–3, October 3, 2013, http://www.dsc. nrsc.gov.in/DSC/Drought/indexjsp?include1=homelink5_b1.jsp&&include2= homelink5_ b2.jsp

169. NSPA. Case study of Drought in Pakistan using Satellite data. National Space Agency of Pakistan. Islamabad, Pakistan; **2013,** 1–3, September 11, 2013, http://www.suparco.gov.pk/ pages/abstract.asp?paperid=67.htm

170. Pakissan; Agri Overview: Introduction to Rice. **2013,** 1–3, September, 9, 2013, http://www. pakissan.com/english/agri.ovreview/introduction. to.rice.shtml

171. Pal, D. K.; Bhattacharyya, T.; Srivastava, P.; Chandran, P.; and Ray, S. K.; Soils of the Indo-Gangetic Plains, India: Their Historical Perspectives and Management. **2006,** 1–2, September 20, 2013, http://ww.idd.goth/18wcss/techprogarm/P11378.htm

172. Pal, S.; Indian cotton production. Current Scenerio. *Ind. Text. J.* **2011,** 65, 42–65.

173. Pal, S. S.; and Gangwar, B.; Nutrient management in oilseed based copping system. *Fert. News.* 2004. 49, 37–45.

174. Parashar, D. C.; Gupta, K. P.; and Bhattacharyya, S.; Methane emissions from paddy fields in India-An update. *J. Radio Space Phys.* **1997,** *26,* 237.

175. Pasricha, N. S.; and Brara, M. S.; Role of mineral fertilizer to increase wheat production. *Fert. News.* **1999,** *44,* 39–43.

176. Pathak, H.; Bhatia, A.; Prasad, S.; Singh, S.; Kumar, S.; Jain, M. C.; and Kumar, U.; Emission of nitrous oxide from rice-wheat systems of India-Gangetic plains of India. *Environ. Monit. Asses.* **2002,** 77, 163–178.

177. Pathak, H.; Li, C.; Wassman, R.; and Ladha, J. K.; Simulation of nitrogen balance in the rice-wheat systems of the Indo-Gangetic plains. *Soil Sci. Soc. Am. J.* **2006,** *70,* 1612–1622.

178. Pathak, H.; Malang. A.; Bhatia, A.; Prasad, S.; Jain, N.; and Singh. J.; Mitigation of Nitrous oxide and Methane emissions from rice-wheat system of the Indo-Gangetic Plain. **2013,** 1–7, September 13 2013, http://www.coalinfo.net.cn/coalbed/meeting/22203/ papers/agriculture/ AG009.pdf

179. Planning Commission, Government of India. Report of the Task force on Grasslands and Deserts. New Delhi: Government of India; **2012,** 1–83.

180. Prasad, R.; Rice-wheat cropping systems. *Adv. Agron.* **2005,** *86,* 255–339.

181. Prasad, S.; and Kar, S. K.; Quaternary alluviation of the Ganga plain. *Spec. Publ. Paleontol. Soc. India.* **2005**, *2*, 225–243.

182. Qureshi, A. L.; Khero, Z. I.; and Lashari, B. K.; Optimization of Irrigation Water Management: A Study of Secondary Canal, Sindh, Pakistan. Sixteenth International Water Technology Conference IWTC, Istanbul, Turkey; **2012**, 1–15.

183. Rachie, K. O.; and Peters, L. V.; The Eleusines. A review of the world literature. International Crops Research Institute for the Semi-arid Tropics. Patancheru 502 324, India. **1977**, 178 p.

184. Ramakrishna, A.; Gowda, C. L. L.; and Johansen, C.; Management factors affecting legumes production in the Indo-Gangetic plains. In: Legumes in Rice-Wheat cropping systems of the Indo-Gangetic Plains. Johansen, C.; Duxbury, J. M.; Virmani, S. M.; Gowda, C. L. L.; Pande, S.; and Joshi, P. K.; eds. Hyderabad, India: International Crops Research Institute for the Semi-arid Tropics; **2000**, 156–165.

185. Rana, I.; Pakistan extremely vulnerable due to climate change. Tribune. **2013**, 1–2, September 11, 2013, http://www.tribune.com.pk/ story/559868/ pakistan-extremely-vulnerable-due-to-climate-change/

186. Randhawa, H. A.; Water development for irrigated agriculture in Pakistan: Past trends, returns and future requirements. Food and Agricultural organization of the United Nations. Rome Italy. **2005**, 1–22, August 9, 2013, http://www.fao.org/docrep/005/ac623e/ ac623e0i.htm

187. Randhawa, M. S.; History of Agriculture in India. Indian Council of Agricultural Research; **1983**, *3*, 243.

188. RGI. Indian Demography. Registrar General of India: Planning Commision: Demography. New Delhi, India: Government of India; **2012**, 1–83.

189. Rhind, P. M.; Terrestrial Biozones-A rough guide to world's major terrestrial ecosystems: Plant formation in the Upper Gangetic Plains bio-province. Upper Gangetic Plain Moist Deciduous Forest. **2010**, 1–34, September, 20, 2013, http://www.terrestrial-biozones.net/ Paleotropic%20Ecosystems/Upper20%Gangetic%20Ecosystems.html

190. SAARC. Drought SAARC Disaster management Centre, Kathmandu, Nepal. **2012a**, 1–20, http://www.saarc-sdmc.nic.in/pdf/ drought.pdf

191. SAARC. Drought. Kathmandu, Nepal: SAARC Disaster Management Centre; **2012b**, 147–158, http://www.saarc-sec.org/userfiles/ Large%20Publications/CCNDPPE/16-CCNDPPE-Chapter%20XII%20-%20Drought.pdf

192. SAARC. Drought. Kathmandu, Nepal: SAARC Disaster Management Centre; **2012c**, 1–28, October 3, 2013, http://www.saarc-sdmc.nic.in/pdf/publication/sdr/chapter4.pdf

193. Saif, S.; Major Crops of Pakistan. **2013**, September, 8, 2013, http://www.saif113sb.hubpages.com/hub/OR-CROPS-OF-PAKISTAN

194. Saini, H. S.; Climate change and its future impact on the Indo-Gangetic Plains (IGP). *E-J. Earth Sci. India.* **2010**, *1*, 138–147.

195. Shaheen, A.; Naeem, M. A.; Jilani, G.; and Shafiq, M.; Restoring the Land Productivity of Eroded Land through Soil Water Conservation and Improved Fertilizer Application on Pothwar Plateau in Punjab Province, Pakistan. Plant Production Science. **2011**, *14*, 196–201.

196. Shaikh, S.; Pakistan debates GM-Cotton's success. *Sci. Dev. Net.* **2012a**, 1–3 October 1, 2013, http://www.scidev.net/global/gm/news/pakistan-   debates-gm-cotton-s-success.html

197. Shaikh, S.; Monsanto to carry out new GM maize trials in Pakistan. *Sci. Dev. Net.* **2012b**, 1–3, October 1, 2013, http://www.scidev.net/ global/biotechnology/news/monsanto-to-carry-out-new-gm-maize-trials-in-pakistan.html

198. Sharma, A.; Drought Management in Vertisols of Central India—A review. *Agricu. Rev.* **2002a**, *23*, 134–139.

199. Sharma, R. C.; Use of GPS for farm level mapping of salt affected soils. Asian GPS. Proceedings of "GPS in Agriculture". New Delhi; **2002b**, 1–3.

200. Sharma, S. K.; Gangwar, K. R.; Pandey, D. K.; and Tomar, O. K.; Increasing productivity of rice-based systems for rain fed upland and irrigated areas of India. *Ind. J. Fert.* **2006**, *2*, 29–40.

201. Sharma, G. R.; Misra, V. D.; Mandal, D.; Mishra, B. B.; Pal, J. N.1980. Beginnings of Agriculture: From Hunting and Food Gathering to Domestication of Plants and Animals. Allahabad. 283 p.

202. Sharma, C.; Tiwari, M. K.; and Pathak, H.; Estimates of emission and deposition of reactive nitrogenous species for India Current Science **2008**, *94*, 1439–1446.

203. Sharpe, P.; Sugarcane: Pat and Present. Southern Illinois University. **1998**, 327, October 10, 2013, http://www.siu.edu/~ebl/leaflets/sugar.htm

204. Siddiqui, N. K.; Geological basins of Pakistan. **2011**, 1–3, September 11th, 2013, http://www.en.allexperts.com/q/Geology-1359/2011/1/Geological-Basins-Pakistan-1.htm

205. Sidhu, G. S.; Singh, S. P.; Walia, C. S.; and Singh, R. P.; Degraded Soils in Rice-Wheat areas of Indo-Gangetic Plain and their economic evaluation-A case study of Punjab State, India. Proceedings of 12th International Science Congress, Beijing, China; **2002**, 142–146

206. Sindh Agricultural Department. Sugarcane. **2011**, 1–9, September 9, 2013, http://www.sindhagri.gov.pk/sugar-about.html

207. Singh, A.; Probable agricultural biodiversity heritage sites in India: The upper Gangetic Plains region. *Asian Agri-History.* **2012**, *16*, 21–44.

208. Singh, R. B.; Drought Preparedness and Mitigation. New Delhi, India: National Academy of Agricultural Sciences; **2011**, 1–78.

209. Singh, R. B.; Land use change, diversification of agriculture and agroforestry in Northwest India. **2013**, 1–28, September, 12, 2013, http//www. ncap.res.in/upland_files/workshop/Vol5_chapter7.pdf

210. Singh, H.; and Bansal, S. K.; A review of Crop productivity and Soil fertility as related to Nutrient Management in the Indo-Gangetic plains of India. *Better Crops-South Asia.* **2010**, *93*, 27–29.

211. Singh, A.; and Kaur, J.; Impact of conservation tillage on soil properties in rice-wheat cropping systems. *Agricu. Sci. Res. J.* **2012**, *2*, 30–41.

212. Singh, H. P.; Sharma, K. L.; Ramesh, V.; and Mandal, U. K.; Nutrient mining in different agroclimatic regions of Andhra Pradesh. *Fert. News.* **2001**, *46*, 29–44.

213. Snyder, C. S.; and Bruulsema, T. W.; Nutritive efficiency and effectiveness in North America. Indices of Agronomic and Environmental Benefit. Norcross, Georgia, USA: International Plant Nutrition Institute; Report No 07076, **2007**, 1–4 p.

214. Solomon, S.; and Singh, G. B.; Sugar diversification: Recent developments and future prospects. In: Sugarcane Agro-industrial Alternatives. Singh, G. B.; and Solomon, S.; eds. New Delhi: Oxford and IBH Publishers Inc.; **2005**, 523–541 p.

215. Srivastava, M.; Sharma, S. D.; and Kudrat, M.; Effect of crop rotation, soil temperature and soil moisture on CO_2 emission rate in the Indo-Gangetic plains of India. *Int. J. Agricu. Forest.* **2012**, *2*, 117–120.

216. The World Bank. India: Issues and Priorities for Agriculture. **2012**, 1–4, September 10, 2013, http://www.worldbank.org/en/news/feature/2012/ 05/17india-agriculture-issues-priorites.htm

217. Thomas, S. M.; and Palmer, M. W.; The montane grasslands of the Western Ghats, India: Community Ecology and Conservation. *Comm. Ecol.* **2007**, 8, 67–73.

218. Timsina, J.; and Connor, D. J.; Productivity and management of rice-wheat cropping systems: Issues and challenges. *Field Crops Res.* **2001**, *69(2)*, 93–132.

219. TNAU. Nutrient Management: Sugarcane. **2008**, 1–4, July 25, 2012, http://agritech.tnau.ac.in/agriculture/agri-nutrient mgt-sugarcane.html

220. Tripathi, V.; Agriculture in the Gangetic plains during the First Millennium B.C. In: History of Agriculture in India up to 120 A.D. History of Science, Philosophy and Culture in Indian Civilization. New Delhi, India: Concept Publishing; **2008**, 348–365.

221. UAS. Research Report-Ragi (Finger millet). **2006**, 1–4, October 20, 2009, http://uasbng.kar. nic.in/ragi.htm

222. Uma, H. R.; Madhu, G. R.; and Habeeb, M.; Regional migration for inclusion: A study of Agricultural labourers from North Karnataka. *Int. J. Sci. Res. Publ.* **2013**, *3,* 1–7 p.

223. Van Der Maesen, L. J. G.; Pigeon Pea: Origin, History, Evolution and Taxonomy. In: Pigeon-pea. Nene, Y. L.; Hall, S. D.; and Sheila, V. K.; eds. United Kingdom: CAB International; **1990**, 15–46 p.

224. Venkataraman, C.; Habib, G.; Kadamba, D.; Shrivatsava, M.; Leon, J. F.; Crouzille, B.; Boucher, O.; and Streets, D. G.; Emissions from open biomass burning in India: Integrating the inventory approach with high-resolution Moderate Resolution Imaging Spectrometer (MODIS) active-fire and land cover data. *Global Biogeochem. Cycles.* **2006**, *20,* 1–12.

225. Velayutham, M.; Sustainable productivity under rice-wheat cropping system-issues and Imperatives of Research. In: Sustainable Soil Productivity Under Rice-Wheat System. Biswas, T. D.; and Naransamy, G.; ed. Indian Society of Soil Science. **1997**, *18,* 1–6.

226. Verma, R. S.; Sugarcane Production Technology in India. New Delhi: International Book Distributors; **2004**, 628 p.

227. Vishnu-Mittre; Plant economy in ancient Navdatoli-Maheshwar. In: Technical Report on Archaeological Remains. Sankalia, H. D.; ed. Department of Archaeology and Ancient Indian History, Deccan College and University of Pune, Publication 2, **1961,** 13–52.

228. Vishnu-Mittre; Discussions in Early History of Agriculture. Philosophical Transactions of Royal Society of London. **1976**, *B275,* 141 p.

229. Vishwanathan, B.; and Kavi Kumar, K. S.; Rural migration in India: How important is weather variability. *Climate.* **2012**.

230. Watt, G.; A Dictionary of Economic Products of India. Delhi, India: Cosmo Publications; Reprinted, **1889,** 257 p.

231. Wikipedia. Terai. **2011,** 1–7, September 10, 2013, http://www.en.wikepdia.org/wiki/Terai

232. Wikipedia. Deccan Plateau. **2012,** 1–5, September 10, 2013, http://www.en.wikepedia.or/wiki/decanplateau/

233. Wikipedia. Geography of South India. **2013a,** 1–8, September 13, 2013, http://www.en.wikepedia.org/wiki/geography_of_southindia

234. Wikipedia. Geography of India. **2013b,** 1–7, October 1, 2013, http://www.en.wikipedia.org/wiki/physical_features_of_India

235. Wikipedia. Climatic regions of India. **2013c,** 1–4, October 4, 2013, http://www.en.wikepedia. orgwiki/Climate_regions_of_India

236. World Bank. Agriculture in South Asia. Thimphu Statement on Climate Change. **2011,** 1238, February 10, 2012, http://www.saarc-sec.org/userfiles/ ThimphuStatementonClimateChange-29April2010.pdf

237. Zafar, Y.; Development of Agriculture Biotechnology in Pakistan. **2007,** 1–2, September 12, 2013, http://www.ncbi.nim.nih.gov/pubmed/17955999.htm

CHAPTER 7

STEPPES OF NORTHEAST CHINA: NATURAL RESOURCES, ENVIRONMENT, CROPS, AND PRODUCTIVITY

CONTENTS

7.1   INTRODUCTION

7.1.1   GEOLOGICAL ASPECTS

Briefly, geological history of Northern China has been immensely influenced by tectonic activity of Asian plate and Himalayas. Movement of Indian plate in Southern China and Pacific plate in the east has affected geomorphology of main land China (NAS, 1992). The collision of Indian plate with Central Asia began about 50 million years ago. It has doubled the thickness of crust that supports Tibet and Himalayan region. Mountain region traced in Western China area is result of tectonic movement and earthquakes. Formation of Mongolian plateau in North is supposedly a secondary effect of tectonic movement. Alluvial deposits of major rivers such as Huang He, Yangt Ze, and others have induced formation of Northern plains.The grasslands of North China extend from Manchuria, across Inner Mongolian steppe, Loess plateau of Shanxi and Gansu, and up to Tibetan plateau. Geological features of North China plains could be classified into (a) high plains of Inner Mongolia; (b) Daxinganling, Yinshan, and Helenshan of Inner Mongolia; and (c) Deseret basins (Xinshi, 1990).

7.1.2   PHYSIOGRAPHY AND AGROCLIMATE OF STEPPES

China is a large country and spreads into 9.6 million km². It stretches for 5,026 km from east to west within the main Asian landmass. Physiographical and cultural diversity is immense. It is host to large human population of 1.3 billion. China is bordered by sea in the east and south. There are several countries that border China in the west and north. China can be divided into 5 physiographical regions namely Eastern China (Northeast Plain, North plain, and Southern Hills), Xinjian-Mongolia, Tibetan Plateau, Western and Northern region of China that has sunken basins (Gob and Taklamkan). Southern and Eastern plains are fertile and regions of immense agricultural value (Figure 7.1). Tibetan plateau is mountainous and elevations range from 1,000 to 3,000 m.a.s.l. Sichuan basin is ringed by mountains and is one of the intensely farmed zones of China. Yunnan-Guizhou plateau in the Southwest has an average altitude of 2,000 m.a.sl. Landscape below the Yangtze river is rugged and covers regions in Shanxi, Hunan, and Jiangxi.

In the Northwest of Tibetan plateau, the Tarim basin is a vast stretch of landscape that extends to 1,500 km east–west and 600 km north–south. Average elevation of the basin is 1,000 m.a.s.l. Northeast of Tibetan plateau is mountain region called Altun Shan Qilian. Further north is Inner Mongolian plateau. Loess region found in between Quinling and Inner Mongolian plateau is very large stretch of arid and semiarid plains. It has yellowish eroded and silty soil that supports cropping at subsistence levels.

Agricultural zones and natural vegetation within the Northeast China form the center peace of discussions in this chapter. However, information about North Plains that encompasses Hubei, Henan, Beijing, Shanxi, and Shandong has also been included. In some cases, data pertaining to entire nation are referred mainly to provide a better perspective. Northeast China is also known as "Manchuria." It encompasses provinces such as Heilongjiang, Jilin, and Liaoning. These three provinces are known commonly as *"Dongbei Sansheng."* Another term used is *"Guangdong,"* which means east of pass. The Northeast China considered in this chapter includes Inner Mongolia. North China plain is another area discussed in this chapter. The North China plains extend into Inner Mongolia, Hebei, Shanxi, and Henan (see Figure 7.1). Agricultural landscape is dominated by wheat and other cereals that form large patches of prairie vegetation.

China is served by several major rivers. The total length of rivers is supposedly more than 4,12,000 km. Majority of rivers flow from west to east and drain into Yellow sea, South China Sea, and Eastern Sea. Yangt Ze is the longest river with 1.8 million km² catchment. Huang He or Yellow river supports one of the densest populated regions and intense farming zones in China. It traverses through North China and drains into sea in Shandong province. The Heilongjiang also called Amur (in Russia) is an important source of irrigation to farmers in Northeast plains. In South China, Pearl river is 2,114 km long. Together with its three tributaries, it is a major source of irrigation water to farms in South China plains.

FIGURE 7.1   Physiography of China.
Source: USDA Joint Agricultural Weather facility, Washington, D.C. USA
Note: Steppes region and parts of Northern Plains shown above are dealt in greater detail in this chapter.

7.1.3   AGROCLIMATE

Basic agroclimate and precipitation patterns in China are regulated by monsoon winds and Pacific oscillations. China experiences wide variation in climatic conditions that is attributable to difference in latitude, longitude, altitude, topography, and vegetation pattern. Climatic pattern depends on monsoon winds created by heat absorption by land mass and ocean. East Asian monsoon carries moist air and causes precipitation that affects entire nation. Siberian cold front affects winter temperature and precipitation pattern. In the Northeast plains, Inner Mongolia and Heilongjiang provinces experience subarctic climate. Average temperature in Northeast plains is very low. It ranges from −30 to 18°C. Southern plains experience subtropical or tropical climate. However, in Hainan and Yunnan provinces, average temperature is 17°C–28°C. Considering entire nation, temperature in China ranges from −52°C in Mohe, Heilongjiang to 48°C in Xingjiang. Annual sunshine hours range from 1,100 h in Sichuwan to 3,400 hrs in parts of Qinghai.Annual precipitation ranges from 20 mm in parts of Xinjiang to 2,000 mm in Guangdong, Guangxi, and Hainan.

Several climatic factors affect crop production in the plains of Northeast and North China. The intensity and period for which they affect the crop growth and yield formation too differ. In order to assess their quantitative influence on crops and decide on various agronomic procedures, currently simulation models and computer based decision support systems are being utilized. Such computer based decisions consider a series of climatic factors and make a make comprehensive forecast (Zhang et al., 2011b).

Northeast steppe of China is a cold belt, with temperature reaching <0°C for 5 months from November till next March. The region is warmer during April to October. Ambient temperature reaches 22°C during June to September (Loucks and Jinaguo, 1990). In areas such as Songnen steppes, annual mean temperature is 5°C and precipitation is 350–460 mm (Guo, 2010).Precipitation pattern is skewed and depends on monsoon winds. Majority of rainfall, up to 70 percent of total, occurs during June to September. The monthly precipitation for July, August, and September is around 80–95 mm. Annual precipitation of Northeast China ranges from 350 to 600 mm.

North China plain is classified as semiarid to arid with regard to agroclimatic conditions. Precipitation in the North China plains ranges from 200 to 400 mm. Northern Xingjian is a dry tract. It receives only 200–240 mm annually. Moisture is a limiting factor for growth and productivity of crops. Sunshine and photosynthetic radiation is sumptuous. Radiation is generally higher in the western region and decreases as we move to east. The temperature increases from Northeast to southwest of plains. The growing period ranges from 170 to 190 days in the east, and total heat radiation (>10°C) received is about 2,000°C. Agroclimate in the western part of North China is warmer and growing period is 250 days. Accumulated heat units reach 3,000–4,000°C. In Xinjiang, the agroclimate is arid and warmer. Total heat units received ranges from 5,600 to 5,700°C (See Xinshi, 1990).

Considering the context of this chapter, agroclimate in China could be grouped into four zones. They are (a) deserts and mountains in the west including Tibet and Mongolia. Here temperature ranges from −4°C to 12°C and precipitation is about 200 mm. Area is dominated by grasslands and man-made pastures. (b) North-eastern region experiences relatively higher precipitation at 600–1,000 mm annually and temperature ranges from 2 to 14° C. Crops grown in this agroclimatic region are wheat, rice, maize soybean, and vegetables. (c) Subtropical regions of south experiences temperature >14° C and precipitation ranges from 1,000–12,000 mm annually. (d) In the Southeast Coast and plains, precipitation is much high at 2,000 mm annually and temperature ranges from 18 to 24°C. Such tropical and subtropical climate allows 2–3 crops a year (Agribenchmark Network, 2009).

In general, based on accumulated temperature units (>10°C), cropping zones in entire China could be classified as Cold – 800–1,500°C; Cold temperate – 1,500 to <3,000°C; Warm temperate – 3,200 to <4,500°C; Northern subtropical region – 4,500–5,000°C; Southern subtropical region – 5,000–7,000°C; Quasi Tropical – 7,000–8,000°C; and Tropical >8,000°C (Thomas, 2008). Reports by Dong et al. (2009) suggest that in the spring wheat zone, accumulated temperature units range from 1,600 to 3,400°C. In the winter wheat zone, it ranges from 3,500 to 4,500°C. During 1980s, single crop per year was changed to three crops per year. Further, winter wheat zone moved north words as accumulated heat units were higher.

7.1.4 CLIMATE CHANGE EFFECTS

Temperature increase is among easily accepted effects of climate change in the Chinese agrarian zones. Since 1980, temperature fluctuation has been perceived more frequently. Yang et al. (2010) believes that it is more important to understand the effects of temperature fluctuations on crops. Firstly, we ought to understand the effects of temperature increase on crop production trends during past 2–3 decades and equally important is the forecasts on weather pattern and its consequences on crops during next few decades. They have tried to study the effect of increased temperature on crop yield, mainly rice, maize, and wheat in two segments, first from 1950 to 1980, next from 1980 to 2007. They have analyzed effect of temperature increase on cropping fronts possible in northern zones of China. According to Yang et al. (2010), during past 50 years, warming of climate has consistently caused northward movement of cropping front. Climate change has moved the cropping fronts of winter wheat and double rice-cropping to shift northwards. First, compared with cropping boundaries during 1950–1980, northern limits of the two different cropping systems has shifted northwards significantly. The northern displacement in provinces such as Shaanxi, Hubei, Beijing, and Liaoning has been conspicuous. The enhanced temperature has also resulted in higher grain productivity in provinces such as Hunan, Hebei, Anhui, and Liaoning. Gain in grain yield has been reported to be 56 percent in case of double cropping zones and 27 percent in case of single crop regions. A few other studies have shown that during 1980–2007, northern limits for wheat shifted considerably in Inner Mongolia, Hebei, and Liaoning. It also increased wheat grain yield by 25 percent compared to 1950–1980 period. Incase of maize, stable yields were more common as the crop shifted southwards. It was attributed to changes in temperature and rainfall pattern. Rainfall affected cropping fronts of both wheat and maize. Forecast for 2011–2050 by Zhao and Guo (2013) suggests that considering average daily temperature of 20°C, it is believed that northern boundaries of double and triple crop systems would shift further northwards. Multiple cropping systems may replace single crop systems. The increase in multiple cropping systems and displacement of single or double crop zones could be gradual.

There are several case studies about possible climate change effects on grassland biomes found in China. For example, Guo (2010) states that among four global change models, Songnen meadows may experience a rise in temperature by 2.8–7.5°C. Precipitation may increase by 10 percent by next century. Salinity in the cropland and natural grasslands too may increase. Drought, along with salinity will supposedly reduce biomass production in the Steppes of Northeast China. Climate change may also affect N dynamics and CO_2 flux.

7.1.5 NATURAL VEGETATION OF STEPPES OF NORTH EAST CHINA

Historical data from some parts of Northeast China indicate that during past 1,000 years, the steppes have experienced fluctuation in flora in response to natural factors. Pollen and charcoal analysis suggest abundance of *Pinus* and *Artemisia*. Several other species of trees such as Huglans, Alnus, Carpinus, Thalictrum, several species of Oleaceae and Rosaceae occurred in the plains during 800 A.D.–1500 A.D. (Li et al., 2013b). Modern vegetation is predominantly composed of *Pinus koraieusis, Thuja korainsis, Betula polyphyla, Juglans mandhsurica, Quercus mongolica, Ulmus propinqua, Tilia amurensis, Fraxinus mandhusurica, Acer species, Poplus davidania, Corylus mandhusurica, and Syringa reticulate.*

Grasslands and natural pastures cover an area of 400 million ha equating to 41.7 percent of total land area of China. They stretch 4,500 km from North east to Southwest (Guo, 2010). China is host to over 32,000 species of higher plants. There are over 7,000 woody species out of which 2,800 are tall tree species (Beijing Government, 2013). Natural vegetation of Northeast China could be easily grouped into two zones namely, Northern steppe zone and easterly forest steppe zone. There are reports that group these two regions and desert areas in Inner Mongolia into Temperate Steppes of China (Zhao,1990). According to Haibo et al. (2012), vegetation pattern can be quantitatively classified into forests and their transition zones, semihumid meadow steppes, and semiarid grassland. The steppes of Northeast are known to receive relatively slightly more precipitation than several other dry zones. Steppes soil is rich in organic matter and moderately fertile. Hence, it supports diverse species of forage grass species, crops, shrubs, and forest trees. The characteristic meadow or steppes species is sheep grass (*Aneurolepidium chinensis*). The grasslands are dense in many regions within Inner Mongolia. However, surrounding urban settlements, the grasslands have become sparse, soils are degraded, and biomass generation is of lower order.

Let us consider description about natural vegetation in the Northeast China. Fund and Hogan (2013) state that, in the Heilongjiang-Amur region, natural vegetation has broad-leaved deciduous species such as Quercus (oak) and mixture of hardwood species. Some parts of the region are prone to floods. This region supports swamps, grasses, and legumes. As we move southwards from Amur valley, vegetation is composed predominantly of *Quercus mongolica, Acer species, Telia amurensis, Ulmus propinua*, Manchuria Ash (*Fraxinus mandshurica*), *Betula dahurica, Crataegus pinnatifida*, and Amur rose (*Rosa davurica*). The Chinese red pine stand is also encountered frequently in Northeast China.

Xinshi (1990) has grouped the natural grassland of North and Northeast China into steppe, meadow, desert, and sparse forest brush. The steppes could be further classified and studied individually. Let us consider a few types of steppes.

(a) Typical dry steppes are traced in Inner Mongolia. Here, annual precipitation is low at 200–250 mm. Plant community is primarily drought-tolerant. Soils are Chernozems with a calcic horizon. Fresh grass production reaches 2,000–3,000 kg ha^{-1} per season.

(b) Temperate Steppe is formed by drought- and low temperature–tolerant grasses. It is predominant in Northeast China from the Songnen plain to Hulunbeir plateau. It is also encountered in Inner Mongolia.

(c) Meadow steppe is distributed in the eastern zone and in high desert plains. The meadow region experiences subhumid climate. Precipitation ranges from 350 to 500 mm annually. Plant community is composed of *Aneruolelidpium chinensis, Stipa baicalensis*, and several drought-tolerant grasses. Dry matter production is low at 150–500 g m^2 year^{-1} (See Tsuiki et al., 2005; Xinshi, 1990).

(d) Desert Steppe is found west of typical steppe of Inner Mongolia. Annual precipitation is low at 150–200 mm and scattered. Plant community is mostly composed of xerophytes and drought-tolerant species. *Stipa* species and super xeric shrubs are common. Soils are light, sandy, and chestnut in color. Average productivity of grass is just 300–700 kg ha^{-1}.

(e) Brush steppe is found in Loess and Sands east of Desert Steppes. The general climate is warm. Annual precipitation is 300–450 mm. Natural vegetation consists of dense shrubs. *Caragana intermedia* and *Heydiayarum scoparium* are examples of shrubs. *Artmesia* community is scattered in the entire Brush steppe zone. Brush steppe supports relatively higher biomass production.

(f) High-Frigid steppe is widely distributed in Tibet. Natural vegetation is composed of *Stipa purpurea*. Temperature is low at 5–7°C.

Meadows could be further classified into Forest meadows, Flood plain meadows, Lowland meadows, and Alpine meadows.

(a) Forest meadows are composed of grasses and legumes. About 30–50 different understory plant species could be easily traced. Meadow forest vegetation is tall and dense. It produces 30–50 t forage ha^{-1} (Xinshi, 1990).

(b) Flood plain meadows occur on the river banks. They harbor grasses, legumes, and sedges. Shrubs and trees such as Poplar and willow are common. During recent years, flood meadows have been briskly converted low crop fields.

(c) Lowland meadows occur in fringes of alluvial zones. They are prone to salt accumulation. Major plant species encountered in lowland meadows are *Achantherium splendens, Glycyrrhiza uralensis, Sophora alopecuriodes, Phragmites communis, Althagi sparsifolis*, and *Karelinai capsica*.

(d) Alpine meadows are traced in mountainous regions. It encompasses grasses, sedges, and forbs. *Kobresia* is the dominant species of alpine meadows.

Desert vegetation is found widely distributed in China. Deserts are characterized by sparse xeric vegetation. Grazing occurs in regions with vegetation. High frigid deserts are also traced in China. Microphanerophytes that are leafless are common. *Haloxylon ammodendron* and *H persicum* are examples. These species thrive on sandy soils and could offer some herbage all through the year. Several other types of deserts that occur in China are Sagebrush desert, Succulent halophytic deserts, Shrub desert, ephemeral Deserts, and Frigid desert. Overall, major problem in the steppes region is the low biomass turnover leading to insufficient forage supply to grazing animals. In addition, procedures that thwart soil degradation, erosion, and maintain natural biodiversity are needed. During past 5 decades, climate change effect on natural vegetation in the Northeast has been notable (Haibo et al., 2011). Programs that restore natural vegetation have been initiated at several places in Northeast and the results have been encouraging (Li et al., 2013c)

Northeast and North China plains are endowed with well spread out wetlands. Wetlands include regular lowland rice fields, natural marshes, fen, peat land, and swamps. Total wetland zone in China is estimated at 6.6×10^5 km^2, equivalent to 6.5 percent of total land area in China. These wetlands are rich in flora and fauna. Zhou et al. (2006) report that wet vegetation is diverse and encompasses over 2,000 species of plants. There are over 200 plant species that are endangered with extinction. Dominant wetland species in Northeast China are *Calamagostis angustifolia* and *Carex species*.

The Amur-Heilongjiang river basin is an important area in terms of natural vegetation, its diversity, and biomass generation. It hosts several forest trees species, under story plant species, scrub vegetation, and expanses of crops such as wheat, maize, barley, and soybean. Forests in the river valley are rich in Korean Pine (*Pinus koriensis*), mixed broadleaf trees, boreal forest species, and several other cold-tolerant species. Tree species dominant in the forest steppe are *Quercus mongolica* (Mongolian oak), *Betula dahurica* (Birch), shrubs such as *Lespedeza bicolor, Corylus hegterophylla, Rhamnus dahuricus*, and *Rosa dahurica* (VICARR, 2013a). The Amur-Heilongjiang region is also exposed to environmental vagaries just like any other biome. Its forest cover and biodiversity is put to test because of human requirements of firewood, crop production, and industrial infrastructure. This is in addition to climate change effects. Climate change effects can be severe. Forecasts suggest that climatic fluctuations may cause drought and floods. Higher ambient temperature may affect flora, its diversity, and cropping pattern (VICARR, 2013b, c).

7.1.6 AGRICULTURAL DEVELOPMENT IN CHINA

Description about evolution of agriculture and its development in China in the following paragraphs are drawn mainly from Bonjean (2010). Archeological evidences suggest that *Homo sapiens* resided in China's mainland sometime 60,000 years ago. However, evidences related to agricultural evolution suggest that, like other regions of earth, it began during early Neolithic period around 10000 B.C. Several archeological sites dating to 7th millennium B.C. have shown permanent dwelling and agricultural cropping. An (1984) suggests based on evidences of pottery, grains, and other artifacts, that agricultural activity might have begun in China during 14000–10000 B.C. Clearly, seeds for agricultural prairies were sown during early Neolithic period. Earliest of the crops that were domesticated in the hills and plains of China were Proso millet (*Panicum milleaceum*) and Foxtail millet (*Setaria italic*) (Fuller et al., 2009). Cereal production was in vogue in the Yellow river (Huang He) basin during 8000–3000 B.C. Human migration aided transfer of Neolithic agricultural technology to other locations in Yangt Ze valley. Rice, wheat, barley, millets, and soybean were among the most important crop species that sustained Chinese human population and domestic animals during past several millennia. Crops such as Foxtail millet and Proso millet served as staple cereal diets right until twentieth century. However, their cultivation zones have dwindled in the recent period (Kipple and Ornelas, 2000). Perhaps, this trend with minor millets is similar to other areas of the world where wheat, rice, or maize are dominant in the landscape. Rice (*Oryza sativa*) was domesticated in Yunnan and Yangt Ze valley sometime during 6500 B.C. There are several Neolithic sites in Southern China dating from 6500 B.C. to 3200 B.C that prove that rice was grown and consumed by the China's Neolithic population. Since then, rice cultivation has been in vogue. Landraces of rice ruled the landscape right up to mid-twentieth century. During past 5 decades, wetland rice belt is dominated by semidwarf high-yielding varieties. During past 10–12 years, hybrid rice is spreading fast into rice belt of China. Wheat (*Triticum aestivum*) and barley (*Hordeum vulgare*) were derived from "Fertile Crescent" in West Asia and introduced into China's prairie region during 3rd millennium B.C. Maize was derived from Meso-America during medieval period about 1550 A.D. Spanish and Portuguese traders brought maize into China and Fareast. Maize spreads rapidly into the plains of North and Northeast China because of its ability to offer higher quantity of grain and forage.

Like most other agricultural zones, spread of domesticated crop species during Neolithic period occurred through human migration and new settlements that developed. During ancient period, several techniques including use of plow, line sowing, and systematic harvesting of crops were in vogue in China. During medieval period, several species of edible vegetables, fruit crops, and spices were introduced into China. Migration of traders, conquering armies, and religious apostles were instrumental in bringing several new crop species into China (Bray, 1984). These eventually allowed development of larger agricultural zones and supported villages, towns, and cities with enhanced population levels. During medieval times, like many other agrarian regions on earth, China too had its share of crop introduction from other regions, especially from Americas. During past 6 decades, China's cropping zones spread into several new areas and production techniques were intensified. In the Northeast, maize, wheat, and soybean cultivation was intensified using fertilizers, irrigation, and high-yielding genotypes. During 1970s–1990s, productivity of major crops was improved markedly.

7.1.7 HUMAN MIGRATION, POPULATION, AND AGRICULTURAL PRODUCTION IN NORTHEAST CHINA

Northeast China is a moderately populated zone within China. It hosts a human population of 1,07,400,000 equivalent to 8 percent of 1.3 billion population of the entire nation. Much of the settlements in Northeast occurred during eighteenth and nineteenth century due to human migration through a socio-economic movement called "Chuang Guandong," which means "venture into east of pass." The native agricultural community in Northeast is "Han Chinese." Immigration of agricultural community into Northeast was aided by lift of ban during Qing dynasty. During this period, large number of people moved from Hebei and Sandong provinces and developed farm lands. Since formation of People's Republic of China, there have been well directed programs to develop the "Great Northeast Wilderness." Currently, Hans, Manchus, Mongols, Koreans, and Huis have agricultural settlements in the Northeast China (Wikepedia, 2013). Northeast plains support several nomadic tribes that practice pastoralism and ethnic groups with permanent settlements and farms. Manchus, Ulchs, Hezhen, Sushen, Xianbei, and Mohe are few examples of ethnic tribes that practice crop production.

Northeast China, also known popularly as Manchuria has large population of agricultural farms and workers. Farmers are concentrated more in the relatively warmer southern region of Manchuria. Farmers in the Huang He region grow maize, wheat, and millets. Soybean and flax are other popular crops. Farm animals are reared in plenty utilizing natural meadows and pastures. Northern portion of this region experiences cold climate. It is relatively sparsely populated. Farmers find poorly drained black soils difficult to work on and reach high grain productivity. The region is supported by river based irrigation through, Heilongjiang-Amur.

It seems human population and number of settlements in the China's forested steppes began to increase several centuries ago. Initially, fertile Chernozems were brought under crop production. During the next phase, excessive migration into steppes occurred. This led to overgrazing of grasslands and overcultivation of agricultural zones. Land degradation and deterioration of soil fertility depreciated the subsistence grain yield. It seems that the nomadic and grazing stretches known for centuries were erased due to rampant human settlement and crop production trends (Sheehy, 1990; Loucks and Jianguo, 1990). Further, it has been stated that historically, rainfall pattern, fertile soils, availability of fuel (forest wood), and productive range lands have dictated human migration and settlements in Northeast China (Loucks and Jianguo, 1990).

7.2   NATURAL RESOURCES OF NORTH CHINESE STEPPES

7.2.1   SOILS OF NORTHEAST CHINA

Broadly, soil types traced in China's steppes are classifiable from east to west into Forest Black Chernozems that are rich in organic matter; Meadow steppe Chernozems, Steppe Kastanozems, Semidesert Brown Calcareous soils, and Deserts sandy soil (Tianjie, 1990). Black soil of Northeast China's steppes is among most fertile regions with regard to agricultural crop productivity. About 70 percent of this region is utilized for crop production. For the past 100 years or more, productivity of crops has been relatively higher in Black soil region. Yet, during recent years, continuous cultivation has affected soil fertility and quality. Therefore, improving agricultural land scape, by adopting soil fertility conservation practices and implementing integrated nutrient management practices seem essential (Wen and Liang, 2001). Black soil in Northeast is prone to crusting due to rainfall. It affects seedling emergence. Compactness of soil that develops due to repeated tillage, reduces infiltration, seepage and percolation of water and nutrients (Bu et al., 2008). Major constraints to crop production in these soils are related to deficiencies of N, P, K, and micronutrients such as B, Mo, and Zn. Soil acidity, toxicity due to Al, Mn, and Fe is also traced in some parts of China that possess Alfisols and Inceptisols.

7.2.2 SOIL TILLAGE SYSTEMS IN THE STEPPES

Conventional tillage using heavy plow, disks, and harrows followed by preparation of land surface into ridges, flat beds, or flat plots are among the most common procedures followed in the Northeast China for over 100 years now. This system has sustained crop productivity and helped farmers to achieve their yield goals. However, in many regions, it is said that conventional tillage accelerates soil oxidative procedures, soil microbial component increases, and CO_2 emanates leading to decrease in SOC. Soil quality gets deteriorated due to loss of SOM, if conventional tillage is repeated too often (Ding et al., 2011). The Northeast China steppes are among important food generating systems of the world. Soil fertility, quality, and grain productivity levels need to be sustained at appropriate levels in order to satisfy the local demands. Therefore, Li et al. (2013a) state that devising appropriate tillage systems is essential, if soil degradation that has become rampant in certain parts of Northeast China has to be thwarted. The wheat-maize system suffers loss of soil fertility and quality.They believe that adopting no-tillage helps in reducing loss of SOC, topsoil, and mineral nutrients. In addition to adoption of mulching, formation of raised beds is essential, if soil degradation has to be minimized. Preparation of ridges helps in improving soil aggregation.

No-tillage systems are relatively recent to farmers in Northeast China, especially if we consider the long agricultural history of this region. Conventional tillage with heavy disks, chisel plow, moldboard plow (MP), and ridge tillage (RT) are in vogue since several decades (He et al., 2010). Conventional tillage causes severe disturbance to soil profile, induces oxidative conditions, improves soil microbial activity, and hence leads to loss of massive quantities of soil-C as CO_2. Soil erosion, especially loss of topsoil and nutrients could be deleterious to productivity (Liang et al., 2007, 2009). Fan et al. (2012) believe that tillage and its interaction with crop rotations is important with regard to preservation of soil fertility and productivity. They have evaluated different tillage procedures and compared the consequence with no-tillage (NT) systems. For example, soybean–maize rotation provided slightly more grain yield under NT compared to MP or RT. In the Black soil zone of Northeast China, NT was 22 percent more profitable compared to RT or MP. Firstly, it minimized expenditure on plowing. Next, it helped in preserving soil fertility. Fan et al. (2012) have concluded that no-tillage and maize–soybean rotation is more profitable than other combinations of tillage and crop rotations. No doubt, the steppes in Northeast China are mostly filled up with maize–soybean cropping system.

He et al. (2010) believe that excessive tillage and cultivation have induced soil deterioration in North China. Farmers here do adopt no-tillage systems in order to reduce disturbance to soil and control loss of topsoil. Long-term effects of no-tillage need to be understood. Field trials conducted for over 11 years with wheat–maize rotation system in Hebei province indicate that no-tillage improves soil organic matter, available N, and available-P compared to conventional tillage. Percentage macroaggregates in no-tillage were higher by 8 percent than in conventional tillage plots. The soil macroporosity was also improved due to adoption of no-tillage. Grain yield of wheat increased by 3.1 percent and that of maize by 1.2 percent in fields under no-tillage compared to conventional tillage systems (He et al., 2010). No-tillage helped in conserving soil moisture better than fields that were disced.

No-tillage systems are prone to support weed growth, since weed seeds/seedlings do not get destroyed. Volunteer plant from the previous crop is another problem connected with no-tillage system. Hence, famers are advised to weed the plots immediately after seeding or prior to it. Herbicide sprays are also done early to thwart weed growth. Mulching is yet another procedure used often along with no-tillage plots. Mulching is known to regulate soil microclimate, especially temperature, reduce weed growth, and avoid soil erosion. During recent years, rice farmers in Northeast China are using plastic films as mulches to regulate soil temperature, avoid coldness, suppress weeds, and control erosion to a certain extent (Gu et al., 2012).

7.2.3 SOIL FERTILITY AND CROP PRODUCTIVITY

Farmers in Northern plains apply fertilizer-N to wheat crop based on recommendations of agricultural agencies, traditional methods, maximum yield goals, and more importantly based on soil test and plant analysis values. Soil test values allow farmers to judge general soil fertility, so that fertilizer needs could be classified based on yield goals. Plant analysis, chlorophyll meter readings, and soil fertility variation maps help in channeling fertilizer-N more accurately at various stages of crop growth. Soil N fertility versus plant growth/grain yield curves is also used. Farmers also use plant critical-N concept to supply fertilizers at different stages (Chen and Zhu, 2013).

Knowledge about absorption rates and total requirement of N to achieve a yield target is important. It allows us to apply fertilizer-N in quantities as needed by the crop at different stages. Application of split dosages of fertilize-N too could be accurately decided. Yue et al. (2012) have opined that generally, farmers have overestimated N requirements of wheat crop grown in Northeast China. They have applied fertilizer-N in excess. The excess N that accumulates in the soil is vulnerable to loss via leaching, seepage, emissions, and so on. Analysis of previous data on N removal by wheat crop indicates that N requirement across the North China ranges from 20.8 kg N t^{-1} grain in fields not provided with fertilizer-N and 25.7 kg N t^{-1} grain if the crop is fertilized. For a grain yield ranges from 4.5 t ha^{-1} to 6.5 t ha^{-1}, N requirement decreased from 27.1 kg N t^{-1} grain to 24.3 kg N t^{-1} grain. This was attributed to higher harvest index and decreases in N concentration in grains with increasing grain harvest (Yueet al., 2012). Further, for grain yield ranges from 6.5 t ha^{-1} to 10 t ha^{-1}, N requirement t^{-1} grains deceased from 24.5 kg N t^{-1} grain to 22.2 kg N t^{-1} grain. Clearly, there is dilution of N concentration of grains as we enhance yield ha^{-1}. Grain N concentration decreases from 2.41 to 2.2 percent. Similar evaluation made on maize (*Zea mays*) during 2006–2009 has shown that it requires 15.3 kg N, 2.9 kg P, and 8.3 kg K per t grain, at 60–70 percent yield potential known for North China plains (Zhang et al., 2012b). We should keep accurate estimates of N requirement trends in the Steppes and accordingly channel fertilizers based on intensity of cropping and yield goals. It helps in avoiding N loss and emission in addition to minimizing on cost of fertilizers.

Zhao et al. (2006) state that overfertilization of N is a common problem traced in wheat–maize belt of North China plains. They evaluated a series of fields during 2003–2009 for soil-N still unused in the soil profile after harvest of crop. Soil-N accumulated in the soil profile at 90–200 cm depth. Soil nutrient movement out of root zone was the major avenue through which excess N was lost. Application of excess fertilizer-N therefore reduces N-use efficiency. In fact, there are estimates and suggestions that fertilizer-N input to intensively cropped wheat or maize could be reduced. In case of wheat, we could reduce N input by 80 kg N ha^{-1} and still answer crop's need accurately (Zhao et al., 2006). Several other reports suggest that fertilizer-N supply to wheat could be kept below 180 kg N ha^{-1} without risk of lowering grain yield. There are recently developed improved N$_{min}$ techniques that allow us to synchronize N need of crop with supply, accurately. It improves N-use efficiency. Following is a typical N supply schedule adopted during wheat culture, by farmers in the Steppes of Northeast China:

Year	Fertilizer Nitrogen Supply (kg N ha^{-1})			
	Before Sowing	Seedling	Tillering	Total
1999–2000	0	16	38	54
2000–2001	0	18	46	64
2001–2002	0	37	61	98
2002–2003	0	30	36	66

Source: Zhao et al. (2006)

Studies on N balance indicate that farmers adopting conventional N supply recommendation tend to supply more than that needed by crop. Conventional N supply stipulates 300 kg N ha^{-1}, whereas optimized N is only 54 kg N ha^{-1} for the yield goal set. Zhao et al. (2006) state that in fields under "No-N supply" N supply derived from N$_{min}$ before sowing is59 kg N ha^{-1}, that derived from apparent N mineralization is 72 kg N ha^{-1} totaling 131 kg Nha^{-1}. In case of fields under "Conventional N supply," fertilizer-N supplied is 300 kg N ha^{-1}, that derived from N$_{min}$ before sowing is 205 kg N ha^{-1} and apparent N$_{min}$ is 72 kg Nha^{-1} totaling 577 kg N ha^{-1}. Whereas, within fields under "Optimized-N system" fertilizer-N supply is nil, that supplied by N$_{min}$ before sowing is 116 kg N ha^{-1} and apparent N$_{min}$ is 72 kg N ha^{-1} totaling 242 kg N ha^{-1}. Clearly, N channeled to root zone under Optimized N system is just sufficient to achieve the yield goals.

Balanced nutrition of wheat crop using all three major nutrients, namely N, P K, and micronutrients seems essential, if the aim is to achieve high grain productivity. During, past few decades, farmers in Northeast China have applied N at optimum rates, small amounts of P, and ignored K supply. Soils are believed to contain sufficient levels of K. However, depletion rates of P and K are high if cropping intensity is high. Incessant cropping also causes uneven depletion of soil nutrients leading to imbalance. Jianhua et al. (2000) have stated that farmers in Hebei plains have fertilized wheat with fertilizer-N ignoring P and K. Hence, based on field experiments, they suggest that in addition to N and P, K should be supplied to wheat crop. Soil, once thought rich in K, has now become deficient. Regardless of N level, 150 kg P_2O_5 and 112 kg K_2O is essential to derive 3.2 t grain ha^{-1}. According to Jianhua et al. (2000), corn sown in rotation too suffers soil nutrient imbalance, hence these nutrients should be refurbished considering soil tests and yield goals.

Knowledge about spatial and temporal distribution of major elements in the different soil types encountered in Northeast and other parts of China is essential (Huang, 2009; Zhang et al. 2013a; and Zhu et al. 2013). It helps farmers in deciding cropping pattern and yield goals. It also helps them to channel fertilizers accurately within fields using fertility maps. In fact, synchrony between soil nutrient availability, supply pattern, and crop's demand/uptake trend is most needed. In such cases, soil fertility map that depicts spatial distribution of soil nutrient and its availability will be useful. Based on several field trials in Northeast China, Liping (2009) has reported that accumulation of nutrients in soil may vary. Hence, it is advisable to supply nutrients such as N in split portions at different stages of the crop. This way, we can maintain optimum availability of nutrients at all stages of crop.

Over all, application of all three major nutrients, N, P, and K is essential. They ought to be available to crops in balanced proportions. In addition, reports by IPNI (2009) suggest that it is the interaction of N, P, and K with soil moisture that decides crop growth and productivity. For example, field trials with summer maize in Hebei province suggest that optimum combination would be conventional irrigation plus supply of medium levels of major nutrients. Such a combination provided best nutrient recovery from soil.

7.2.4 SOIL ORGANIC MATTER AND CARBON SEQUESTRATION

The terrestrial ecosystem in Northeast China, like many other regions of this nation supports vegetation that includes coniferous forests, deciduous forests, cropland, that is, agricultural prairies, natural grasslands, and wasteland with sparse vegetation. According to Wang et al. (2002), carbon in vegetation and for the entire region of Northeast China is 2.81 PgC (10^{15} g C) and 26.4 PgC, respectively. There are spatial variations in the distribution of organic C in the ecosystem and fluxes of C are also significant. Following are few examples that depict carbon density and total C stock in soil:

Soil Type	Area	Carbon Density	Carbon Stock
	($\times 10^4$ km^2)	Mg C ha^{-1}	($\times 10^{15}$ g C)
Chernozems	5.98	186.7	1.12
Meadow Chernozems	3.87	154.9	0.60
Castanozems	2.32	93.7	0.22
Meadow Castonzems	0.67	128.3	0.09
Black soils	0.75	165.9	0.12
Meadow soils	1.34	174.8	0.23
Steppe Eolian soil	3.51	27.0	0.09
Total Northeast China	124	212.7	26.43

Source: Wang et al. (2002)

Black soils found in Northeast China are rich in organic-C. Yu et al. (2006) have reported that organic-C storage was 646.2 Tg C. The net C emission from these soils was 1.3 T g C year^{-1} under current agricultural crop management systems. Simulations using CENTURY model have suggested that climate change and elevated CO_2 may increase precipitation and SOC content. In addition, it is believed that long-term organic manure input will further enhance SOC content. A different study on estimation of C storage with an aim to understand the C balance in soil states that total storage of C in the three states of Northeast China is 1243.48 $\times 10^6$ g C (soil layer 0–30 cm depth). The three states namely Heilongjiang, Jilin, and Liaoning are experiencing negative SOC balance (Qui et al., 2005). They are losing excessive SOC than that sequestered through various agronomic procedures. Careful management of soil, its C content, sequestering more C, and reducing emissions are essential in all regions. Paddy soils in the Northeast are supposed to sequester more C than upland fields (Yang et al., 2012; Yan et al., 2013).

Farmers in Northeast China are often prone to use both organic and inorganic manures to enhance productivity of cereal-based rotations. Indeed, there are several reports about influence of inorganic fertilizer mixes that contain N, P, K, and organic manures on wheat productivity in the steppes. Such inputs are known to enhance soil organic matter, reduce erosion, improve soil microbial activity, and at the same time increase crop productivity. For example, on a sandy loam, SOC improved by 12.2 t C ha^{-1} if organic manure is supplied; by 7.8 t C ha^{-1} if organic manure and complex fertilizer containing N, P, and K is applied; and by 3.7 t C ha^{-1} if only inorganic fertilizers are used. Wheat grain productivity ranged from 5.2 to 7.5 t ha^{-1} if both inorganic and organic manures were added. In comparison, application of organic manure alone or inorganic fertilizer alone gave low grain yield (Cai and Qin, 2006). Similar trends were noticed with long-term trials during 1987–2005 with crops such as rice, maize, soybean, vegetables, and cotton grown in the plains (Zhong et al., 2010; Lou et al., 2011; and Duan et al., 2012). Long-term trials (1989–2007) with wheat grown on sandy loams in Henan has shown that microbial population increased significantly, if organic matter is provided and it improved soil quality (Gong et al., 2009). The microbial population increase ranged from 5 to 24,000 CFU due to FYM supply.

Management of organic manure is an important aspect during crop production on Black soils of Northeast. Productivity of major crops such as wheat, maize, and soybean was regulated by manure, but in many regions, manure applied with irrigation was critical. Maize and wheat crops may just need appropriate manures/fertilizers, but soybean grain yield was enhanced only if manure was applied with optimum irrigation (Liu et al., 2004). Soil organic matter and tillage practices are important in the Northeast Plains of China, because these procedures have direct influence on soil moisture retention. Precipitation retention and its use efficiency are affected SOM content (Zhang et al., 2013b).

7.3 ENVIRONMENTAL CONCERNS IN CHINA'S CROP BELTS

7.3.1 SOIL EROSION IN THE STEPPES OF NORTHEAST CHINA

Soil degradation in China's Northeast is noticed relatively frequently. Several natural and man-made factors induce soil degradation. Major soil degradation processes are (a) increase in sand fraction called "sandization"; (b) deterioration of physical properties of soil such as aggregates, loss of soil structure, and so on; (c) decline of soil organic fraction leading to low buffering capacity and retention of nutrients; (d) increase in salinization; (e) accentuation of soil erosion and loss of topsoil (Shiet al., 1990; Tianjie, 1990). Loss of grass cover is among the most important factors that induces soil erosion and degradation. Overgrazing, conversion to cropland and incessant tillage, fallows without cover crops, and high wind/rainfall causes erosion. Some of these phenomena, including increase in soil erosion, loss of topsoil, reduction in grass cover, and fallows without crops could be easily observed using satellites and correction measures could be devised. In the Northeast China, soil erosion and ground cover have been studied in great detail using both ground estimates and remote sensing data. In China, extent of soil erosion is classified using "soil erosion index." Following is an example of classification of soil erosion (in tonkm^{-2}) (See Shi et al., 1990):

   (a) Gentle erosion if soil erosion due to water is <240 and due to wind is also <240;
   (b) Light erosion if soil erosion due to water is 200–2,500 and that due to wind is 200–2,250;
   (c) Medium erosion if soil erosion due to water is 2,500–5,000 and that due to wind is 2,250–4,500;

(d) Strong erosion if soil erosion due to water is 5,000–8,000 and that due to wind is 4,500–9,000

(e) Grave erosion if soil erosion due to water is 9,000–15,000 and that due to wind is 9,000–18,000

(f) Severe erosion if soil erosion due to water is >15,000 and that due to wind is >18,000

According to Jianyu (2010), at the present erosion rates, black soil region of Northeast China may forfeit 1.0 million ha of cropland within next 50 years. Louckes and Jianguo (1990) have stated that overgrazing, lack of ground cover, and fallows without cover crop have induced desertification of large areas in Northeast China, especially in Inner Mongolia. Desertification of forest Steppes is of course due to lumbering and lack of forest cover. Incidentally, conversion of forest steppes into agricultural prairies is another important cause of soil erosion and land degradation. Establishment of shelterbelts and sand stabilization is necessary to restore soil productivity. Enhanced human population density due to migration, increase in livestock population, overgrazing, and conversion of land to urban purposes too have induced, land degradation (Cao, 1984; Hays, 2008).

Black soil zone of Northeast China has undergone rampant erosion leading to lowered crop yield. Field trials that simulated erosion of different intensities were conducted to assess the effect on soybean grain production. Soybean grain yield decreased exponentially with increase in soil erosion. The grain yield decrease was related to loss of soil fertility aspects such as organic matter, N and P. Supply of inorganic fertilizer could overcome deleterious effects of erosion (Wang et al., 2009). In many locations, erosion of SOM, especially particulate organic matter depended on slope of field (Cheng et al., 2010).

Black soil region of Northeast China also suffers land degradation due to gully erosion. Gully erosion reduces agricultural cropping zones and crop productivity. Gully erosion is characterized mainly by land cuts and side wall failures in trenches that expand. Gully erosion intensity and spread are currently monitored using GPS and satellite pictures of cropland. Ephemeral gullies measured during summer suggest that gully formation is marked during late spring. Soil loss is high reaching 4.0 t ha^{-1} year^{-1}, which is higher than threshold by 2.0 t soil loss ha^{-1} year^{-1} (Wu et al., 2013). According Liu et al. (2010), soil degradation due to gully erosion is severe particularly in the black soil area of Northeast China. Gully erosion has depreciated soil organic matter, nutrients, and topsoil of this area. Reports suggest that SOM erosion rates have ranged from 1.2 t ha^{-1} to 78 t ha^{-1} depending on slope of the region, gully width, and intensity of precipitation events (Liu et al., 2010).In some locations, average annual decrease in SOM was 0.5 percent. In 50 years, SOM in cropland has decreased by half of its original value. Soil erosion control measures such as adoption of no-tillage, contour planting, mulching, and intercropping have helped in reducing soil erosion in few locations (Liu et al., 2011; Yang et al., 2012; and MWR, 2013).

The steppes of Northern China have also undergone soil degradation, and to a certain extent, desertification has occurred. Several types of rehabilitation measures have been adopted to thwart expansion of desert ecosystem (Liu and Zhang, 2010). Several programs that conserve steppe vegetation and cropland are in place since 1960s. They have aimed at reduction in exploitation of natural vegetation, regulating livestock grazing and soil conservation in crop fields. Hong-feng et al. (2009) state that black soil region in Northeast China actually experiences soil degradation due to several factors acting at different intensities and for different length of period. Therefore, they propose adoption of integrated practices that includes erosion control using principles of biology, soils, crop production techniques, soil conservation engineering, irrigation, and so on.

Regarding NO_3 leaching, Zhou et al. (2013) have reported that, it is perceptible in the intensively cultivated zones of North and Northeast plains of China. Black soils are prone to NO_3 leaching, but depending on inherent soil-N status and fertilizer-N supply. For example, NO_3-N leaching increased from 10 to 12 kg N ha^{-1}in unfertilized check fields to 25–28 kg NO_3-N ha^{-1} at 200 kg N ha^{-1} supply and then to 50 kg NO_3-N loss if fertilizer-N supply was further increased to 350 kg N ha^{-1} in wheat fields. The NO_3-N leached in maize fields increased from 25 kg NO_3-N ha^{-1} at 50 kg fertilizer-N supply and to 150 kg NO_3-N ha^{-1}if maize fields were supplied 450 kg fertilizer-N (Zhou et al., 2013). Incidentally, fertilizer usage is among highest known in the North and Northeast plains of China. The high yield goals of 8 t grain ha^{-1} maize and 5–6 t wheat grains ha^{-1} necessitate application proportionately higher quantities of fertilizer-N. It seems in all counties of North China about 20,000 t of fertilizer-N is applied to croplands per county per year (Zhou et al., 2013). Obviously, such high input induces high NO_3 leaching. Nitrate leaching fluxes are accentuated during monsoon cropping season, beginning from June and it reaches a peak in September. Nitrate-N leaching occurs through surface leaching, seepage, and topsoil loss. Annual NO_3 leaching loss observed in main fields ranged from 31 to 53 kg N ha^{-1}. Peak loss of soil-N occurred from June to August (Zhou et al., 2013). The NO_2-N emissions were related to rate of NO_3-N leaching. Fields with pronounced NO_3-N leaching generally experienced lower levels of NO_2-N emission.

7.3.2 GASEOUS EMISSIONS FROM THE STEPPES

Gaseous emissions such as nitrous oxide, nitric oxide, gaseous nitrogen, carbondioxide, and methane induce greenhouse gas effects as they get released to atmosphere. They also reduce fertilizer-N efficiency, since a share of fertilizer-N applied is lost to atmosphere. Therefore, Guo et al. (2013) have argued that timing and quantity of application of organic manures and inorganic fertilizer-N are crucial. The seasonal weather pattern, soil conditions, and nutrient transformation rates affect the extent of soil-N lost as nitrous oxide. In Northeast China, Black soils moderately rich in organic matter called Chernozems are utilized to grow maize and wheat. It is a usual practice to supply higher ranges of fertilizer-N during spring. This practice supposedly enhances N_2O emissions and reduces fertilizer-use efficiency. Soil-N emissions as N_2O fluxes were marked immediately after each application of inorganic fertilizer to the crop field. Hence, Guo et al. (2013) hypothesized that it could be advantageous to supply inorganic fertilizer (N) during autumn, when maize is to be sown or it is a standing crop. Results suggested that applying organic manure to the crop grown in autumn and reducing inorganic fertilizer application to both spring wheat and then maize that succeeds is very useful in reducing N_2O emissions into atmosphere.

In a crop field, both crop and soil emit N_2O, although crops are known to generate very low quantities of N_2O (Chen et al., 2002a). Field trials in Northeast China suggest that application of fertilizer-N (NO_3) induces N_2O emissions. Fluxes of N_2O are higher if fertilizer-N inputs are greater in quantity and atmospheric conditions are warmer. In case of soybean fields, N_2O emitted by crop alone was 6–11 percent of total N_2O emission recorded for entire crop-soil system. Maize crop emitted 8.5–16 percent of total N_2O emission by crop-soil system. It is worth noting that proportion of applied fertilizer-N lost via plant N_2O emission was very low at 0.8–1.9 percent of total N lost as N_2O (Chen et al., 2002a). Therefore, fertilizer-N loss and reduction in fertilizer-N use efficiency is regulated by emissions from soil as N_2O.

Studies at Harbin State Agroecological Experiment Station, Harbin, China suggest that N_2O emissions are affected by fertilizer-N inputs and crop species that dominate the field. Quite often, N_2O emission in Black soil region is erratic and highly variable. Field trials indicate that, in an un-planted fallow field, application of fertilizer-N increased N_2O emission from 141 g N_2O-N ha^{-1} to 570 g N_2O-N ha^{-1}. In maize fields, N emissions increased from 209 to 882 g N_2O-N ha^{-1}. Approximately 75 percent of N_2O emission was derived from fertilizer-N. Presence of crop often significantly increased N_2O emission. Seasonal variations in soil moisture and temperature too affected N_2O emission (Ni et al., 2012).

Tian et al. (2012) have arrived at some interesting facts and conclusions regarding fertilizer-N usage, its effect on GHG emissions, C sequestration, and grain increase in North China plains. Their study aimed at deciphering the effects of N inputs from 1949 when fertilizer-N usage began in China till 2008. They used a C–N coupled model to understand the effects on C sequestration in soil due to N supply. Tian et al. (2012) state that fertilizer-N supply increased by 4 g N m^{-1} year^{-1} per decade since 1950 and has now reached 21.1 g N m^{-1} year^{-1}. Fertilizer-N supply is highest in North China and Yangt Ze, where cropping is intense. Double or triple crops year^{-1} are common and yield expectations are commensurately high from the agricultural prairies. Results indicate that N emissions increased as fertilizer-N supply increased. During past decade, N emissions ranged from 222G g N year^{-1} in 1999 to 369 G g N year^{-1} in 2010. In terms of global warming potential, fertilizer-N supply increased C sequestration. It firstly stimulated biomass formation and then its recycling into cycle. It increased CO_2 sink in the cropping ecosystem. According to Tian (2012), firstly, crop growth productivity in Chinese plains increased phenomenally due to N supply. It induced biomass and SOC component in the ecosystem. It seems to be that sequestration of C in soil peaked during 1990s and until 2005. Overuse of fertilizer has induced GHG emissions (e.g., N_2O). Primary suggestion is to reduce fertilizer-N usage by 40–60 percent and avoid overuse. This will help in avoiding GHG emissions commensurately.

Methane emissions are common in rice production zones. Agricultural crop production, especially lowland flooded ecosystems, livestock, and firewood burning generate CH_4 emissions. China has large area under lowland paddy cultivation that is prone to emit significant quantities of CH_4. Chinese agriculture, in addition, generates large quantity of crop residues and firewood that is periodically burnt. This procedure too emits CH_4. Estimates on CH_4 emissions from agricultural activity increased from 16.4 T g $year^{-1}$ in 1990 to 19.3 Tg $year^{-1}$ in 2006 (Fu and Yu, 2010). We should note that among all GHG emissions, CH_4 has 25–30 times greater global warming potential (IPCC, 2007). Regarding CH_4 emissions from rice belt of China, it is said that it harbors 20 percent of global rice belt and produces 31 percent of global rice grains. High fluxes of CH_4 have been reported from many locations of rice belt of China. No doubt, CH_4 emissions from rice fields vary with several natural and man-made factors (Zhang et al., 2011a). Models and periodic estimates suggest that Northeast China's rice growing area that extends into 1.44 million ha generates 0.48–0.58 T g CH_4 annually.

7.3.3 DROUGHTS AND HEAT WAVES

Agroclimatic changes that occur due to monsoon winds and pacific oscillations have proportionate changes in precipitation pattern, floods, droughts, and heat wave conditions. Such environmental changes do affect crop production. Phenomenon of drought is detrimental to natural vegetation, crop production trends, and general economic conditions. To quote a few examples, during 2012, rainfall pattern was skewed and 40–50 percent above normal during mid-season. It created floods and crop damage was severe. Standing crops of maize were affected severely. According to Zhang (2004), during 1949–2010, agrometeorological disasters such as floods, droughts, and heat wave have depreciated China's grain production. Drought alone has decreased crop yield by 40 percent in some seasons. Maize grain yield, in particular, decreased perceptible during all three seasons of 1989 due to drought. Estimates suggested that during 2009, drought affected large stretches of maize fields of Heilongjiang, Liaoning, Jilin, and Shanxi. Erratic temperature too affected crop production. For example, during past decades, periodic heat wave condition has affected steppes and other important cropping zones of China (Zhang, 2004; Garriss, 2013).

It is interesting to note that during past 2 decades, Northeast plain of China has experienced warming. Ambient temperature has increased by 1.0–2.5°C due to GHG emissions and global warming (Fang et al., 2003). This has stretched growing period by few more days. Farmers have changed cropping systems in response to weather pattern. As a consequence, grain productivity increased marginally in Northeast. Global warming and GHG emissions have also affected the natural vegetation in Northeast plains. Satellite and AVHRR observations suggest that growing season for vegetation has increased by 2.2 percent during past 2 decades. Fang et al. (2003) state that response of Hulun Bur steppes has been remarkable with regard to GHG emissions and global warming effects. Mean vegetation growth has increased by 7.8percent during past 2 decades. Reports by Niu and Wan (2008) suggest that reaction of natural vegetation and individual plant species that encompasses the natural prairies to global warming is varied. It helps certain species to compete better than others and spread rapidly in the area. Generally, of the six plant species studied, all of them showed increase in biomass in response to global warming. The competitive ability of species such as *Pennesitum centrasiaticum* was suppressed. Competitiveness of *Artmesia capillaris* and *Stipakrylovii* was enhanced.

Detailed knowledge about response of major cereals to global warming and marginal increase to growing season temperature is useful. Rice is important among crops of China. Its response to global warming has been studied by various agricultural agencies. It seems, rice grain yield has increased due to warming. Increase in ambient temperature by 1°C upwards could increase rice grain yield by 4–6percent(Zhang et al., 2012a). In Heilongjiang, rice cropping zones have moved northwards in response to warming. Therefore, it is believed that we can negotiate global warming effects on crops carefully by adopting suitable crop breeding strategy, adjusting time of sowing, and cropping systems. In addition, cropping region too could be altered to take appropriate advantage of global warming. In case of soybean, Zheng et al. (2011) have opined that global warming may advance seed filling time and lengthen seed maturity period. This seems to enhance grain yield. Soybean sowing date could be adjusted to suit ambient temperature changes. Ju et al. (2013) opine that global warming may bring in both negative and positive effects on agricultural cropping zones and their productivity. Alterations in temperature, precipitation pattern, and extremes of drought, floods, and outbreaks of disease/insects need to be negotiated appropriately. Global warming may increase evapotranspiration and increase water requirements of crops, therefore, soil moisture may deplete rather quickly. There are several adaptive measures that avoid all of the ill effects at least to a certain extent. Adjusting cropping zones, sowing time, augmenting irrigation, breeding suitable genotypes, and timely pesticide spray may all help in avoiding ill effects of global warming. This leaves only positive effects, if any, to express.

7.3.4 WATER RESOURCES OF NORTHEAST CHINA

Agricultural crop production in China thrives on water derived from different sources. Crop production occurs under rainfed, dryland agricultural systems, where in water supply to crops is limited. Crop production under irrigation support is also widespread considering that several major rivers, large number of lakes, dams, and other water bodies supply water. China has large share of groundwater source that again supports crop production. Annual precipitation in China is 645 mm year^{-1} and total water received via precipitation is 6192 km^3 year^{-1}. Surface water accounts for 2712 km^3 year^{-1}, groundwater 828 km^3 year^{-1}, and overlap is 727 km^3 year^{-1}. Together, internal renewable water resource of China is 2813 km^3 year^{-1} (Aquastat, 2013). Northern China is relatively less endowed with water resources, but has 64percent of nation's farm land. This region therefore has larger share of dry arable cropping zones that support maize and wheat. In contrast, Central and Southern China has better water resources, where in rice crop that requires higher quantity of water dominates the landscape (Vassilos et al., 2008). Vassilos et al. (2008) state that demand for irrigation water has increased in the North China, but droughts and water deficits are getting more frequent. This leads to overexploitation of water resources and to water crisis. Water allocation to agriculture has to increase. Currently, 3,50,000 million cubic meters of water is utilized by agricultural crop production, 1,00,000 million cubic meters by industry, 50,000 by urban/domestic regions, and small share of 5–10,000 million cubic meters is lost as overflow in different regions (Vassilos et al.,2008).

According to Fuqiang and Heping (2006), agricultural water management needs to be improved enormously. It involves saving of agricultural water and improving its production efficiency. For example, water use efficiency (WUE) in China is 45 percent while it is 80 percent in USA. Average water production efficiency (WPE) is 1.0 kg m^{-3} in China compared to 2.0 kg m^{-3} in USA. Clearly, there is a wide gap of efficiency with regard to water and crop production that needs to be decreased. It is believed that China's agricultural zones still adopt and use old water management systems that, perhaps, need to be upgraded. Reports suggest that in North China efforts to increase irrigation use efficiency has been in place since several years. Advanced irrigation systems such as sprinkler and drip irrigation systems are being introduced.

Agricultural crop production in eastern half of China is supported by some form of irrigation, in addition to precipitation. Agricultural cropping zones in North and Northeast China possess perennial irrigation supported by riverine systems, canals, groundwater, and lakes. About 50 percent or more of the farms in Northeast are irrigated yet, irrigation deficits in this region range from 0 to 50 mm annually. Greatest water deficiency is felt in North-west dry/arid zones, wherein deficiencies range from 700 to 800 mm (Thomas, 2008).

Crop wise, much of rice belt in China is irrigated. About 92percent of rice production zones are supported by irrigation. Arable crops too are supported by irrigation to different extents. For example, 80 percent area of wheat, 58 percent area of cotton, 47 percent area of maize, 45 percent of oilseed crops, 23 percent area of soybean, and 20 percent area of sweet potato are irrigated using some form of water resource (Wang et al., 2006).

Wheat-maize rotation is among most important cropping patterns in the Steppes of North China. Together, these two cereals require >850 mm water. However, known long-term (20 year) average precipitation in this region of steppes is only 450–650 mm. In addition, precipitation pattern is skewed. Nearly 70 percent rainfall occurs during maize phase of the rotation extending from July to September. Precipitation that occurs during wheat phase satisfies only 25–40 percent crop's water requirement. Hence, farmers in this region of China try to augment irrigation, plus adopt a range of agronomic techniques that enhances water use efficiency of wheat crop (Zhang et al., 2005). Groundwater is the major source of supplemental irrigation in the North China Plains (Yang and Zehnder, 2001). Tolerance of wheat genotypes to paucity of water varies. Water deficiency that occurs during critical stages such as stem elongation, spikelet formation and seed fill are most deleterious to grain yield. Xu and Zhao (2001) point out that despite consistent occurrence of water deficits during wheat production, annual grain yield averaged for past 25 years has increased. This phenomenon has been attributed to genetic gain in traits such as size of spikelet, number and weight of grains per spikelet. Harvest index has shown perceptible increase within the wheat genotypes that dominate the North China plains. Generally there was an in improvement of water-use efficiency from 10 to 15 kg mm^{-1} ha^{-1}for winter wheat and from 14 to 20 kg hg mm^{-1} ha^{-1} for maize. Mulching with crop residues was highly effective in preventing evaporation of soil moisture (Chen et al., 2002b).Timing of irrigation appropriately during active growth stage and milking in case of maize and tillering to seed fill in case of wheat was highly effective, in enhancing productivity. Following is the general trend in water supply (mm) to crops noticed in zones that support wheat-maize rotation:

	Crop/Year	Rainfall	Irrigation	Total
1999–2000	Winter wheat	77	319	396
	Summer maize	339	0	339
2000–2001	Winter wheat	125	310	435
	Summer maize	214	40	254

Source: Zhao et al. (2006)

7.4   CROPS AND CROPPING SYSTEMS OF STEPPES OF NORTHEAST CHINA

7.4.1   MAJOR CROPS OF NORTHEAST CHINA

Wheat and barley were domesticated in Fertile Crescent region of West Asia. Archeological excavations at Donguanmiao in Henan Province indicate that wheat was utilized by the natives as early as 5000 B.C. Carbon dating of grains of wheat and barley suggest it was cultivated in North-western China by 2650 B.C. Grains from several other excavation sites suggest both wheat and barley cultivation and usage was in vogue between 3000 and 2000 B.C. in Xingjian, Anhui, and nearby regions. It is believed that domesticated wheat was introduced into China via Afghanistan. Wheat was also introduced into China through the famous silk route from Turkmenistan. Several other pathways for introduction of wheat and barley into China are possible. They are via Mongolia, through South China Sea, and migrations from Southeast Asia (Bray, 1984). Wheat was also brought and introduced into China via Indus region, that later was transported right up to Korea (Bonjean and Angus, 2001). It seems wheat and barley were preferred less compared to millets such as foxtail millet (Bonjean, 2010). Initially, winter wheat and barley were grown in North China. Winter wheat/barley followed by summer millets, peas or legumes were preferred. Cultivation of both Spring and Winter wheat occurred during Han dynasty's regime (206–220 B.C.). Wheat cultivation stretched into several areas of China during medieval period. By seventeenth century about 50 percent of Chinese population consumed wheat/barley (Bonjean, 2010; He and Bonjean, 2010).

Wheat cultivation thrives under subtropical climate. Spring wheat is predominant in the Northern and Northeast region of China. Spring wheat is sown during February/March and harvested by mid-July/early August. Winter wheat that is predominant in eastern and southern plains is sown in October. Winter wheat crop forms panicles during March/April and the mature grains are harvested by May/June. Wheat crop thrives on Inceptisols, Alfisols, Alluvial soils, and Coastal sands (See Krishna, 2014; Figure 7.2). The wheat-based prairie has experienced changes in the composition of cultivars, sometimes gradually but at times with drastic consequences to the agro eco system. Wheat crop production area in China is classified into 10 zones. They are (a) Northern wheat zone, (b) Yellow and Huai river valley's facultative wheat zone, (c) Middle and Low Yangtze valleys Autumn sown Spring wheat, (d) South-western Autumn sown Spring wheat belt, (e) Southern Autumn sown spring wheat zone, (f) Northeastern Spring Wheat zone, (g) Northern Spring wheat zone, (h) North-western Spring wheat zone, (i) Qinghai-Tibetan Plateau Spring wheat zone, (j) Xingiang Winter wheat zone (He et al., 2001). In the context of this chapter, we should note that wheat production in Northeast and Northern plains of China is mostly intense and plains are highly productive with grain yield above 4.5 t grain ha^{-1}. The wheat genotype that dominates the steppes and one that provides grains to human populace in this region has undergone periodic changes. According to He et al. (2001) between 1949 and 2000, Chinese wheat belt has experienced 4–6 changes in the dominant cultivar that extends into cropping zones and provides most of the grain requirement. Interestingly, each change in genotype has increased yield by 10 percent over previous levels. Major progress in the composition of wheat belt is related to grain yield, rust resistance, earliness, and lodging resistance. Plant stature of the wheat crop has been shortened by 30 cm, from 115 to 85 cm height. Semi dwarfs have of course replaced tall landraces in most regions. Several landraces selected specifically for resistance to lodging has been released into larger areas (He et al., 2001). Land races with good adaptation to agroclimate, tall in stature, and good resistance to diseases have been released despite low yield of 1.5 t ha^{-1}. For example, wheat landraces such as Nexiang, Abbondanza, and Funa derived from Italy are found in Northeast China. Since 1950s, wheat hybrids have been preferred in the Chinese wheat belt. Hybrids occur in 30–68 percent area of various wheat zones. Clearly, steppes have been manipulated drastically with regard to composition of wheat genotype.

FIGURE 7.2   Spring and winter wheat producing regions of China.
Source: USDA, Joint Agricultural Weather Facility, Washington, D.C.; http://www.fas.usda. gov/pecad2/highlights/2002/06/wheat/chiwht.gif.

Some of the earliest records about maize cultivation and usage as food crops emanate from books written during Ming Dynasty dating 1505 A.D. Annals of Ying Zhou dated 1511 A.D. mentions maize cultivation in Henan. Maize brought from India and Burma was cultivated in provinces of Henan, Anhui, Jiangzu, and Fujian during 1525–1535 A.D. European travelers too brought maize grains through ports. Maize was accepted quickly in the plains of Central China. In the South, Yunnan supported maize cultivation in large areas of over 50,000–90,000 ha during mid-sixteenth century (Bonjean, 2010). For along stretch of three centuries, from sixteenth to nineteenth century, maize grown in China was mostly landraces that evolved through simple selection and natural hybridization. Flint and waxy corns were common. During 1920s, several other maize genotypes were introduced into China's cropping zones. Popular maize cultivars such as Golden Queen, Italian white, and several Dent corns were noteworthy. During recent past, beginning 1960s, maize hybrids and high-yielding composites have dominated the agrarian zones. These maize genotypes have evolved into one of the most intensely cultivated zones in the Northeast China, where maize productivity is very high at 7–9 t grain ha^{-1}. Northeast plains of China harbors a most intensely cultivated maize belt. It is comparable to "Corn belt of USA" (See Krishna, 2012; Figure 7.3). Fertile black soils, high fertilizer input, congenial precipitation pattern, and irrigation facilities allow farmers to review yield goals upwards of 7–8 t grain ha^{-1} plus 15–20 t forage ha^{-1}. Plains of Northeast China contribute 40 percent of total maize grain harvest of China. Another 40 percent of maize grains are harvested in the valleys of Yangt Ze river.

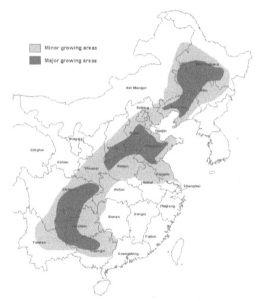

FIGURE 7.3 Maize producing regions of China.
Source: USDA, Joint Agricultural Weather Facility, Washington D.C., USA

Rice was domesticated in Southwest China and Assam in the North-eastern India. Most widely cultivated species *Oryza sativa* is supposedly derived from wild species complex of *O. rufipogon* and *O. nivara*. Archeological sites in valleys of Huang He and Yangt Ze in Central China clearly indicate use of rice during 3rd millennium B.C. (Bray 1984; You, 1988). Bonjean (2010) states that, initially Chinese natives might have used wild rice for a long stretch of time, since it is believed that domestication involving selection for nonshattering, color, taste, and seed dormancy was gradual (Fuller et al., 2009; Bonjean, 2010; He and Bonjean, 2010). Later, it seems clear that several hundreds of landraces and populations were selected and perpetuated during the period since Neolithic times till twentieth century. Plant characteristics such as maturity period, grain length, glutinous endosperm, color of kernels, taste, cooking quality, that is, governed the cultivation and dominance of a particular landrace of rice in a given region within China (Jin et al., 2008). The wetland type of rice spreads to different regions in Northeast China, Korea, and Japan during 1st millennium B.C. Mostly, *japonica* types that were cold-tolerant compared to *indica* types derived from subtropics became acclimatized in the North and Northeast plains of China (An, 1999). Preference for rice landraces changed periodically based on plants ability to resist diseases, withstand water shortages, maturity period, and grain quality. During medieval period, that is, twelfth-fifteenth century, *indica* types called "Champa varieties" named after Assam state in present India, dominated the Chinese wet land region, especially in the Central and Southern China. Since mid-1900, selection of rice genotypes has been methodical. It has been based on innumerable traits preferred by the Chinese population. Higher yield goal, rice grain quality, and economic advantages have been prime aims during selection. Hybridization with dwarf genotypes resulted in creation of large number of dwarf and semidwarf genotypes with high yield potential and disease resistance. During the period 1970–2000, semidwarf cultivars have dominated the China's Riceland. However, since 1998, hybrids with even greater yield potential of 8–10 t grains, 20–22 t forage ha^{-1} and resistance to disease/pests have been invading the rice landscape.

Sorghum was domesticated in Ethiopian highlands. It has secondary centers of origin and regions of high genetic biodiversity in West Africa, Southern Africa, and Southern Indian plains. Sorghum was introduced into Indian peninsula via Mozambique, that later moved into China's Southern plains via Yunnan and Sichuan (See Bonjean, 2010). Sorghum from Africa might have also entered Chinese agrarian zones via "silk route." Sorghum became popular in the Northeast during Qing dynasty. Sorghum is known as Kaoling in China. Two species of Sorghum namely *S. propinquun* and *S. nitidum* were traced in archeological sites of Henan and Shanxi. The cultivated sorghum *S. bicolor* evolved later through hybridization. Sorghum cultivation spreads into larger areas during 1920 and 1930s. It occupied about 4.7 percent of cropland. Currently, sorghum is predominant in Northeast Plains and adjoining regions in semiarid region of loess.

Millets such as Proso millet (*Panicum milleaceum*) and Fox tail millet (*Setaria italica*) are among the earliest of the cereals domesticated and grown for grains/forage by native Chinese. These two millets covered areas in Northeast and North China plains. Archeological sites in the Yellow river valley suggest that Proso millet was utilized by native Chinese during 5th millennium B.C. Millets were grown in Loess plateau during early Neolithic period. Proso millet was domesticated from wild strain *P. ruderdale* found in Central China. Foxtail millet was derived from wild progenitor *S. viridis* (See Bonjean, 2010; Krishna, 2014). We should note that millets were the main stay for Chinese natives for 4 millennia. Wheat, maize, and sorghum arrived into Chinese agrarian zones rather much later into the river valleys and plains (Yang, 1997; Toussant-Samat, 1997). Millets mostly dominated the

Huang He valley, Loess, and Northeast China. Domesticated millets spread from China to several areas in India, West Asia, Russia, and Southern Europe (Jones and Liu, 2009). Replacement of millets and consequent loss of acreage in China perhaps began during nineteenth century. Currently, major cereals are wheat, maize, and rice that have displaced large areas of millet in Northeast and North China.

Soybean is native to Central and Eastern China, where it was domesticated and utilized as a food grain since several millennia. Soybean production was localized to China's agrarian zones for a rather long stretch of time. The cultivated species of soybean is *Glycine max*. Wild ancestor is *Glycine soja*. Evidences for earliest cultivation, storage, and use of soybean in China's plains date to 5000 B.C. Soybean cultivation spreads to other regions of China and adjacent areas in Korea, Japan, and Philippines. Soybean cultivation in China became significant during medieval times. Soybean is leguminous crop with capability to fix atmospheric-N by establishing symbiotic association with *Bradyrhizobium*. Therefore, farmers supply relatively lower levels of fertilizer-N to soybean. In addition, soybean cultivation in rotation with crops such as wheat or maize in North and Northeast China helps in enhancing soil-N fertility. Cereals grown after soybean derive a certain quantity of N benefits from previous soybean. It may range from 30 to 80 kg N ha^{-1}. Currently, soybean production in China is intense in Jilin, Liaoning, parts of Heilongjiang in Northeast steppes. In North China plains, particularly in Hebei, Shandong, and Shanxi provinces, soybean is major crop (Figure 7.4).

FIGURE 7.4 Soybean growing regions of China.
Source: USDA Joint Agricultural Weather Facility, Washington, D.C. USA
Note: Darker zone in the map suggests intensive cultivation zones of soybean. Lighter zones are moderate or low input soy bean regions.

7.4.2 GM CROPS IN CHINA'S STEPPES AND OTHER REGIONS

Initially, landraces of different crops that were adapted to subsistence farming dominated the Chinese agricultural landscape. Large-scale changes in plant genetic composition occurred in cropland of China between 1950 and 1970s due to introduction of improved composites and hybrids of major cereal crops such as wheat, rice, maize, sorghum, and millets. Soon, new cultivars of legumes and oilseeds too invaded the vast plains that support crop production. During 1970s, large-scale use of high yield varieties and hybrids allowed farmers to review the grain yield goals upwards. This necessitated higher inputs of fertilizers, pesticides, and irrigation. Landraces of cereals that had adapted to different climatic zones faded as new genotypes that responded to fertilizers took over the agricultural landscape. Most of the new varieties and hybrids were generated through classical genetic procedures of pollination, formation of recombinants and repeated selection. However, during past decade, China's cropping zones have been experiencing a different kind of invasion that may affect genetic composition of cropping expanses. In this case, crop genotypes that are genetically tailored using DNA hybridization and other molecular techniques have been introduced into crop fields. They replace while genotypes that were developed using simple plant selection and classical genetic procedures. The Chinese agricultural agencies have invested large sums in developing GMO (Genetically Modified Organisms) crops. China, it seems to be the first country to introduce GMO-rice. GMO-cotton with ability to tolerate insect attack was introduced in North China during 2000 (Hayes, 2013). However, there were opinions expressed, which said insects may develop resistance to introduced BT genes. Aspects such as diffusion and movement of introduced genes in the plant community, particularly other closely related crop species and wild types need to be clarified or at least we need an educated guess about time period and consequences if genes introduced through GMO go astray. Measures that curtail untoward effects need to be in place. Considering that agricultural crop production is an all important aspect of China and cropping expanses are really very large and care is needed. Periodical checks about any ill effects of GMO crops either to natural plant community, soil biotic flora, or livestock and human population are needed.

7.4.3 CROPPING SYSTEMS

The agrarian region of China is predominant along the major river valleys, plains, undulated regions, and coasts. Cultivation of crops is highly diversified and relatively intense in Northeast, South, and South-eastern sections of the country. Cereals such as wheat, maize, and rice; legumes such as soybean dominate the landscape. Wheat monocrops, intensively cultivated wetland rice monocropping zones, and high productivity maize zones in Northeast are among most common cropping systems. Farmers in China enhanced land use-efficiency and utilized soil and fertilizers efficiently by adopting several different cropping systems. Cropping systems that

include two or three crops in a year and long-term sequences involving a few different crop species are also in vogue in the Northeast plains (see Figure 7.5).

Following is the list of most common multiple cropping systems adopted in
North and Northeast plains of China:

Major Cropping Systems Traced in Northeast and North China	
Region	Northeast/North
Provinces	Heilongjiang, Inner Mangolia, Jilin, Liaoning and Gansu.
Multi crop Rotations	Rice-vegetable, rapeseed-vegetable, vegetable-maize; maize-soybean, potato-soybean, rapeseed-maize vegetable-vegetable-vegetable.
Region	North China Plain and Central region.
Provinces	Beijing, Hebei, Henan, Ningxia, Qinghai, Shaanxi, Shandong, Shanxi.
Multi crop Rotations	Winter wheat-rape seed, winter wheat-maize, winter wheat–rice, winter wheat-vegetable, rice-rapeseed, rape seed –maize, vegetable-maize, rice-vegetable, vegetable-vegetable-vegetable, maize-soybean, winter wheat-soybean, soybean-hay, winter wheat-potato, potato-soybean, winter wheat-cotton, rice-soybean.

Source: Qui et al. (2003)

Note: Multiple crops such as rice-rice-alfalfa, rice-rice-winter wheat, rice-rice-rapeseed,
potato-vegetable-vegetable, and winter wheat–rice are common in other regions of China
such as Anhui, Jiangsu, Shanghai, Zhejiang, and Guangdong.

China hosts 9 percent of total arable crop area and 14 percent of total cereal
area of the world. The cropping zones together contribute about 24 percent of total
grains harvested in the world (Qui et al., 2003). Total area of each cropping system
such as triple crops, double crops, and single crops varies. Qui et al. (2003) studied
crop cover in China using fine resolution maps from Landsat Thematic Mapper. It
suggested that about 1.3 million km² of cropland was covered by single crops and it
is equivalent to 60 percent of total cropped area. Single crop rotations are concentrated in the north and northeast regions of China and they are intensively cultivated.
About 30 percent of total crop area was utilized for double cropping and 10 percent
for triple cropping systems.

Following is the area covered by major cropping systems practiced in the China
mainland (in 1,000 km²):

Triple Crops
Total land area under triple crops in China = 140
Rice-rice-alfalfa = 50; rice-rice-rapeseed = 42; rice-rice-wheat = 20; vegetable-vegetable-potato = 19; rice-rice-vegetable = 11; vegetable-vegetable-vegetable = 2;

Double Crops
Total land area under Double crops in China = 360
Wheat-maize = 110; wheat-rapeseed = 66; rice-rice = 45; rice-wheat = 27; rice-oat = 22; wheat-soybean = 16; potato-maize = 14; maize-soybean = 11; potato-potato = 9; wheat-potato 7.5; wheat-cotton= 6.0; rice-vegetable = 4.3; potato-soybean = 3.6; rapeseed-maize = 3.5; wheat-vegetable = 3.3
Single Crops
Total area under Single crops in China = 810
Maize = 150; wheat =140; oats = 82; rice = 76; soybean = 74; potato = 65; cotton = 56; rapeseed = 35; millet = 33; vegetable = 22; sorghum = 20; sugarcane = 12; Beets = 10; Alfalfa = 5.7; Tobacco = 0.1, Hay = 0.1

Source: Qui et al. (2003)

Based on number of harvests in a year and altitude, cropping systems followed in Northeast China could be classified as follows: Nil harvest = Rangeland; single harvest =wheat or maize; 1.5 harvest = Rice/wheat or maize/wheat; 2 harvests = rice/rice; 2.5 harvests = rice/rice/wheat; 3 harvests = rice/rice/sweet potato or rice/rice/rice (Thomas, 2008).

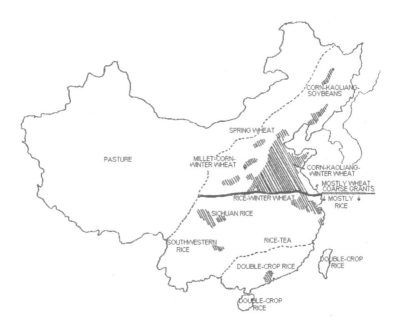

FIGURE 7.5 Agrarian regions and cropping systems of China.
Source: USDA, Foreign Agricultural Service, Washington, D.C.

Wheat–maize double cropping system is among important cropping system in Northeast China, especially in Hebei, Shaanxi, Liaoning, and Heilongjiang. Hebei plains support large stretches of maize–wheat rotation. Field evaluations during past decade suggest that farmers in Hebei reap wheat yield ranging from 3,375 kg grain ha^{-1} to 9,000 kg grain ha^{-1} with an average of 6,556 kg ha^{-1}. Maize harvest ranged from 3,250 kg grain ha^{-1} to 11,250 kg grain ha^{-1} with an average of 7,546 kg gain ha^{-1}. Together, in a year, farmers harvested 14,105 kg grain ha^{-1} (Liang et al., 2011; Carberry et al., 2013). Total grain yield in farmer's plots was lower and only 78 percent of that recorded in Agricultural Experimental Stations (19,455 kg grain ha^{-1}). Simulations using computer models indicated that potential yield possible with wheat–maize double cropping is 24,145 kg grain ha^{-1}. Hence, it leaves a large yield gap between grain harvest realized in Farmer's fields and Experimental stations. Grain yield gap between simulations and farmers' harvest is still larger. Farmers reap only 60 percent of potential yield forecasted by simulations. Liang et al. (2011) suggest that it should be possible to reach the grain harvest levels known for experimental farms. Farmers could improve the productivity by adopting irrigation. Irrigation at stem elongation and seed-fill period seems crucial. Fertilizer supply could be split and applied in 3–4 installments.

Maize based prairies are taller and wide spread across all agrarian regions of China mainland. Maize belt can be grouped into Summer maize region, North-eastern maize region, and Southwest maize region (Langemeier and Hu, 2012). Summer maize region includes provinces in the North China plains such as Hebei, Henan, Shandong, and Shanxi. Typically, farmers in this region adopt two crops per year, namely wheat and maize. The wheat–maize belt receives 600–900 mm precipitation annually and is predominantly rain-fed. The North-eastern region is popularly called "Spring Corn region." It includes agrarian regions of provinces such as Heilongjiang, Jilin, Liaoning, and Inner Mongolia. Cropping intensity is high. Major cropping systems are continuous maize that covers 50 percent of Northeast region (Figure 7.5). The other major rotation is maize–soybean that spreads into 40 percent of Northeastern China. Maize belt here is mostly rain-fed. Annual precipitation is about 400–800 mm. However, maize belt in Inner Mongolia is provided with supplemental irrigation. Irrigation supplements coincide with critical stages such as silking and grain-fill. The southwest maize region spreads into the steppes in Shaanxi, Sichuan, and Yunnan. Maize region is predominantly confined to mountainous and undulated tracts. Maize is rotated with potato, peas, rapeseed, and wheat (Langemier and Hu, 2012). Continuous maize is also common. In Southern plains, three crops of maize year^{-1} actually give a vast monocropping stretch. The maize belt is predominantly rain-fed and it receives 800–1,200 mm year^{-1}.

Intercropping with agroforestry tree species such as *Leucana, Faidherbia,* and *Acacia* is a useful practice. Jujube (*Zizyphus zuzuba var spinosa*) is indigenous to North China plains. It is a useful tree species that is commonly intercropped with crops such as wheat, soybean, cotton, or vegetable (Tianchang et al., 1998). Jujube is a successful agroforestry species in North and Northeast China, especially in the provinces of Henan, Shandong, Hubei, and Liaoning. Farmers in this zone adopt different planting models that include spacing jujube at 3–4 m and up to 15 m apart in a field with cereal or other crops in between the rows (see Figure 7.6; Peng, 1991; Thianchang et al., 1998). Trees are planted north–south and the canopy is mended to provide good diffusion of light to the intercropped cereals. Jujube provides about 4,500 kg fruits ha^{-1} in the intercropped system, but the yield depends on planting density and soil fertility. Generally, land-use efficiency and productivity of steppes is higher, if crops such as jujube are intercropped.

FIGURE 7.6   Jujube trees and cereal intercropping.
Note: Jujube trees and cereals such as wheat or maize are planted at 1:4 or 5 rows. Such planting pattern helps in reducing soil erosion due to wind/water, improves crop residue recycling and enhances biomass production.

7.4.4   LAND USE AND ITS CHANGE

Currently, land resource in China is predominantly used to support man-made pastures or natural grasslands that extend into 41.7 percent of total land area of the nation. About 12.7 percent land area is covered by agricultural field crops, 1.4 percent area by Orchards and Plantations, 19.0 percent by forests, 1.8 percent by water bodies, and 23.5 percent by others (Agribenchmark Network, 2009). About 25 percent of land area in China is consumed by infrastructure items such as roads, urban settlements, industries, and agriculturally noncongenial deserts and scrubland. China has 135 million ha of land under field crops. If double cropping systems are considered as total area under food crops reaches 172 million ha. Among food crops, cereals dominate the landscape. Rice, maize, and wheat together constitute 55 percent of cropland in China. Among legumes, Non-GMO soybean dominates the cropping zones (Agribenchmark, Network, 2009). Native grassland is an important land use pattern in China. About 40 percent of land area, that is, 400 million ha are under grasslands and pastures (Wang and Ba, 2008). Native grasslands are predominant in Northeast. Grazing and natural succession of flora do affect various parameters such as biomass accumulation, nutrient cycling, ecosystematic functions, and land use pattern. However, during recent decades, there has been a massive change from grasslands to cropland. Crops such as maize and rice have spread into erstwhile natural steppes.

Studies that deal with impact of land use pattern and its changes on C dynamics, especially storage and loss from soil ecosystem seems very useful. Lian et al. (2012) have studied C dynamics in at least five different land use systems in Northeast China. They grouped study area into pasture, natural meadow, corn field, temperate savannah, and bush wood shrubland. Total C storage ranged from 5,958 g C m^{-2} in natural meadows to 11,992 g C m^{-2} in corn fields. Soil organic fraction in different types of land use accounted for over 85 percent of C stored in the ecosystem. Carbon stored in bush wood was least compared to cropland (maize) that possessed 2,032 g C m^{-2}. Carbon stored in maize roots accounted for 20 percent of total C in crop field (Lian et al., 2012). We should expect changes in C storage with each change in land use, say from natural grassland to crops, or bush to crops, and so on.

During past 3 decades, cropping intensity and productivity in Northeast China has increased perceptibly. Large patches of natural meadows have been converted to maize, rice, and soybean. Further, even within the erstwhile cropland, traditional cropping patterns have undergone enormous changes. Traditional crops such as millet, wheat, and legumes have been replaced by high input maize and rice farms. The wetland rice, in particular has gained in area and productivity. Historically, rice production zones were confined to south. The rice growing regions have now shifted from traditional southern parts of North China plains into northern colder zones. Rice production technologies that make them adapt well to colder north have been introduced. During past 3 decades, beginning from 1985, wetland rice production zones in Northeast China have increased from 3.7 percent of total cropland in China to 8.3 percent. About 10 percent of China's rice grain harvest is derived from three provinces of Northeast. Indeed, rice production has become popular in Liao He, Songhua, and lowlands of Nen river in Heilongjiang region. In case of maize, area in Northeast has increased from 25.2 to 27.3 percent of total China's maize belt. Intensively cultivated maize farms of Northeast contribute about 30.5 percent of total maize grain generated in China. During the same period, that is, beginning of 1985, horticultural crop production zones have depreciated in Liaoning from 3,90,500 ha to 3,69,900 ha, in Jilin it has increased from 60,800 ha to 1,02,800 ha and in Heilongjiang from 22,800 to 52,100 ha.

Amur-Heilongjiang river valley is an important area that has supported massive forests, shrubland, crops, and pastures that allow nomadic pastoralism. Wasteland

and cold swamps support vegetation with grass and dicots. Amur-Heilongjiang area is actually shared by Russia, China, and Mongolia. Following is the land use distribution (in million ha) of cropland, forests, pastures, and wasteland in the countries that share Amur-Heilongjiang river basin:

Nation	Cropland	Forests	Nomadic Pasture	Wasteland	Total
Russia	7.9	119	14.8	16.7	178.7
Mongolia	0.2	1.6	21.5	0.5	28.6
China	26.3	39.0	10.3	9.4	95.1

Source: VICARR (2013d)
Note: values are in million ha

7.5   CROP PRODUCTIVITY IN NORTHEAST CHINA

During past 6 decades, China's agricultural prowess has improved enormously. It has been aided by several measures such as land development, expansion of agricultural area, introduction of improved cultivars and hybrids, soil conservation, fertilizer inputs, advanced agronomic procedures, irrigation, pesticides, mechanized harvest, and postharvest techniques. Productivity of Agricultural prairies of China has improved. The per capita grain production has increased from 200 kg in 1949 to 350 kg in 1983, 400 kg in 1990, then to 385 kg in 2011 (Zhang, 2011b). During 1950–1990, population increased by 2.5 fold, but grain production improved by 4.5 fold (Zhang, 2011b). We should note that, demography, especially intensity of human inhabitation varies with region. It is particularly high in Northeast, North, East, and South of China. The crop production intensity too coincides with high input farming and better grain output in these same regions (see Figure 7.7). Northwest China is arid with low fertility soils and is also relatively feebly populated. Grain productivity is commensurately low in Northwest China.

Majority of grain output in China occurs during autumn harvest. During 2007, grain output was 510 million t year^{-1}. Then, in 2009, grain production reached a record 530 million t year^{-1}. Major share of grains is contributed by three cereals, wheat, rice, and maize. Together, these three cereals have contributed over 70 percent of grain harvests since past 2–3 decades. For example, in 2011, out of 512 million ton grain harvest, 440 t grains were derived from the three major cereals. Rice production was 204 million t, maize 114 million t, and wheat 102 million t during 2011 (FAOSTAT, 2013). During past decade, productivity of crops such as sugarcane hovered around 89 million t, potatoes 70 million t, and sweet potatoes 105 million t.

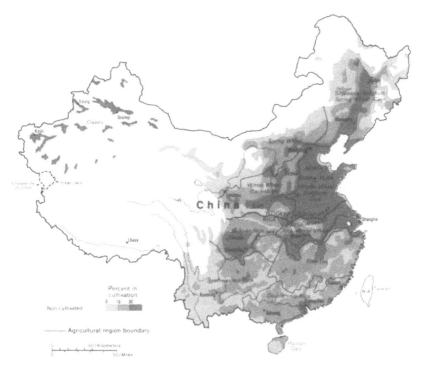

FIGURE 7.7   Agricultural cropping pattern and intensity in China.
Note: Crop production is predominant in Eastern half of the country. Intensity of cropping depicted by darker region is relatively high in East, North, and Northeast China.
Source: USDA, Joint Agricultural Weather Facility, Washington D.C.

There are several ways to assess and forecast roughly the crop growth and grain yield pattern in different regions. During recent years, crop growth monitoring systems (CGMS) are used to estimate grain yield. For example, in Hebei plains, CGMS considers weather parameters, soil conditions, and crop growth to arrive at appropriate yield forecasting. Crop production data derived from remote sensing are often tallied with ground truth data and calibrated wherever necessary (USDA-FAS, 2003). The CGMS considers various static and dynamic traits of soils and crops. Such forecasts made for large expanses in the steppes are useful during macroeconomic analysis of grain production trends and in regulation of cropping pattern (Fei et al., 2012; Mingwei et al., 2008). Forecasts about the cropping intensity in a given area are useful (see Figure 7.7).

Wheat, maize, and rice are three staple cereals of China. They also nourish large number of farm animals with forage. Total cereal grain production in China is about 400 million year^{-1}. The productivity of cereals in China is averaged at 4.4 t grain ha^{-1} compared to 5.4 t ha^{-1} in USA, 2.0 t ha^{-1} in India and 1.8 t ha^{-1} in Russia. During 1949–2000, average wheat productivity of fields increased from 0.7 t grain ha^{-1} to 3.86 t grain ha^{-1} (He et al., 2001). Until 1980, China has been a major importer of wheat grains to augment domestic production that often fell short. During 1980s, China imported, on an average, 13 million t wheat grains year^{-1} (Hsu et al., 2003). We should note that demand for wheat is almost inelastic and is based on requirements of large human population of 1.3 billion. Per capita consumption of wheat has been steady at 200 kg per person since 1983. Since mid-1995, China embarked on increase of wheat production. China is gaining self-sufficiency with regard to food grains. It was specifically aimed at satisfying wheat grain requirements of those provinces that experienced deficiency. During 1998, wheat harvest reached 123.3 t million grain, which is double the grain production of 55.2 million t during 1997. Wheat imports decreased to 1.2 million t during 2003 compared to 15.8 million t in 1980 (Hsu et al., 2003). Bumper harvests during 2000–2005 caused fall in grain price and government subsidies decreased. During 2005, wheat production has declined because area planted has decreased in response to relatively high grain price. We should realize that wheat area planted, fertilizer, and irrigation inputs are generally in response to economic benefits and demand for grains. The influence of alteration in cropping expanse and soil management procedures on grain yield too needs attention. Reports by USDA Agricultural Weather Facility suggest that spring wheat is localized in the North and Northeast of China. Spring wheat accounts for 15 percent of total wheat production in China. Major spring wheat areas are in Heilongjiang (26.5%), Jilin (1.8%), Liaoning (4.9%), Inner Magnolia (24.4%), Ningxia (6.4%), and Gansu (18.8%).

Winter wheat cropping zones dominate agrarian regions of Eastern and South-Central plains. Major winter wheat producing regions occur in Shandong (21.2%), Henan (20.4%), Hebei (11.2%), Jiangsu (7.9%), Anhui (7.0%), Sichuan (7.4%), Hubei (4.1%), Shanxi (3.1%), and Yunnan (1.4 %.). During past 2 decades, since 1994, there has been a gradual decrease in wheat cropping area from a peak of 32.2 million ha in 1997 to 26 million ha in 2003. Total wheat grain harvest peaked at 122 m t during 1997 and gradually declined to 96 m t in 2003.

Maize production is fairly wide spread in China. Its production is intense in Northeast and Northern plains, and equally so in Southern China. The fringes of maize belt as shown in Figure 7.3 supports moderately intense maize cropping zone. Maize is a dominant crop in the landscapes of provinces such as Jilin (13.3%), Shandong (12.8%), Heilongjiang (10.9), Hebei (9.9%), Henan (8.3), Liaoning (7.5%), Sichuan (6.1), and Inner Mongolia (5.3%). The maize belt in China extended into 19 million ha during 1988 and it has increased gradually to 26 million ha in 2007. Consequently, production of maize grains has increased from 78 million ton in 1988 to 122 million ton 2007. The productivity of maize crop has improved from 3.8 t grain ha^{-1} in 1988 to 5.3 t grain ha^{-1} in 2007 (USDA-FAS, 2003).

Rice harvest during past decade has steadily increased from 177 m t grain year^{-1} in 2001 to 204 m t grain year^{-1} in 2011. Productivity has increased significantly from

3.5 t grain ha^{-1} in 1990s to 5–6 t grain ha^{-1} in 2005, based on location and fertility/irrigation practices.

Soybean is a staple pulse crop for Northeast China. The steppes actually hosts large expanse of soybean that gets rotated with cereals such as wheat or maize. China imports a certain quantity of soybean to supplement domestic production. Hence, emphasis to expand soybean production and enhance productivity is high. During past 2 decades, soybean production zones increased from 8.2 million ha to 9.5 million ha in 2010. Soybean grain productivity has increased from 1.25 t ha^{-1} in 1988 to 1.7 t ha^{-1} in 2010 As a result, total soybean grain harvest has improved from 11.8 million t in 1988 to 16.2 million t in 2010.

KEYWORDS

- **Agricultural cropping**
- **Agroclimate**
- **Black soil**
- **Fertilizer-N**
- **Global warming**
- **Manchuria**
- **Soil degradation**
- **Steppes**

REFERENCES

1. Agribenchmark Network. Country profile: China. **2009**, 1–5, http://www.agribenchmark.org/cashcrop/sector-country-farmersinformation/country-profile-china htm
2. An, Z.; Early Neolithic culture in Northern China. Archaeology. **1984**, *10*, 934–936.
3. An, Z.; Origin of Chinese rice cultivation and its spread eastward. *Beijing Cultural Relics.* **1999**, *2*, 63–70.
4. Beijing Government. Physical Geography of China. *Beijing Int.* **2013**, 1–3, http://www.ebejeing.gov/cn/ BeijingInfo/BJInfoTips/BeijingHistory19511209.htm
5. Aquastat. China: Water Resources. Rome, Italy: Food and Agricultural Organization of the United Nations; **2013**, 1–3, October 18, 2013, http://www.fao.org/nr/aquatsta/
6. Beijing Government. Physical Geography. *Beijing Int.* **2013**, 1–3, October 16, 2013, http://www.ebeijing.gov.cn/BeijingInfoTips/ BeijingHistrory/1951209.htm
7. Bonjean, A. P.; Origin and historical diffusion in China of major native and alien cereals. In: Cereals in China. He, Z.; and Bonjean, A. P. A.; eds. Mexico (CYMMIT), Mexico: International Centre for Maize and Wheat; **2010**, 1–14 p.
8. Bonjean, A. P.; and Angus, W. J.; The World Wheat Book: A History of Wheat Breeding. Paris: Lavoisier-Intercept; **2001**, 1136.
9. Bray, F.; Agriculture. In: Science and Civilization in China. Needham, J.; ed. Taipei: Caves Book Ltd; **1984**, 724 p.

10. Bu, C. R.; Wu, S. F.; Zhang, X. C.; and Cai, Q. C.; Development process of Crust in Black soil of Northeast China. Ying Yang Sheng Tai Xue Bao. **2008**, *19,* 357–362.

11. Cai, Z. C.; and Qin, S. W.; Dynamics of Crop yields and Soil organic carbon in a Long term fertilization experiment in the Huang-Huai-Hai plain of China. Geoderma **2006**, *136,* 708–715.

12. Cao, 1984; The structure function and the way of transforming ecosystems of Inner Mongolia. Proceedings of International Symposium on Integrated control of land Desertification. 41–44 p.

13. Carberry, P.; et al. Scope for improved eco-efficiency varies among diverse cropping systems. Proceedings of National Academy of Science. **2013**, 110, 8381–8386.

14. Chen, X.; Cabrera, M. L.; Zhang, L.; Wu, J.; Shi, Y.; Yu, W. T.; and Shen, S. M.; Nitrous oxide emission from Upland crops and Crop-soil systems in Northeast China. *Nutr. Cycl. Agroecosyst.* 2002a, *62,* 241–247.

15. Chen, S. Y.; Zhang, X. Y.; Hu, C. S.; and Liu, M. Y.; Effect of mulching on growth and soil water dynamics of summer corn field. *Agricul. Res. Arid Areas.* 2002b, *20,* 54–57.

16. Chen, P.; and Zhu, Y.; A new method for winter wheat critical nitrogen curve determination. *Agron. J.* **2013**, 1–2, October 10, 2013, https://dl.scincesocieties.org/publications/aj/articles/0/0/agronj2013.0257

17. Cheng, S.; Fang, H.; Zhu, T.; Zheng, J.; Yang, X.; Zhang, X.; and Yu, G.; Effects of soil erosion and deposition on soil organic matter dynamics at a slope in the field in Black soil region of Northeast China. *Soil Sci. Plant Nutr.* **2010**, *56,* 521–529.

18. Ding, X.; Zhang, B.; Zhang, X.; Yang, X.; and Zhang. X.; Effects of tillage and crop rotation on soil microbial residues in a rain fed agroecosystem of Northeast China. *Soil Tillage Res.* **2011**, *114,* 43–49.

19. Dong, J.; Liu, J.; Tao, F.; Xu, X.; and Wang, J.; Spatio-temporal changes in annual accumulated temperature in China and the effects on cropping systems, 1980s to 2000. *Climate Res.* **2009**, *40,* 37–48.

20. Duan, Y.; Xu. A.; He, X.; Li, S.; and Sun, X.; Long term pig manure application reduces the requirement of chemical phosphorus and potassium in two rice-wheat sites in subtropical china. *Soil Use Manag.* **2012**, *27,* 427–436.

21. Fan, R.; Zhang, X.; Liang, A.; Shi, X.; Chen, X.; Bao, K.; Yang, X.; and Jia, X.; Tillage and rotation effects on crop yield and profitability on a Black soil in Northeast China. *Can. J. Soil Sci.* **2012**, *92,* 463–470.

22. Fang, X.; Yu, W.; and Ma, Y.; Response of the grassland in Hulun Buir Steppe of Northeast China to Global Warming based on NOAA/AVHRR NDVI. Proceedings of Symposium on Ecosystem Dynamics and Remote Sensing. *Appl. Semi-Arid Arid Land.* **2003**, *4890,* 287.

23. FAOSTAT. Grain Production in China. Rome, Italy: Food and Agricultural Organization of the United Nations; **2013**, October 20, 2013, http://www.faostat.org

24. Fei, T.; Chen, Z.; Wu, W.; and Huang Qing; Yield estimation of winter wheat in North China Plain by using crop growth monitoring system (CGMS). Proceedings of First international Conference on Agro-Geoinformatics Shanghai, China, **2012**, 1–4.

25. Fu, C.; and Yu, G.; Estimation and Spatiotemporal analysis of Methane emissions from Agriculture in China. *Environ. Manag.* **2010**, *46,* 618–632.

26. Fugiang, T.; and Hu, H.; A general review of the capacity development for agricultural water management in China. **2006**, 89–97, October 17, 2013, www.fao.org/docrep/009/a0415e/A0415E10.htm

27. Fuller, D. O.; Qin, L.; Zheng, Y.; Zhao, Z.; Chen, X.; Hosoya, L.; and Sun, G. P.; The domestication process and domestication rates in rice: spikelet bases from the lower Yangt Ze. *Science.* **2009**, 3231607–1610.

28. Fund, A.; and Hogan, M.; Northeast China Plains: Deciduous forests. **2013**, 105, October 20, 2013, http://www.eoeaorth.org/view/article /154891

29. Garriss, E. B.; Floods and Droughts causing major crop problems in China. **2012**, 1–4, http://www.financialsense.com/ contributors/evelyn-browning-garriss/floods-droughts-crop-problems-china

30. Gong, W.; Yan, X.; Wang, J.; Hu, T.; and Gong, Y.; Long term manure and fertilizer effects on soil organic matter fractions and microbes under wheat-maize cropping system in Northern China. *Geoderma.* **2009**, *149*, 318–324.

31. Gu, X.; Liang, Z.; Huang, L.; Ma, L.; Wang, M.; Yang, H.; Liu, M.; Lv, H.; and Lv, B.; Effects of plastic film mulching and plant density on rice growth and yield in Saline sodic soil of Northeast China. *J. Food Agric. Environ.* **2012**, *10*, 560–564.

32. Guo, J.; Songnen Meadow Steppes of Northeast china response to Global Climate Change. Ministry of Education, Northeast Normal University, China. **2010**, 1–32, http://grassnet_info/Datien/Grassnet%/presentation/guo%20-%20Grassland %20and%20water%20fluxes.pdf

33. Guo, Y.; Luo, L.; Chen, G.; Kuo, Y.; and Xu, H.; Mitigating Nitrous oxide emissions from a maize-cropping Black soil in Northeast China by a combination of reducing chemical N fertilizer application and applying manure in Autumn. *Soil Sci. Plant Nutr.* **2013**, *59*, 392–402.

34. Hays, J.; Deforestation and Desertification in China. **2008**, 1–17, October 18, 2013, http://www.factsanddetails.com/china.php?itemid=389

35. Hays, J.; Crops in China: Grain, Imports, Exports and GM Crops. **2013**, 1–19, October 17, 2013, http://www.facts and details.com/ china .php?itemed =345.html

36. Haibo, D.; Zhengafang, W.; and Ming, L.; Inter-decadal changes of vegetation transition zones and their responses to climate in Northeast China. *Theor. Appl. Climatol.* **2011**, *106*, 179.

37. Haibo, D.; Zhengfang, W.; Ming, L.; Shengwei, Z.; and Xiangjun, M.; Quantitative division of vegetation ecotones in Northeast China. *Appl. Ecol. Environ. Res.* **2012**, *10*, 319–332.

38. He, Z.; and Bonjean, A. P. A.; Cereals in China. Mexico: International Maize and Wheat Centre; **2010**, 129 p.

39. He, J.; Li, H.; Kuhn, N. J.; Wang, J.; and Zhang, X.; Effect of Ridge tillage, No-tillage and conventional tillage on Soil temperature, Water use and Crop performance in cold and semi-arid areas of Northeast China. *Aust. J. Soil Res.* **2010**, *48*, 737–744.

40. He, Z. H.; Rajaram, S.; Xin, Z. Y.; and Huang, G. Z.; A History of Wheat Breeding in China. Mexico: International Maize and Wheat Centre; **2001**, 1–95.

41. Hong-feng, B.; Guang, Y.; Lian, X.-S.; and Jing, J.; The Degradation, prevention and treatment of Black Soil in Jilin Province. *WSEAS Transac. Inform. Sci. Appl.* **2009**, *6*, 95-105

42. Hsu, H.; Lohmar, B.; and Gale, F.; Surplus wheat production brings emphasis on quality. China Agriculture in Transition/WRS-01-02. Economic Research Service of United Stated Department of Agriculture. **2003**, 17–25.

43. Huang, S.; Temporal and Spatial Variability of Soil Nutrients Under the Collective Contract System, 2007. Norcross, Georgia, USA: International Plant Nutrition Institute; **2009**, 1–2, October 19, 2013, https://www.ipni.net/far/farguide.nsf/$webindex/article= 0661EF9448257 3E00035864D73CB563!open document

44. IPCC. Climate Change 2007 Mitigation of Climate Change. Contribution of the Working Group III to the Fourth Assessment Report of the Inter-Governmental Panel on Climate Change. United Kingdom: Cambridge University Press; **2007**, 63–67.

45. IPNI. Interactive Effects of Nitrogen, Potassium, and Water on Maize Growth in Hebei. Norcross, Georgia, USA: Internation Plant Nutrition Institute; **2009**, 1–2, October 19, 2013, http://www.ipni.net/far/farguide.nst/$webindex/article/article= 8A24981B 482574900135F82E10A221A!open

46. Jianhua, G.; Zhu, X.; and Zhingheng, L.; Responses to phosphorus and potassium application in a wheat corn rotation in Hebei province. *Better Crops Int.* **2000**, 14, 3–5.

47. Jianyu, J.; Vast Zone of Black Soil will Disappear in Northeast China in 50 Years. **2010,** 1–3, October 28, 2010, http://www.globaltimes.cn/ china/ news/ 2010-11/588449.html

48. Jin, H.; et al. Genetic control of Rice Plant architecture under domestication. *Nat. Genet.* **2008,** *40,* 1365–1369.

49. Jones, M. K.; and Liu, X.; Origins of Agriculture in East Asia. *Science.* **2009,** *324,* 730–731.

50. Ju, H.; Velde, M.; and Li, X.; The impacts of climate change on agricultural production systems in China. *Climate Change.* **2013,** *120,* 313–324.

51. Kipple, K. F.; and Ornelas, K. C.; Cambridge World History of Food. Cambridge University Press; **2000,** *1,* 234–258.

52. Krishna, K. R.; Maize Agroecosystem: Nutrient Dynamics and Productivity. Waretown, NJ, USA: Apple Academic Press Inc; **2012,** 1–42.

53. Krishna, K. R.; Agroecosystems: Soils, Climate, Crops, Nutrient Dynamics and Productivity. Waretown, New Jersey, USA: Apple Academic Press Inc; **2014,** 87–118.

54. Langemeier, M.; and Hu, X.; A comparison of United States and Chinese Crop Production Systems. West Lafayette, Indiana: Centre for Commercial Agriculture, Purdue University; Internal Report. **2012,** 1–23.

55. Li, H.; et al. Permanent raised beds improved soil physical properties in an annual double cropping system. *Agron. J.* **2013a,** https://www. agronomy.org/publications/aj/abstracts/0/0/ agronj2013.0169

56. Li, J.; Mackay, A. W.; Zhang, Y.; and Li, J.; A 1000 year record of vegetation change and wild fire from Maar lake Erlongwan in Northeast China. *Quat. Int.* **2013b,** *290,* 313–321.

57. Li, S.; Yan, J.; Liu, X.; and Wan, J.; Response of vegetation restoration to climate change and human activities in Shaanxi-Gansu-Ningxia region. *J. Geograph. Sci.* **2013c,** *23,* 88–85.

58. Lian, P.; Zeng, D.; Liu, J.; Ding, F.; and Wu, Z.; Impact of land use change on carbon stocks in Meadow steppes of Northeast China. *Appl. Mech. Mater.* **2012,** *108,* 262–268.

59. Liang, W.; Carberry, P.; Wang, G.; Lu, R.; Lu, H.; and Xia, A.; Quantifying the yield gap in wheat-maize cropping systems of the Hebei plain, China. *Field Crops Res.* **2011,** *124,* 180–185.

60. Liang, A. Z.; Zhang, X. P.; Fang, H. J.; Yang, X. M.; and Drury, C. F.; Short-term effects of tillage practices on organic carbon in clay loam soil of Northeast China. *Pedosphere.* **2007,** *17,* 619–623.

61. Liang, A. Z.; Zhang, X. P.; Yang, X. M.; McLaughlin, N. B.; Shen, Y.; and Li, W. F.; Estimation of total erosion in cultivated Black soils in Northeast China from vertical profiles of soil organic carbon. *Eur. J. Soil Sci.* **2009,** *60,* 223–229.

62. Liping, Y.; Soil Nutrient Availability and Nutrient Uptake in Northeast China as Influenced by Different Fertilization Pattern. Norcross, Georgia, USA: International Plant Nutrition Institute; 2009, 1–2, October 19, 2013, https://www.ipni.net/ far/farguide.nsf/$webindex/article=E B317CD482576D60003C64ECA2F9E7D6?opendocument

63. Liu, X.; Herbert, S. J.; Jin, J.; Zhang, Z.; and Wang, G.; Responses of photosynthetic rates and yield/quality of main crops to irrigation and manure application in the Black soil area of Northeast China. *Plant Soil.* 2004, *261,* 55–60.

64. Liu, Z.; and Zhang, X.; Steppe Degradation and Rehabilitation in China. Desertification and Its Control in Northern China. 2010, 299–350.

65. Liu, X. B.; et al. Soil degradation: A problem threatening the sustainable development of Agriculture in Northeast China. *Plant Soil Environ.* 2010, *56,* 87–97.

66. Liu, X.; Zhang, S.; and Zhang, X.; Soil erosion control practices in Northeast China: A mini review. *Soil Tillage Res.* 2011, *117,* 44–48.

67. Lou, Y.; Xu, M.; Wang, W.; Sun, X.; and Liang, C.; Soil Organic Carbon fractions and Management index after 20 years of manure and fertilize application on vegetables. *Soil Use Manag.* **2011,** *27,* 163–169.

68. Loucks, O.; and Jianguo, W.; The Northeast. In: Grasslands and Grassland Sciences of Northern China. Washington, D.C. USA: National Academy of Sciences; **1990**, 55–88, October 18, 2013, http://www.images.nap.edu/openbook/ 030904684X/ gifmid/55. gif

69. Mingwei, Z.; Qingbo, Z.; Zhongxin, C.; Jia, L.; Yong, Z.; and Chingfa, Y.; Crop discrimination in Northern China with double cropping systems using Fourier analysis of time series MODIS data. *Int. J. Appl. Earth Observ. Geo-Inform.* **2008**, *189*, 1–10.

70. MWR. Soil and Water Conservation in China. Ministry of Water Resources, The Republic of China. **2013**, 1–4, October 18, 2013, http://www.mwr.gov.cn/english/swcc.html

71. NAS. Grasslands and Grassland Sciences in Northern China. Washington, D.C. USA: National Academy of Sciences; **1992**, 230 p.

72. Ni, K.; Ding, W.; Zaman, M.; Cai, Z.; Wang, Y.; Zhang, X.; and Zhou, B.; Nitrous oxide emissions from a rain fed-cultivated Black soil in Northeast China: Effect of fertilization and maize crop. *Biol. Fertil. Soils.* **2012**, *48*, 973–979.

73. Niu, S.; and Wan, S.; Warming changes plant competitive hierarchy in a temperate steppe in Northern China. *J. Plant Ecol.* **2008**, *2*, 103–110.

74. Peng, E.; Intercropping with Jujube in China. Chinese Academy of Forestry. The Archives of the Rae fruit Council of Australia. **1991**, 1–3, October 19, 2013, http://www.rfcarchives.org. au/Next/Fruits/Jujube/JujubeIntrcrops5-91.html

75. Qui, J.; et al. Mapping single, double and triple crop agriculture in China at 0.5 x 0.5 by combining county scale census data with a remote sensing—derived land cover map. *Geocarto Int.* **2003**, *18*, 3–13.

76. Qui, J.; Wang, L.; Tang, H.; Hing, L.; and Changsheng, L.; Studies on the situation of soil organic carbon storage in Croplands in Northeast of China. *Agric. Sci. China.* **2005**, *4*, 101–105.

77. Sheehy, D. P.; Using deferred rotation grazing to improve natural rangelands of East-Central Inner Mongolia. In: Grasslands and Grassland Science in Northern China. Washington D.C. USA: National Academy of Science; **1990**, 613–620.

78. Shi, P.; Research on soil Erosion in Inner Mongolia of China by Remote Sensing. In: Grasslands and Grassland Sciences of Northern China. Washington, D.C. USA: National Academy of Sciences; **1990**, 137–140, October 18, 2013, http://www. images.nap.edu/ openbook/030904684X/gifmid/55.gif

79. Thianchang, Z.; Huimin, Z.; Yong, G.; Zilu, G.; and Jun; Intercropping with Zizyphus zuzuba a Traditional Agroforestry Model. IDRC Documents on Agroforestry Systems. **1998**, 1–4, October 19, 2013, http://www.archive.idrc.ca/library/document/090916/ chap14_e.html

80. Thomas, A.; Agricultural demand under present and future climate scenerios in China. *Global Planet. Change.* **2008**, *60*, 306–326.

81. Tian, H.; et al. IOP Science. **2012**, 1–12, October 19, 2013, http://www.iopscience.io-porg/1748-9326/7/4/044020/article

82. Tianjie, L.; Application of Remote Sensing to the Investigation, Mapping, and Monitoring of Soil Cover in the Grassland of Inner Mongolia. Beijing: Proceedings of the International Symposium on Grassland Vegetation Science Press; **1990**, 83–89 p.

83. Toussaint-Samat, M.; Histoire Naturelle et Morale de la Nourriture. Paris: Larousse-in Extenso; **1997**, 958 p.

84. Tsuiki, M.; Wang, Y.; Tsutsumi, Y. M.; and Shiyomi, M.; Analysis of Grassland Vegetation of the Southwest Heilongjiang Steppe. *Acta Botanica Sinica.* **2005**, *47*, 1–8.

85. USDA-FAS. China: 2002/03 Wheat Update. United States Department of Agriculture, Foreign Agricultural Service, Washington D.C. USA; **2003**, 1–3, October 19, 2013, http://www. fas.usda.gov/pecad2/highlights/2002/06/wheat/ncp.htm

86. Vassilos, R.; Franke, T.; and Sanchez, J.; Peoples Republic of China: Agricultural Situation Water Situation in the North China Plain. USDA Foreign Agricultural Service GAIN Report No: CH8049, **2008**, 1–16.

87. VICARR. Northeast China Plain deciduous forests. Virtual Information Centre for Amur River Region. **2013a,** 1–3, October 16, 2013, http://www.amur-heilong.net/http?02-ecosystem/02 15Northeasternchinaplainforests.htm

88. VICARR. Global climate change observed and predicted. Virtual Information Centre for Amur-River Region. **2013b,** 1–3, October 17, 2013, http://www.amuur-heilong.net/http/01-climate-water/0106/GlobalClimateChange.html

89. VICARR. Climatic Fluctuations: Floods and Droughts. Amur-Heilong river basin. **2013c,** 1–5, November 2, 2013, http://amurheilong.net /http/07 _landuse_chap.html

90. VICARR. General Land Use Trends in Amur-Heilongjiang River Basin. **2013d,** 1–5, October 16, 2013, http://www.amur-heilong.net/http /07landuse _chap.html

91. Wang, D.; and Ba, L.; Ecology of Meadow steppe in Northeast China. *Rangeland J.* **2008,** *30,* 247–254.

92. Wang, J.; Huang, J.; Blanke, A.; Huang, Q.; and Rozelle, S.; The development challenges and management of groundwater in rural China. In: The Agricultural Groundwater Revolution: Opportunities and Threats to Development. Giordano, M.; and Vilholth, K. G.; eds. United Kingdom: CAB International; **2006,** 37–61.

93. Wang, Z.; Liu, B. Y.; Wang, X. Y.; Gao, M.; and Liu, G.; Erosion effect on the Productivity of Black soil in Northeast China. *Sci. China Ser.-Earth Ser.* **2009,** *52,* 1005–1021.

94. Wang, S.; Zhou, C.; Liu, J.; Tian, H.; Li, K.; and Yang, X.; Carbon storage in Northeast china as estimated from vegetation and soil inventories. *Environ. Pollut.* **2002,** *116,* S157–S165.

95. Wen, D.; and Liang, W.; Soil fertility quality and agricultural sustainable development in the Black soil region of Northeast China. *Environ. Dev. Sust.* **2001,** *3,* 31–43.

96. Wikepedia. Geography of China. **2013,** 1–14, October 16, 2013, http://en.wikipedia.org/wiki/ Geography_of_China

97. Wu, Y.; Liu, b.; Yun, X.; and Yonguang, Z.; Gully erosion at the black soil region of Northeast China. Paper Presented at Annual Meeting of the Soil and Water Conservation Society, Saddle Brook; Tampa, Florida, USA; **2013,** 1–2, October 28, 2013, http://www. citaion.allacademic. com/meta/p202368_index.html

98. Xinshi, Z.; Northern China. In: Grasslands and Grassland Sciences of North China. Washington, D.C. USA: National Academy of Sciences; **1990,** 39–52, October 18, 2013, http://www. images.nap.edu/openbook/030904684X/gifmid/39.gif

99. Xu, F. A.; and Zhao, B. Z.; Development of crop yield and water use efficiency in Fengqqiu in the NCP of China. *Acta Pedologia Sinica.* **2001,** *38,* 491–497.

100. Yan, X.; et al. Carbon sequestration efficiency in Paddy soil and upland soil under long term fertilization in Southern China. *Soil Tillage Res.* **2013,** 130, 43–51.

101. Yang, Z.; Food storage in pre-Qin days. *Agric. Archaeol.* **1997,** *3,* 2–8, (in Chinese)

102. Yang, X.; et al. Adaptation of Agriculture to global warming in Northeast China. *Climate Change.* **2007,** *84,* 45–58.

103. Yang, X.; Ren, W.; Sun, B.; and Zhang, S.; Effects of contrasting soil management regimes on total and labile soil organic carbon fractions in a loess soil in China. *Geoderma.* **2012,** *178,* 49–56.

104. Yang, X.; Liu, Z.; and Chen, F.; The possible effects of Global Warming on Cropping Systems in China.1. The possible effects of climate Warming on Northern Limits of Cropping Systems and Crop yields in China. *Scientia Agricultura Sinica.* **2010,** *43,* 329–336.

105. Yang, H.; and Zehnder, A.; China's regional water scarcity and implications for grain supply and trade. *Environ. Planning.* **2001,** *33,* 79–95.

106. You, X.; Historical growth and values of rice agriculture. *Agric. Archaeol.* **1988,** *4,* 406–408, (in Chinese).

107. Yu, G.; Fang, H.; Gao, L.; and Zhang, W.; Soil organic carbon budget and fertility variation of Black soils in Northeast China. *Ecol. Res.* **2006,** *21,* 855–867.

108. Yue, S.; et al. Change in nitrogen requirement with increasing grain yield for winter wheat. *Agron. J.* **2012,** *104,* 1687–1693.

109. Zhang, J.; Risk assessment of Drought disaster in the Maize-growing region of Songliao Plain, China. *Agric. Ecosyst. Environ.* **2004,** *102,* 133–153.

110. Zhang, X.; Chen, S.; Liu, M.; Pei, D.; and Sun, H.; Improved Water use efficiency associated with cultivars and Agronomic Management in the North China Plain. *Agron. J.* **2005,** 97, 783–790.

111. Zhang, W.; et al. Actual response and Adaptation of rice cropping system to Global warming in Northeast China. *China Agric. Sci.* **2012a,** *45,* 1265–1273.

112. Zhang, Y.; Hou, P.; Gao, Q.; Chen, X.; Zhang, F.; and Cui, Z.; On Farm estimation of Nutrient requirements for spring corn in North China. Agron. J. **2012b,** *104,* 1436–1444.

113. Zhang, Y.; Wang, Y. Y.; Su, S. L.; and Li, C. S.; Quantifying methane emissions from rice paddies in Northeast China by integrating remote sensing mapping with a biogeochemical model. *Biogeosciences.* **2011a,** *8,* 1225–1235.

114. Zhang, S.; Zhang, Z.; and Jian, S.; Effect of different management systems on soil water in Black soil of Northeast China. *Adv. Mater. Res.* **2013b,** *613,* 2912–2915.

115. Zhang, S.; Zhang, X.; Liu, X.; Liu, W.; and Liu, Z.; Spatial distribution of soil nutrient at depth in black soils of Northeast China: A case study of soil available nutrients. *Nutr. Cycl. Agroecosyst.* **2013a,** *95,* 319–331.

116. Zhang, J.; Zheng, Y.; and Wang, X.; Decision support System for Quantitative calculation of crop climatic suitability in Hebei Province. *Adv. Inform. Commun. Technol.* **2011b,** *344,* 381–389.

117. Zhao, X.; Survey of the temperate steppe of China. In: Grasslands and Grassland Sciences of North China. Washington, D.C. USA: National Academy of Sciences; **1990,** 235–238, October 18, 2013, http://www.images.nap.edu/openbook/ 030904684X /gifmid/55.gif

118. Zhao, R.; et al. Fertilizer and Nitrogen balance in a wheat-maize rotation system in North China. *Agron. J.* **2006,** *88,* 938–945.

119. Zhao, J.; and Guo, J.; Possible trajectories of agricultural cropping systems in China from 2010 to 2050. *Am. J. Climate Change.* **2013,** *2,* 191–197.

120. Zheng, H. F.; Chen, L. D.; and Han, X. Z.; Response and adaptation of soybean systems to climate warming in Northeast China: Insights gained from long term field trials. *Crop Pasture Sci.* **2011,** *62,* 876–882.

121. Zhong, W.; et al. The effects of mineral fertilizer and organic manure on soil microbial community and diversity. *Plant Soil.* **2010,** *326,* 511–522.

122. Zhou, M.; Butterbach-Bahl; and Zhu, B.; Nitrous Oxide Emissions and Nitrate Leaching from a Rain-Fed Wheat-Maize Rotation in the Sichuan Basin, China. Paper Presented at Land Use and Water Quality Conference: Reducing effects of agriculture. Hague, The Netherlands; **2013,** 1–22 p.

123. Zhou, D.; Gong, H.; Luan, J.; Hu, J.; and Wu, F.; Spatial pattern of water controlled wetland communities on the Sanjian Floodplain, Northeast China. *Commun. Ecol.* **2006,** *7,* 223–234.

124. Zhu, X.; Han, B.; and Zhao; Spatial and Temporal variability of soil nutrients in the Black soil are of Northeast China. *J. Food Agric. Environ.* **2013,** *11,* 1386–1389.

CHAPTER 8

AGRICULTURAL PRAIRIES AND MAN—WITHER BOUND

CONTENTS

8.1   INTRODUCTION

Natural vegetation that has evolved through the ages varies widely in terms of composition and dominance of plant species. Natural vegetation can be identified into forests, shrubland, prairies, wetlands, swamps, wasteland with scanty vegetation cover, and a few other variants. Development of a particular type of vegetation depends on variety of factors related to topography, soils type and its fertility, climate, and plant diversity available in the region. In the context of this chapter, we are concerned with "Agricultural Prairies" that are grouped as wetlands (e.g., rice), and arable cropping expanses that support variety of crops.

The inception, sustenance, and perpetuation of agricultural prairies are almost entirely a human effort. Agricultural prairies world-wide are termed as marvels or living wonders generated through persistent human endeavor (see Krishna, 2003). Agricultural Prairies are generated to negotiate natural factors and yield food grains for human population. Agricultural prairies differ from natural prairies with regard to several aspects related to interaction with natural factors, botanical composition, genetic nature of individual plant species, human involvement in mending soils and their fertility, utilization of water resources, and upkeep of prairie vegetation and its perpetuation. Overall, Agricultural Prairies are predominantly the result of human endeavor directed toward answering his need for food grains and few other items.

As stated earlier, agricultural evolution began around 12000 B.C. in the Fertile Crescent area of Middle East. There are many suggestions that seeding and crop production was independently discovered by Neolithic human beings, in different continents and locations during different periods of the history. For example, human population in West Africa residing in the upper reaches of Niger might have domesticated and cultivated a set of crops such as pearl millet, sorghum, cowpeas, and initiated agriculture practices. Agriculture might have been independently invented in Indo-Gangetic plains during 3–5th millennia. Similarly, in China, agriculture was supposedly initiated in the valleys of Yangt Ze and Huang He during Neolithic period (An, 1984, 1999; Kipple and Ornelas, 2000; Fuller et al., 2009; and Bonjean, 2010).

Agricultural Prairies, in particular, that are predominantly filled with cereals and other food crops were obviously initiated just around the period crop production was invented by humans. Agricultural Prairies are basically biomes where in crops are identified as prominent species. Their creation weaned humans from hunting-gathering and to a large extent reduced practice of nomadic farming and even pastoralism. Agricultural prairies, as we perceive, visualize, and utilize are predominantly products of constant mending of soils, selection of crops, and their adaptation to agroclimate. This fact is unlike natural prairies that have developed and perpetuated almost entirely due to interacting natural factors. Human interference in natural prairies has been least, if any.

Human intelligentsia has created different types of agricultural prairies such as those based on dryland/arable crops species (e.g., wheat, maize sorghum, millets);

wetland crops (rice, sugarcane); mixtures of cereals; and other crops (maize/cotton, wheat/soybean, sorghum/pigeonpea); parklands that encompass cereal/pulse plus interspersed trees (e.g., millets/groundnut plus Faidherbia trees in Senegal, millet plus Kejri trees in Rajasthan); Tropical forests with understory of prairie cereals (e.g., Sorghum in Guinea Savannahs of West Africa), and so on. Human intelligentsia discovered, domesticated, and cultivated several plant species during Neolithic period, across several hot spots located on earth. In each case, selection of crops that fill the prairies was based on traits useful to human beings, mainly to serve his food requirements—carbohydrates, proteins, and fat; fodder; fabric; fuel, and so on.

The evolution, domestication, and selection of crops and their genotypes have major impact on diversity, spread, and intensity of agricultural prairies. Currently, among various types of vegetation, agricultural prairie vegetation dominates. In most areas, natural vegetation has been removed to develop agricultural prairies. Wasteland with scanty vegetation too has been converted to agricultural prairies. Forests have also given way for agricultural crops in many locations. A few of the examples are (a) Conversion of natural forest in European plains; (b) Conversion of arable regions in East coast of India into low land rice ecosystem; and (c) Conversion of natural prairies of Canada and Northern USA into Wheat-based agricultural prairie during nineteenth century. The change of natural vegetation of Cerrados to develop soybean-based biome during 1970s is a very recent event yet it is a prominant example.

Today, human population in almost every region of earth is predominantly dependent on food grains/products derived from agricultural prairies. In some regions, human dependence on nourishment derived via food grains generated by agricultural prairies is almost obligate. In many other regions, agricultural prairie–related activity *per se* is the main occupation of human population. It is almost the only source that offers them food grains and exchequer to buy goods and services.

8.1.1 NEOLITHIC HUMANS AND GIFT OF DOMESTICATED PRAIRIE CROP SPECIES

Neolithic human dwellers across several locations on earth have played a remarkable role in domesticating various crops that today are the chief sources of food grains, forage, fuel, and other useful items to mankind. They also dispersed many of these crops into several locations through migrations, disseminated earliest known agricultural technologies, and as well guided the development of agricultural prairies to a certain extent. Further, ingenuity in selecting several different types of landraces and periodic introductions of new crops from other geographic locations has offered us with posse of crops to cultivate. We have no detailed reasons about their preference to cereals and crop mixtures of prairie. Tall or short statured food grain crops with relatively shorter duration to maturity, seems to have been selected more often, over others with different morphogenetics such as trees, shrubs, or creepers. Following paragraphs offers evidences for domestication, dissemination, and consolidation of agricultural prairies by Neolithic human settlers and those who colonized earth in later stages of history:

Wheat and barley were among the earliest of the crops domesticated by human species. Human dwellers in the "Fertile Crescent" of West Asia are credited with selecting nondehiscent panicles and perpetuating a few different types of *Triticum* species, such as hexaploid *T. aestivum*, diploid *T durum*, einkorn wheat, and so on (Kipple and Ornelas, 2000; Newman and Newman, 2006; and Palmer et al., 2009). Human movement toward different directions into European plains and Asia induced its dispersal and spread during Neolithic age. There are instances of independent and repeated domestication events, for example, in Indo-Gangetic belt and Steppes in Russia. Wheat cropping zones were intensified around citadels, villages, and towns during ancient period. Human involvement in land (plowing) and soil management (ridging), interculture procedures (hoeing, weeding), and harvest (grain separation, winnowing, de-husking) of wheat are noteworthy. Large expanses of wheat/barley took shape during medieval times (Arias, 1995). Higher demand for grains/forage caused by enhanced population of human beings and domestic animals, development of storage facilities, advent of trade routes, and transport vehicles, allowed farmers in several regions of West Asia, Europe, and Asia to develop large prairies that are exclusively aimed at wheat/barley production (see Bonjean, 2010; CIMMYT, 2011).

Rice (*Oryza sativa*) was domesticated and propagated by Neolithic human population residing in Yunnan, Assam, and swamps along the river Brahmaputra (Chen, 1989; Zhimin, 1999). Its cultivation practices were later spread to Central Chinese river valleys and Indo-Gangetic plains. Neolithic migratory trends were responsible for spread of *Oryza sativa* into Southern Indian Plains and Coasts. Several types of rice genotypes have been cultivated in these regions (Vishnu-Mittre, 1976; Lu and Chang, 1980; Randhawa, 1980; Fuller et al., 2004; and Manasala, 2006). Human ingenuity has played a great part in devising a series of land, water, and crop management procedures that allow production of rice under submerged conditions and as upland arable crop. Ploughing, flooding/stagnating with water, puddling, application of manure/fertilizers, development of high tillering, and semidwarfs with shorter maturity period are among excellent efforts by human intellect. The wetland prairies have currently spread into vast area in India, Pakistan, Southeast Asia, and Fareast. Rice-based prairies are also conspicuous in wetlands of New Orleans, Florida, and in the Caribbean. Wetland rice-based prairies flourish in West African coastal regions (Harlan, 1973; Somada et al., 2012). Wetland rice is often alternated with arable crops such as oilseeds, pulses, and vegetables. Wetland rice belts in Asia are among intensely maintained agricultural prairies. Their productivity is held high commensurate with demand for grains/forage. We should note that rice culture involves large populations of Asian farmers, relatively higher quantities of inputs like fertilizer and irrigation. Rice-based prairie is relatively inefficient in terms of use of natural resources, fertilizers, and irrigation.

Maize is native crop of Meso-America. Maize was domesticated by Central Mexican tribes (Piperno and Flannery, 2001). Maize was disseminated into North American plains by Indian tribes who migrated into several locations within Great Plains during ancient period, around 500 B.C. to second century A.D. (Hudson, 1994; Gallinat, 1999). Maize production zones later spread to Northeast region of America. Human ingenuity selected maize genotypes of diverse traits related to grain, its color, cooking quality, and palatability. Crop-related traits such as adaptability, duration to maturity, compatibility with pulses, and legumes were also important (Hudson, 1994). Maize found its way into Southern Europe in fifteenth century. Maize based prairies were established in Italy and Central European regions (Andrews, 1993; Brandolini and Brandolini, 2009). Maize had spread into almost all agrarian regions of earth by medieval times. Human migration, trade practices, conquests, and travel aided the spread of maize prairies. Maize based prairies in North America also called "Corn Belt" are highly intensively cultivated prairie. Its productivity ranges from 7 to 11 t grains plus 12–15 t forage per hectare. Maize has been grown in mixed culture with beans and vegetables in West African humid tropics, Egypt, and Arabia, since medieval times (McCann, 2000; USDA, 2008). There are also, maize prairies that receive low or no worthwhile inputs, for example, in hilly regions of South America, dryland of West Asia, India, and Northeast China (see Krishna, 2012).

Sorghum based prairies were initially generated by Neolithic settlers in North Africa, particularly in the Ethiopian highlands and adjacent regions of Sudan, around third millennium B.C. Sorghum-based prairies dominate the landscape in Sudanian and Guinean regions of West Africa (Harlan and DeWet, 1972; Smith and Frederickson, 2000; CGIAR, 2004). Sorghum prairies in West Africa are taller, since they are composed of landraces and improved cultivars that are taller and reach 5–6 m in height. Sorghum-based prairies are common to Vertisol plains of Peninsular India (Krishna, 2010a). During past 2 decades, sorghum based prairies in India have experienced shrinkage in expanse, but hybrids that encompass the taller prairies seem to generate greater quantity of grains/forage that compensates for decreased area. Human preference has induced generation of diverse sorghum genotypes and their cultivation. Genetic diversity for traits related to plant stature, panicle shape, size, grain color, nutritional content, and palatability is a noteworthy effort of farmer's agricultural acumen and ingenuity.

Pearl millet based prairies that tolerate the harsh, drier conditions of Sahelian West Africa were generated by human dwellers in the upper reaches of river Niger. Pearl millet landraces offered tall prairies (>5 m) mixed with legumes. Pearl millet prairies developed in response to short rainy period in the West African semiarid tropics. Pearl millet genotypes of diverse stature, canopy, duration, panicle size, and grain types were selected by West African farming community and propagated. Pearl millet, like other cereals has spread to other continents and established itself into prairies that tolerate drought and yet provide food grains/forage. Northwest region of India is yet another region where pearl millet has been nurtured by farmers since Neolithic period. Here, it has developed into prairies of relatively medium height. Genetic selection and diversity preserved in the fringes of Thar Desert is relatively shorter. Pearl millet prairies have also found a niche in the drier tracts of Peninsular Indian Vertisol and Alfisol zones. Pearl millet prairies in South India are relatively more productive than those traced in West Africa or Northwest India. It is attributable to fertilizer/irrigation inputs and higher planting density.

Finger millet was domesticated by Neolithic dwellers in highlands of Eastern Africa. Neolithic migration to Central Africa and other locations in the continent aided spread of finger millet based cropping zones (Hilu and Dewet, 1976). Southern Indian Neolithic settlements utilized finger millet. They induced its spread all across the plains and hills of South India. Here, human ingenuity has resulted in development of several cultivars of finger millet that have served the local populations with grains. Finger millet belts developed by farmers in South India are aimed to withstand dryland conditions, low soil fertility, and intermittent drought (Rachie and Peters, 1977; Devraj et al., 1995; Fuller et al., 2004; Krishna, 2010a; and Krishna, 2014a).

Foxtail millet is a native crop of Central Asia and Russian Steppes. It later spread to other regions such as Indian plains, West Africa, and southern Europe (Baltensperger, 1996; Smekalova, 2009). Setaria-based prairies were relatively prominent grain generators until mid-1900. During recent decades, foxtail millet belts have been gradually replaced by better yielders such as maize and wheat.

Legumes too form large prairies. Legumes were domesticated in several regions, mainly to serve the protein requirements of human populace and farm animals. Legumes have developed into either short or tall prairies depending on the stature of crop species and its genotype. For example, soybean in Cerrado is an excellent demonstration of legume prairie that is short in stature. Whereas, many of the leguminous trees that supply fodder and crop residue (manure) have developed into tall prairies. They are usually cultivated as intercrops, strip crops, and wind rows. Leguminous agroforestry species add to soil-N and organic matter. They are often traced in agricultural prairies as intercrops. Let us now consider few important legume grain crops:

Cowpea is an important legume in several regions of Earth. Cowpea cultivation began in West Africa sometime during third millennium B.C. (Ng, 1995; Fuller, 2003). Cowpea spread to several regions in Africa and Asia through Neolithic migration trends (Fuller and Harvey, 2006). However, its cultivation is now predominant in Africa and Asia (Krishna, 2010a; Krishna, 2014c). Cowpea cultivation generates a short statured leguminous prairie, if it is grown as monocrop. Currently, cowpea is grown in West African savannahs as a monocrop that adds to grain and forage harvests. Here, it is also grown as an intercrop with pearl millet or sorghum.

Pigeonpea-based prairies were initiated in the hills and plains of South India during late Neolithic period (2nd millennium B.C.). Pigeonpea-based prairie is a staple pulse source to large population in India. Pigeonpea-based prairies are also traced in Caribbean region and Eastern Africa. Pigeonpea is a leguminous crop.

Therefore, these prairies add to soil-N fertility, in addition to supplying grain/fodder (Nene and Hall, 1990; Van Der Maesen, 1990; Fuller et al., 2004; Fuller and Harvey, 2006; Ikissan, 2006; Krishna, 2010a).

Lentils were domesticated by native tribes resident in Indo-Gangetic plains during 3rd millennium B.C. Lentils were also domesticated independently and grown in large patches in Northern Syria, Israel, and Iraq during 4th millennium B.C. It has developed into large prairie vegetation in West Asia, Central Europe, and Russian steppes. Lentil based prairies are often alternated with cereals such as wheat or barley (Al-Issa, 2006; Yadav et al., 2007; and Erskine, 2009).

Soybean is native to Chinese agrarian zones where it was domesticated during Neolithic age. This crop was disseminated by migrants to other regions in Southeast Asia. Soybean prairies were confined within China for a long period. Ancient trade routes aided its dispersal to other continents. Soybean based prairies, currently flourish in American agrarian regions. Soybean culture dominates the landscape in Cerrados of Brazil and Cardoba region of Argentinean Pampas. Soybean based prairies developed into one of the intensive monocropping zones of North American plains during 1930s–1970s. The soybean based short, leguminous prairies are often intercropped or rotated with crops such as wheat, maize, and cotton. American farmers have applied their ingenuity and resources to develop soybean belts that actually generate legume grains to human population residing in other continents (Singh, 2010; Soy info Centre, 2011a, 2011b, 2011c, 2011d).

Groundnut was domesticated in South America, within regions encompassed by Peru, Bolivia, and Gran Chaco. Groundnut crop was held within South America for a long stretch of period until medieval times (fifteenth century). Groundnut production spread rapidly to other regions such as West Africa, India, and China during Medieval period through Spanish travelers. Currently, groundnut-based short prairies are flourishing in United States of America (Texas, North Carolina, Virginia, Georgia, Alabama, and Florida); Sahelian region (Senegal, Burkina Faso, Niger and Nigeria); India (Gujarat, Andhra Pradesh, Karnataka, and Tamil Nadu); China (Shandong and Shanxii); and Australia (Queensland). Groundnut is a dryland crop in South Asia and thrives on subsistent inputs in many locations within West Africa. Groundnut prairies add to soil-N fertility, because it is a leguminous crop with ability to fix atmospheric-N (see Hammons, 1994; Weiss, 2000; and Krishna, 2008).

Sunflower is a native species of Southern USA. It was first domesticated in Tennessee region around 1,000 B.C. (Lentz, et al., 2006). Sunflower has spread into many agrarian regions. Its expanses are traced in South America, Eastern Europe, Africa, and Asia (see Krishna, 2014b). Sunflower based prairies are conspicuous in the Vertisol regions of Peninsular India.

Sugarcane originated in Southeast Asia. It was domesticated in the Indo-Gangetic plains around 3rd millennium B.C. *Saccharum barberi* was domesticated in India, while *S. edule* and *S officinarum*, it seems were domesticated in New Papua Guinea. Sugarcane spread into different nations in Middle-east, especially Mesopotamia, and North Africa during tenth century A.D. Spaniards introduced sugarcane into

American tropics (Sharpe, 1998). Sugarcane cultivation became a major occupation in the Caribbean during eighteenth–twentieth century. Sugarcane crop production and related industries were primarily manned by slaves drawn from South Asian nations. It is a clear case of crop-induced migration and slavery of human being.

Cotton was domesticated by inhabitants of Indus Valley during 3rd millennium B.C. It was also independently domesticated in Nile valley during the same age. Mexican archeological sites suggest that cotton was domesticated around 5,800 B.C. It seems ancient dwellers of Nazca and Chaco region used cotton fabrics. Like other crops, cotton cultivation techniques spread to different regions through Spanish travelers and merchants. Cotton cultivation is predominant in Great Plains of USA, Egypt, Pakistan, and India (see Wayne and Cotherne, 1999; Clayton, 2011; Krishna, 2014e).

At this stage, we may have to enlist a few facts about the array of crops that have been domesticated, grown into vast expanses (prairies) and are useful to human beings in various ways. It seems over 7,000 plant species have been identified as useful to humans. However, only 300 species have been cultivated regularly. Further, only 30 crop species have been accentuated into large prairie vegetation and they answer the food necessities of current population of human species. A step further, there are indeed only very few top cereal crops and legumes that dominate the global agricultural land scape (Wood et al., 2000). One of the consequences is lopsided development of prairies. Agricultural Prairies are endowed much less with regard to plant genetic diversity, ability to explore soil, biomass accumulation, and productivity compared to natural trees/shrubs. Yet, human preferences have led to overemphasis and development of prairie vegetation, based mostly on few cereals. Commonly, they are termed as grain or field crops. Tree crops grown for fruits are dominant only in certain geographic locations of earth, but generally are a less preferred botanical species, currently.

8.1.2 AGRICULTURAL PRAIRIES, HUMAN MIGRATION, AND SETTLEMENTS

Agricultural Prairies have affected human behavior, his sustenance, and perpetuation in variety of ways. Firstly, with the domestication of crops and establishment of crop fields/expanses, it seems clear that food habits were affected. Even before this, the way human beings found or generated food got changed. The hunting-gathering systems faded with time. At certain point of time before elimination of hunting/gathering systems, both crop production strategies and hunting coexisted (Morrison, 2007). In greater detail, agricultural prairies that got established in the plains and open ecosystems-induced cave-dwelling humans to change shelter systems and migrate to plains, closer to food crops. First open sedentary villages with predominantly agrarian occupation began appearing in early Neolithic age (Godwin, 1965; Klein and Klein, 1988; and Piperno and Flannery, 2001). Food habit included more of vegetarian ingredients. Crops such as cereals, legumes, tubers, fruits, and leafy vegetables replaced flesh eating habits. Agricultural prairies also induced large-scale changes in daily routine of the Neolithic human. It made humans to sow, nurture, and harvest food crops generated in fields, in preference to pastoralism and hunting errands. Agricultural prairies that generated grains allowed him to store his food more securely. Forage harvest helped him to find food items for his domestic cattle and other species.

Apart from above changes in human behavior and his daily routine, agricultural prairies of modern times has affected his long range migratory systems, finding locations that suit crop production and secure settlements. In the following paragraphs, let us consider several examples that convincingly prove that agricultural prairies are instrumental in inducing human migrations, within short distances, say in search of new crop fields, more fertile, and amenable locations and even long range migrations into different continents. We may also note that gregarious nature of humans has in turn helped several domesticated and natural species of plants to disperse into different continents and establish themselves in that region. Human migration and plant dispersal on several occasions has been a mutually related phenomenon—an interactive effect.

8.1.3 HUMAN MIGRATION DRIVEN BY CROPS AND CROP PRODUCTION TRENDS

During past two centuries, wheat has been the major cereal crop in the Great Plains of North America. Wheat-based prairie has induced consistent migratory response by Europeans. Parton et al. (2007) point out that for over 150 years, European migrants and settlers entering the Great Plains have converted a sparsely cropped zone into an intensively cultivated wheat belt. It is said that Great Plains yielded totally to axe that cleared the woodland and converted it into agricultural prairie. Wheat was the forerunner of crops grown in the plains. Wheat cropping zones increased most rapidly during 1900–1940. Migration for short distances from farms to nearby metros and vice-versa too was common in the wheat belt, during past century. Parton et al. (2007), further state that human migration in and out of agriculture zones toward urban locations and vice-versa exceeded 50,000 persons on many occasions. Such migrations indicate that agricultural trends, crops, and exchequer derived from them were major causes of human movement. Irrigation that ensured regular returns induced farmers to stay within settlements. Demographic aspects such as age of the farmers also influenced movement into new farming zones. It seems during past 4–5 decades, agricultural technology, cropping pattern, and environmental effects related to prairies have also affected human migration (Parton et al., 2007).

During 1900–1920, well before the advent dust bowls, United States of America actually experienced large-scale expansion of cropping zones aided by migrants from European nations. This period is termed the "Golden Age of Agriculture in

United States of America." In the Northern plains comprising Dakotas, parts of Montana, Minnesota, and Nebraska, gross income of wheat/barley farms were high. This led to migration of 2,50,000 of pioneer Europeans into Dakotas. Further, incentives such as larger farm holdings, and better price for grains, induced human migration and settlements in Dakota (Library of Congress, 2013). Wheat belt in Dakota increased from 4.3 million acres to 9.1 million acres.

Canadian Prairies is a prominent wheat belt in North America. Earliest of human intrusions into Canadian prairies for crops and livestock production, it seems began in late 1800s. Human migration intended to produce wheat in the prairie ecozone began in 1870s. Construction of railways induced over 2,00,000 European settlers to migrate to prairies and initiate wheat farming. By 1916, Canada was a leading wheat exporter in the world. Wheat belt was supported by migrant Europeans. However, in due course as mechanization and economic activity other than agriculture gained primacy, farm workers began migrating to urban locations. Population of migrant workers in wheat farms then decreased from 60 percent in 1916 to 10 percent during 1980s. Further, it is interesting to note that entire Canadian prairies supported only 15 crop species with wheat as the dominant one. Clearly, wheat farming was the major inducer of European migration into Canadian praries. However, during recent times, secondary activities that include mining, wood, chemicals, and food processing have weaned away human population from wheat production (Environment Canada, 1999).

Maize is an important crop in the Great Plains and adjacent regions of North America. Initially, during ancient period spanning from first to seventh century A.D., native Indians moved into plains from Mexico along with posse of maize genotypes that had been preferred by them. Maize cultivation induced migrations as far as Northeast. Mexican farming community and maize crop spread to different locations in South America (Vigouroux et al., 2008; see Krishna, 2012). European migration into Argentinean Pampas helped in expanding maize cultivation zones. Farmer migrations were also induced with introduction of soybean in the Pampas. Droller (2013) states that between 1857 and 1920, nearly 5.5 million European settlers moved into Pampas to grow native crops like maize and those brought by them. Further, farm-related migration into Argentina it seems peaked immediately after the World War I.

Human migrations from West Asia into Russian steppes peaked during seventeenth and eighteenth century. Expansion of agricultural prairies into forested zones was the lure. Wheat, barley, and lentils were the chief crops that attracted migrants. Migrations within the steppes were also in vogue and they were guided by availability of fertile soils (Stebelsky, 1974, 1983; Chibiliyov, 1984; and Masalsky, 1897). Fertile Chernozems held the sway regarding human migration.

Since several centuries, crops have caused migrations of farm owners and workers. Migrations of human population in the agrarian zones have been attributable to variety of reasons related to cropping zones, climatic conditions, productivity, and economic advantages. A few of migratory events have been exclusively connected with a crop species, or geographic location and human preferences/food habits. Let us now consider Sugarcane Plantation-related migrations of Asian farm workers. During early 1800s, British called the migrant sugarcane workers, who for all purposes were enslaved to work in sugarcane fields as "indentured labourers." Such indentured farm workers moved from Asian and Southeast Asian locations into various British colonies in Fiji, Natal, Burma, Malaysia, Jamaica, Trinidad, Suriname, French Guiana, Mauritius, and so on. (Hugh, 1993; Lai, 1993; Laurence, 1994; National Archives of United Kingdom, 2010). Movement of Asian indentured labor into Caribbean, Mauritius, and Fiji is noteworthy. Indentured farm workers have also been forced to migrate into French colonies, into Hawaii in North America, and Pacific islands. Sugarcane plantations in these regions indeed induced human migration and latter settlements. It is said, that sugarcane plantations in Caribbean, like cotton farms in Central Plains, has almost caused slavery of farm workers during eighteenth and nineteenth century. This is a clear example of how crops have coevolved, and induced migrations of human population across continents. They have affected various aspects of human behavior including his gregarious nature. We should also note that such migrants have induced spread of sugarcane based wet prairies in these locations.

During recent times, sugarcane cultivation in Gangetic plains, Maharashtra, Gujarat, Karnataka, and Tamil Nadu in India is intense and labor requirement to maintain the wet prairies is fairly high. Deshingkar and Farington (2006) and Deshinger (2011) point out that, migratory trend of sugarcane workers comes handy in maintaining the tall prairies. We may note that about 40 percent of sugarcane laborers in Peninsular India are constituted by temporary migrants from relatively poorer farming zones. Further, it is said migrations are stringently controlled by crop stages and cycles. For example, in Maharashtra, farm workers from poor regions of Maratwada move in to sugarcane belt in November and leave after April that coincides with sugarcane cutting period. Incidentally, in Southern Brazil, sugarcane is a cash crop that offers better wages to migrants. Sugarcane based farms in the Cerrados were major reasons for labor movement into Sao Paulo region during 1950s.

Cotton is another crop that has induced human migrations from across different continents. It has also caused periodic short journeys from different locations into its production zones. Human migration into Central Plains from European ports to cultivate cotton in the Central Great Plains is a glaring example of how a crop species, its related economics and infrastructural advantages can induce large-scale migration. European migration into North American regions in Tennessee, Kansas, North Texas Highlands, and Arkansas to support cotton cultivation is an example of long distance migration of human beings, which is induced by a crop species. Currently, we find that small distance local migrations from cotton farms to urban locations and vice versa is too common. Cotton cultivation trends, wages, and facilities related to crop affects farm workers movement. We should also note that, cotton crop failures in Oklahoma during 1934–36 did induce migration out of that zone (McCleman and Smit, 2006).

Currently, in Peninsular Indian Vertisol belt, cotton is a cash crop, grown intensively with high inputs and the resultant exchequer is commensurately impressive. Labor requirements are generally high during Kharif season. Migrant farm workers literally cultivate the crop from seeding till harvest and send it to market. Each of the southern States, hire about 3,00,000 migrant cotton farm workers each season. A large share of migrants is young and prone to settle permanently in the cotton belt (Deshingkar and Akter, 2009).

Crop-induced migratory behavior of human beings is rampant in South Asia since many millennia. Perhaps it began during Neolithic period when human migration seeking new pastures, fertile locations, and congenial dwelling zones was common. During Medieval period, there are instances of entire village and its farming community moving out of their locations in Gangetic belt and settling in other locations. It seems availability of land and water resources induced farmers to shift from a location to other one (Rath, 2013). During recent times, human migration induced by rice-wheat based Agricultural Prairies of Indo-Gangetic belt is noteworthy. In Nepal, hill country landlords migrate to plains to own and develop farms. The farm workers regularly move from Sivalik hills to rice-wheat belt in Tarai region to cultivate the crops and earn lively hood (Hatlebakk, 2007). It seems, all through the Gangetic Belt, it is most often crop controlled migration of farm workers that maintains the rice-wheat prairies. Farm workers from neighboring states move into Uttar Pradesh and Bihar. Migrants enter prairie zone at the onset of seeding time, stay until harvest, and return to original locations. According to Dehsingkar and Akter (2009), crop production-related migration has been in vogue since several centuries. However, during recent period, lasting past 5–6 decades, agricultural prairies have developed unequally in Indian plains. There are pockets of less developed zones, from where poor farm workers tend to periodically migrate to well-established zone in the Gangetic plains. The so called "Green revolution areas" that support intensive production of rice and wheat seem to offer higher wages and facilities. Hence, they induce migration of agricultural labor from other areas close by. For example, in the peak season of cropping, about 0.8 million migrant farm workers enter the prairies of Bihar and Uttar Pradesh. In fact, in case of Punjab, it is said about 23 percent of farm labor in the wheat belt is drawn from migrants. They usually stay in that crop zone for the season of 50–90 days and leave to their original locations. Similarly, it seems over 5,00,000 farm workers migrate to rice fields of Burdwan in West Bengal, India at the onset of rice transplanting season, stay till harvest, and move back to original locations (Rogaly et al., 2002). A very clear case of stringent control of human behavior related to crops and their production procedures, almost at the beck, and call of crops.

Soybean based prairies in North America, especially in the mid-west states of Indiana, Illinois, and Ohio have experienced migration of farm settlers and migrants from Europe. Similarly, large-scale expansion of soybean production since early1970s has caused massive migration into the Cerrados of Brazil. In fact, Cerrados has experienced change in vegetation from a tropical shrubland with diverse vegetation into almost vast stretch of soybean monocrops or cereal-soybean intercrops. This activity has been largely possible due to active migration of farm workers from different parts of Brazil. Soybean based economy that is largely supported by export market is a clear incentive to farm migrations (see Chapter 2; Schmidt, 2009).

There are several reports about sugar beet production in the Central Plains that has induced periodic farmer migration from Mexico. It seems during and after math of World War II, farm workers in sugar beet fields were in shortage. Mexican migrants, known as *Braceros* were recruited in the plains to perpetuate sugar beet-based prairies (Douglas Hurt, 2008). During 1942–1945, *Braceros* from Mexico migrated and settled in areas as far as Wyoming and Dakotas. They helped in raising sugar beet, potato, corn, and wheat farms.

In addition to crops and performance of agricultural prairies, we ought to realize that farm infrastructure related to irrigation, processing, transport of grains, and market prices too affect human migratory trends. For example, almost each and every irrigation dam that assures supply of water does induce farmers to migrate into that zone. Agricultural prairies that develop near the rivers, dams, lakes, and in regions with good groundwater resources do attract human migration. In many of the Great Plains locations, construction of rail heads, transport facilities, and better market value for farm products has induced migration into that region. For example, rail transport into Dakotas and Midwest induced farm workers to move into Northern Plains (Library of Congress, 2013).

Groundnut is an important crop in Senegal. Currently, "Groundnut Basin" has developed into Parkland with interspersed *Faidherbia* and other tree species. Groundnut in Senegal has attracted migrations from Northern Senegal since mid-nineteenth century. Several ethnic groups such as *Fulani* and *Wolofs* mixed with natives such as Peul Habobe and Mandingue to work in the groundnut prairies, during 1960s and 70s. Migrant labor in groundnut basin increased rapidly. For example in 1970s, ethnic *Wolof* migrants accounted for 44 percent of farm worker population and 17.5 percent were migrant *Fulanis*. Early migrants such as *Mandigue* were only 7 percent of farm work force. During 1980s, when groundnut-based economy was slack they migrated to regions in Guinea Savannahs (Badiane, 2001). Predominantly, pastoralists such as *Serere* too moved away to North Senegal from groundnut basin during drought years.

Land resource is among the important factors that have induced migration either into or out of an agricultural zone. Agricultural expansion, in general, involves bringing new areas into crop production. It needs migration of humans to those new areas. Migrations westward in the Great Plains of America occurred as long as there was new land to occupy. White farmers went devastating natural prairies to convert them to cropland. Such migration became conspicuous, if the previous location was afflicted by dust bowl conditions, soil erosion and loss of topsoil. Always, better land induced human movement toward them (Sears, 1941). Historically, expansion of agrarian regions has involved waves/groups of human migration into areas. Then, forest/shrub clearing and initiation of crop production systems has followed. For example, in Cerrados and Central Brazil, migration of Indian tribes from Andean regions and crop diffusion occurred due to availability of new fertile soil. Human flow between locations within the continent was important for rapid development of agricultural cropping (Feltran-Barbieri, 2012). During more recent times, human migration into Cerrados for crop production is a good example. Land resource became easily available in Cerrados, immediately upon clearing the natural vegetation of shrubs and grass. Farmers and personnel from different agricultural companies moved into Cerrados to culture soybean. Here, human migration is effected by land resource in conjunction with a crop species (soybean) (Rezende, 2013). Migration into Argentinean Pampas has been guided by the vast expanse of land that could be utilized for farming. Migration was brisk prior to and aftermath of World War 1. Europeans migrated to Pampas in search of agricultural expanses (Droller, 2013). Crop production exercise affects soil fertility. Soil fertility below threshold has often induced human migration out of a zone. "Shifting agricultural" is among important methods that induces farmer to move to new areas. It is deterioration of land resource that induces farmers to move out of a previous location. It is practiced in all continents in different forms.

8.1.4 AGRICULTURAL PRAIRIES, CLIMATE CHANGE, AND FARMER MIGRATIONS

Agricultural cropping induces changes in climate. Transformation of natural vegetation to crops and adoption of different cropping systems do affect climate. We should note that climate change and its effects on agricultural prairies and human population is not a new phenomenon (Pongratz and Reick, 2012). Human beings dependent on farms, crops, and prairie vegetation have consistently adapted themselves to such changes through ages. Migration due to congenial locations is among the most popular adaptive mechanisms since Neolithic age. Migrations due to climate change have occurred periodically in ancient times, medieval era, and right till recent period. Migration is an important human behavior affected by climate-change-related impacts on agricultural prairies. For example, Lockeretz (1978) states that favorable precipitation pattern and climatic conditions in the plains induced migration into Great Plains. It coincided with the "Golden Age of Agriculture in USA". On the other hand, human vulnerability to drought, crop loss, reduced grain productivity, lack of timely availability of resources, and paucity of exchequer also drives them to migrate, but out of that zone (McLeman and Smit, 2006). The influence of climate change on agricultural prairies is varied in terms of aspects of farm operation it affects, productivity, and economic advantages accrued. Extremes of weather patterns, or even minor changes in soil water and nutritional status at critical stages of crop could result in depreciation of grain harvests. There are also reports suggesting possible increased grain harvest due to climate change, for example due to enhanced temperature (Yang et al., 2010). Crop production zones may shift in response to climate effects. Crop production strategies and time table of various operations too may alter. We ought to expect proportionate alterations in cropping pattern with each extreme swing in climatic pattern. For example, enhanced temperature and glacial melt in Northern plains may induce shift of cereal cultivation zone (IRSNDSU, 2013). Such factors may induce human migration. In Northern Florida, extension of cold front south words may necessitate further shift of cereal, groundnut and citrus plantations south words into Central Florida Spodosol zones (Grierson, 1995; see Krishna, 2003). In response, farmers migrate proportionately. McLeman and Smit (2006) state that climate change and prevailing conditions were among important factors that induced European settlers to migrate to various locations in Northern Great Plains during eighteenth and nineteenth century. In Northeast China, higher temperature forecasted due to climate change effects may actually shift wheat/soybean cropping zones north words into Inner Mongolia (Zheng, 2011; Zhang et al., 2012). In India, adaptation to climate change has direct consequences on agricultural labor, its requirements, and migrations that may ensue. As such in many locations, procedures such as seeding, inter-culture, weeding, harvest, and postharvest procedures have specific influence on movement of agricultural laborers. Great Plains may experience warming of agricultural zones. It has consequences on crop species, sowing time, and productivity in Northern plains. Glacial melt and water resources too may dictate certain changes in agricultural operations. Together, these factors may affect farm labor movement into Northern plains and *vice versa* (Library of Congress, 2013).

It seems climate change effects in Sahel may require human migration, plus planting of crop genotypes that tolerate shorter precipitation period and drought. It may also require alterations in planting time and cropping systems. In West Africa, farm workers located in Saharo-Sahelian region, closer to arid zones move into Sudanian and Sudano-Guinean region based on precipitation pattern, pearl millet/cowpea sowing time, and harvest season. They return to their original locations in Sahara once the crop season culminates. Such migrations need further careful adjustments.

In addition to above adaptations to climate change mediated effects on agriculture, there are certain climatic patterns that are drastic and bring about severe distortions in agricultural prairies and their productivity. Let us consider a few of them in the following paragraphs.

8.1.5 DROUGHT, DUST BOWLS, AND LOW PRODUCTIVITY INDUCED MIGRATION

Dust bowls of Central Great Plains of USA induced rampant movement of erstwhile well settled farmers and workers, in different directions. The primary cause was inability of farm community to grow crops and preserve topsoil fertility. Wind erosion resulted in soil fertility loss and inability to culture crops. In some locations of Great

Plains, crops grown yielded low and exchequer was severely affected resulting in loss of farm jobs. Farmers and workers moved away from droughts/dustbowl zones. It seems during 1929–1932, farmers from Oklahoma, Kansas, and Northern Texas, that is, "Dust Bowl zone" moved away in large numbers to California in the hope of finding farm jobs. Flight of migrants in search of cropping zones was so great, that in many camps, over 20 percent of migrants were from Oklahoma. Migrants from any region of Central Great Plains then were called "Okie Migrants" (Fanslow, 2013). According to Gregory (1989), 1930s decade witnessed migration of 3,00,000 farm workers out of United States South west. Dust Bowl migrants did disperse crops, techniques, and their folklore to different parts. All these movements/migrations were of course generated by crops on which the human population depends, so convincingly. In addition to immediate dependence on crops for food grains, economic aspects of farming too generated migrations. For example, during recession after World War I, need to intensify farming using mechanization, resulted in loss or reduction in exchequer. It forced many farmers to abandon independent farms. They migrated in search of jobs and better profits from farm-related activity elsewhere in California or Northeast (Fanslow, 2013). These are examples of migrations induced by crop-related activities.

Droughts and dust bowls created by intense dust storms are among important factors that affects agricultural prairies in African continent. Let us consider a few examples. In Ethiopia, migration has been a major human behavioral response to droughts. During 1980s, large-scale migration of famers out of Ethiopia and Somalia was primarily in response to droughts, impending famines, and problems related to gainful farm employment and hunger (McLemon and Smit, 2006). Migration of farm workers in and out of West African Sahel, especially in Burkina Faso and Niger is a common phenomenon. Migration out of Sahel to Guinean region is related to climate and agricultural activity (Henry et al., 2010). *Harmattan* is an important dust storm that affects soil, crop establishment, and productivity in West African Saharo-Sahel and Sahelian region. Crop loss does affect movement of farm workers to Guinean zones. In the general course, farm workers do transit periodically from fringes of Sahara, to Sudano-Sahelian regions, Guinea Savannahs, and into wet tropics of West Africa. They return with culmination of cropping season.

Droughts and dust bowls in the Gangetic Plains have created near famine conditions. Crop loss due to delayed onset of monsoon, erratic precipitation pattern, or lack of timely rainfall events have often led to migratory trends in India. Drought prone Eastern Gangetic plain often induces movement of farm workers. Migratory response of Bihar farmers in response to severe droughts of 1930s is well known. Droughts and resultant famines have occurred periodically in the Southern Indian plains. For example, droughts that occurred in Madras province of South India during 1870s affected crop production severely. Farmers' crop production strategies failed to varying extents. Prairies were decimated leaving soil surface to deterioration. It led to massive dislocation of farmers and they were forced to migrate to towns as the most plausible strategy. Since grain harvests were negligible, even the market yards moved out of drought affected zone (Morrison, 2012). During 1970s, Droughts in the Vertisol plains of Peninsular India resulted in abandonment of cotton farms and migration out of that zone to southern locations. For example, droughts that ravaged districts of Bijapur, Raichur, Koppal, and Kalburgi in North Karnataka, India, have affected crops and human sustenance. About 90 percent of rural population depends on crop production activity. Hence, most farm workers migrated out of drought zone and reached nearby cities and towns (Nagaraja et al., 2010).

Floods that inundate large patches of arable cropping zones are detrimental to productivity. Farmers are induced to migrate to distances where influence of flood water is not perceived. Flood-induced agricultural migrations are known to occur in all continents. Let us consider an example from Gangetic Delta in South Asia that is rampantly affected by floods (Bhattacharya and Werz, 2012). Flood on farms and migrations they induce are manifested in different ways in this Delta zone. Firstly, floods erode the river banks and destroy cropland. This induces farmers to shift to locations away and those placed at higher altitude. Frequent floods makes the farmers to find a different location altogether. In the Delta, salt and saline incursions due to sea water surge is also common. Saline incursion either through underground seepage or flooding initiates farmers to abandon the field/location and move to other areas. Salt water surge may also destroy standing crops and their productivity. Most commonly, as floods occur and water recedes, farmers experience a perceivable loss of topsoil fertility.

8.1.6 AGRICULTURAL PRAIRIES THAT ENSLAVE FARMERS

There are many facets of human slavery and reasons for slavish tendencies. Several types of slavery too could be identified within the human race, since ages. Within the context of this chapter, we are concerned with only slavery caused by crops, slavery to crops, their causes and resultant effect on human beings. Above all, for a very long stretch of time, since domestication of crops and establishment of agricultural prairies, we ought to acknowledge that man has slaved to perpetuate crops. Agricultural prairies (crop species) have induced a large section of human race to spend their physical energy, time, and intellect to further their own cause of dispersal and establishment in different terrains. Agricultural Prairies have definitely enslaved a large share of human population to work for their cause. This phenomenon, perhaps, began when human dwellers in permanent Neolithic settlements slowly changed from hunting/gathering and nomadic pastoralism to permanent agriculture, and higher density of human population necessitated total involvement in crop husbandry or related activity.

Examples of Human Slaves recruited to work in Crop fields and Prairies:

Braceros from Mexico who migrated either short or long distance from their present locations in the vicinity of Southern Plains were enlisted to work in prairies. They were mostly skilled and worked in prairies that supported sugar beet, corn, and wheat. These Mexican farm workers were asked to leave the farms once the daily work schedule was over. They were segregated from land lords and other elite. However, Braceros were praised for their work ethic, ability to pick up farming skills, use of machinery, harvesting, and postharvest processing, rapidly despite migration from far off locations. *Braceros* were also praised for their ability to slave in the fields for long period for low wages (Douglas Hurt, 2008).

In Brazil, large sugarcane farms owned by powerful landlords also harbors several bonded slaves who mend the sugarcane prairies. Slaves in Brazilian Sugarcane prairies are generally traced in remote locations within Cerrados. Typically, enslavement of sugarcane workers in Brazil begins with a contractor known as *Gato*. Gatos recruit workers (slaves) who need employment in sugarcane farms placed remotely in Cerrados. Currently, Brazilian sugarcane zone has over 100,000 slaves involved in various activities of sugarcane production, marketing, and other related activities (Campbell, 2008; Gervasio, 2013).

The sugarcane-based tall prairies in West Indies were among important attractors of slave labor. The recruitment of slaves in sugarcane farms was predominantly due to economic disparities between land lord and laborer (Frank, 1926). Major farm-related activities such as clearing fields, holing, planting, weeding, and cane cutting were usually done by Negro recruits. Slavery, it seems confined more to sugarcane farming and did not spread into other agricultural activities. Slavery and recruiting slaves was generally economically advantageous to sugarcane farm owner until late eighteenth century. Negro slaves from Gold Coast and Ivory Coast kept sugarcane farms running for over a century in West Indies. In Barbados, slaves from India were common. They usually filled up the farms vacated by Negro slaves who left for different locations in Southern Plains.

Cotton-based prairies of Great Plains are major factors that induced migration and slavery. Phillips and Roberts (2011) state that cotton prairie driven economy in Alabama induced large-scale movement of farm settlers and slaves. Cotton cultivation trends clearly created two types of labor arrangements. First, it created slave laborer that stayed in the fringes of cotton farms. Slaves were bonded to work for farmers/land lords. Second type was termed share croppers who derived a portion of economic advantages derived from cotton production. Cotton cultivation in Alabama was a success during nineteenth and twentieth century. Further, it is said that, importation of African slaves from West African coasts into cotton farms of Alabama highlighted the importance of slaves and their daily labor for maintenance and productivity of cotton prairies (Phillip and Roberts, 2011). Cotton is very labor-intensive crop and requires abundant hours of drudgery. Deployment of African slaves, therefore became necessary. The population of enslaved Africans in the cotton production zones increased with time. It seems between 1820 and 1860, population of enslaved African cotton workers increased by 20 percent in Tennessee. The black soil zone (cotton belt) in Tennessee demanded more of African slaves. Soon ratio of African slaves in Tennessee cotton belt increased from 30 to 50 percent

(Phillip and Roberts, 2011). Slaves in cotton farms usually picked seeds from 50 lbs of cotton per person. This aspect of cotton processing affected the profitability for a long time until the development of Whitney Cotton Gin. This invention induced spread of cotton fields and farm workers in the Great Plains. This is a clear case of mechanization and engineering that has generally come to rescue of farm workers/ slaves in overcoming drudgery. Further, due to factors such as "Great Exodus," World War I and World War II, slave labor got scarce. African slaves had moved out of cotton farms into occupations other than farming. This necessitated invention of as many agricultural implements, gadgets and contraptions. Tractors, and mechanized boll picking machines although expensive became necessities to run cotton farms. Again, agricultural engineering and mechanical devices came to rescue of cotton farm owners. They could reduce recruitment of slaves. It helped in meeting higher demand in the after math of wars (Fite, 1984; Phillips and Robert, 2011).

Slavery in agricultural prairies of Southern USA was wide spread during nineteenth century. General notion that large cotton/wheat farms had several hundreds of slaves was not true everywhere. About 88 percent of farms had slaves working in some aspect of farming such as seeding, weeding, cotton picking, separating bolls, harvesting grains wheat farms, transporting, and so on. During antebellum (1830–1860), slaves numbering 50 or fewer worked in cotton/cereal farms in Alabama. There were also small farms that had only few slave laborers (Africans in America Resource Bank 2013). It seems by 1870s, there were over 4 million African slaves drawn from West Central Africa who were mending prairies and plantations in the Southern Great Plains. Whatever be the cause, origin of slaves from different continent, country, or district, religious affiliations and their occupation inside farms, at the bottom line, we have to identify that crop was the center of activity. All humans involved in the farm and who derive lively hood from the crop are slaves-"slaves to crops." Since, it is the crop that offers the exchequer or food grains as the case may be. We have to realize that farm owner, worker, and migrant slave, every one strives physically, spends time, and intellectual ability to further the crop production. The group in *toto* slaves (works) to further the cause of crop production.

During nineteenth and early twentieth century, Caribbean region that grows sugarcane adapted and employed African slaves to mend the plantations. African migrant workers were the main source of labor in sugar plantations. Sugarcane was the major crop that attracted slave labor, but other species such as indigo, coffee, and rice also induced slavery. The slave population in the sugarcane prairies of Caribbean was initially high because it was a cash crop and earned good value upon export. Sugarcane plantations in Jamaica employed 60–150 slaves for each form. About 250 slave laborers were monitored by four or five farm managers (Fogel, 1994) During 1800s, mechanization, reservoirs built by irrigation engineers, and improved sugarcane crushing mills reduced drudgery and slave number. Further, slaves had migrated to other locations in North America and for different kind of jobs, instead of localizing in sugarcane farms. Abolition of slavery in Caribbean sugarcane farms reduced recruitment of African laborers. Here is again an example, where in farm managers, workers and imported migrant slaves, all toil hard to just sustain wet sugarcane prairies. It is sugarcane, after all that drives humans to slavery and its own perpetuation, in return for sugar.

During seventeenth and eighteenth century, it seems movement of Europeans into Brazilian farming zones affected the indigenous Indian tribes and their farming trends. Firstly, it uprooted them from original locations. The Indian tribes had to make way for larger companies. New cropping procedures made many of the natives seek employment in the farms and this induced slavery to farm masters, sometimes perpetually (Feltran-Barbieri, 2012).

Debt bondage and regular adoption of slavery during farming is in vogue in Indian subcontinent since several centuries. It occurs in several variations, but all are directed to develop, sustain, and cultivate crops in farms and accomplish crop-related activities. In many areas, enslavement in farms takes the form of caste hierarchy and economic impoverishment (See Cambell, 2008; Hjejle, B. 2011; and Knight, 2012).

Domestication of crops and invention of farming procedures led more humans to wean themselves from hunting/gathering system and become farmers. Since late Neolithic period, a large population in the prairies have become totally involved with crop production. Human population engaged in crop production increased progressively until recent history. Farming enslaved a large share of population for almost two millennia. Draught animals reduced human hardship in farm work to a certain extent. Then, mechanization was introduced. During past 150 years, farmers and farm worker population engaged in fields have progressively reduced due to variety of reasons. In particular, efficiency of farm operations has increased enormously during past century. Historically, efforts to reduce drudgery and slavish tendencies in farms began with invention of plow that could easily open furrows and allow farm labor to sow seeds. Invention of mechanical automotive tractors during mid-1800s immensely reduced farm work force required to toil in the field. Steam or petrol powered gadgets further reduced human energy requirement. It was farm engineers to the rescue. Almost every gadget or farm contraption powered by electricity or fuel oils has definitely reduced human labor hours in the farm. Many of the tasks have been accomplished with greater accuracy, better quality, and efficiency by using automated farm implements and gadgets. During recent past, chemical pesticides/herbicides have enormously helped in reducing labor/drudgery in farms. Large tractors with several tines, coulters, nozzles to apply liquid fertilizers, and pesticides/herbicides have reduced farm drudgery. Combine harvesters and many of the postharvest machines, after all, reduce slavish occupations. They reduce requirement of human labor in farm operations. As a proof, we should note that during eighteenth and nineteenth century, a large section of population (82–90%) in Great Plains of North America depended and was engaged in farming. Farm workers drawn from different regions were entirely occupied with crop production. However, during recent times, mechanization and automation has reduced farm laborers to <7.5 percent of total human population in USA (see Chapter 1). In the Pampas of

Argentina, initially, migration of farm workers into the plains changed the demography. However, with the inception of large mechanized farms, labor requirements per unit area decreased. Later, farm workers began to migrate away from farms to urban locations and farm worker number reduced from 30 to 10 percent of total population (see Chapter 3; Droller, 2013). Clearly, mechanization has offered relief to many farmers from drudgery and enslavement. Let us consider yet another example. In West African Sahel, human population is expected to cross 100 million by 2020. About 60 percent of it is deemed to be engaged in farm work (Jalloh et al., 2013a, b). Farm mechanization is low and manual drudgery is more. Since farm products offer 35 percent of region's GDP, reductions in farm labor population are not forecasted. Subsistence farming techniques may further make it difficult for farmers to introduce mechanization of any kind because of prohibitive cost. We may agree that reduction in farm workers is a common phenomenon across different prairies. Yet, in certain regions, as in India, farm dependent population is 50–60 percent and farm workers constitute 30–32 percent. Again, enhanced mechanization, automation, and use of chemicals may reduce drudgery and farm slavery.

During recent times, GPS guided precision farming techniques are becoming popular. Such satellite guided farming procedures may enormously reduce farm labor requirements. Literally, they lead us to robots/drones with computer aided decisions, electronic controls, and variable-rate applicators. They make development, maintenance, and harvest of agricultural prairies easier to accomplish. Farm operations such as seeding accurately, application of fertilizers, their combinations, pesticides, herbicides, harvesting, preparation of soil/yield maps could all be accomplished with perceptible reductions in farm labor requirements. Overall, introduction of GPS guided vehicles, soil fertility mapping systems, robotics, and electronically controlled fertilizer/chemical dispersers and irrigation systems seem to hold the key for an efficient agricultural prairie in future.

Some of the above examples depict the way agricultural prairies harness human energy and intellectual abilities to further their own cause. The above examples pertain mainly to human labor in the field and field-related activity. However, agricultural crops as a botanical species derive advantage from humans in many other ways. Aspects such as genetic improvement, seed production, dispersal, and preservation of specific germ plasm are some examples. Classical and Molecular genetic procedures are adopted just to see that crops thrive better, without being susceptible to pests/diseases or abiotic stresses. Their ability to produce foliage and seeds (propagules) is enhanced using genetic crossing and selection. These aspects involve human toil, in locations other than field but are related directly to perpetuation and productivity of agricultural prairies. In a way, crops covertly utilize human energy, intellect and skills to further their own cause on the surface of earth. Humans, help crops to overcome undue competition by other crops/vegetation, and to withstand interference by several biotic and abiotic factors.

8.2 AGRICULTURAL PRAIRIES OF THE WORLD: PRESENT SITUATION

Agricultural crops have originated in different locations across all continents. The center of origin and region of high genetic diversity occur in different regions of the world. These domesticated species have been introduced into several regions that are close-by or far away from center of origin. Crops' ability to adapt to climate, soils, and other natural factors have affected the spread of each crop. Human migration is among important reasons for spread of crops. At present, agricultural prairies established based on a crop occur in regions that are not necessarily the region of origin (Beddow et al., 2010). Following is a list of crops and top five nations where their prairies currently are highly productive and occupy large areas:

Wheat originated in West Asia, currently top five regions are in China, India, United States of America, Russian Federation and France. They contribute 51 percent of global wheat grain harvest and offer 310 m t grains annually.
Corn originated in Central Mexico; currently top five regions are in United States of America, China, Brazil, Mexico and Argentina. They contribute 523 m t maize grains annually equivalent to 71 percent of global maize grain harvest;
Rice originated Eastern India; at present top five regions that produce rice are China, India, Indonesia, Bangladesh and Vietnam. Together they contribute 457 m t rice grains annually equivalent to 71 percent of global rice harvest.
Barley originated in Ethiopia; at present it is grown more in Russian Federation, Germany, Canada, France and Turkey. Together they offer 58 m t grains annually equivalent to 42 percent of global barley grain harvest.
Soybean originated in Central China; currently top five nations that generate soybean grains are United States of America, Brazil, Argentina, China and India. Together they contribute 200 m t grains annually equivalent to 92 percent global soybean harvest.
Potatoes originated in South America. At present its cultivation is predominant in China, Russian Federation, India, Ukraine and United States of America. They contribute 171.5 m t tubers annually equivalent to 54 percent of global harvests.
Cotton originated in Central Mexico; currently top regions that cultivate cotton are in China, United States of America, India, Pakistan and Uzbekistan. They contribute 52 m t annually equivalent to 74 percent of global cotton harvest.

Source: Beddow (2010)
Note: At present, most crops form prairies and dominate landscape in regions entirely different and often far away from origin. Crops have been dispersed to areas where they adapt, flourish, and are needed. Human beings have of course played a major role in dispersal of these crops, also helped in intensifying their cultivation.

During past 5 decades, since 1960s, global cropland has increased steadily with regard to some categories such as cereals, oilseeds, and vegetables. Prairies based on pulses and roots crops have increased in area, but at a lower rate than cereals.

Prairies allocated to fiber crops like cotton has decreased marginally. Plantation vegetation that offers fruits has increased marginally (Table 8.1). Perhaps, even before domestication of field crops, Neolithic population had clear option to accentuate and propagate fruit trees in preference to other types of food sources. However, preference for horticultural crops was relatively feeble. Even now, only 4–7 percent (<25 m ha) of global agricultural cropping zone is covered by fruit crops.

TABLE 8.1   Status of agricultural prairies during past 5 decades (1961–2010)

	Area (in million ha)						
Year	**Cereals**	**Oilseeds**	**Pulses**	**Root Crops**	**Vegetables**	**Fiber**	**Fruit Trees**
1961	648	113	64	47	24	39	25
2010	701	250	73	55	53	36	47
Increase in Area (per decade)	10.6	26.4	1.8	1.6	5.8	-0.6	4.4
Share of Cropland (%)	58	21	6.0	4.5	4.3	3.0	4

Source: Beddow et al. (2010)

Note: Development of Agricultural Prairies, in other words, cropland began long ago since Neolithic period. The rate of expansion of crop area has been influenced by variety of natural and man-made factors. Whatever be the fluctuations in area recorded for each category of crops, currently, cereal-based prairies dominate the global landscape, followed by oilseed crops. Prairie vegetation as a source of food is highly preferred. Area occupied by prairie vegetation is lopsided and accounts for 95 percent of cropland. In comparison, fruit trees and plantations (excluding forests) that yield energy occupy almost negligible proportion of area on earth's surface. Fruit trees have not been preferred as a staple source of nourishment by human race—why?

8.3   AGRICULTURAL PRAIRIES: A COMPARATIVE STUDY

Human endeavor, ingenuity, and his agricultural acumen have allowed him to develop different types of agricultural prairies. Development of agricultural prairies has been influenced by geographic location, topography, soils, climate, crop species, infrastructure, human preference, and economic advantages. At times, demand for crop product has been enormous and it induced farmers in a location to develop larger expanse of a particular crop, or intensify it, so that it is exported to region where there is greater demand for it. For example, Soybean-based prairie in Cerrados has been developed by clearing vast stretches of natural shrub vegetation. Enlarged soybean prairie has resulted due to excessive demand for soybean and its products in Europe, Africa and China. Wheat/maize based prairie in Pampas is intensified to satisfy export demand for cereals to Eastern Europe and other places.

Currently, we encounter prairies that are large but thrive on soils with low fertility, meager water resources, and are subsistent in terms of seed and fertilizer inputs. Resultant grain/forage outputs may just satisfy human population *in situ.* There are agricultural prairies based on crops such as maize, wheat, legumes, and oilseeds that yield moderately. They thrive on fertile soils, yet are supplied periodically with fertilizers, herbicides, and water. Productivity of such prairies is commensurately moderate, but may offer a certain amount of surplus for farmers to sell and derive economic advantages. Economic advantage is among major factors that drive farmers to intensify the agricultural prairies. Few of the examples are Wheat in Northeast China, Rice in Southeast Coast of India, Corn Belt of USA. These prairies are supplied high quantities of fertilizers, water, and pesticide. Resultant grain/forage output is commensurately high. Yet, we should note that potential grain/forage is higher than current productivity and it leaves option to intensify the prairies further. As stated earlier, human ingenuity has played a key role in developing agricultural prairies of different intensities. Let us compare them with regard to geographic location, topography, soil type, manure supply, precipitation patter and irrigation, yield goals, and productivity.

Agricultural Prairies maintained at low intensity that offer proportionately low grain/forage productivity are well distributed, across different agrarian zones of the world. Farmers supply low levels of input, if any, especially fertilizers and pesticides, but prefer to select crop species/genotypes that tolerate low or subsistent levels of inputs. Farmers reap harvests that are low. Major factors that drive farmers to develop such low productivity prairies are inadequate natural resources. Hence, farmers try to match crops and yield goals appropriately. Here, soils are infertile. Nutrient availability to roots is restricted. Soil moisture at critical stages of crop growth is inadequate. Economic returns are poor. A few examples of agricultural prairies that thrive as low productivity biomes are Pearl millet/cowpea farming zones in Sahelian regions of Niger, Burkina Faso, and Senegal; Groundnut in Senegal's "Groundnut Basin"; Wheat and Barley in Dry steppes of Southern Russia; Wheat and Barley in Syria, Iran, and Iraq; Finger millet and legume intercrops in South Indian rain-fed drylands; Sorghum in southern African drylands; Wheat/sorghum in dry regions of Pampas; Groundnut in Peninsular Indian drylands; and Pearl millet in Northwest India, and so on.

Low intensity agricultural prairies are distributed in all continents. They are traced in plains, undulated stretches, and coastal zones. As stated earlier, they are often localized to areas with poor quality soils. Soils with inappropriate texture, structure, and nutrient availabilities are utilized. Precipitation is generally inadequate and not well distributed during crop season. Cropping season is restricted due to short rainfall period, frost, cold climate, or dry season with high temperature and drought. Fertilizer inputs are low ranging from nil to 20 kg N. It is crop residue recycling that sustains fertility and organic fraction of soil. Nutrient cycles in such prairies operate at low rates. Productivity of grain/forage is low and usually <1.0 t grain per hectare plus 3–4 t forage per hectare (Table 8.2). Intensification of such agricultural prairies is possible through removal of constraints, for example improving water resources, application of fertilizers, use of improved cultivars, and so on.

Moderately intense agricultural prairies are again traceable in different continents. Agricultural prairies based on Maize/Soybean in America are supplied with fertilizers (80–100 kg N, 60 kg P, and 100 kg K), organic manures (10–15 t per hectare FYM), irrigated, and sprayed with pesticides. Farmers aim at higher grain/forage yield ranging from 4 to 5 t maize grain per hectare plus 7–8 t forages. Soybean produces 2–3 t grains per hectare plus 5 t forage per hectare. Nutrient recycling in the ecosystem is commensurate (Table 8.2). Soil deterioration due to exhaustion of nutrients, erosion, gaseous emissions and net removal of biomass/nutrients via grains/forage needs periodic corrections.

Intensely cultivated agricultural prairies occur in geographic locations that possess apt soils of high fertility, congenial weather pattern, and optimum water resources for production of crops. Intensification of crop production is usually effected using high quantity of fertilizer timed shrewdly by splitting the dosages. Fields are irrigated more frequently. Crop genotypes with high grain/forage yield potential are only selected. Pesticide and herbicide needs are greater. Soil deterioration due to incessant cropping, gaseous emissions of CO_2, CH_4, N_2O, and NO_2, soil erosion, removal of nutrients through grain/forage is common. Grain productivity in case of maize ranges from 7 to 11 t ha^{-1}, rice 8–10 t ha^{-1}, wheat 6–7 t ha^{-1} sorghum 5 t ha^{-1}, soybean 3.2 t ha^{-1}, and so on. (Table 8.2).

TABLE 8.2 Comparison of agricultural prairies

Agricultural Prairie	Crop Duration	Nutrient Supply	Water Requirement	Grain Yield	
	days	kg ha−1	mm season−1	kg ha−1	(kg ha−1 day−1)
Pearl millet in sub-Sahara1	80–100	20–40 N	450	800–1200	8–12
Groundnut Basin of Senegal2	140	nil—40 N	550–600	900–1200	8–10
Wheat in Ukraine/Russian Steppes3	140	nil–80 N, 70 P, 70 K	450–550	1500–2500	10–16
Finger millet/legume in South India4	110–130	20 N, 8 P, 20 K	350–550	1330–2225	13–22
Wheat in dryland Prairies of West Asia7	130–140	40 N, 20 P, nil	450–500	1550–2400	12–18

TABLE 8.2 *(Continued)*

Agricultural Prairie	Crop Duration	Nutrient Supply	Water Requirement	Grain Yield	
	days	kg ha−1	mm season−1	kg ha−1	(kg ha−1 day−1)
Wheat belt of Argentina5	130–140	40 N, 20 P, nil	550–600	3500–4400	26–31
Soybean in Cerrados of Brazil6	130	nil, 80 P, 90 K	650–700	2650–3200	22–24
Intensive Rice belt in South India8	120–130	180 N, 105 P, 90 K	1200–2400	4200–6500	34–52
Maize in Northeast China's Steppes9	140	240 N, 60 P, 120 K	550–900	7000–8,300	48–60
Corn Belt of North America10	130–140	180 N, 60 P, 100 K	700	6500–7500	55–62

Source: [1]Buerkert et al. (2000, 2002); Bationo et al. (2003, 2004); [2]Coura Badiane (2001); Krishna (2008); [3]FAO (2005); [4]Krishna (2010b); Dubey and Shrivas (1999); Selvi et al. (2004); [5]Bono et al. (2004); Alvarez et al. (2009); [6]Benites (2010); Franca et al. (2010); FAO (2008); [7]Krishna (2003); [8]Janaiah (2003); Parihar et al. (1999); Pandey et al. (2008); Cassman, (1999); [9]Carberry et al. (2013); Liang et al. (2011); [10]Duvick and Cassman (1999); Egli (2008).

Note: Grain yield calculated is based on one crop season. For many prairies, there may not be a second season, and crops are not on the field all year round. Hence, we should not assume that some of these prairies will fetch grain yield at the same rate for the whole year. In comparison, a fruit tree crop whose productivity is measured per day is for the whole year.

Let us consider a concept known as agrosphere that has relevance to crop production systems. Agrosphere is a conglomerate of agricultural biomes, including prairies on earth (Krishna, 2003). It excludes natural vegetation of forests, prairies, swamps, and wastelands. Agrosphere is exclusively concerned with the way we develop, mend, sustain, and harvest agriculture crops on earth. Agrosphere extends from depths of crop root tips in the soil phase till the tip of above-ground shoot system of the crop. The Agrosphere thickness varies with the crop and its genotype cultivated in a location. Therefore, agrosphere thickness on earth surface fluctuates enormously. Agrosphere productivity also differs based on crop, environment, and management factors.

A Fruit Tree	Maize and Rice
Above-Ground Canopy: Tree height 7–10 m	Above-Ground Canopy: Cereal crop height 1 m (rice)—3 m (maize)
Soil Phase: Rooting depth 5–8 m	Soil Phase: Rooting depth 1 m (rice)—1.5 m (maize)
Agrosphere Height: 12–18 m	Agrosphere Height: 2–4.5 m

FIGURE 8.1 Agrosphere in regions with fruit tree crops versus those supporting semidwarfed cereals (rice/wheat) or taller prairies of maize.

Note: Agrosphere thickness is equal to space occupied by vegetation from root tips to above-ground foliage/canopy tip. Larger and thicker (height) canopy of a fruit tree crop with profuse foliage allows better photosynthetic ability, biomass production, fruit formation, and residue recycling. Deeper well distributed roots help in exploring greater volume of soil. It helps in absorption of higher quantity of nutrients at higher rates. In comparison, short canopy despite high number of tillers forms relatively lower levels of biomass/grains. Further, cereal roots are shallower and nutrient scavenging ability is relatively restricted.

The agrosphere is restricted when we grow genetically dwarfed cereals on a large scale. Such a biome with cereal roots and above-ground stem, foliage and panicles put together may extend for 1.0–1.5 m into soil and at best 1.3–3 m in height with stem and foliage. Even tall landraces of sorghum or pearl millet grown in Africa do not cross a height barrier of 5–7 meters. Essentially, we are curtailing the photosynthetic zone (height) and nutrient exploration in soil (rooting depth) (Figure 8.1). Now, the trend to curtail the height of cereal hybrids and varieties further to 1.5 m or less despite increased tillering does not offer higher photosynthetic area. Potential light interception gets curtailed. Diffused light is lessened to a great extent because canopy height is itself shortened in a genetically dwarfed cereal. A fruit tree crop, on the contrary, has foliage extending to 6–8 m height at the least and explores soil to more than 5 m depth. Annual accumulation of biomass and edible fruits is much higher than best cereals that make up the prairie vegetation. Even if total produce for all three seasons possible in many of the tropical regions are considered, best total biomass accumulation by prairies fall short of a season's harvest of tree crop harvest. These are limitations of agricultural prairies that have been overlooked since long.

Let us consider root systems of different Prairie crops. The nature of crop roots, their architecture, spread in soil phase, depth to which they penetrate, soil volume they explore, extent of secondary and tertiary roots, and physiological ability of roots decide quantity and rate at which soil moisture and dissolved nutrients are explored and absorbed by the crop. A crop genotype with large and efficient root system is advantageous. Similarly, a tree crop with deep root system that is well spread is better, since it explores larger soil volume and absorbs greater quantity of mineral nutrients. Let us consider a few examples: in case of maize, root growth is confined to 0–90 cm depth. Water and nutrient recovery by maize is greatest from top layers of soil phase of agrosphere. First 0–30 cm explored by roots provides 40 percent of moisture. Next, 30–60 cm depth offers 30 percent moisture, 60–90 cm depth provides 20 percent and rest 10 percent of moisture and nutrients is derived from depths beyond 90 cm. Root distribution beyond 90 cm is feeble. Therefore, agrosphere in case of maize is confined well within top 1.0 m of soil phase (Sharp and Davies, 1985; Nakamoto et al., 1992; Hsiao, 2012a; and Krishna, 2013). In other words, maize-based prairies are efficient in exploiting soil only to a depth of 1.0 m or less.

Wheat roots are fibrous and localize in shallow upper layers of soil phase of ecosystem. They penetrate and explore up to a depth of 90–120 cm depending on soil type, nutrient, and moisture distribution (Hsiao, 2012b). Tall land races may produce roots that penetrate slightly deeper layers of soil up to 1.2–1.5 m depth (Hsiao, 2012a, b). Rice roots are localized at upper layers of soil up to 1.0 m (Hsiao, 2012c). These roots are able to thrive in the submerged condition. Rice crop requires higher quantity of water about 900–2200 mm per season to mature. Its water use efficiency and even energy efficiency is much lower than drought-tolerant dryland crops like sorghum or pearl millet. It is glaring; we ought to note that water used for rice crop per season suffices to produce arable cereals (sorghum, maize, finger millet) for two or three seasons. Further, rice genotypes currently in vogue around the world are semidwarfs with restricted height. Its ability to utilize agrosphere height is restricted. These rice genotypes tiller profusely, produce many leaves and their photosynthetic ability in relation to land races or many other previously cultivated genotypes is high. Yet, utilization efficiency of sunlight is curtailed because of shortened height. Agrosphere thickness in rice production regions is relatively smaller than tall prairies formed by sorghum, maize, or pearl millet and much small compared to a fruit tree crop. Production potential of rice hybrids grown in experimental fields are 10–12 t grain per hectare plus 12–15 t forage per hectare.

Groundnut, an important oilseed crop develops into short statured prairies. The leguminous roots are however deep and penetrate the soil phase up to 1.5–2.0 m depth (Shorter and Patanathoi, 1997; Krishna, 2008). Groundnut foliage is profuse. Its biomass accumulation and pod formation is relatively efficient. It is capable of improving soil-N status through symbiosis with Bradyrhizobium.

Soybean also forms a short prairie. Soybean roots reach a depth of 1.5 m (Hsiao, 2012d). They form symbiotic association with Bradyrhizobium that have the ability to fix atmospheric-N. Soybean plant branches and produces profuse foliage. Biomass productivity ranges from 1.2 t grain per hectare to 3.2 t grain per hectare plus forage 3–5 t ha^{-1} (USDA-ERS, 2012). The agrosphere height which is relevant to nutrient acquisition and photosynthetic carbon fixation is restricted compared to tall cereal/leguminous prairies.

Pigeonpea is a staple pulse crop in Peninsular India. Its rooting depth varies with genotype. Perennial pigeonpea grows to 5 m in height. Roots are profuse and reach deeper layers of soil, up to 4 m. The vast expanses of pigeonpea in the Vertisol belt is composed of medium maturity short statured genotypes. Rooting depth of roots is 2–3 m. Pigeonpea explores greater depth of soil phase of the agrosphere compared to other small sized legumes such as horse gram or cowpea (Anandraj, 2000; ICRISAT, 2007; Krishna, 2010a; 2014d). Productivity of pigeonpea is 462 kg ha^{-1} in Myanmar, world average is 689 kg ha^{-1} and in West Indies it is 1.5 t grain per hectare (FAOSTAT, 2008).

Cotton is a deep-rooted prairie crop. Its roots penetrate 1.5–2.0 m in depth and are efficient in scavenging water and nutrients. Cotton forms canopy with profuse leaves. Therefore, photosynthetic area is also higher than many other short statured prairies. Sugarcane is a tall prairie crop with deep roots. It explores soil and absorbs water/nutrients at much higher rate than smaller sized cereal genotypes. Sugar cane forms a tall prairie at 3–5 m height and photosynthetic surface (foliage) is greater compared to short statured prairies.

Let us consider a few perennial fruit tree crops. Trees are taller by 3–4 m over field crops. They branch and form leaves profusely. Photosynthetic radiation interception and biomass formation rates are significantly high compared to annual cereal based prairies (Figure 8.1). Fruit crops maintained on fertile soils with regular fertilizer and water supply offer over 15–60 t fruits per year^{-1}. Tree crops scavenge nutrients from 0 to 5 m depth of soil profile. However, roots that absorb nutrient/water may localize at certain depths in some species. Tree roots generally absorb water/nutrients from great depths of up to 3–4 m. Trees recycle nutrients efficiently when leaves senesce. Trees also allow development of understory crops that enhance land use efficiency, photosynthetic efficiency and nutrient recovery rates enormously compared to a short statured cereal. In the normal course, biomass, and fruit productivity of tree crops is greater than a cereal/legume crop. Fruit crops such as grapes grown at higher planting density offer 15–30 t fruits per hectare per season! (Nilnond, 1998; Shikamany, 2010). Fertilizer usage per year is lower than a set of two cereal/legume crop seasons. For example, fruit tree such as grape is supplied with proportionately high quantity of nutrients that gets converted. In two seasons, rice crop that offers 4.2 t grains/forage per season may consume 120–160 kg N, 60 kg P, and 120 kg K ha^{-1} season^{-1}. Arable sorghum is provided with 80–120 kg N, 40–60 kg P, and 80–100 kg K ha^{-1} each season and it offers only 3.5–5.0 t grains ha^{-1} plus forage. Maize grown in the "Corn Belt of USA" consumes over 160 kg N, 80 kg P, and 120 k ha^{-1} season^{-1} and offers 9–11 t grain ha^{-1} plus forage (See Krishna, 2012; IPNI, 2008). However, its average productivity in many regions of the world is less than 4.7 t grain ha^{-1} (USDA-NAGS, 2013). Mango orchards regularly offer 10–15 t fruits plus foliage per hectare, depending on planting density (FAOSTAT, 2011). Farmers in low input, dry belts of South Asia harvest 5.2–12.7 t mango fruits ha^{-1} (ICAR, 2002; Deng and Janssens, 2004; FAOSTAT, 2004; NHB, 2009). The average productivity of citrus orchards in Florida (USDA-NAGS, 2013), Brazil and China is 30–40 t fruits ha^{-1} plus foliage that recycles organic matter/nutrients (FAO-STAT, 2010a). This is comparatively way too higher than best annual grain yield from a cereal prairie.

The cereal crop is endowed with relatively shallow roots system. Its ability to scavenge inherent soil nutrients and that applied as fertilizers is restricted to upper horizon of soil. The loss of nutrients to lower layers of soil profile, and possible groundwater contamination is high. Fertilizers unused and held in soil profile is also vulnerable to loss via emissions, percolation seepage, and so on. Soil nutrients held unused has to be efficiently scavenged by the succeeding crop and within stipulated period. Now, compare it with a tree crop that can extract soil nutrients from greater depth and volume of soil. It offers very little unused fertilizer-based nutrients to be lost via emissions. It adds to biomass of lower layers of soil, when the deeper roots decompose. The carbon sequestration component of a tree plantation is surely several times more than a cereal crop. The carbon sequestration is higher because of foliage that senesce, foliage/twigs that are recycled and periodic pruning. Root biomass of a tree crop that accumulates each season and adds to soil-C is several times greater than that derived from cereal crop (See Beyer et al., 2002; Krishna, 2002).

The development of understory crops is a great advantage in horticultural locations. Land use efficiency is increased. It allows for greater biomass formation, grain/forage production, and C sequestration in the multiple cropping systems. Tree crops may also negotiate drought better compared to cereal. Tree's ability to store more of moisture in its tissues, ability to translocate water in times of scarcity helps it to survive the drought period. While a frail cereal may wilt at the same intensity of drought. Overall, we may have to compare several aspects of cereal prairies and fruit tree crops and count on the advantages. A monotonous low productivity cereal prairie may not be worthwhile option in many regions of earth. A multiple cropping system with fruit trees, cereal/legume understory that is rotated periodically seems a better option. Productivity of multiple cropping systems could be 4 to 5-folds greater than a cereal/legume prairie. A thicker agrosphere with trees and understory vegetation is more efficient in land use compared to cereal (dwarfs/semidwarfs) based prairie.

What did we gain by preferentially accentuating short statured cotton, instead of perennial short tree cotton? Why did we go far short statured genotypes of staple cereals like wheat and rice to spread all over the agricultural belts? Why did we adopt short season, dwarf statured pigeonpea hybrids Why did we accentuate short-statured annual crops for oilseed production (e.g., groundnut, sunflower, brassicas, sesame, etc.)? There are perennial oil sources like palms, olives, jujube, and so on. There are many tree crops that could be used as protein sources, instead of excessively depending on short statured annual legume-based prairies. At the bottom line, a tree crop accumulates greater amount of biomass and sequesters C more than most short prairies.

8.4   WHEAT, RICE, AND MAIZE—"THREE MASTERS"

A few edible crops that generate highly palatable food grains and vegetables literally dictate terms with other crop species in terms of their ability to spread on the surface of earth. Human preference, palatability, and few other advantages of cultivating these crops seems the crux of the situation. Crops, in this case literally take help from human species for their dispersal, spread, development, and perpetuation on earth. There are innumerable examples that suggest that pioneer plant population and well established natural flora has been just removed callously and in its place crop—based prairies have sprung up in no time. Such activities have been conspicuous since historical times. Removal of naturally acclimatized species and replacement with crop species was rampant in spurts depending on human needs. Citadels with higher human population and well diversified activities required expansion of cropping zone. During more recent times, removal of natural vegetation in Cerrados and Amazonia during 1970s, to plant the cleared stretches with soybean/maize is a good example for how prairie crops take human support to spread and perpetuate.

Covertly, these few crop species are also inducing man to look after them and their well-being in almost every corner of the earth. This seems to be a nature's mission that began some 10,000 years ago and that continues with great gusto—we have no idea where it leads to! A covert and seemingly natural process that makes agricultural crops overcome competition from other botanical species, so that they spread and perpetuate with relative ease and have a stringent control over food and energy supply to humans and several domesticated animal species. In other words, agricultural grain crops—cereals, legumes and other groups have literally dominated the global land mass through a partnership with human kind. Today there seems to be very few alternatives to choose.

Man seems to be working incessantly for over 9000 years on few botanical species for his sustenance-why did he do so? Mind you!, there are merely three or four grass species and their genotypes plus a few legumes that have garnered most land mass and cropping zones on the surface of the earth—Rice, Maize, and Wheat plus few legumes (See Wood et al., 2000; WDI, 2013; FAOSTAT, 2013; USDA-ERS,

2013; Figure 8.2). Human preferences like palate, ease of processing, and knowledge of production techniques plus the ever troubling economy seem to have played key roles in generation and spread of agricultural prairies based on less than 10 crop species. To a great extent availability of genetic diversity in crops and even assortment of natural resources seem to have been overlooked during evolution of agricultural prairies.

The three major cereals and select legumes seem to induce humans to prefer them excessively so that they could overcome any competition from other species in the wide open natural setting. They literally fight other botanical species using human kind as a proxy. In simple terms, area of cultivation and seed production of many a domesticated crop species is improved preferentially by human race. Several of the major obstacles to spread and perpetuation of crops, that otherwise affect in natural settings are removed when these crop species are produced in fields. Excessive focus on few species or their genotypes has of course reduced chances of several genotypes of different crop that may have otherwise continued to perpetuate in the open. Human selection pressure for specific gene sets of course has influenced genetic diversity.

FIGURE 8.2   Human evolution and cereals—maize, wheat and rice.
Note: Did human evolution that took several thousand years finally lead him to utter dependence on three major cereals and few legumes on earth. Was he covertly utilized to spread maize, wheat, and rice germ plasm and develop their prairies in every possible nook and corner. Why is he spending so much of his time, ingenuity, and resources just to see these few crops flourish in the name of serving himself with food?

The three master cereals, namely maize, wheat, and rice occupy a very large share of agricultural land on earth. They together occupied over 555 m ha of cultivable land on earth to contribute 2176 m t grains (FAO STAT, 2013; USDA-ERS, 2013). Maize, wheat, and rice together cover up 78 percent of cereal cropping zone on earth. They attract and enslave a large population of farm workers in each continent. We have no definite reason to explain why these three cereal species drives human species located in different geographical conditions and variable human dietary practices to still depend immensely on them. To quote an example, it seems during 1985–1999, more of agricultural land in Northeast China was brought under maize, rice, and wheat in preference to others. Cereal production zones were extending even into hitherto uncongenial northern region. The three cereals occupy 82 percent of cropping zone available in Northeast China (Motoki, 2002).

Wheat based prairies are distributed all over the earth between 50°N and 50°S latitude. Wheat is a versatile crop. Wheat genotypes show high adaptability to soil and agroclimatic variations that occur in different continents. Man's preference to cultivate and consume wheat is matching. It is so great that every day, wheat crop is sown and harvested in one or the other region on the surface of earth. Seasonal variations do not matter. Wheat-based prairies are conspicuous in China, India, USA, Russia, and France. It masters the human diet in temperate world. Globally, wheat prairie spreads into 225 m ha. Human beings depend immensely on wheat products. We consume about 681 m t grains each year (FAOSTAT, 2013; OKState, 2013; and USDA-ERS, 2013). Human population resident in wheat belts have consistently toiled to perpetuate its genotypes. Wheat productivity varies. Global average productivity is 3.2 t grain ha^{-1}. Farmers who strive hard in the intensive cropping zones have achieved up to 8.9 t grain ha^{-1}. Wheat farms worldwide employ workers at high rates to accomplish various agronomic procedures. The share of humans directly dependent on wheat farming and related activities that involve processing, transport marketing is quite high in different regions.

Rice is the dominant cereal in the wet prairie regions of Asia, parts of West Africa, and a few other locations such as Southern USA. It covers over 161 million ha of farm land. The greatest concentration of wet prairies filled with rice occurs in China, India, and other Southeast Asian regions. Global average productivity is 4.2 t grain ha^{-1} plus forage. Human population harvests and consumes about 678 m t of rice grains each year (FAOSTAT, 2010b; OKState, 2013). Hence, dependence on rice prairies is high. In the tropical Asia, rice crop masters the human dietary preferences. Rice grains are staple source of carbohydrates for over 2 billion human beings, most of whom are located in Southeast Asia. The wetland rice zones are generally intensively cultivated using high fertilizer inputs (120–180 kg N, 40–60 P$_2$O$_5$, 120 kg K K$_2$O ha^{-1}). Rice is among inefficient users of water resources. Irrigation needs are high at 900–2,200 mm compared to 450–500 mm for a dryland sorghum crop in the same region. Farm operations in wetlands such as nursery development, transplanting, weeding, fertilizer supply, harvesting, postharvest processing, and marketing are labor intensive. Rice farms do enslave a sizeable portion rural human population.

Maize is a versatile crop species that has been perpetuated with great care by human beings. Again, maize dominates agricultural landscapes, rather intensely in some parts of the globe (e.g., Corn Belt of USA; Northeast Chinese Prairies) (See Krishna, 2012). Maize spreads into all agrarian zones between 50°N and 50°S latitude. It occupies 167 m ha and offers 817 m t grains annually at an average productivity of 5.12 t grain ha^{-1}. Intensively cultivated zones produce 11.3 t grain ha^{-1}. Maize crop masters the human diets in Americas, Africa, and Asia. Human beings situated throughout different continents depend on maize grains immensely. Maize prairie consumes a large farm work force and drives them to dependence.

Let us consider a different view of our agricultural prowess. We realize that over billion people engaged in agricultural farm work are toiling hard on few cereal species, especially three master cereals. Human beings are perpetuating these three botanical species in the entire agrarian landscape. It is a great bargain that operates in nature. Food grains (a few morsels of food) for each human being (farm labor) ensures development, perpetuation, and preservation of cereal germ plasm worldwide. Now, who is the master, prairie crops or human beings?

Humans nurture the prairies of these three crops with great interest, care, and costs in terms of energy, time, and physical inputs (fertilizers, water). Since, these three cereals and other food crops occupy a very large expanse of possible cultivable area on earth; errors in judgment could have devastating consequences. Farmers tend to spread crops, genotypes, and new agronomic procedures rather rapidly. Therefore any advantage or error/loss is quickly felt by the prairies and human beings. At times, human dependence on maize, wheat, and rice may become obligate on performance of their prairies. It is not a good situation to forecast. There are an assortment of other cereals, millets, and tubers to support human diets that need accentuation in the prairies. A change in food habit to say other cereals, millets, tubers, fruits, and a good extension service advising people to relish other crops seem to hold the key. Instead, we are placing great efforts on genetic manipulation, selecting the same cereal species for adaptation into diverse agroclimates. We are also expanding and intensifying the same botanical species using higher quantity of fertilizers and water, and so on. This trends adds monotony of vegetation, induces greater susceptibility to pests, diseases, and related disasters. A crop like rice, a vast monotonous expanse of wetland is among high greenhouse gas emitting species. It affects soil structure and microbial flora rather immensely each season. Fauna too may get affected if they have to encounter monotonous vegetation all through the ecosystem. We may ultimately develop an "Agrosphere" that is monotonously filled with just semi dwarfs and hybrids of three or few crops and that is highly vulnerable to massive disasters. In terms of grain/forage productivity, it seems for now, that best potential of these cereals still do not match an assortment of fruit tree crop, an understory cereal plus and/or floor species. Cereal genotypes with dwarfing genes were opted in 1970s. This further reduced the canopy size (height) and photosynthetic potential of the agrosphere. Further, crop's smaller root systems exploited limited volume of soil. It seems such cereal genotypes were amenable for farm operations. Instead of devising machinery to suit the crop, cereal crops were manipulated genetically, with tedious procedures. It continues even now with tedious molecular procedures.

8.5 FUTURE COURSE OF AGRICULTURAL PRAIRIES

Agricultural Prairies have appeared and flourished on Earth's surface since early Neolithic period. Their initiation, development, sustenance, productivity, and perpetuation are predominantly a human effort. Agricultural prairies are constantly in interaction with nature and its several variations. They occur as small or large stretches of crops in different continents. They negotiate varied weather patterns. Agricultural prairies thrive luxuriantly in diverse soil types, with a wide range of fertility levels and constraints. The grain/forage productivity or any other product for which the prairies are raised is an important factor for all human beings. Farmers try to invest resources, time, capital, and adopt series of agronomic procedures based on geographic locations, product, and economic advantages that they reap.

Agricultural prairies have come to stay. Thus far, there are no immediate alternatives to food grains generated by these prairies, especially, the large quantity of grains that they supply to humans on earth each season and year after year. Right now, Agricultural prairies seem to flourish in every nook and corner of the globe between latitudes 50°S–50°N. The intensity and food grain/forage generated varies depending on geographic aspects, soil fertility, agroclimatic parameters, and inputs. Agricultural prairies are obligate necessities for human nourishment. Their performance now and in future, with regard to food grains and other commodities of value to humans is a matter of great concern to all agriculturists, policy makers and the general populace too. Prairies that thrive on adverse soils and agroclimate, as in Sahel or other arid zones may offer less than sufficient food grains/forage per hectare. Intensification of such low productivity Agricultural prairies is needed. However, it requires higher supply of soil nutrients, water and crop management-related inputs. Moderately intense prairies are wide spread. Forecasts suggest their expansion and intensification where ever possible. High-intensity crop production zones occur in large patches in different continents. For example, Corn Belt in USA or Rice belts in Southeast Asia may be consistently held so, or even further intensified using fertilizers and/or water. The grain yield potential of many crop species is much higher than present levels. Yield gaps could be reduced in future. Regular farming procedures in any region is accompanied with gradual loss of fertility, soil quality, exhaustion of water resources, appearance of disease/pests, and a general decline in agroenvironment. This aspect has to be corrected timely by farmers.

Agricultural Prairies are dynamic in terms of crop species. Maize is gaining in area because of its superior grain/forage productivity. Similarly, rice too is preferred where ever feasible. When maize replaces a cereal such as sorghum, productivity, nutrient cycling, and C sequestration trends are marginally positively enhanced. Whereas when rice replaces an arable maize or sorghum, water requirements double. Low land rice emits greenhouse gasses at higher rates. Such wetland rice prairies are widely distributed all across South and Southeast Asia. Consider a replacement of rice with arable cereal legume like wheat/sorghum/maize/ cowpea/soybean or pigeonpea. Two crops of arable cereal can be managed with same quantity of water.

As stated earlier, rice is relatively inefficient and needs 900–2,200 mm, but arable cereals need 550–600 mm water season^{-1}. Irrigation requirements get halved if arable crops are preferred in place of rice. Fertilizer-based nutrient use efficiency too is better with maize/sorghum genotypes grown in Southeast Asia. In future, arable crops could be retained in preference to wetland rice. Perhaps, it is worthwhile to change food habit to maize/sorghum flour compared to rice grains/flour. Water efficient rice genotypes or drought-tolerant versions of rice are not easy to generate. It could be tedious, difficult, and not possible at all in some cases because of complex physiological genetics. Upland paddy is a clear alternative to improve water use efficiency. Perhaps we can adapt to palatability and amend our requirements much easily, than to toil hard on the genetics of crop. Agricultural extension agencies hold the key. They have to advise that it is helpful to feed on dryland crops than rice.

Since, 1970s, short stature rice/wheat prairies have dominated agricultural landscape all over the world. Such semidwarfs, with erect stem, higher tiller number, panicle, and seed number/weight were rapidly accepted by farmers. They were fertilizer responsive genotypes with better partitioning (harvest index) and importantly were amenable to combine harvesters, when grown in large fields/vast expanses. However, we should note that shortening cereal genotype reduced the photosynthetic height of Agrosphere. Genetic selection for tallness, high number of tiller, and leaves could have led us to higher C-fixation and biomass. Hopefully, in future, taller cereals with thick stem (nonlodging), better foliage, and high grain productivity are also preferred more frequently than at present. During past 5 decades, spread of dwarf and semidwarf cereals has reduced potential photosynthetic height of the agrosphere. Carbon sequestration in nature has been proportionately curtailed. We have to grow taller prairies with better foliage and photosynthetic ability time^{-1} area^{-1} to enhance biomass.

We may note that short statured rice hybrids potentially yield not beyond 9–12 t grain plus 8–10 t forage ha^{-1}. Similarly, maize hybrids yield 9–11 t gains plus 12 t forage ha^{-1}. Semidwarf wheat offers 7–8 t grain plus 5–8 t forage ha^{-1}. Tall prairies of sugarcane offer 40–50 t cane ha^{-1} year^{-1}. Tree crops are efficient in biomass and fruit formation. Depending on planting density, they provide 15–30 t fruits ha^{-1} plus crop residue. Cereals could be interspersed or strip cropped with fruit trees. It results in higher land use efficiency. Nutrient scavenging could also be higher. Soil nutrients vulnerable to loss via seepage, emissions and drainage could be reduced.

Land use change from short statured cereal/legume genotypes to fruit tree crops should be considered. Agricultural prairies could be curtailed. Their spread is already lopsided. The dependence on food grains from short prairies of cereals/legume is very high. It needs to be regulated. Fruits could be used to satisfy human nutritional requirements. It requires a change in food habit which seems easier. Perhaps a large expanse of fruit tree plantation could have negotiated dust bowls, droughts, floods and climate change a bit better than the frail cereal/legume prairie. It is a debatable point that could be studied using simulations and actual field situations. In the drought prone areas (e.g., Sahel) often trees survive the rigor better.

Several aspects such as effect of thunderstorms, unfavorable precipitation pattern, cold front, droughts, and pests need to be weighed carefully between cereal/legume prairie and tree plantations. Do we have to propagate agricultural prairies beyond thresholds into areas with an accumulation of unfavorable conditions? We have already done so to a certain extent. Short statured cereal prairies face many disadvantageous compared to a fruit tree crop. Adaptation to fruits and their variations in taste is important. Agricultural extension agencies professing cultivation of trees, in strips, or allocating larger share of area to tree crops in a farm is a worthwhile option. It allows us to fix more biomass per unit time and area and helps in harvesting more of carbohydrates, proteins and minerals in the fruits. We may derive better or equal benefits by adopting a mixture of Agriculture prairies plus much larger fruit tree vegetation.

Agricultural prairies, in the near future could be entirely managed using electronically controlled tractors, implements, machines, seeders, sprayers, harvesters, and processers. Right now, most combine harvesters have electronic controls and are GPS guided. Farm labor requirement is expected to be reduced enormously. Farm operation could be accomplished with greater versatility and accuracy. Soil fertility maps and GPS-guided variable dispensers, make it easy to regulate crop production intensity, fertilizer supply, and yield goals.

Our ability to compute the influence of various factors and variations and arrive at forecasts was limited. However, currently, there are innumerable computer simulations that could help us contemplate changes in land use, cropping systems and yield goals better. We can apply computers to judge and forecast effects of various environmental changes more accurately. Computer simulations could be effectively deployed to review and re-review our decisions regarding crops, fertilizers, cropping systems, and production goals. Intensification of agricultural prairies with due consideration to environmental concerns seems a good option.

Why not alter composition of agricultural prairies a bit drastically? Agricultural Prairies mixed with Tree Plantations seems plausible. Multistoried cropping systems are better in terms of land use, nutrient and irrigation use efficiency. Strip cropping of fruit crops with cereals could be useful. We have to modify food habits to include more fruits. Right now, it is monotony of single or few cereal crops. We ought to realize that short prairies developed since past 5 decades have affected plant genetic diversity beyond threshold in many locations. Genetic diversity stays inside cold stores of different agricultural agencies—if not used effectively.

Over all, agricultural prairies on the surface of earth have spread into vast expanses of food generating systems. The spread and intensification of Agricultural Prairies needs to be regulated effectively without untoward influence on soil, environment and crop productivity. Human food habits especially his ability to opt out of excessively small number of cereals is crucial. A highly narrowed food grain preference could potentially hold humans at ransom. A disastrous season can affect food security. An assortment of tree crops and cereal/legumes plus other useful species seems better for future. Agricultural drudgery is an aspect that needs greater attention. Precision farming, GPS-guided agricultural drones, computers, and satellite guided crop production trends hold the key for an efficient food generating system in the future. Agricultural drones, robots, computer guided GPS vehicles/variable applicators, and electronically controlled harvest processors should take over and dominate farming in the prairies and plantations in future.

KEYWORDS

- **Canadian Prairies**
- **Famine**
- **Groundnut Basin**
- **Indo-Gangetic plains**
- **Sorghum**

REFERENCES

1. Africans in America Resource Bank. Conditions of Antebellum Slavery: Judgement day-pat 4. **2013**, 1–7, http://www.pbs.org/wgbh/aia/part4/4pc2956.html
2. Al-Issa, Y.; Lentils and Chickpea. Syria: National Agricultural Policy Centre; Commodity Brief No. 7, **2006**, 1–12.
3. Alvarez, C. R.; Taboada, M. A.; Gutlerrez, F. H.; Bono, A.; Fernandez, P. L.; and Prystpa, P.; Top soil properties as affected by tillage systems in Rolling Pamapsregion of Argentina. *Soil Sci. Society Am. J.* **2009**, *73*, 1242–1250.
4. An, Z.; Early Neolithic culture in Northern China. *Archaeology.* **1984**, *10*, 934–936.
5. An, Z.; Origin of Chinese rice cultivation and its spread eastward. *Beijing Cult. Relics.* **1999**, *2*, 63–70.
6. Beijing Government. Physical Geography of China. Beijing International. **2013**, 1–3, http://www.ebejeing.gov/cn/ BeijingInfo/BJInfoTips/BeijingHistory19511209.htm
7. Anandraj, D.; *Cajanus cajan.* Agriculture, Man and Ecology. Bangalore, India. **2000**, 1–3, May 11, 2012, http://ecoport.org/ep.htm
8. Andrews, J.; Diffusion of Meso-American food complex to Southern Europe. *Geograph. Rev.* **1993**, *83*, 194–303.
9. Arias, G.; Mejoramiento Genetico y Production de Cebada Cervecera en America el sur. Regional Office of the Flood and Agricultural Organization for Latin America and Caribbean. **1995**, 60.
10. Badiane, C.; Senegal's trade in Groundnuts: Economic, Social and Environmental Implications. **2001**, 1–14, November 30, 2013, http://www. american.edu/ted/senegal-groundnut.htm
11. Baltensperger, D. D.; Foxtail and Proso Millet. In: Progress in New Crops. Janick, J.; ed. Virginia, USA: ASHS Press Alexandria; **1996**, 182–190.
12. Bationo, A.; Mokwunye, U.; Vleck, P. L. G.; Koala, S.; Shapiro, B. I.; and Yamaoh, C.; Soil fertility management for sustainable land use in West African Sudano-Sahelian zone. In: Soil Fertility Management in Africa-A Regional Perspective. Gichuru, M. P.; et al. eds. Cali, Colombia: Tropical Soil Biology and Fertility/Centro Internacional Agricultura Tropicale; **2003**, 438.

13. Bationo, A.; et al. Cropping Systems in the Sudanain-sahelian Zone: Implications on Soil Fertility Management. Bamako, Mali: Syngenta Workshop; **2004,** 209.
14. Beddow, J. M.; Pardey, P. G.; Koo, J.; and Wood, S.; The changing landscape of global agriculture. In: Shifting Patterns of Agricultural Production and Productivity Worldwide. Ames, Iowa, USA: The Midwest Trade Research and Information Centre, Iowa State University; **2010,** 7–38.
15. Benites, V. D. M.; Sustainable soil fertility management of crop systems in the Brazilian Cerrado. Proceedings of a workshop "Better Soils for Better Life: Protecting Our Future Through Soil Conservation." Jacobs University Brehman and Julius Kuhn Institute; **2010,** 1–3, June 10, 2013, http://www.jacobs-univefrsity.de/SES/better_soils_better_life.htm
16. Beyer, L.; Pingpank, K.; and Sieling, K.; Soil organic matter in temperate arable land and influence on crop production. In: Soil Fertility and Crop Production. Krishna, K. R.; ed. Enfield, New Hampshire, USA: Science Publishers Inc.; **2002,** 189–288.
17. Bhattacharyya, A.; and Werz, M.; Climate Change, Migration and Conflict in South Asia. Centre for American Progress. **2012,** 1–93, September 14, 2013, http://www.org/downloads/bhattacharyya-Werz-Climatechange-Migration-Conflict-in-South-Asia.pdf
18. Bonjean, A. P.; Origin and historical diffusion in China of major native and alien cereals. In: Cereals in China. He, Z.; and Bonjean, A. P. A.; eds. Mexico (CYMMIT), Mexico: International Centre for Maize and Wheat; **2010,** 1–14.
19. Bono, A.; Paepe, D. J.; and Alvarez, E. A.; In-Season Wheat Yield Prediction in the Semi-Arid Pampas of Argentina Using Artificial Neural Networks. **2004,** 133–150, http://www.novapublishers. com/catalog/product_info.php?products_id=272234
20. Brandolini, A.; and Brandolini, A.; Maize introduction, evolution and diffusion in Italy. *Maydica.* **2009,** *54,* 233–242.
21. Buerkert, A.; Bagayoko, M.; Bationo, A.; and Romheld, V.; Soil fertility and crop production in semi-arid and sub humid West Africa. In: Farming in West Africa: Issues, Potentials and Perspectives. Graef, F.; Lawrence, P.; and Von Oppen M.; eds. Stuttgart, Germany: Verlag Ulrich; **2000,** 27–57.
22. Buerkert, A.; Piepho, H. P.; and Bationo, A.; Multi-site time trend analysis of soil fertility management effects on crop production in Sub-Saharan West Africa. *Exp. Agric.* **2002,** *38,* 163–183.
23. Cambell, J.; A Growing Concern: Modern Slavery and Agricultural Production in Brazil and South Asia. **2008,** 131–141, http:// www.du.edu/korbel/hrhw/researchdigest/slavery/agriculture.pdf
24. Carberry, P.; et al. Scope for improved eco-efficiency varies among diverse cropping systems. Proceedings of National Academy of Science. **2013,** *110,* 8381–8386.
25. Cassman, K. G.; Ecological intensification of cereal production systems: Yield potential, soil quality and precision farming. Proceedings of National Academy of Sciences AMBIO. **1999,** 31, 132–140.
26. CGIAR. Sorghum. **2004,** 1–3, May 19, 2010, http://www.cgiar.org/compact/research/sprghum.html
27. Chen, W. H.; Several problems concerning the origin of rice growing in China. *Agric. Archaeol.* **1989,** *12,* 63–69.
28. Chibiliyov, A.; Steppe and Forest Steppe. In: The Physical Geography of Northern Eurasia Shahagedanova, M.; ed. New York: Oxford University Press; **1984,** 248–266.
29. CIMMYT. People and Production Affected by Wheat Within Each Wheat Mega-Environment Targeted by Wheat CRP. Mexico: International Maize and Wheat Centre; **2011,** 1–3, July, 2012, http://www.Wheatatals.cimmyt.org
30. Clayton, B. D.; King Cotton: A Cultural, Political and Economic History Since 1945. MS, USA: University of Mississippi Press; **2011,** 440.

31. Coura Badiane. Senegal's Trade in Groundnut: Economic, Social and Environmental Implications. TED case studies number 646. **2001,** 1–32, August 3, 2013, http://www.Senegalcasestudy.htm

32. Deng, Z.; and Janssens, M.; Shaping the Future Through Pruning the Mango Tree. **2004,** 1–6, 9, August 28, 2012, http://www.tropen.uni-bonn.de/new_website/englische_seiten/Research/Research_%20project

33. Deshingkar, P.; and Akter, S.; Migration and Human Development in India. United Nations Development Programme: Human Development Reports. UNDP, New Delhi Research Paper No 2009/13, **2009,** 1–48.

34. Deshingkar, P.; and Farrington, J.; Rural labour markets and migration in south Asia: Evidence from India and Bangladesh. **2006,** 1–31, September 14, 2013, http://openknowledge.worldbank.org/bitstream/handle/10986/9199/wdr2008_0011.pdf?sequence=1.htm

35. Deshingkar, Priya; *Migration, Remote Rural Areas and Chronic Poverty in India.* Working Paper. London: ODI and Manchester: CPRC, University of Manchester. England, United Kingdom. **2011,** 1–8, http://www.sro.sussex.ac.uk/12007

36. Devraj, D. V.; Shafer, J. G.; Patil, C. S.; and Balasubramaniam, K.; Watgal excavations: An interim report. *Man Environ.* **1995,** 20, 57–74.

37. Douglas Hurt; The Great Plains During World War II: Agriculture. Lincoln, Nebraska: The University of Nebraska Press; **2008,** 1–110.

38. Droller, F.; Migration and Long-run Economic Development: Evidence from settlements in the Pampas. **2013,** http://blogs.brown.edu/fdroller/files/2013/01/DROLLER_JMP.pdf

39. Dubey, O. P.; and Shrivas, D. N.; Response of finger millet genotype to nitrogen. *Ind. J. Agron.* **1999,** *44,* 564–566.

40. Duvick, D. N.; and Cassman, K. G.; Post Green Revolution trends in yield potential of temperate maize in North Central USA. *Crop Sci.* **1999,** *39,* 1622–1630.

41. Egli, D. B.; Comparison of Corn and soybean yields in USA-Historical trends and future prospects. *Agron. J.* **2008,** *100,* S78–S79.

42. Erskine, W.; Meulbhaeur, F. J.; Sarkar, A.; and Sharma, B.; Lentils. Oxford, England: CAB International Inc.; **2009,** 457.

43. Environment Canada. The Ecological Frame Work of Canada: Prairies Eco zone. **1999,** 1–2, http://ecozones.ca/english/ index.html

44. Fanslow, R. A.; The migrant experience. American Folklore Centre. Washington, USA: Library of Congress; **2013,** 1–3.

45. FAO. Fertilizer Use by crops in Ukraine. Rome, Italy; Food Agricultural Organization of the United Nations; Internal Report **2005,** 1–56.

46. FAO. The Cerrados. Rome, Italy: Food and Agricultural Organization of the United Nations; **2008,** 1–6, May 23, 2013, http://www. fao.org /docrep/004/Y2638E/y2638e08.htm

47. FAOSTAT. Mango Statistics. Rome Italy: Food and Agricultural Organization of the United Nations, United Nations; **2004,** October 25, 2012, http://www.faostat.org

48. FAOSTAT. Soybean Statistics. Rome, Italy: Food and Agricultural Organization of the United Nations; **2008,** 1–3, December 8, 2013, http://www.faostat.org pp

49. FAOSTAT. Citrus Fruit production. Food and Agricultural Organization of the United Nations, Rome, Italy. **2010a,** 1–3, August 29, 2012, http://faostat.org.fao.org/site/567/DesktopDefault. aspx?PagelID=567#ancor

50. FAOSTAT. Rice Production Statistics. Rome, Italy: Food and Agricultural Organization of the United Nations; **2010b,** December 5, 2013, http://www.faostat.org. htm

51. FAOSTAT. Mango Production Statistics. Rome, Italy: Food and Agricultural Organization of the United Nations; **2011,** September 22, 2013, http://www.faostat.org/site/567/desktopDefault.asp?PgelID=567#ancor

52. FAOSTAT. Maize, Wheat and Rice World Statistics. Rome, Italy: Food and Agricultural Organization of the United Nations; **2013,** 1, December 3, 2013, http://www.faostat.org

53. Feltran-Barbieri, R.; The Other Side of the Agricultural Frontier: A Brief History on Origin and Decline of Indian Agriculture in Cerrado. **2012,** 1–16, May 25, 2013, http://www.social sciences.scielo.org/pdf/s_assoc/v5nse/sasoc_aO2.pdf

54. Fite, G. C.; Cotton Fields No More: Southern Agriculture, 1865–1980. Lexington, Kentucky, USA: University Press of Kentucky; **1984,** 376.

55. Fogel, R.; Slavery in the New World. Without Consent and Contract: The Rise and Fall of American Slavery. Washington, D.C. USA: W.W. Norton & Company; **1994,** 21–23.

56. Franca, J. G.; Peres, J. R.; and Anjos, J. R.; The Brazilian Cerrados and its implications for investment to produce surplus food. **2010,** 1–37, June 15, 2013, www.rsis.edu.sg/nts/events/docs/ICAFS-PPT-Jose_Geraldo.pdf

57. Frank, P.; Slavery in British West Indies Plantations in the Eighteenth Century. *J. Negro History.* **1926,** *11,* 584–668.

58. Fuller, D. Q.; Korisettar, R.; Venkatasubaiah, P. C.; and Jones, M. K.; Early plant domestications in Southern India: Some preliminary Archaeological results. *Vegetat. Hist. Archaeol.* **2004,** *13,* 115–129.

59. Fuller, D. O.; African Crops in Prehistoric South Asia: A review In: Food, Fuels and Fields. Neumann, K.; and Butler, A.; eds. Progress in African Studies. Germany: Heinrich-Barth Institute; **2003,** 239–271.

60. Fuller, D. O.; and Harvey, E. L.; The Archaeobotany of Indian Pulses: Identification, processing and evidence for cultivation. *Environ. Archaeol.* **2006,** *11,* 218–246.

61. Fuller, D. O.; et al. The domestication process and domestication rates in rice: spikelet bases from the lower Yangt Ze. Science **2009,** 3231607–1610.

62. Gallinat, W. C.; Maize: gift from Americas first peoples. In: Chiles to Chacolate: Food the Americas gave the World. Foster, N.; and Cordell, L. S.; eds. Tuscon, USA: University of Arizona; **1999,** 47–60.

63. Gervasio, G. R.; Technological change and Agricultural growth in the Brazilian Cerrado: A Theoretical Analysis. Social Science Research Network. **2013,** 1–2, May 23, 2013, http://www.papers.ssrn.com/sol3/papers.cfm?abstract_id=444320

64. Godwin, H.; The Beginnings of Agriculture in Northwest Europe. In: Essays on Crop Plant Evolution. Hutchinson, J.; eds. United Kingdom: Cambridge University Press, Cambridge; **1965,** 1–22.

65. Grierson. 400 years of Florida Citrus. *Citrus Ind.* **1995,** 76, 28–36.

66. Gregory, J. N.; American Exodus: The Dust Bowl Migration and Okie Culture in California. New York, USA: Oxford University Press; **1989,** 489.

67. Hammons, R.; The origin and History of Groundnut. In: The Groundnut Crop: A Scientific Basis for Improvement. Smartt, G.; ed. New York: Chapman and Hall; **1994,** 22–35.

68. Harlan, J. R.; Genetic Resources of some major field crops in Africa. In: Genetic Resources in Plants-their Exploration and Conservation. Frankell, O. H.; and Bennet, E.; eds. Philadelphia, Pennsylvania, USA; **1973,** 19–32.

69. Harlan, J. T.; and DeWet, J. M. J.; A Simplified classification of cultivated Sorghum. *Crop Sci.* **1972,** 12, 172–176.

70. Hatlebakk, M.; Economic and Social structures that may explain the recent conflicts in the Terai of Nepal. Nepal: Norwegian Embassy; **2007,** 243, November 24, 2013, http://www.norway.org.np/NR/rdonlyres/0993F5660B3548A98F819167B4FD596C/72944/http_wwwcmi.pdf

71. Henry, F.; Millions Face Starvation in West Africa. **2010,** 1, August 10, 2013, http://www.gaurdian.co.uk/world/2010jun/21/millions-face-starvation-westafrica.htm

72. Hilu, K. W.; and DeWet, J. M. J.; Evolution of Eleusinian Coracana Subsp Coracana. (Finger Millet). *Am. J. Bot.* **1976,** 66, 330–333.

73. Hjejle, B.; Slavery and Agricultural Bondage in South India in 19th Century. *Scand. Econ. Rev.* **2011,** *15,* 71–126, November 30, 2013, DOI:10.1080/03585522.1967.10414353

74. Hsiao, T. C.; Herbaceous Crops. Rome, Italy: Food and Agricultural Organization of the United Nations; **2012a,** 1–42, http://www.fao.org/docrep/016/i2800e/i2800e07.pdf

75. Hsiao, T. C.; Maize. In Herbaceous Crops Food and Agricultural Organization of the United Nations. Rome, Italy; **2012b,** 1–28, http://www.fao.org/docrep/016/i2800e/i2800e07.pdf

76. Hsiao, T. C.; Rice. In: Herbaceous Crops. Rome, Italy: Food and Agricultural Organization of the United Nations; **2012c,** 1–32, http://www.fao.org/docrep/016/i2800e/i2800e07.pdf

77. Hsiao, T. C.; Soybean. In: Herbaceous Crops. Rome, Italy: Food and Agricultural Organization of the United Nations; **2012d,** 1–46, http://www.fao.org/docrep/016/i2800e/i2800e07.pdf

78. Hudson, J. C.; Making the Corn Belt: A Geographical History of Middle-Western Agriculture. Bloomington, Indiana, USA: Indiana University Press; **1994,** 260.

79. Hugh, T.; New System of Slavery. London: Hansib Publishing; **1993,** Check pages…

80. ICAR. Hand Book of Agriculture Directorate of Information and Publication of Agriculture, Indian Council of Agriculture. New Delhi, India; **2002,** 876.

81. ICAR. Pulse Crops. **2006,** 1–2, August 10, **2012,** http://www. icar.org/.in/drought/drought1. htm

82. ICRISAT. Pigeonpea Hybrids. **2007,** 1–5, August 10, 2012, http://www.icrisat.org/NewsEvents/hybrid_pigeonpea_trigger_pulse_revolution.htm

83. Ikissan. Redgram. **2006,** 1–3, August 13, 2012, http://www.ikisasn.com/links/ap_redgram-histiry.shtml

84. IPNI. Maize planting and Production in the World. International Plant Nutrition Institute, Norcross, Georgia, USA. **2008,** 1–8, August 23, 2012, http://www.ipni.net/ppi-web/nchina.nsf/$webindex/3FA09D72EC945883482573BB0030EB2A

85. IRSNDSU. Northern Great Plains-1880 to 1920. Institute for Regional Studies North Dakota State University. Fargo, North Dakota, USA. **2013,** 1–2, February, 20, 2013, memory.loc.gov/ammem/award97/ndfahtml/paz_ag.html

86. Jalloh, A.; et al. Overview. In: West African Agriculture and Climate Change. Washington, D.C.: International Food Policy Research Institute; **2013a,** 1–30.

87. Jalloh, A.; et al. Summary and Conclusions. In: West African Agriculture and Climate change. Washington, D.C.: International Food Policy Research Institute; **2013b,** 383–391.

88. Janaiah, A.; Hybrid Rice in Andhra Pradesh: Findings of a Survey. Economic and Political Weekly. **2003,** *38,* 2513–2516 p.

89. Kipple, K. F.; and Ornelas, K. C.; Cambridge World History of Food. Cambridge University Press **2000,** *I,* 234–258 p.

90. Klein, R. M.; and Klein, D. J.; Fundamentals of Plant Science. New Delhi, India: Harper and Row Publishers; **1988,** 617 p.

91. Knight, S.; Debt-bondage slavery in India. *Glob. Dialog.* **2012,** *14(2),* 1–8.

92. Krishna, K. R.; Nutrient dynamics in agro-environments. In: Soil Fertility and Crop Production. Krishna, K. R.; ed. Enfield, New Hampshire, USA: Science Publishers Inc.; **2002,** 387–410 p.

93. Krishna, K. R.; Agrosphere: Nutrient Dynamics, Ecology and Productivity. Enfield, New Hampshire, USA: Science Publishers Inc.; **2003,** 257–278.

94. Krishna, K. R.; Peanut Agroecosystem: Nutrient Dynamics and Productivity. Oxford, United Kingdom: Alpha Science International Inc.; **2008,** 1–17.

95. Krishna, K. R.; Agroecosystems of South India: Nutrient Dynamics, Ecology and Productivity. Boca Raton, Florida, USA: Brown Walker Press Inc.; **2010a,** 175–239.

96. Krishna, K.; Finger millet cropping zones of South India: Nutrients, ecology and productivity. In: Agroecosystems of South India: Nutrient Dynamics and Productivity. Boca Raton, Florida, USA: Brown Walker Press Inc.; **2010b**, 279–312.

97. Krishna, K. R.; Maize Agroecosystem: Nutrient Dynamics and Productivity. Apple Academic Press Inc., Waretown, New Jersey, USA, **2012**, 1–40.

98. Krishna, K. R.; Maize Agroecosystem: Nutrient Dynamics and Productivity. Waretown, New Jersey, USA: Apple Academic Press Inc.; **2013**, 185–206.

99. Krishna, K. R.; Finger Millet in Asia and Africa. Waretown, New Jersey, USA: Apple Academic Press Inc.; **2014a**.

100. Krishna, K. R.; Sunflower agroecosystem. In: Agroecosystems: Soils, Climate, Crops, Nutrient Dynamics and Productivity. Waretown, New Jersey, USA: Apple Academic Press Inc.; **2014b**, 191–198.

101. Krishna, K. R.; Cowpeas in Africa and Asia. In: Agroecosystems: Soils, Climate, Crops, Nutrient Dynamics and Productivity. Waretown, New Jersey, USA: Apple Academic Press Inc.; **2014c**, 139–146.

102. Krishna, K. R.; Horse Gram Cropping Zones of Asia and Africa. Waretown, New Jersey, USA: Apple Academic Press Inc.; **2014d**, 169–174.

103. Krishna, K. R.; Cotton Cropping Zones. Waretown, New Jersey, USA: Apple Academic Press Inc.; **2014e**, 219–232.

104. Lai, W. L.; Indentured Labor, Carribbean Sugar: Chinese and Indian Migrants to the British West Indies 1838–1918. Baltimore, USA: John Hopkins University Press; **1993**, 370.

105. Laurence, K.; A Question of Labour: Indentured Immigration into Trinidad and British Guiana. 1875–1917. St Martin Press; **1994**, 224–242.

106. Lentz, D.; Robert, B.; and Victor, S.; New Evidence for Sunflower (Helianthus Annuus) Origins, Ecological Niche Model for Distribution in Mexico. Botanical Society of America, USA; **2006**, 1–2, April 12, 2012, http://www.2006botanyconference.org/engine/ search/index.php?func=detail&aid=881

107. Liang, W.; Carberry, P.; Wang, G.; Lu, R.; Lu, H.; and Xia, A.; Quantifying the Yield gap in Wheat-maize Cropping systems of the Hebei plain, China. Field Crops Research **2011**, 124, 180–185.

108. Library of Congress. Northern Great Plains, 1880–1920. **2013**, 1–4, November 26, 2013, http://memory.loc.gov/ammem/award97/ndfahtml/

109. Lockeritz, W.; The Lessons of the Dust Bowl. American Scientist. **1978**, *66*, 560–569.

110. Lu, J. J.; and Chang, T. T.; Rice in its temporal and spatial persepectives. In: Rice Production and Utilization. Luh, B. S.; ed. Westport, Connecticut, USA; **1980**, 1–74.

111. Manasala, P. K.; A New look at Vedic India. **2006**, 1–17, http://asiapacificuniverse.com/pkm/vedicindia.html.html

112. Masalsky, V. I.; Gullies in the Black Soil Region of Russia: Their Extent, Development and Impacts. Russia: St Petersburg; **1897**, 146, (in Russian).

113. McCann, J.; Maize and Grace: History Corn and Africa's New Landscapes, 1500–1999. **2000**, 1–29, November 22, 2013, http://ruafrica.rutgers.edu/ events/media/0405_media/maize_and_grace_jamesmccann.pdf

114. McLeman, R.; and Smit, B.; Migration as an adaptation to climate change. *Climate Change.* **2006**, 76, 31–53.

115. Morrision, K. D.; Forager and forager-traders in South Asian Worlds: Some thoughts from the last 10,000 years. In: The Evolution of Human Populations in South Asia. Petraglia, M. D.; and Allchin, B.; eds. Springer, Dordrecht, Netherlands; **2007**, 233–278.

116. Morrison, K. D.; Naturalizing Disaster: From Drought to Famine in Southern India. **2012**, 1–15, http://cis.uchicago.edu/outreach/ summerinstitute/2012/documents/sti2012-morrison-naturalizing-disaster-southern-india.pdf

117. Motoki, Y.; Changes in the Structure of Agricultural land use in the Northeast China, with special reference to role of Rice paddy production. **2002,** 1–4, December 5, 2013, http://www-basin.nies.go.jp/project/lugec/ReportIIX_E/014_Motoki02.pdf

118. Nagaraja, B. C.; Somashekara, R. K.; and Kavitha, A.; Impact of Drought on Agriculture: Challenges facing poor farmers of Karnataka, South India. **2010,** 1–10, November 29, 2013, http://climsec.prio.no/papers/climate%20change-norway%20final%20paper.pdf

119. Nakamato, T.; Matasuzaki, A.; and Shimado, K.; Root spatial distribution of Field grown maize and millets. *Jpn. J. Crop Sci.* **1992,** *61,* 304–309.

120. National Archives of United Kingdom. Forced Labour. **2010,** 1–26, November 24, 2013, http://www.nationalarchives.gov.uk/pathways/blackhistory/ india/forced.htm

121. Nene, Y. L.; and Hall, S. D.; The Pigeonpea. Hyderabad, India: International Crops Research Institute for the Semi-arid Tropics; **1990,** 490.

122. Newman, C. W.; and Newman, R. K.; A brief history of Barley foods. Cereal Foods World. **2006,** *51,* 1–7.

123. Ng Cowpea. Vigna Unguiculata. In: Evolution of Crop Plants. Smartt, J.; and Simmonds, N. W.; eds. London: Longman Scientific and Technical Publishers; **1995,** 326–332.

124. NHB. Indian Horticultural Data Base. New Delhi: National Horticultural Board; **2009,** 485 p.

125. Nilnond, S.; Grape Production in Thailand. Rome, Italy: Food and Agricultural Organization of the United Nations; **1998,** 1–9.

126. OKState. World Wheat, Corn and Rice. Oklahoma State University, Stillwater, Oklahoma. **2013,** 1, December 3, 2013, http://www.nue.okstae.edu/ Crop_Information/World_Wheat_production.htm

127. Palmer, S. A.; Moore, J. D.; Clapham, A. J.; Rose, P.; and Allaby, R. G.; Archaeogenetic evidence of Ancient Nubian Barley evolution from six to two indicates local adaptation. The McDonald Institute for Archaeological Research. United Kingdom: University of Cambridge, Cambridge; **2009,** 1–13.

128. Pandey, M. P.; Srinivasa Rao, K.; and Saha, S.; Agro-economic analysis of Rice-based cropping systems. *Ind. J. Fertil.* **2008,** *4,* 39–47.

129. Parihar, S. S.; et al. Energetics, Yield, water use and Economics of based cropping system. *Ind. J. Agron.* **1999,** *44,* 205–209.

130. Parton, W. J.; Guttmann, M. P.; and Ojima, D.; Long term trends in population, farm income and crop production trends in the Great Plains. *BioScience.* **2007,** *59,* 737–747.

131. Phillips, K. E.; and Roberts, J.; Cotton. Encyclopedia of Alabama, USA. **2011,** 1–4, http://www.encyclopediaofalabama.org/face /Article.jsp?id=h-1491

132. Piperno, D. R.; and Flannery, K. V.; The earliest Archaeological maize (*Zea mays*) from Highlands of Mexico. New Mass Spectroscopy Dates and Their Implications. **2001,** 1–2, November 22, 2013, http://www.pnas.org/content/98/4/2101.abstract

133. Pongratz, J.; and Reick, C.; Agriculture: Is plowing up the Climate. Max Plank Research Report. **2012,** December 3, 2013, http://.mpg.de/788203/w005_Environment_Climate_076_082.pdf

134. Rachie, K. O.; and Peters, L. V.; The Eleusine: A Review of the World Literature. International Crops Research Institute for the Semi-arid Tropics, Patancheru, India; **1977,** 178.

135. Randhawa, M. S.; A History of Agriculture in India. New Delhi: Indian Council of Agricultural Research; **1980,** 243.

136. Rath, B.; Migration and its Impact on Agriculture: Economic Fabric in Early India. Reseapro. **2013,** 1–3, November 11, 2013, http://www.resapro.com/2013 /08/migration-and-agriculture/

137. Rezende, G. C.; Technological Change and Agricultural Growth in the Brazilian Cerrado: A Theoretical Analysis. Social Science Research Network. **2013,** 1–2, May 23, 2013, http://www.ssm.com/absstract=444320

138. Rogaly, B.; Coppard, D.; Ratique, A.; Rana, K.; Sengupta, A.; and Biswas, J.; Seasonal migration and welfare/illfare in Eastern India: A Social Analysis. *J. Dev. Stud.* **2002,** *38.5,* 89–114.

139. Schmidt, N.; The Cerrados Biome in Central Brazil-natural Ecology and Threats to Its Diversity. **2009,** 12, June 5, 2013, http://www.goek.tu-freiberg.de/oberseminar/OS_09/Nadja_Schmidt.pdf

140. Sears, A. B.; The desert threat in the Southern Great Plains: The historical implications of soil erosion. *Agric. Hist.* **1941,** 15, 1–11.

141. Selvi, D.; Santhy, P.; Dakshinamurthy, M.; and Maheshwari, M.; Microbial population and biomass in rhizosphere as influenced by continuous intensive cultivation and fertilization in an Inceptisol. *J. Ind. Soc. Soil Sci.* **2004,** 52, 254–257.

142. Sharpe, P.; Sugarcane: Past and Present. Illinois, USA: Southern Illinois University Press; **1998,** 439 p.

143. Sharp, R. E.; and Davis, W. J.; Root growth and water uptake by Maize plants in drying soil. *J. Exp. Bot.* **1985,** *36,* 1441–1446.

144. Shikamany, S. D.; Grape production in India. Rome, Italy: Food and Agricultural organization of the United Nations; **2010,** 1–9.

145. Shorter, S. L.; and Patanathoi, A.; *Arachis Hypogea. L.* PROSEA Hand Book on CD-ROMS. Thailand: Jan Kops House; **1997,** 1–9.

146. Singh, G.; The Soybean: Botany, Production and Uses. England: CAB International; **2010,** 494.

147. Smekalova, N.; Interactive Agricultural atlas of Russia and neighbouring countries. **2009,** 1–3, September 19, 2012, http://wwwagroatlas.ru/encontent/ related/Setaria_italica/

148. Smith, W. C.; and Frederickson, R. A.; Sorghum: Origin, history, technology, and production. Wiley Series in Crop Science. New York: Wiley Inc.; **2000,** 840.

149. Somada, E. A.; Guei, R. G.; and Nguyen, N.; Overview: Rice in Africa. Liberia: West Africa Rice Centre; **2012,** 1–9, December, 8, 2013, http://www.africarice.org/publications/nerica-comp/module%201_Low.pdf

150. Soy Info Centre. History of Soybeans and Soybean Foods in South America (1882–2009). **2011a,** October 2, 2012, http://www. soybeaninfocenter.com/books/132

151. Soy Info Centre. History of Soybeans and soybean foods in Africa. **2011b,** October 2, 2012, http://www.soybeaninfocenter.com/books/134

152. Soy Info Centre. History of soybeans and Soybean food in Australia. **2011c,** October 2, 2012, shttp://www.soybeaninfocenter.com/books/138

153. Soy Info Centre. History of Soybeans and Soybean Food in Australia. **2011d,** October 2, 2012, shttp://www.soybeaninfocenter.com/books/140

154. Stebelsky, I. V.; Environmental deterioration in the Central Russian Black soil region: The Case of Soil Erosion. *Can. Geograph.* **1974,** 18, 232–249.

155. Stebelsky, I. V.; Agriculture and Soil erosion in the European forested steppes. In: Studies in Russian Historical Geography. Bater, J. H.; and French, B. A.; eds. London: Academic Press Inc.; **1983,** 45–63.

156. USDA. Mahindi and Milho in Africa. In: Plants and Crops. Beltsville, Maryland, USA: United States Department of Agriculture; **2008,** 1–3, November 20, 2013, http://www.nal.usda.gov/research/maize/chapter 3.html

157. USDA-ERS. Pigeonpea Statistics. United States Department of Agriculture. **2012,** 1–3, http://www.ers.usda.gov/data-products/cajanuscajan-data-aspx

158. USDA-ERS. World Wheat Supply and Disappearance. **2013,** 1–2, December 3, 2013, http://www.ers.usda.gov/data-products/wheat-data-aspx

159. USDA-NAGS. World Crop Production Summar. Beltsville, Maryland, USA: United States Department of Agriculture; **2013,** 1–17, January 3, 2014, http://www.fas.usda.gov.htm

160. Van Der Maesen, L. G. J.; Pigeonpea. Origin, history, evolution, taxonomy. In: Pigeonpea. Nene, Y. L.; and Hall, S. D.; eds. United Kingdom: CAB International; **1990,** 15–46.

161. Vigouroux, Y.; Glaubitz, J. C.; Matsuoka, Y.; Goodman, M. M.; Sanchez, J. G.; and Doebley, J.; Population structure and genetic diversity of new world maize races assessed by DNA microsatellites. *Am. J. Bot.* **2008,** *95,* 1240–1253.

162. Vishnu-Mittre, A.; Discussions in early history of agriculture. *Philos. Transac. Royal Soc. London.* **1976,** B275, 141.

163. Wayne, S. C.; and Cothern, J. T.; Cotton: Origin, History, Technology and Production. New York, USA: John Wiley and Sons; **1999, 850.**

164. Weiss, E. A.; Oil Seed Crops. London, United Kingdom: Blackwell Scientific Company; **2000,** 487.

165. Wood, S.; Sebastian, K.; and Scherr, S. J.; Agroecosystems. A Pilot Analysis of Global Ecosystems. Washington, D.C.: International Food Policy Institute: **2000,** 185.

166. WDI. World bank development indicators. The World Bank. Washington D.C. USA; **2013,** 1–5, http://www. wdi. worldbank.org/table/3.2

167. Yadav, S. S.; McNeil, Y. D.; and Stevenson, P. C.; Lentil: An Ancient Crop for Modern Times. Heidelberg: Springer Verlag Inc.; **2007,** 461.

168. Yang, X.; Liu, Z.; and Chen, F.; The possible effects of global effects on cropping systems in China. 1. Possible effects of climate warming on Northern limits of cropping systems in China. *Scientia Agricultura Sinica.* **2010,** *43(79),* 79–95.

169. Zhang, W.; Chen, J.; Xu, X.; Che, C. Q.; Deng, A. X.; Qian, C. R.; and Dong, W. J.; Actual response and adaptation of rice cropping system to Global warming in Northeast China. *China Agric. Sci.* **2012,** *45,* 1265–1273.

170. Zheng, H. F.; Chen, L. D.; and Han, X. Z.; Response and adaptation of soybean systems to climate warming in Northeast China: Insights gained from long term field trials. *Crop Pasture Sci.* **2011,** *62,* 876–882.

171. Zhimin, A.; Origin of Chinese Rice Cultivation and Its Spread East. **1999,** 1–11, September 2, 2012, http://http-server.carleton.ca/-bgordon/rice/papers/ zhimin99.htm

INDEX